Selected Titles in This Se

MW00912230

(*Continued in the back of this publication*)

Operads: Proceedings of Renaissance Conferences

CONTEMPORARY MATHEMATICS

202

Operads: Proceedings of Renaissance Conferences

Special Session and International Conference on
Moduli Spaces, Operads, and Representation Theory/
Operads and Homotopy Algebra
March 1995 / May–June 1995
Hartford, Connecticut/Luminy, France

Jean-Louis Loday
James D. Stasheff
Alexander A. Voronov
Editors

American Mathematical Society
Providence, Rhode Island

Editorial Board

Dennis DeTurck, managing editor

Andy Magid   Michael Vogelius

Clark Robinson   Peter M. Winkler

This volume features proceedings from the special session "Moduli spaces, operads, and representation theory" held at Hartford, CT, on March 4–5, 1995, and at a conference "Operades et algebre homotopique" held at the Centre International de Rencontres Mathematiques at Luminy, France, May 29–June 2, 1995.

1991 *Mathematics Subject Classification.* Primary 08A05, 17Axx; Secondary 16S80, 17B69, 18Gxx, 55Pxx, 55Sxx, 81T40.

Library of Congress Cataloging-in-Publication Data

Operads : proceedings of renaissance conferences / Jean-Louis Loday, James D. Stasheff, Alexander A. Voronov, editors.

    p. cm.—(Contemporary mathematics, ISSN 0271-4132 ; 202)

    "Special session and international conference on moduli spaces, operads, and representation theory/operads and homotopy algebra, March 1995/May–June 1995, Hartford, Connecticut/Luminy, France."

    Includes bibliographical references and index.

    ISBN 0-8218-0513-4 (alk. paper)

    1. Ordered algebraic structures—Congresses. 2. Moduli theory—Congresses. 3. Representations of algebras—Congresses. I. Loday, Jean-Louis. II. Stasheff, James D. III. Voronov, Alexander A. IV. Series: Contemporary mathematics (American Mathematical Society) ; v. 202.

QA172.064 1997

512′.2—dc21

96-37049

CIP

Contents

Preface

"Operads" are mathematical devices which model many sorts of algebras (such as associative, commutative, Lie, Poisson, alternative, Leibniz, etc., including those defined up to homotopy, such as A_∞-algebras). The notion of an operad appeared in the seventies in algebraic topology (J. Stasheff, J. P. May, J. M. Boardman, R. M. Vogt), but there has been a renaissance of this theory due to the discovery of relationships with graph cohomology, Koszul duality, representation theory, combinatorics, cyclic cohomology, moduli spaces, knot theory, and quantum field theory.

This renaissance was recognized at a special session "Moduli spaces, operads, and representation theory" of the AMS meeting at Hartford, CT, on March 4–5, 1995, and at a conference "Opérades et algèbre homotopique" held at the Centre International de Rencontres Mathématiques at Luminy, France, from May 29 to June 2, 1995. Both meetings drew a diverse group of researchers, as we hope these proceedings reflect, though not all the speakers are represented here.

We have arranged the contributions so as to emphasize certain themes around which the renaissance of operads took place: homotopy algebra, algebraic topology, polyhedra and combinatorics, and applications to physics. We begin the collection with two short papers which drop out of this classification. The first one is the paper "Definitions: operads, algebras and modules" by Peter May, which recalls basic notions of operad theory. The other paper, "The pre-history of operads", by one of us describes how operads had been used before they were created.

Jean-Louis Loday, Jim Stasheff, and Alexander A. Voronov

Contemporary Mathematics
Volume **202**, 1997

Definitions: operads, algebras and modules

J. P. MAY

Let $\mathscr{S}$ be a symmetric monoidal category with product $\otimes$ and unit object κ.

DEFINITION 1. An operad $\mathscr{C}$ in $\mathscr{S}$ consists of objects $\mathscr{C}(j)$, $j \geq 0$, a unit map $\eta : \kappa \to \mathscr{C}(1)$, a right action by the symmetric group Σ_j on $\mathscr{C}(j)$ for each j, and product maps

$$\gamma : \mathscr{C}(k) \otimes \mathscr{C}(j_1) \otimes \cdots \otimes \mathscr{C}(j_k) \to \mathscr{C}(j)$$

for $k \geq 1$ and $j_s \geq 0$, where $\sum j_s = j$. The γ are required to be associative, unital, and equivariant in the following senses.

(a) The following associativity diagrams commute, where $\sum j_s = j$ and $\sum i_t = i$; we set $g_s = j_1 + \cdots + j_s$, and $h_s = i_{g_{s-1}+1} + \cdots + i_{g_s}$ for $1 \leq s \leq k$:

$$
\begin{array}{ccc}
\mathscr{C}(k) \otimes (\bigotimes_{s=1}^{k} \mathscr{C}(j_s)) \otimes (\bigotimes_{r=1}^{j} \mathscr{C}(i_r)) & \xrightarrow{\gamma \otimes \mathrm{id}} & \mathscr{C}(j) \otimes (\bigotimes_{r=1}^{j} \mathscr{C}(i_r)) \\
\downarrow{\scriptstyle \mathrm{shuffle}} & & \downarrow{\gamma} \\
& & \mathscr{C}(i) \\
& & \uparrow{\gamma} \\
\mathscr{C}(k) \otimes (\bigotimes_{s=1}^{k} (\mathscr{C}(j_s) \otimes (\bigotimes_{q=1}^{j_s} \mathscr{C}(i_{g_{s-1}+q})))) & \xrightarrow{\mathrm{id} \otimes (\otimes_s \gamma)} & \mathscr{C}(k) \otimes (\bigotimes_{s=1}^{k} \mathscr{C}(h_s)).
\end{array}
$$

1991 *Mathematics Subject Classification.* Primary 18-02, 55-02; Secondary 18C99, 55U99.
The author was supported in part by NSF Grant #DMS-9423300.

(b) The following unit diagrams commute:

(c) The following equivariance diagrams commute, where $\sigma \in \Sigma_k, \tau_s \in \Sigma_{j_s}$, the permutation $\sigma(j_1, \ldots, j_k) \in \Sigma_j$ permutes k blocks of letter as σ permutes k letters, and $\tau_1 \oplus \cdots \oplus \tau_k \in \Sigma_j$ is the block sum:

$$
\begin{array}{ccc}
\mathscr{C}(k) \otimes \mathscr{C}(j_1) \otimes \cdots \otimes \mathscr{C}(j_k) & \xrightarrow{\sigma \otimes \sigma^{-1}} & \mathscr{C}(k) \otimes \mathscr{C}(j_{\sigma(1)}) \otimes \cdots \otimes \mathscr{C}(j_{\sigma(k)}) \\
\gamma \downarrow & & \downarrow \gamma \\
\mathscr{C}(j) & \xrightarrow{\sigma(j_{\sigma(1)}, \ldots, j_{\sigma(k)})} & \mathscr{C}(j)
\end{array}
$$

and

$$
\begin{array}{ccc}
\mathscr{C}(k) \otimes \mathscr{C}(j_1) \otimes \cdots \otimes \mathscr{C}(j_k) & \xrightarrow{\mathrm{id} \otimes \tau_1 \otimes \cdots \otimes \tau_k} & \mathscr{C}(k) \otimes \mathscr{C}(j_1) \otimes \cdots \otimes \mathscr{C}(j_k) \\
\gamma \downarrow & & \downarrow \gamma \\
\mathscr{C}(j) & \xrightarrow{\tau_1 \oplus \cdots \oplus \tau_k} & \mathscr{C}(j).
\end{array}
$$

The $\mathscr{C}(j)$ are to be thought of as objects of parameters for "j-ary operations" that accept j inputs and produce one output. Thinking of elements as operations, we think of $\gamma(c \otimes d_1 \otimes \cdots \otimes d_k)$ as the composite of the operation c with the $\otimes$-product of the operations d_s.

Let X^j denote the j-fold $\otimes$-power of an object X, with Σ_j acting on the left. By convention, $X^0 = \kappa$.

DEFINITION 2. Let $\mathscr{C}$ be an operad. A $\mathscr{C}$-algebra is an object A together with maps

$$\theta : \mathscr{C}(j) \otimes A^j \to A$$

for $j \geq 0$ that are associative, unital, and equivariant in the following senses.

(a) The following associativity diagrams commute, where $j = \sum j_s$:

$$
\begin{array}{ccc}
\mathscr{C}(k) \otimes \mathscr{C}(j_1) \otimes \cdots \otimes \mathscr{C}(j_k) \otimes A^j & \xrightarrow{\gamma \otimes \mathrm{id}} & \mathscr{C}(j) \otimes A^j \\
\text{shuffle} \downarrow & & \downarrow \theta \\
& & A \\
& & \uparrow \theta \\
\mathscr{C}(k) \otimes \mathscr{C}(j_1) \otimes A^{j_1} \otimes \cdots \otimes \mathscr{C}(j_k) \otimes A^{j_k} & \xrightarrow{\mathrm{id} \otimes \theta^k} & \mathscr{C}(k) \otimes A^k.
\end{array}
$$

(b) The following unit diagram commutes:

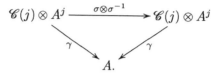

(c) The following equivariance diagrams commute, where $\sigma \in \Sigma_j$:

$$\mathscr{C}(j) \otimes A^j \xrightarrow{\sigma \otimes \sigma^{-1}} \mathscr{C}(j) \otimes A^j$$
$$\underset{\gamma}{\searrow} \qquad \underset{\gamma}{\swarrow}$$
$$A.$$

DEFINITION 3. Let $\mathscr{C}$ be an operad and A be a $\mathscr{C}$-algebra. An A-module is an object M together with maps

$$\lambda : \mathscr{C}(j) \otimes A^{j-1} \otimes M \to M$$

for $j \geq 1$ that are associative, unital, and equivariant in the following senses.
   (a) The following associativity diagrams commute, where $j = \sum j_s$:

$$(\mathscr{C}(k) \otimes (\bigotimes_{s=1}^{k} \mathscr{C}(j_s))) \otimes A^{j-1} \otimes M \xrightarrow{\gamma \otimes \mathrm{id}} \mathscr{C}(j) \otimes A^{j-1} \otimes M$$

with the left vertical map labelled shuffle, the right side descending via λ to M and back up via λ, and the bottom map

$$\mathscr{C}(k) \otimes (\bigotimes_{s=1}^{k-1}(\mathscr{C}(j_s) \otimes A^{j_s})) \otimes (\mathscr{C}(j_k) \otimes A^{j_k-1} \otimes M) \xrightarrow{\mathrm{id} \otimes \theta^{k-1} \otimes \lambda} \mathscr{C}(k) \otimes A^{k-1} \otimes M.$$

(b) The following unit diagram commutes:

$$\kappa \otimes M \xrightarrow{\cong} M$$
$$\eta \otimes \mathrm{id} \downarrow \qquad \nearrow \lambda$$
$$\mathscr{C}(1) \otimes M.$$

(c) The following equivariance diagram commutes, where $\sigma \in \Sigma_{j-1} \subset \Sigma_j$:

$$\mathscr{C}(j) \otimes A^{j-1} \otimes M \xrightarrow{\sigma \otimes \sigma^{-1} \otimes \mathrm{id}} \mathscr{C}(j) \otimes A^{j-1} \otimes M$$
$$\underset{\lambda}{\searrow} \qquad \underset{\lambda}{\swarrow}$$
$$M.$$

Maps of operads, of algebras over an operad, and of modules over an algebra over an operad are defined in the evident ways: all structure must be preserved.

VARIANTS 4. (i) *Non-Σ (or non-symmetric) operads.* When modelling non-commutative algebras, it is often useful to omit the permutations from the definition, giving the notion of a non-Σ operad. An operad is a non-Σ operad by neglect of structure.

(ii) *Unital operads.* The object $\mathscr{C}(0)$ parametrizes "0-ary operations". When concerned with unital algebras A, the unit "element" $1 \in A$ is defined by a map $\kappa \to A$, and it is sensible to insist that $\mathscr{C}(0) = \kappa$. We then say that $\mathscr{C}$ is a unital operad. For types of algebras without units (e.g. Lie algebras) it is sensible to set $\mathscr{C}(0) = 0$ (categorically, an initial object).

(iii) *Augmentations.* If $\mathscr{C}$ is unital, the $\mathscr{C}(j)$ have the "augmentations"

$$\epsilon = \gamma : \mathscr{C}(j) \cong \mathscr{C}(j) \otimes \mathscr{C}(0)^j \to \mathscr{C}(0) = \kappa$$

and the "degeneracy maps" $\sigma_i : \mathscr{C}(j) \to \mathscr{C}(j-1)$, $1 \le i \le j$, given by the composites

$$\mathscr{C}(j) \cong \mathscr{C}(j) \otimes \kappa^j \longrightarrow \mathscr{C}(j) \otimes \mathscr{C}(1)^{i-1} \otimes \mathscr{C}(0) \otimes \mathscr{C}(1)^{j-i} \overset{\gamma}{\longrightarrow} \mathscr{C}(j-1),$$

where the first map is determined by the unit map $\eta : \kappa \longrightarrow \mathscr{C}(1)$.

EXAMPLE 5. Assume that $\mathscr{S}$ has an internal Hom functor. Define the endomorphism operad of an object X by

$$\mathrm{End}(X)(j) = \mathrm{Hom}(X^j, X).$$

The unit is given by the identity map $X \to X$, the right actions by symmetric groups are given by their left actions on $\otimes$-powers, and the maps γ are given by the following composites, where $\sum j_s = j$:

$$\mathrm{Hom}(X^k, X) \otimes \mathrm{Hom}(X^{j_1}, X) \otimes \cdots \otimes \mathrm{Hom}(X^{j_k}, X)$$

$$\downarrow \text{id} \otimes (k\text{-fold} \otimes \text{-product of maps})$$

$$\mathrm{Hom}(X^k, X) \otimes \mathrm{Hom}(X^j, X^k)$$

$$\downarrow \text{composition}$$

$$\mathrm{Hom}(X^j, X).$$

Conditions (a)-(c) of the definition of an operad are forced by direct calculation. In adjoint form, an action of $\mathscr{C}$ on A is a morphism of operads $\mathscr{C} \to \mathrm{End}(A)$, and conditions (a)-(c) of the definition of a $\mathscr{C}$-algebra are also forced by direct calculation.

EXAMPLE 6. The operad $\mathscr{M}$ has $\mathscr{M}(j) = \kappa[\Sigma_j]$, the coproduct of a copy of κ for each element of Σ_j; the maps γ are determined by the formulas defining an operad. An $\mathscr{M}$-algebra A is a monoid in $\mathscr{S}$ and an A-module in the operadic sense is an A-bimodule in the classical sense of commuting left and right actions $A \otimes M \longrightarrow M$ and $M \otimes A \longrightarrow M$.

EXAMPLE 7. The operad $\mathcal{N}$ has $\mathcal{N}(j) = \kappa$; the maps γ are canonical isomorphisms. An $\mathcal{N}$-algebra A is a commutative monoid in $\mathcal{S}$ and an A-module in the operadic sense is a left A-module in the classical sense. If we regard $\mathcal{N}$ as a non-Σ operad, then an $\mathcal{N}$-algebra A is a monoid in $\mathcal{S}$ and an A-module in the operadic sense is a left A-module in the classical sense. A unital operad $\mathcal{C}$ has the augmentation $\varepsilon : \mathcal{C} \longrightarrow \mathcal{N}$; an $\mathcal{N}$-algebra is a $\mathcal{C}$-algebra by pullback along ε.

There are important alternative formulations of some of the definitions. First, there is a conceptual reformulation of operads as monoids in a certain category of functors. Assume that $\mathcal{S}$ has finite colimits. These allow one to make sense of passage to orbits from group actions.

DEFINITION 8. Let Σ denote the category whose objects are the finite sets $n = \{1, \cdots, n\}$ and their isomorphisms, where 0 is the empty set. Define a Σ-object in $\mathcal{S}$ to be a contravariant functor $\mathcal{C} : \Sigma \longrightarrow \mathcal{S}$. Thus $\mathcal{C}(j)$ is an object of $\mathcal{S}$ with a right action by Σ_j; by convention, $\mathcal{C}(0) = \kappa$. Define a product $\circ$ on the category of Σ-objects by setting

$$(\mathcal{B} \circ \mathcal{C})(j) = \coprod_{k, j_1, \dots, j_k} \mathcal{B}(k) \otimes_{\kappa[\Sigma_k]} ((\mathcal{C}(j_1) \otimes \cdots \otimes \mathcal{C}(j_k)) \otimes_{\kappa[\Sigma_{j_1} \times \cdots \times j_k]} \kappa[\Sigma_j]),$$

where $k \geq 0$, $j_r \geq 0$, and $\sum j_r = j$. The implicit right action of $\kappa[\Sigma_{j_1 \times \cdots \times j_k}]$ on $\mathcal{C}(j_1) \otimes \cdots \otimes \mathcal{C}(j_k)$ and left action of Σ_k on $(\mathcal{C}(j_1) \otimes \cdots \otimes \mathcal{C}(j_k)) \otimes_{\kappa[\Sigma_{j_1 \times \cdots \times j_k}]} \kappa[\Sigma_j]$ should be clear from the equivariance formulas in the definition of an operad. The right action of Σ_j required of a contravariant functor is given by the right action of Σ_j on itself. The product $\circ$ is associative and has the two-sided unit I specified by $I(1) = \kappa$ and $I(j) = \phi$ (an initial object of $\mathcal{S}$) for $j \neq 1$.

A trivial inspection gives the following reformulation of the definition of an operad.

LEMMA 9. Operads in $\mathcal{S}$ are monoids in the monoidal category of Σ-objects in $\mathcal{S}$.

Similarly, using the degeneracy maps σ_i of Variant 4(iii), if Λ denotes the category of finite sets n and all injective maps, then a unital operad is a monoid in the monoidal category of contravariant functors $\Lambda \longrightarrow \mathcal{S}$.

These observations are closely related to the comparison of algebras over operads to algebras over an associated monad that led me to invent the name "operad".

DEFINITION 10. Define a functor $C : \mathcal{S} \longrightarrow \mathcal{S}$ associated to a Σ-object $\mathcal{C}$ by

$$CX = \coprod_{j \geq 0} \mathcal{C}(j) \otimes_{\kappa[\Sigma_j]} X^j,$$

where $\mathcal{C}(0) \otimes_{\kappa[\Sigma_0]} X^0 = \kappa$.

By inspection of definitions, the functor associated to $\mathscr{B} \circ \mathscr{C}$ is the composite BC of the functors B and C associated to $\mathscr{B}$ and $\mathscr{C}$. Therefore a monoid in the monoidal category of Σ-objects in $\mathscr{S}$ determines a monad (C, μ, η) in $\mathscr{S}$. This leads formally to the following result; it will be expanded in my paper "Operads, algebras, and modules" later in this volume (which gives background, details, and references for most of the material summarized here).

PROPOSITION 11. *An operad $\mathscr{C}$ in $\mathscr{S}$ determines a monad C in $\mathscr{S}$ such that the categories of algebras over $\mathscr{C}$ and of algebras over C are isomorphic.*

There is also a combinatorial reformulation of the definition of operads that is expressed in terms of "$\circ_i$-products".

DEFINITION 12. Let $\mathscr{C}$ be an operad in $\mathscr{S}$. Define the product

$$\circ_i : \mathscr{C}(p) \otimes \mathscr{C}(q) \longrightarrow \mathscr{C}(p+q-1)$$

to be the composite

$$\mathscr{C}(p) \otimes \mathscr{C}(q)$$

$$\downarrow \mathrm{id} \otimes \eta^{i-1} \otimes \mathrm{id} \otimes \eta^{p-i}$$

$$\mathscr{C}(p) \otimes \mathscr{C}(1)^{i-1} \otimes \mathscr{C}(q) \otimes \mathscr{C}(1)^{p-i}$$

$$\downarrow \gamma$$

$$\mathscr{C}(p+q-1).$$

These products satisfy certain associativity, unity, and equivariance formulas that can be read off from the definition of an operad. Conversely, the structure maps γ can be read off in many different ways from the $\circ_i$-products. In fact, just the first one suffices. By use of the associativity and unity diagrams, we find that the following composite coincides with γ:

$$\mathscr{C}(k) \otimes \mathscr{C}(j_1) \otimes \cdots \otimes \mathscr{C}(j_k)$$

$$\downarrow \circ_1 \otimes \mathrm{id}$$

$$\mathscr{C}(k+j_1-1) \otimes \mathscr{C}(j_2) \otimes \cdots \otimes \mathscr{C}(j_k)$$

$$\downarrow \circ_1 \otimes \mathrm{id}$$

$$\vdots$$

$$\downarrow \circ_1 \otimes \mathrm{id}$$

$$\mathscr{C}(k+j_1+\cdots+j_{k-1}-(k-1)) \otimes \mathscr{C}(j_k)$$

$$\downarrow \circ_1$$

$$\mathscr{C}(j_1+\cdots+j_k).$$

We deduce that operads can be redefined in terms of $\circ_i$-products. This leads to another useful variant of the notion of an operad. If we are given Σ_j-objects $\mathscr{C}(j)$ for $j \geq 1$ and $\circ_i$-products that satisfy the associativity and equivariance laws, but not the unit laws, that are satisfied by the $\circ_i$ operations of an operad, we arrive at the notion of an "operad without identity" (analogous to a ring without identity). Such structures arise naturally in some applications related to string theory.

THE UNIVERSITY OF CHICAGO, CHICAGO, IL 60637
E-mail address: may@math.uchicago.edu

Contemporary Mathematics
Volume **202**, 1997

The pre-history of operads

Jim Stasheff

Peter May will tell you some of the influences that led to his orig-
ination of the concept of operad. I will try to supply some of the
pre-history: before there were operads in name, there was an example. Before Peter May introduced the concept of an operad and used
it thoroughly to describe the homotopy type of an infinite loop space,
I constructed the family of associahedra $\{K_n\}$ in order to character-
ize from a homotopy point of view spaces of the homotopy type of a
(based) loop space. With hindsight, we can say that the associahedra
form a non-symmetric operad (equivalently, the free Σ-spaces $\Sigma_n \times K_n$
form an operad) and the characterization is as (topological) algebras
over that non-symmetric operad. Let me recall the origins in more
detail.

For me, it all begins with Poincaré. As we teach our students, if we
define based loops as maps

$$\lambda : [0, 1] \to X$$

$$0, 1 \mapsto x_0,$$

then Poincaré's loop composition is not associative but does give associativity on homotopy classes. By the time of Serre's thesis [**8**] (1951),
the set of based loops was considered as a topological space ΩX and
the multiplication as a continuous map (that is, we had an H-space)
with a homotopy for associativity:

$$I \times (\Omega X)^3 \to \Omega X,$$

which was continuous because it was induced by the familiar parameter
homotopy.

1991 *Mathematics Subject Classification.* Primary 55P99, 55P35 Secondary
55P45, 18G99, 17B37, 18G55.

Research supported in part by the NSF throughout most of my career, currently
under grant DMS-9504871.

When I was a graduate student around 1958, John Moore suggested I look at the problem of determining when a cohomology class of a loop space ΩX was a 'suspension' or loop class, i.e. came from a cohomology class of X. It was known that it was necessary but not sufficient for the class to be primitive and that there was a relation to homotopy associativity [2].

In pursuing this question, I was led to work of Sugawara [12], who had a *recognition principle* for characterizing loop spaces up to homotopy type. In the course of simplifying his criteria by eliminating the use of homotopy inverses, I was led to consider higher homotopies for associativity, e.g. the pentagon

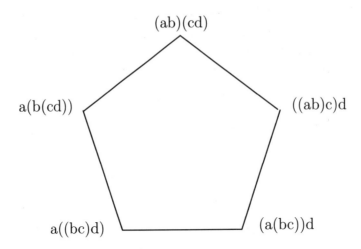

While I continued this work as a student in Oxford, Frank Adams visited and discussed his work with Mac Lane on PACTs and PROPs. With a key idea from Adams, I created the polytopes now known as *associahedra* [11]. The name is due to Gil Kalai, a geometric combinatorialist [6, 5].

The associahedron K_n can be described now as a convex polytope with one vertex for each way of associating n ordered variables, that is, ways of inserting parentheses in a meaningful way in a word of n letters. For $n = 5$, here is a portrait I copy (literally) from Masahico Saito.

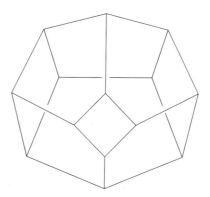

From a slightly different perspective, the 6-fold cyclic symmetry is more manifest, as in this portrait I learned from John Harer a very few years ago. (It also appears in Kapranov's paper on the permutoasso-ciahedron [4]). In fact, the associahedra can be realized so that the geometric symmetry is that of the dihedral group [6].

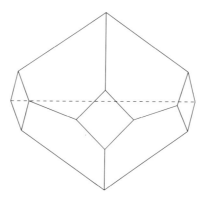

To describe all the cells of K_n, the language of planar rooted trees is helpful, as Adams indicated to me. The cells of K_n are all of the form $\Pi_i K_{n_i}$. In particular all the facets (cells of codimension 1) are of the form $K_r \times K_s$. If we let K_n be indexed by the *corolla*

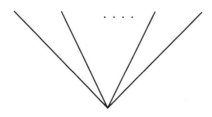

with n branches, then the facets of, e.g., K_4 are labelled as follows:

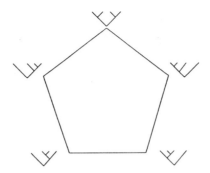

In general, the facets are labelled by grafting the s-corolla to a leaf i of the r-corolla. Denote the grafting operation by

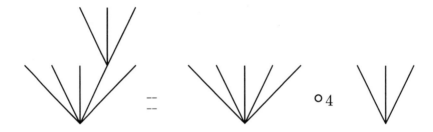

This makes the set of rooted planar trees into the non-symmetric **tree operad**. The corresponding inclusion of facets

$$\circ_i : K_r \times K_s \hookrightarrow K_{r+s-1}$$

makes the set of associahedra into a non-symmetric operad. Cells of lower dimension are indexed by trees obtained by iterated grafting and have the product form indicated above. (Perhaps because of my emphasis originally on the $\circ_i$ inclusions of facets and/or the lack of symmetric group actions, the structure of an operad here went unrecognized for some time after May's introduction of the concept. The extension to a full operad can readily be achieved by free generation: replace K_n by $\Sigma_n \times K_n$ with the obvious action on the $\circ_i$.)

Jumping ahead in time, there was an 'open' problem in combinatorial geometry from at least 1978 until solved in 1984 by Haiman as to whether there was a *convex polytope* realizing the associahedron. As I originally constructed them, the K_n were convex but curvilinear, although linearized by Milnor (unpublished) early on. Thanks to Kapranov, who bridged the two communities, the combinatorialists realized that the solution predated the question. (A particular realization, related to symplectic moment maps and toric varieties, as a truncation of the standard simplex was presented (on the cover of their book [9]) by Shnider and Sternberg; it is described in more detail in an appendix to my talk in these proceedings where it is related to a corresponding realization of the cyclohedra [**10**].)

The main result of my dissertation with regard to the associahedra was a simplification of Sugawara's recognition principle; morover, it brought to the forefront the relevance of operad structures.

THEOREM   *A connected space Y (of the homotopy type of a CW-complex) has the homotopy type of a loop space ΩX for some X if and only if there exist maps*

$$K_n \times Y^n \to Y$$

which fit together to make Y an algebra over the operad $\mathcal{K} = \{K_n\}$.

In fact, such a space X was constructed as a quotient of $\coprod K_n \times Y^n$.

Because I was dealing here with associativity only, planar trees sufficed - there was no room nor need for symmetry. I next looked at when ΩX was the loop space on an H-space X. Obviously this required homotopy commutativity. Again generalizing work of Sugawara [**13**], I introduced the hexagon which in turn was generalized to higher dimensions in Milgram's permutahedra.

From this point, the subject began to diverge. Another hexagon appears which together with the pentagon is crucial in Mac Lane's first coherence theorem [**7**] in category theory. My attempts to describe even two-fold loop spaces by specific cell complexes became too complicated, although at this conference we have with hindsight seen that things could have been done that way [**3**]. For infinite loop spaces, Boardman and Vogt [**1**] introduced what is now known as the linear isometries operad on R^∞.

Thus we come finally to the dawning of the age of operads, for which history I'll turn things over to Peter May, but recent reincarnations of the first topological operad of associahedra will appear in my second talk along with the remarkable appearance of operads in mathematical physics.

References

1. J. M. Boardman and R. M. Vogt, *Homotopy invariant algebraic structures on topological spaces*, Lecture Notes in Math., vol. 347, Springer-Verlag, 1973.
2. A.H. Copeland Jr., *On H-spaces with two non-trivial homotopy groups*, Proc. AMS **8** (1957), 184–191.
3. R. Fox and L. Neuwirth, *Braid groups*, Math. Scand. **10** (1962), 119–126.
4. M. M. Kapranov, *The permutoassociahedron, Mac Lane's coherence theorem and asymptotic zones for the KZ equation*, J. Pure and Appl. Alg. **85** (1993), 119–142.
5. C. Lee, *Some notes on triangulating polytopes*, 3. Kolloquium über Diskrete Geometrie, Institut für Mathematik, Universität Salzburg, pp. 173–181.

6. _____, *The associahedron and triangulations of the n-gon*, Europ. J. Combinatorics **10** (1989), 551–560.

7. S. Mac Lane, *Natural associativity and commutativity*, Rice Univ. Studies **49** (1963), 28–46.

8. J.P. Serre, *Homologie singulière des espaces fibrés*, Ann. of Math. **53** (1951), 425–505.

9. S. Shnider and S. Sternberg, *Quantum groups - from coalgebras to Drinfeld algebras*, Internatioonal Press Publications, Cambridge, MA and Hong Kong, 1993.

10. J. Stasheff, *An operad-chik looks at configuration spaces, moduli spaces and mathematical physics*, Operads: Proceedings of Renaissance Conferences (J.-L. Loday, J. Stasheff, and A. A. Voronov, eds.), Amer. Math. Soc., in this volume.

11. J. D. Stasheff, *On the homotopy associativity of H-spaces, I*, Trans. Amer. Math. Soc. **108** (1963), 275–292.

12. M Sugawara, *On a condition that a space is group-like*, Math. J. Okayama Univ. **7** (1957), 123–149.

13. _____, *On the homotopy-commutativity of groups and loop spaces*, Mem. Coll. Sci. Univ. Kyoto, Ser. A Math. **33** (1960/61), 257–269.

DEPARTMENT OF MATHEMATICS, UNIVERSITY OF NORTH CAROLINA, CHAPEL HILL, NC 27599-3250, USA

E-mail address: jds@math.unc.edu

Contemporary Mathematics
Volume **202**, 1997

Operads, algebras and modules

J.P. MAY

There are many different types of algebra: associative, associative and commutative, Lie, Poisson, etc., etc. Each comes with an appropriate notion of a module. As is becoming more and more important in a variety of fields, it is often necessary to deal with algebras and modules of these sorts "up to homotopy". I shall give a very partial overview, concentrating on algebra, but saying a little about the original use of operads in topology.

The development of abstract frameworks in which to study such algebras has a long history. As this conference attests, it now seems to be widely accepted that, for many purposes, the most convenient setting is that given by operads and their actions. While the notion was first written up in a purely topological framework [**19**], it was thoroughly understood by 1971 [**12**] that the basic definitions apply equally well in any underlying symmetric monoidal (= tensor) category.

The definitions and ideas had many precursors. I will indicate those that I was aware of at the time.

- Algebraists such as Kaplansky, Herstein, and Jacobson systematically studied algebras defined by different kinds of identities.
- Lawvere [**15**] formalized algebraic theories as a way of codifying different kinds of algebraic structures.
- Adams and MacLane [**16**] developed certain chain level concepts, PROP's and PACT's, with a view towards understanding the algebraic structure of the singular chain complex.
- Stasheff [**23**] introduced A_∞ spaces and constructed their classifying spaces, using associahedra and what in retrospect was an example of an operad.

1991 *Mathematics Subject Classification.* Primary 18-02, 55-02; Secondary 18C99, 55P35, 55S12.

This talk is largely based on Parts I and II of [**14**] to which the interested reader is referred for details and more complete references.

The author was supported in part by NSF Grant #DMS-9423300.

- Milgram [22] proved an approximation theorem for iterated loop spaces, using what are now known as permutahedra.
- Boardman and Vogt [3] switched from algebraic to topological PROP's and used PROP's to prove a recognition principle in infinite loop space theory.
- Beck [2] pointed out the relevance of monads to infinite loop space theory.
- Dyer and Lashof [6] systematized homology operations for iterated loop spaces as analogs of Steenrod operations in the cohomology of spaces.

It was my extraordinary good fortune to have had close mathematical contact with all of the people whom I have mentioned. I learned from Steenrod at Princeton and from Jacobson at Yale. Remarkably, all of the rest were at Chicago, or were there when the relevant work was done, or were regular visitors there.

I wanted a notion that carried the combinatorial structure familiar to me from a paper that I had written [18] that gave a general algebraic approach to Steenrod operations. The diagrams in the definition of an operad action are generalizations of diagrams that were used there to prove the Cartan formula and Adem relations. I wanted the notion to be so intimately related to monads that one could easily go back and forth between locally and globally defined structures. I had a view towards using monads to obtain a new approximation theorem for iterated loop spaces and towards using this approximation theorem together with a monadic bar construction to prove a new recognition principle for infinite loop spaces.

To these ends, I consciously sacrificed all-embracing generality: many types of algebras defined by identities are deliberately excluded. The name "operad" is a word that I coined myself, spending a week thinking about nothing else. Besides having a nice ring to it, the name is meant to bring to mind both operations and monads. Incidentally, I persuaded MacLane to discard the term "triple" in favor of "monad" in his book "Categories for the working mathematician" [17][1], which was being written about the same time. I was convinced that the notion of an operad was an important one, and I wanted the names to mesh.

What I did not foresee was just how flexible the notion would be, how many essentially different mathematical contexts there are in which it would play a natural role, how many philosophically different ways it could be exploited. Things have gone so far that I feel quite incompetent to give a thoroughgoing survey. It would be a little like surveying the use of groups in mathematics. Maybe I will feel a little more competent by the end of the conference.

---

[1] To quote MacLane (op cit, p. 134), "The frequent but unfortunate use of the word 'triple' in this sense has achieved a maximum of needless confusion, what with the conflict with ordered triple, plus the use of associated terms such as 'triple derived functors' for functors which are not three times derived from anything in the world. Hence the term *monad*."

1. The definitions of operads, algebras, and modules

To work. Let $\mathscr{S}$ be any symmetric monoidal category, with product $\otimes$ and unit object κ. In this talk, I will focus on algebraic examples and their relationship with some basic examples in topology, but there is no point in specializing the general definitions. Details are given in the handout[2]. While there are perhaps more elegant equivalent ways of writing them down, my original explicit versions of the definitions still seem to be the most convenient, especially for concrete calculational purposes.

An operad $\mathscr{C}$ consists of objects $\mathscr{C}(j)$ with actions by the symmetric groups Σ_j, a unit map $\eta : k \longrightarrow \mathscr{C}(1)$ and product maps

$$\gamma : \mathscr{C}(k) \otimes \mathscr{C}(j_1) \otimes \cdots \otimes \mathscr{C}(j_k) \to \mathscr{C}(j)$$

that are suitably associative, unital, and equivariant. The $\mathscr{C}(j)$ are thought of as parameter objects for j-ary operations that accept j inputs and produce one output.

A $\mathscr{C}$-algebra A is an object together with Σ_j-equivariant maps

$$\mathscr{C}(j) \otimes A^j \to A$$

that are suitably associative, unital, and equivariant, where A^j denotes the j-fold $\otimes$-power of A with $A^0 = \kappa$.

An A-module M is an object M together with Σ_{j-1}-equivariant maps

$$\mathscr{C}(j) \otimes A^{j-1} \otimes M \to M$$

that are suitably associative, unital, and equivariant.

These are the non-negotiable ingredients, but there are variants. If we drop all reference to symmetric groups, we obtain "non-Σ" versions of the concepts. If we insist that $\mathscr{C}(0) = \kappa$, then we call $\mathscr{C}$ a "unital operad". The resulting map $\kappa = \mathscr{C}(0) \longrightarrow A$ encodes the unit elements of algebras over such operads (and is not to be confused with the unit map $\eta : \kappa \longrightarrow \mathscr{C}(1)$, which encodes the identity operation present on any kind of algebra). Unital operads come with "augmentation maps"

$$\varepsilon : \mathscr{C}(j) \longrightarrow \kappa$$

and "degeneracy maps"

$$\mathscr{C}(j) \longrightarrow \mathscr{C}(j-1), \quad 1 \le i \le j.$$

Thinking of elements of the $\mathscr{C}(j)$ as operations, we think of $\gamma(c \otimes d_1 \otimes \cdots \otimes d_k)$ as the composite of the operation c with the tensor product of the operations d_s. When $\mathscr{S}$ has an internal Hom functor, so that

$$\mathscr{S}(X \otimes Y, Z) \cong \mathscr{S}(X, \mathrm{Hom}(Y, Z)),$$

---

[2]Reproduced in expanded form earlier in this volume.

this is made precise by the endomorphism operad $\mathrm{End}(X)$ of an object X. We set

$$\mathrm{End}(X)(j) = \mathrm{Hom}(X^j, X).$$

The unit is given by the identity map $X \to X$, the right actions by symmetric groups are given by their left actions on tensor powers, and the maps γ are given by composition of tensor products of maps. The identities that define an operad are then forced by direct calculation. An action of $\mathscr{C}$ on A can be redefined in adjoint form as a morphism of operads $\mathscr{C} \to \mathrm{End}(A)$. The identities that define a $\mathscr{C}$-algebra are then also forced by direct calculation.

There is a notion of a monoid and of a commutative monoid in any symmetric monoidal category $\mathscr{S}$. These are objects A with unit maps $\kappa \longrightarrow A$ and product maps $A \otimes A \longrightarrow A$ such that the obvious diagrams commute. There are also notions of left A-modules, which are defined in terms of maps $A \otimes M \longrightarrow M$, and of A-bimodules. These classical notions are encoded in operad actions. Assume that $\mathscr{S}$ has finite coproducts; for a finite set S, let $\kappa[S]$ be the coproduct of a copy of κ for each element of S.

EXAMPLE 1.1. The unital operad $\mathscr{M}$ in $\mathscr{S}$ has $\mathscr{M}(j) = \kappa[\Sigma_j]$. The unit map η is the identity and the product maps γ are dictated by the equivariance formulas. An $\mathscr{M}$-algebra A is the same thing as a monoid in $\mathscr{S}$. The operad action encodes all of the iterates and permutations of the product on the monoid. In terms of elements,

$$\theta(\sigma \otimes a_1 \otimes \cdots \otimes a_j) = a_{\sigma(1)} \otimes \cdots \otimes a_{\sigma(j)}.$$

An A-module M in the operadic sense is an A-bimodule in the classical sense. In terms of elements, given an operad action λ, we define

$$am = \lambda(e \otimes a \otimes m) \quad \text{and} \quad ma = \lambda(\tau \otimes a \otimes m),$$

where e and τ are the identity and transposition in Σ_2. Conversely, given an A-bimodule M, we define

$$\lambda(\sigma \otimes a_1 \otimes \cdots \otimes a_j) = a_{\sigma(1)} \cdots a_{\sigma(j)},$$

where $\sigma \in \Sigma_j$, $a_i \in A$ for $1 \le i < j$ and $a_j \in M$.

EXAMPLE 1.2. The unital operad $\mathscr{N}$ has $\mathscr{N}(j) = \kappa$ for all j. The Σ_j-actions are trivial, the unit map η is the identity, and the maps γ are the evident identifications. An $\mathscr{N}$-algebra A is the same thing as a commutative monoid in $\mathscr{S}$ and an A-module M in the operadic sense is the same thing as a left A-module. We may regard $\mathscr{N}$ as a non-Σ operad. In the non-Σ sense, an $\mathscr{N}$-algebra A is a monoid in $\mathscr{S}$, and an A-module is a left A-module in the classical sense. For any unital operad $\mathscr{C}$, the augmentations give a map $\epsilon : \mathscr{C} \to \mathscr{N}$ of operads. Therefore, by pullback along ϵ, an $\mathscr{N}$-algebra may be viewed as a $\mathscr{C}$-algebra.

2. Monadic reinterpretation of algebras

We recall some standard categorical definitions.

DEFINITION 2.1. Let $\mathscr{S}$ be any category. A monad in $\mathscr{S}$ is a functor C : $\mathscr{S} \to \mathscr{S}$ together with natural transformations $\mu : CC \to C$ and $\eta : \mathrm{Id} \to C$ such that the following diagrams commute:

$$C \xrightarrow{\eta C} CC \xleftarrow{C\eta} C \quad \text{and} \quad \begin{array}{ccc} CCC & \xrightarrow{C\mu} & CC \\ {\scriptstyle \mu C}\downarrow & & \downarrow{\scriptstyle \mu} \\ CC & \xrightarrow{\mu} & C. \end{array}$$

(left diagram: $C \xrightarrow{\eta C} CC \xleftarrow{C\eta} C$ with Id and Id arrows and μ down to C)

A C-algebra is an object A of $\mathscr{S}$ together with a map $\xi : CA \to A$ such that the following diagrams commute:

$$A \xrightarrow{\eta} CA \quad \text{and} \quad \begin{array}{ccc} CCA & \xrightarrow{C\xi} & CA \\ {\scriptstyle \mu}\downarrow & & \downarrow{\scriptstyle \xi} \\ CA & \xrightarrow{\xi} & A. \end{array}$$

(left diagram: $A \xrightarrow{\eta} CA$ with id and ξ down to A)

Taking $\xi = \mu$, we see that CX is a C-algebra for any $X \in \mathscr{S}$. It is the free C-algebra generated by X. That is, for C-algebras A, restriction along $\eta : X \to CX$ gives an adjunction isomorphism

$$C[\mathscr{S}](CX, A) \cong \mathscr{S}(X, A),$$

where $C[\mathscr{S}]$ is the category of C-algebras. Formally, we are viewing C as a functor $\mathscr{S} \to C[\mathscr{S}]$ whose composite with the forgetful functor is our original monad. Thus the monad C is determined by its algebras. Quite generally, every pair $L : \mathscr{S} \to \mathscr{T}$ and $R : \mathscr{T} \to \mathscr{S}$ of left and right adjoints determines a monad RL on $\mathscr{S}$, but many different pairs of adjoint functors can define the same monad.

Now return to our symmetric monoidal category $\mathscr{S}$ and assume that it is cocomplete. We have the following construction of the monad of free algebras over an operad $\mathscr{C}$.

DEFINITION 2.2. Define the monad C associated to an operad $\mathscr{C}$ by letting

$$CX = \coprod_{j \geq 0} \mathscr{C}(j) \otimes_{\kappa[\Sigma_j]} X^j.$$

The unit $\eta : X \to CX$ is $\eta \otimes \mathrm{id} : X = \kappa \otimes X \to \mathscr{C}(1) \otimes X$. The product

$\mu : CCX \to CX$ is induced by the following maps, where $j = \sum j_s$:

$$\mathscr{C}(k) \otimes \mathscr{C}(j_1) \otimes X^{j_1} \otimes \cdots \otimes \mathscr{C}(j_k) \otimes X^{j_k}$$

$$\downarrow \text{shuffle}$$

$$\mathscr{C}(k) \otimes \mathscr{C}(j_1) \otimes \cdots \otimes \mathscr{C}(j_k) \otimes X^{j}$$

$$\downarrow \gamma \otimes \text{id}$$

$$\mathscr{C}(j) \otimes X^{j}.$$

PROPOSITION 2.3. *A $\mathscr{C}$-algebra structure on an object A determines and is determined by a C-algebra structure on A. Formally, the identity functor on $\mathscr{S}$ restricts to give an isomorphism between the categories of $\mathscr{C}$-algebras and of C-algebras.*

PROOF. Maps $\theta_j : \mathscr{C}(j) \otimes_{\kappa[\Sigma_j]} A^j \to A$ that together specify an action of $\mathscr{C}$ on A are the same as a map $\xi : CA \to A$ that specifies an action of C on A. $\square$

Not all monads come from operads. Rather, operads single out a particularly convenient, algebraically manageable, collection of monads.

For the operad $\mathscr{M}$, the free algebra MX is the free monoid in $\mathscr{S}$ generated by X. For the operad $\mathscr{N}$, the free algebra NX is the free associative and commutative algebra generated by X.

Now suppose that $\mathscr{C}$ is a unital operad. In this case, there is a monad that is different from that defined above but that nevertheless has essentially the same algebras. Since $\mathscr{C}$ is unital, a $\mathscr{C}$-algebra A comes with a unit $\eta \equiv \theta_0 : \kappa \to A$. Thinking of η as preassigned, it is natural to change ground categories to the category of objects under κ. Working in this ground category, we obtain a reduced monad $\tilde{C}$. This monad is so defined that the units of algebras that are built in by the θ_0 component of operad actions coincide with the preassigned units η. Formally, for an object X with unit $\eta : \kappa \longrightarrow X$, the reduced monad $\tilde{C}X$ is obtained from CX as an appropriate coequalizer. Informally, elementwise, we identify $\sigma_i(c) \otimes y$ with $c \otimes s_i y$, where $c \in \mathscr{C}(j)$, $y \in X^j$, and

$$s_i = \text{id}^{i-1} \otimes \eta \otimes \text{id}^{j-i} : X^{j-1} \longrightarrow X^{j}.$$

PROPOSITION 2.4. *Let $\mathscr{C}$ be a unital operad. Then a $\mathscr{C}$-algebra structure satisfying $\eta = \theta_0$ on an object of $\mathscr{S}$ under κ determines and is determined by a $\tilde{C}$-algebra structure on A.*

The reduced construction is more general than the unreduced one.

LEMMA 2.5. *For an object X, $CX \cong \tilde{C}(X \coprod \kappa)$ as $\mathscr{C}$-algebras.*

In algebraic contexts, there is not much difference between the two constructions. There, $\mathscr{S}$ will be the category of modules over a commutative ground ring k. If X has an augmentation $\varepsilon : X \longrightarrow k$ such that $\varepsilon\eta = \text{id}$, then $X \cong \text{Ker } \varepsilon \oplus k$

and therefore $\tilde{C}X \cong C(\mathrm{Ker}\ \varepsilon)$. In topological contexts, nothing like this works and there is a huge difference between the two constructions, with the reduced construction being by far the more important one.

Modules also admit a monadic reinterpretation: there is a monad $C[1]$ in the category $\mathscr{S}^2$ such that a $C[1]$-algebra (A, M) is a C-algebra A together with an A-module M. There is a free A-module functor F_A for any $\mathscr{C}$-algebra A, and $C[1](X; Y)$ is the pair $(CX, F_{CX}(Y))$. This construction also has a reduced variant. See [**14**, I§4].

3. The specialization to algebraic operads

Let k be a commutative ring and write $\otimes$ and Hom for $\otimes_k$ and Hom_k. We shall work in the tensor category $\mathscr{M}_k$ of $\mathbb{Z}$-graded differential graded k-modules, with differential decreasing degrees by one. We implicitly use the standard convention that a sign $(-1)^{pq}$ is to be inserted whenever an element of degree p is permuted past an element of degree q. If one prefers the opposite grading convention, one can reindex chain complexes C_* by setting $C^n = C_{-n}$.

I will refer to $\mathbb{Z}$-graded chain complexes over k simply as "k-modules". As usual, we consider graded k-modules without differential to be k-modules with differential zero, and we view ungraded k-modules as graded k-modules concentrated in degree 0. These conventions allow us to view the theory of generalized algebras as a special case of the theory of differential graded generalized algebras.

When the differentials on the $\mathscr{C}(j)$ are zero, we think of $\mathscr{C}$ as purely algebraic, and it then determines an appropriate class of (differential) algebras. When the differentials on the $\mathscr{C}(j)$ are non-zero, $\mathscr{C}$ determines a class of (differential) algebras "up to homotopy", where the homotopies are determined by the homological properties of the $\mathscr{C}(j)$. Recall that a map of k-modules is said to be a quasi-isomorphism if it induces an isomorphism of homology groups.

DEFINITION 3.1. Let $\mathscr{C}$ be a unital operad with each $\mathscr{C}(j)_n = 0$ for $n < 0$. We say that $\mathscr{C}$ is acyclic if its augmentations are quasi-isomorphisms. We say that $\mathscr{C}$ is Σ-free if $\mathscr{C}(j)$ is $k[\Sigma_j]$-free for each j. We say that $\mathscr{C}$ is an E_∞ operad if it is both acyclic and Σ-free; $\mathscr{C}(j)$ is then a $k[\Sigma_j]$-free resolution of k.

By an E_∞ algebra, we mean a $\mathscr{C}$-algebra for any E_∞ operad $\mathscr{C}$. These were called "May algebras" when they were introduced by Hinich and Schechtman [**10**]. If we ignore symmetric groups, we obtain the notion of an A_∞ algebra. These are commutative and non-commutative differential graded algebras up to homotopy. Similarly, there is a class of operads that is related to Lie algebras as E_∞ operads are related to commutative algebras, and there is a concomitant notion of a differential graded Lie algebra "up to homotopy," or L_∞ algebra. Hinich and Schechtman [**11**] called these "Lie May algebras".

An E_∞ algebra A has a product for each degree zero element, necessarily a cycle, of $\mathscr{C}(2)$. Each such product is unital, associative, and commutative up to all possible coherence homotopies, and all such products are homotopic. There is a long history in topology and category theory that makes precise what these "coherence homotopies" are. However, since the homotopies are all encoded in the operad action, there is no need to be explicit. In the last lecture of the conference, I will explain a beautiful new way of thinking about A_∞ and E_∞ algebras which hides the operads in an "operadic tensor product". We will there focus on one particular E_∞ operad, but all such operads give equivalent classes of A_∞ and E_∞ algebras.

One can treat operads as algebraic systems to which one can apply versions of classical algebraic constructions. An ideal $\mathscr{I}$ in an operad $\mathscr{C}$ consists of a sequence of sub $\kappa[\Sigma_j]$-modules $\mathscr{I}(j)$ of $\mathscr{C}(j)$ such that $\gamma(c \otimes d_1 \otimes \cdots \otimes d_k)$ is in $\mathscr{I}$ if either c or any of the d_s is in $\mathscr{I}$. There is then a quotient operad $\mathscr{C}/\mathscr{I}$ with j^{th} k-module $\mathscr{C}(j)/\mathscr{I}(j)$. As observed by Ginzburg and Kapranov [9], one can construct the free operad $\mathscr{F}\mathscr{G}$ generated by any sequence $\mathscr{G} = \{\mathscr{G}(j)\}$ of $\kappa[\Sigma_j]$-modules, and one can then construct an operad that describes a particular type of algebra by quotienting out by the ideal generated by an appropriate sequence $\mathscr{R} = \{\mathscr{R}(j)\}$ of defining relations, where $\mathscr{R}(j)$ is a sub $\kappa[\Sigma_j]$-module of $(\mathscr{F}\mathscr{G})(j)$. Actually, there are two variants of the construction, one unital and one non-unital.

In many familiar examples, called quadratic operads by Ginzburg and Kapranov [9], $\mathscr{G}(j) = 0$ for $j \neq 2$ and $\mathscr{R}(j) = 0$ for $j \neq 3$. Here, if $\mathscr{G}(2)$ is $k[\Sigma_2]$ and $\mathscr{R}(3) = 0$, this reconstructs $\mathscr{M}$. If $\mathscr{G}(2) = k$ with trivial Σ_2-action and $\mathscr{R}(3) = 0$, this reconstructs $\mathscr{N}$. In these cases, we use the unital variant. If k is a field of characteristic other than 2 or 3, we can use the non-unital variant to construct an operad $\mathscr{L}$ whose algebras are the Lie algebras over k. To do this, we take $\mathscr{G}(2) = k$, with the transposition in Σ_2 acting as -1, and take $\mathscr{R}(3)$ to be the space $(\mathscr{F}\mathscr{G})(3)^{\Sigma_3}$ of invariants, which is one dimensional. Basis elements of $\mathscr{G}(2)$ and $\mathscr{R}(3)$ correspond to the bracket operation and the Jacobi identity. We shall see shortly that $\mathscr{L}$ can be realized homologically by the topological little n-cubes operads for any $n > 1$. Note that, in these "purely algebraic" examples, all $\mathscr{C}(j)$ are concentrated in degree zero, with zero differential.

The definition of a Lie algebra over a field k requires the additional relations $[x, x] = 0$ if char$(k) = 2$ and $[x, [x, x]] = 0$ if char$(k) = 3$. Purely algebraic operads are not well adapted to codify such relations with repeated variables, still less such nonlinear operations as the restriction (or p^{th} power operation) of restricted Lie algebras in characteristic p. The point is simply that the elements of an operad specify operations, and operations by their nature cannot know about special properties (such as repetition) of the variables to which they are applied.

As an aside, since in the absence of diagonals it is unclear that there is a workable algebraic analog, we note that a topological theory of E_∞ ring spaces has been developed [20, 21]. The sum and product, with the appropriate version of the distributive law, are codified in actions by two suitably interrelated operads. I may say a bit more about this in my second talk.

Fix an operad $\mathscr{C}$ and a $\mathscr{C}$-algebra A. It is clear that the category of A-modules is abelian. In fact, it is equivalent to the category of modules over the universal enveloping algebra $U(A)$ of A, where $U(A)$ is a certain differential graded algebra.

DEFINITION 3.2. Let A be a $\mathscr{C}$-algebra. The action maps

$$\lambda : \mathscr{C}(j) \otimes A^{j-1} \otimes M \to M$$

of an A-module M together define an action map

$$\lambda : C(A;k) \otimes M = C(A;M) \to M.$$

Thus $C(A;k)$ may be viewed as a k-module of operators on A-modules. The free DGA $M(C(A;k))$ generated by $C(A;k)$ therefore acts iteratively on all A-modules. Define the universal enveloping algebra $U(A)$ to be the quotient of $M(C(A;k))$ by the ideal of universal relations.

Actually, $U(A)$ can be described more economically as a quotient of $C(A;k)$.

PROPOSITION 3.3. *The category of A-modules is isomorphic to the category of $U(A)$-modules.*

It follows that the free $U(A)$-module functor gives us the free A-module functor. This implies the following identifications of $U(A)$ in special cases.

EXAMPLES 3.4. (i) For an $\mathscr{M}$-algebra A, $U(A)$ is isomorphic to $A \otimes A^{\mathrm{op}}$.

(ii) For an $\mathscr{N}$-algebra A, $U(A)$ is isomorphic to A.

(iii) For an $\mathscr{L}$-algebra L, an L-module in our sense is the same as a Lie algebra module in the classical sense, hence $U(L)$ is isomorphic to the classical universal enveloping algebra of L.

In my second talk, I shall construct a tensor product on the category of modules over an E_∞ algebra A. From the universal enveloping algebra point of view, this should look wholly implausible: a $U(A)$-module is just a left module, and, since $U(A)$ is far from being commutative, there is no obvious way to define a tensor product of A-modules, let alone a tensor product that is again an A-module. Moreover, I will give a precise description of $U(A)$ and show that it is quasi-isomorphic to A, as one would hope. Such calculations are often hard to come by, even rationally, as the work of Hinich and Schechtman [**11**] on L_∞ algebras shows.

4. Algebraic operads associated to topological operads

An operad $\mathscr{E}$ of topological spaces consists of spaces $\mathscr{E}(j)$ with right actions of Σ_j, a unit element $1 \in \mathscr{E}(1)$, and maps

$$\gamma : \mathscr{E}(k) \times \mathscr{E}(j_1) \times \cdots \times \mathscr{E}(j_k) \to \mathscr{E}(j)$$

such that the appropriate associativity, unity, and equivariance diagrams commute. For definiteness, we assume that $\mathscr{E}(0)$ is a point.

Via the singular chain complex functor, an operad $\mathscr{E}$ of spaces gives rise to an operad $C_*(\mathscr{E})$ of (differential graded) k-modules for any commutative ring k of coefficients. The operad $\mathscr{E}$ is said to an E_∞ operad if each space $\mathscr{E}(j)$ is Σ_j-free and contractible (a universal Σ_j-bundle), and $C_*(\mathscr{E})$ is then an E_∞ operad in the algebraic sense. Similarly, the chain functor C_* carries $\mathscr{E}$-algebras ($=\mathscr{E}$-spaces) to $C_*(\mathscr{E})$-algebras and carries modules over an $\mathscr{E}$-algebra to modules over the associated $C_*(\mathscr{E})$-algebra.

Taking coefficients in a field k, so that we have a Künneth isomorphism, we can go further and define homology operads.

DEFINITION 4.1. Let $\mathscr{E}$ be an operad of spaces. Define $H_*(\mathscr{E})$ to be the unital operad whose jth k-module is the graded k-module $H_*(\mathscr{E}(j))$, with algebraic structure maps γ induced by the topological structure maps. For $n \geq 0$, define $H_n(\mathscr{E})$ to be the suboperad of $H_*(\mathscr{E}(j))$ whose jth k-module is $H_{n(j-1)}(\mathscr{E}(j))$ for $j \geq 0$; in particular, the 0th k-module is zero unless $n = 0$. The degrees are so arranged that the definition makes sense. We retain the grading that comes naturally, so that the jth term of $H_n(\mathscr{E})$ is concentrated in degree $n(j-1)$, but these operads also have evident "degree zero translates".

If the spaces $\mathscr{E}(j)$ are all connected, then $H_0(\mathscr{E}) = \mathcal{N}$ and $H_*(X)$ is a commutative algebra for any $\mathscr{E}$-space X. If the spaces $\mathscr{E}(j)$ are all contractible, for example if $\mathscr{E}$ is an E_∞ operad, then $H_*(\mathscr{E}) = \mathcal{N}$. Thus, on passage to homology, E_∞ operads record only the algebra structure on the homology of $\mathscr{E}$-spaces, although the chain level operad action gives rise to homology operations, including the Dyer-Lashof and Steenrod operations. It is for this reason that topologists did not formally introduce homology operads decades ago.

In fact, there is a sharp dichotomy between the calculational behavior of operads in characteristic zero and in positive characteristic. The depth of the original topological theory lies in positive characteristic, where passage to homology operads jettisons most of the interesting structure. In characteristic zero, in contrast, the homology operads completely determine the homology of the monads E and $\tilde{E}$ associated to an operad $\mathscr{E}$. Here, for a space X,

$$EX = \coprod \mathscr{E}(j) \times_{\Sigma_j} X^j.$$

For a based space X, $\tilde{E}X$ is the quotient of EX obtained by the appropriate basepoint identifications, and $\tilde{E}X$ has a natural filtration with successive quotients

$$\mathscr{E}(j)_+ \wedge_{\Sigma_j} X^{(j)},$$

where $X^{(j)}$ denotes the j-fold smash power of X. Here the smash product $X \wedge Y$ is the quotient of the product $X \times Y$ obtained by identifying the wedge $X \vee Y$ to a point.

The calculational difference comes from a simple general fact: if a finite group π acts on a space X, then, with coefficients in a field k of characteristic zero,

$H_*(X/\pi)$ is naturally isomorphic to $H_*(X)/k[\pi]$. This leads to the following result.

THEOREM 4.2. *Let $\mathscr{E}$ be an operad of spaces. Let E_H denote the monad in the category of k-modules associated to $H_*(\mathscr{E})$ and let $\tilde{E}_H$ denote the monad in the category of k-modules under k associated to $H_*(\mathscr{E})$. If $char(k) = 0$, then*

$$H_*(EX) \cong E_H(H_*(X)) \quad and \quad H_*(\tilde{E}X) \cong \tilde{E}_H(H_*(X))$$

as $H_*(\mathscr{E})$-algebras for all spaces X, or for all based spaces in the reduced case.

This allows us to realize free algebras topologically. Recall that we have the topological (actually, discrete) versions of the operads $\mathscr{M}$ and $\mathscr{N}$: $\mathscr{M}(j) = \Sigma_j$ and $\mathscr{N}(j) = \{*\}$. For a based space X, $\tilde{M}X$ is the James construction, or free topological monoid, on X, and it is homotopy equivalent to $\Omega\Sigma X$ if X is connected. Similarly, $\tilde{N}X$ is the infinite symmetric product, or free commutative topological monoid, on X, and it is homotopy equivalent to the product over $n \geq 1$ of the Eilenberg-MacLane spaces $K(H_n(X), n)$ if X is connected. Note that the unreduced constructions MX and NX are just disjoint unions of Cartesian powers and symmetric Cartesian powers and are thus much less interesting. At least in characteristic zero, we conclude that

$$H_*(\tilde{M}X) \cong \tilde{M}_H(H_*(X)) \quad and \quad H_*(\tilde{N}X) \cong \tilde{N}_H(H_*(X)).$$

The functors $\tilde{M}_H$ and $\tilde{N}_H$ are the free and free commutative algebra functors on unital k-modules that we previously denoted by $\tilde{M}$ and $\tilde{N}$. Thus the right sides are the free and free commutative algebras generated by $\tilde{H}_*(X)$.

5. Operads, loop spaces, n-Lie algebras, and n-braid algebras

We obtain deeper examples by considering the operads that come from the study of iterated loop spaces. Their homology operads turn out to describe n-Lie algebras and n-braid algebras. Implicitly or explicitly, the case $n = 2$ has received a good deal of attention in the recent literature of string theory. Although the theorems I'm about to state were proven in the early 1970's, their statements came much later, in work of Ginzberg and Kapranov [9] and Getzler and Jones [8].

For each $n > 0$, there is a little n-cubes operad $\mathscr{C}_n$. It was invented, before the introduction of operads, by Boardman and Vogt [3]. Its jth space $\mathscr{C}_n(j)$ consists of j-tuples of little n-cubes embedded with parallel axes and disjoint interiors in the standard n-cube. There is an analogous little n-disks operad defined in terms of embeddings of little disks in the unit disk via radial contraction and translation. These are better suited to considerations of group actions and of geometry, but they do not stabilize over n. There is a more sophisticated variant, due to Steiner [24], that enjoys the good properties of both the little n-cubes and the little n-disks operads. Each of these operads comes with a canonical equivalence from its jth space to the configuration space $F(\mathbb{R}^n, j)$ of j-tuples of distinct points of $\mathbb{R}^n$. The little n-cubes operad, and any of its variants, acts naturally on all n-fold loop spaces $\Omega^n Y$.

Since $\mathscr{C}_1$ maps by a homotopy equivalence to $\mathscr{M}$, we concentrate on the case $n > 1$. When k is a field of characteristic $p > 0$, the homology of a $\mathscr{C}_n$-space, such as $\Omega^n Y$, has an extremely rich and complicated algebraic structure, carrying Browder operations and some of the Dyer-Lashof operations that are present in the homology of E_∞ algebras. I will describe the characteristic zero information and a portion of the mod p information in Cohen's exhaustive mod p calculations [**4, 5**]. (There were earlier partial calculations by Arnol'd [**1**] in the case $n = 2$.) Again, we take k to be a field.

Cohen's calculations have two starting points. One is his complete and explicit calculation of the integral homology of $F(\mathbb{R}^n, j)$, with its action by Σ_j, for all n and j. He used this to define homology operations. The other is my "approximation theorem" [**19**]. It asserts that, for a based space X, there is a natural map of $\mathscr{C}_n$-spaces $\tilde{C}_n X \longrightarrow \Omega^n \Sigma^n X$ that is an equivalence when X is connected. This allowed Cohen to combine the homology operations with the Serre spectral sequence to compute simultaneously both $H_*(\tilde{C}_n X)$ and $H_*(\Omega^n \Sigma^n X)$ for any X.

In characteristic zero, the calculations simplify drastically since calculation of the homology operads $H_*(\mathscr{C}_n)$ determines $H_*(\tilde{C}_n X)$. Cohen showed that each space $F(\mathbb{R}^n, j)$ has the same integral homology as a certain product of wedges of $(n-1)$-spheres. Therefore, the operad $H_*(\mathscr{C}_n)$ can be written additively as the reduced sum $\mathscr{N} \tilde{\oplus} H_{n-1}(\mathscr{C}_n)$ of its suboperads $\mathscr{N}$ and $H_{n-1}(\mathscr{C}_n)$, where the reduced sum is obtained from the direct sum by identifying the unit elements in $\mathscr{N}(1)$ and $H_0(\mathscr{C}_n(1))$. For $n \geq 1$, the algebras over $H_n(\mathscr{C}_{n+1})$ turn out to be the n-Lie algebras and the algebras over $H_*(\mathscr{C}_{n+1})$ turn out to be the n-braid algebras.

DEFINITION 5.1. An n-Lie algebra is a k-module L together with a map of k-modules $[\ ,\ ]_n : L \otimes L \to L$ that raises degrees by n and satisfies the following identities, where $\deg(x) = q - n$, $\deg(y) = r - n$, and $\deg(z) = s - n$.

(i) (Anti-symmetry)
$$[x, y]_n = -(-1)^{qr}[y, x]_n.$$

(ii) (Jacobi identity)
$$(-1)^{qs}[x, [y, z]_n]_n + (-1)^{qr}[y, [z, x]_n]_n + (-1)^{rs}[z, [x, y]_n]_n = 0.$$

(iii) $[x, x]_n = 0$ if $\mathrm{char}(k) = 2$ and $[x, [x, x]_n]_n = 0$ if $\mathrm{char}(k) = 3$.

Of course, a 0-Lie algebra is just a Lie algebra.

For a k-module Y and an integer n, define the n-fold suspension $\Sigma^n Y$ by $(\Sigma^n Y)_q = Y_{q-n}$.

PROPOSITION 5.2. *The category of n-Lie algebras is isomorphic to the category of Lie algebras. There is an operad $\mathscr{L}_n$ whose algebras are the n-Lie algebras, and its "degree zero translate" is isomorphic to $\mathscr{L}$.*

PROOF. For an n-Lie algebra L, $\Sigma^n L$ is a Lie algebra with bracket

$$[\Sigma^n x, \Sigma^n y] = \Sigma^n [x, y]_n.$$

Similarly, for a Lie algebra L, $\Sigma^{-n} L$ is an n-Lie algebra. $\square$

THEOREM 5.3. *If $char(k) \neq 2$ or 3, then, for all $n \geq 1$, the operad $H_n(\mathscr{C}_{n+1})$ defines n-Lie algebras. That is, $H_n(\mathscr{C}_{n+1}) \cong \mathscr{L}_n$.*

For a k-module V, let $L_n V = \Sigma^{-n} L \Sigma^n V$ be the free n-Lie algebra generated by V.

THEOREM 5.4. *Assume that $char(k) = 0$. For any based space X,*

$$(\tilde{C}_{n+1})_H(H_*(X)) \cong H_*(\tilde{C}_{n+1}X) \cong NL_n \tilde{H}_*(X)$$

is the free commutative algebra generated by the free n-Lie algebra generated by $\tilde{H}_*(X)$.

In fact, this can be interpreted as a theorem about n-braid algebras.

DEFINITION 5.5. An n-braid algebra is a k-module A that is an n-Lie algebra and a commutative DGA such that the bracket and product satisfy the following identity, where $\deg(x) = q - n$ and $\deg(y) = r - n$.
   (Poisson formula)

$$[x, yz]_n = [x, y]_n z + (-1)^{q(r-n)} y[x, z]_n.$$

The Poisson formula asserts that the map $d_x = [x, ?]_n$ is a graded derivation, in the sense that

$$d_x(yz) = d_x(y)z + (-1)^{\deg(y)\deg(d_x)} y d_x(z).$$

Batalin-Vilkovisky algebras are examples of 1-braid algebras [7], hence the general case, with non-zero differentials, is relevant to string theory. However, our concern here is with structures that have zero differential.

THEOREM 5.6. *The homology $H_*(X)$ is an n-braid algebra for any $\mathscr{C}_{n+1}$-space X and any field of coefficients. If $char(k) \neq 2$ or 3, then, for all $n \geq 1$, the algebras over the operad $H_*(\mathscr{C}_{n+1})$ are exactly the n-braid algebras. Moreover, the free n-braid algebra generated by a k-module V is isomorphic to NL_nV.*

The last statement could in principle be proven algebraically, but it is much easier to deduce it from the topology, even in characteristic zero. The n-bracket is denoted λ_n and called a Browder operation in the context of iterated loop spaces. The special characteristic 2 and 3 identities (iii) in the definition of an n-Lie algebra are of conceptual interest: they cannot be visible in the operad $H_n(\mathscr{C}_{n+1})$, but they follow directly from the chain level definition of λ_n.

6. Homology operations in characteristic p

When $\mathscr{C}$ is an E_∞ operad, an action of $\mathscr{C}$ on A builds in the kinds of higher homotopies for the multiplication of A that are the source, for example, of the Dyer-Lashof operations in the homology of infinite loop spaces and the Steenrod operations in the cohomology of general spaces. We describe the form that these operations take in the homology of general E_∞ algebras A in this section. Many other examples are known to topologists, such as the Steenrod operations in the Ext groups of cocommutative Hopf algebras and in the cohomology of simplicial restricted Lie algebras.

We take $k = \mathbb{Z}$ and consider algebras A over an integral E_∞ operad $\mathscr{C}$. Let $\mathbb{Z}_p = \mathbb{Z}/p\mathbb{Z}$ and consider the mod p homology $H_*(A; \mathbb{Z}_p)$.

THEOREM 6.1. *For $s \geq 0$, there exist natural homology operations*

$$Q^s : H_q(A; \mathbb{Z}_2) \to H_{q+s}(A; \mathbb{Z}_2)$$

and

$$Q^s : H_q(A; \mathbb{Z}_p) \to H_{q+2s(p-1)}(A; \mathbb{Z}_p)$$

if $p > 2$. These operations satisfy the following properties
  (i) $Q^s(x) = 0$ *if $p = 2$ and $s < q$ or if $p > 2$ and $2s < q$.*
  (ii) $Q^s(x) = x^p$ *if $p = 2$ and $s = q$ or if $p > 2$ and $2s = q$.*
  (iii) $Q^s(1) = 0$ *if $s > 0$, where $1 \in H_0(A; \mathbb{Z}_p)$ is the identity element.*
  (iv) *(Cartan formula) $Q^s(xy) = \sum Q^t(x) Q^{s-t}(y)$.*
  (v) *(Adem relations) If $p \geq 2$ and $t > ps$, then*

$$Q^t Q^s = \sum_i (-1)^{t+i} (pi - t, t - (p-1)s - i) Q^{s+t-i-1} Q^i;$$

if $p > 2$, $t \geq ps$, and β denotes the mod p Bockstein, then

$$Q^t \beta Q^s = \sum_i (-1)^{t+i} (pi - t, t - (p-1)s - i) \beta Q^{s+t-i} Q^i$$

$$- \sum_i (-1)^{t+i} (pi - t - 1, t - (p-1)s - i) Q^{s+t-i} \beta Q^i;$$

here $(i, j) = \dfrac{(i+j)!}{i! j!}$ if $i \geq 0$ and $j \geq 0$ (where $0! = 1$), and $(i, j) = 0$ if i or j is negative; the sums run over $i \geq 0$.

For the proof, one simply checks that one is in the general algebraic framework of a 1970 paper of mine [18] that does the relevant homological algebra once and for all. That paper should have been about operad actions. As I mentioned, however, it was actually written shortly before I invented operads. The point is that $\mathscr{C}(p)$ is a Σ_p-free resolution of $\mathbb{Z}$, so that the homology of $\mathscr{C}(p) \otimes_{\Sigma_p} A^p$ is readily computed, and computation of $\theta_* : H_*(\mathscr{C}(p) \otimes_{\Sigma_p} A^p; \mathbb{Z}_p) \to H_*(A; \mathbb{Z}_p)$ allows one to read off the operations. The Cartan formula and the Adem relations are derived from special cases of the diagrams in the definition of an operad action via calculations in the homology of groups.

Notice the grading. The first non-zero operation is the pth power, and there can be infinitely many non-zero operations on a given element. This is in marked contrast with Steenrod operations in the cohomology of spaces, where the last non-zero operation is the pth power. In fact, Steenrod operations are defined on cohomologically graded E_∞ algebras that are concentrated in positive degrees, in which context the complexes $\mathscr{C}(j)$ of the relevant E_∞ operad must be regraded as cochain complexes concentrated in negative degrees. If we systematically regrade homologically, then Dyer-Lashof and Steenrod operations both fit into the general context of the theorem, except that the adjective "Dyer-Lashof" is to be used when the underlying chain complexes are positively graded and the adjective "Steenrod" is to be used when the underlying chain complexes are negatively graded.

7. A conversion theorem

In characteristic zero, E_∞ operads carry no more homological information than the operad $\mathscr{N}$, and similarly for more general types of algebras.

LEMMA 7.1. *Let $\epsilon : \mathscr{C} \to \mathscr{P}$ be a quasi-isomorphism of operads over a field k of characteristic zero, such as the augmentation $\epsilon : \mathscr{C} \to \mathscr{N}$ of an acyclic operad. Then the map $CX \to PX$ induced by ϵ is a quasi-isomorphism for all k-modules X.*

Taking $\mathscr{P} = \mathscr{N}$ and $\mathscr{P} = \mathscr{L}$, this leads to a proof that, when k is a field of characteristic zero, E_∞ algebras are quasi-isomorphic as E_∞ algebras to commutative DGA's and L_∞ algebras are quasi-isomorphic as L_∞ algebras to differential graded Lie algebras, and similarly for modules. That is, one can convert operadic algebras and modules to genuine algebras and modules without loss of information. The proof is a typical exercise in the use of the two-sided monadic bar construction that I introduced in the same paper [**19**] in which I first defined operads.

THEOREM 7.2. *Let k be a field of characteristic zero and let $\epsilon : \mathscr{C} \to \mathscr{P}$ be a quasi-isomorphism of operads of k-complexes. Then there is a functor W that assigns a quasi-isomorphic $\mathscr{P}$-algebra WA to a $\mathscr{C}$-algebra A. There is also a functor W that assigns a quasi-isomorphic WA-module WM to an A-module M.*

An acyclic operad augments by a quasi-isomorphism to the operad $\mathscr{N}$ that defines commutative DGA's, so the following consequence is immediate.

COROLLARY 7.3. *Let k be a field of characteristic zero and let $\mathscr{C}$ be an acyclic operad of k-complexes. Then there is a functor W that assigns a quasi-isomorphic* *commutative DGA WA to a $\mathscr{C}$-algebra A. There is also a functor W that assigns a quasi-isomorphic WA-module WM to an A-module M.*

The following consequence for differential graded Lie algebras is also immediate.

COROLLARY 7.4. *Let k be a field of characteristic zero and let $\mathscr{J}$ be an operad with a quasi-isomorphism to $\mathscr{L}$. Then there is a functor W that assigns a quasi-isomorphic differential graded Lie algebra WL to a $\mathscr{J}$-algebra L. There is also a functor W that assigns a quasi-isomorphic WL-module WM to an L-module M.*

The second corollary may be of interest in applications to string theory. The motivation for the first corollary originally came from algebraic geometry, where it answered a question of Deligne concerning mixed Tate motives.

The techniques of proof also apply to replace partial algebras over operads by genuine algebras over operads, in any characteristic. This has real force. Using it, Kriz and I were able to carry out a suggestion of Deligne for the construction of categories of both integral and rational mixed Tate motives [13, 14]. However, that is a subject for another talk.

I hope that this has given a bit of a feel for some of the ways that operads work, and I look forward to learning more from all of you during the conference.

BIBLIOGRAPHY

1. V. I. Arnol'd. Cohomology of the group of dyed braids. *Math. Zametki* 5(1969), 227-231.
2. J. Beck. On *H*-spaces and infinite loop spaces, in *Springer Lecture Notes in Mathematics* Vol. 99. Springer, 1969.
3. J. M. Boardman and R. M. Vogt. *Homotopy invariant structures on topological spaces. Springer Lecture Notes in Mathematics* Vol. 347. Springer, 1973.
4. F. R. Cohen. PhD thesis, University of Chicago, 1972.
5. F. R. Cohen, T.J. Lada, and J. P. May. *The Homology of Iterated Loop Spaces. Springer Lecture Notes in Mathematics* Vol. 533. Springer, 1976.
6. E. Dyer and R. K. Lashof. Homology of iterated loop spaces. *American J. Math.* 84(1962), 35-88.
7. E. Getzler. Batalin-Vilkovisky algebras and two-dimensional topological field theories. *Comm. Math. Phys.* 159(1992), 265-285.
8. E. Getzler and J. D. S. Jones. n-Algebras and Batalin-Vilkovisky algebras. Preprint, 1992.
9. V. A. Ginzburg and M. M. Kapranov. Koszul duality for operads. *Duke Math. J.* 76(1994), 203-272.
10. V. A. Hinich and V. V. Schechtman. *On homotopy limit of homotopy algebras*, in *Springer Lecture Notes in Mathematics* Vol 1289. Springer, 1987.
11. V. A. Hinich and V. V. Schechtman. Homotopy Lie algebras. *Advances in Soviet Mathematics*, 16(1993), 1–27.
12. G. M. Kelly. On the operads of J. P. May. Preprint, University of Chicago, January 1972.
13. I. Kriz and J. P. May. Derived categories and motives. *Mathematical Research Letters*, 1(1994), 87–94.
14. I. Kriz and J. P. May. Operads, Algebras, Modules, and Motives. *Astérisque* Vol 233. 1995.
15. F. W. Lawvere. Functional semantics of algebraic theories. *Proc. Nat. Acad. Sci. USA* 50(1963), 869-872.
16. S. MacLane. Categorical algebra. Bull. Amer. Math. Soc. 71(1965), 40-106.

17. S. MacLane. *Categories for the Working Mathematician*. Springer-Verlag, 1976.
18. J. P. May. A general approach to Steenrod operations, in *Springer Lecture Notes in Mathematics* Vol 168. Springer, 1970.
19. J. P. May. *The Geometry of Iterated Loop Spaces. Springer Lecture Notes in Mathematics* Vol. 271. Springer, 1972.
20. J. P. May. E_∞ ring spaces and E_∞ ring spectra. *Springer Lecture Notes in Mathematics* Vol. 577. Springer, 1977.
21. J. P. May. Multiplicative ininite loop space theory. *J. Pure and Applied Algebra*, 26(1982), 1–69.
22. R. J. Milgram Iterated loop spaces. *Annals of Mathematics* 84(1966), 386-403.
23. J. D. Stasheff. Homotopy associativity of H-spaces I, II. *Trans. Amer. Math. Soc.* 108(1968), 457-470.
24. R. Steiner. A canonical operad pair. *Math. Proc. Camb. Phil. Soc.*, 86(1979), 443–449.

THE UNIVERSITY OF CHICAGO, CHICAGO, IL 60637
E-mail address: may@math.uchicago.edu

Contemporary Mathematics
Volume **202**, 1997

Relating the associahedron and the permutohedron

Andy Tonks

Introduction

Recently it was shown by Kapranov [4] that the combinatorics of the permuto-hedra and associahedra can be combined to give a 'hybrid' family of polytopes, the permutoassociahedra. In this short note we put forward a slightly different point of view: the associahedra can themselves be seen as retracts of the permuto-hedra. We construct a natural cellular quotient map from the permutohedron P_n to the associahedron K_{n+1}. In dimension 3 we also give K_5 as the convex hull of a particular subset of the usual vertices of P_4.

1. The quotient map

We begin by recalling the definitions of the permutohedra and the associahedra. See [4] and the references there for more details.

The *permutohedron* [5, 8] (or *zilchgon* [2], or *parallelohedron* [1]) P_n is the convex hull of the $n!$ vertices $(\pi^{-1}(1), \pi^{-1}(2), \ldots, \pi^{-1}(n)) \in \mathbb{R}^n$, for permutations $\pi \in S_n$. As a cellular complex P_n is the realization of the poset $\mathcal{P}_n$ of partitions of $\underline{n} = \{1, 2, \ldots, n\}$. That is, an $(n-r)$-cell of P_n is labelled by a tuple $(A_1, A_2, \ldots, A_r)$ of non-empty disjoint subsets of $\underline{n}$ with $\bigcup A_i = \underline{n}$. A permu-tation $\pi \in S_n$ gives a 0-cell of P_n via $A_i = \{\pi(i)\}$, and the 1-skeleton of P_n is just the Cayley graph of S_n. An r-cell $(A_i)_{i=1}^r$ is isomorphic to the product $P_{a_1} \times P_{a_2} \times \cdots \times P_{a_r}$, where $a_i = |A_i|$, and its boundary consists of those cells given by further partitioning the A_i. Note that P_n is $(n-1)$-dimensional.

The *associahedron* [9, 10] (or *Stasheff polytope*) K_n is the realization of the poset $\mathcal{K}_n$ of bracketings of n variables, or equivalently of rooted trees with n leaves or of certain subdivisions of the $(n+1)$-gon. It has dimension $n-2$. An $(n-r)$-cell of K_n corresponds to a (meaningful) insertion of $r-2$ pairs of parentheses into the expression $x_1 x_2 \ldots x_n$, or to a rooted tree with n leaves and $r-1$ internal nodes. The boundary consists of the cells obtained by inserting further parentheses into the expression. By [3, 6], the associahedron K_n may also be obtained as the convex hull of a particular collection of c_{n-1} points in $\mathbb{R}^{n+1}$, where $c_n = \frac{1}{n+1}\binom{2n}{n}$ is the nth Catalan number. These vertices correspond to the complete bracketings, the binary trees, or the triangulations of the $(n+1)$-gon.

DEFINITION 1.1. Consider a relation $\sim$ on $\mathcal{P}_n$ as follows. For a partition $(A_i)_1^r$ we say that A_{k-1} and A_k are *independent* if there exists $x \in \bigcup_{i>k} A_i$ such that

---

1991 *Mathematics Subject Classification.* Primary 51M20, 05C05.

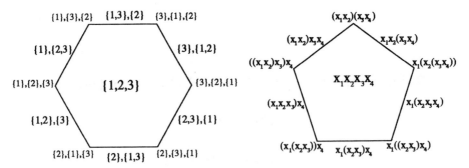

FIGURE 1. The permutohedron and associahedron of dimension 2.

$\max A_{k-1} < x < \min A_k$ or $\max A_k < x < \min A_{k-1}$. Then $\sim$ is the equivalence relation generated by

$$(A_1, A_2, \ldots, A_n) \sim (A_1, \ldots, A_{k-2}, A_{k-1} \cup A_k, A_{k+1}, \ldots, A_n)$$

if A_{k-1} and A_k are independent.

To give the motivation for this definition, consider a composite of $n+1$ variables $x_1 x_2 \ldots x_{n+1}$ which is to be evaluated. There are n composition operations to be performed, and so $n!$ ways of carrying out the evaluation, which we can think of as the vertices of P_n. Similarly we interpret a general face of P_n, given by a partition $(A_i)_{i=1}^r$, as the following evaluation procedure: carry out simultaneously ("in parallel") the composition operations between x_i and x_{i+1} for $i \in A_1$, then on the resulting terms carry out the composition operations indicated by A_2, then for A_3, and so on. An $(n-r)$-dimensional face of P_n gives a procedure for evaluating the composite $x_1 x_2 \ldots x_{n+1}$ in r stages.

To any such evaluation procedure there is an associated tree, with $n+1$ leaves labelled by the variables x_i and at least r internal nodes labelled by the compositions. Thus we have constructed a function from partitions of $\underline{n}$ to trees with $n+1$ leaves:

$$\theta : \mathcal{P}_n \to \mathcal{K}_{n+1}.$$

This respects the poset structures since taking a finer partition gives further parentheses or extra internal nodes. The function is also surjective: for any tree, choose an ordering of the internal nodes which respects the natural partial order. Such an ordering defines a composition procedure and hence a partition which under θ gives the original tree. There is a choice of ordering when two nodes in the tree correspond to terms which are to be composed later; the composition may be carried out first at one node then at the other, or both simultaneously. As in definition 1.1 we call such nodes *independent*. It is clear that θ maps two partitions to the same tree if and only if they are equivalent under the relation $\sim$. Thus $\mathcal{K}_{n+1} \cong \mathcal{P}_n/\sim$ and θ is the quotient map $\mathcal{P}_n \to \mathcal{P}_n/\sim$.

Taking the realization of the map θ gives:

PROPOSITION 1.2. *There is a natural cellular quotient map of* $(n-1)$-*dimensional complexes*

$$P_n \xrightarrow{\;\;\theta\;\;} K_{n+1}$$

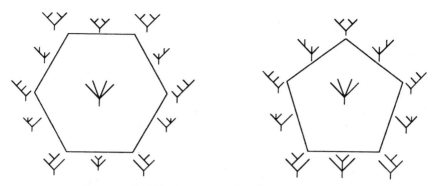

FIGURE 2. The trees associated to P_3 and K_4.

from the permutohedron to the associahedron.

The restriction of θ to the vertices is (essentially) the function from S_n to binary trees used by Loday [7].

In dimension two, θ consists of quotienting one of the edges of the hexagon to give a pentagon. We can see this arising quite naturally in homotopy theory, as follows. We consider the hexagon as the space of paths through the cube: the vertices of the former correspond to the six paths through the edges of the latter, with edges corresponding to the homotopies between paths given by the six square faces. But the cube is in turn the path space of a 4-simplex σ. Five of the faces of the cube correspond to actual homotopies of homotopies of paths, given by the faces of σ. The sixth, however, is the product of the homotopies given by $\sigma(012)$ and $\sigma(234)$. It is this square which corresponds to the "degenerate" edge of the hexagon.

2. Dimension three

Consider the function ϕ given by the restriction of $\theta : P_4 \to K_5$ to the vertices of the permutohedron P_4. We define a right inverse $\iota : K_5 \to P_4$ to ϕ with the property that for any face F of K_5 the vertices $\{\iota(v) : v$ a vertex of $F\}$ are coplanar.

For eight of the vertices $v \in K_5$ there is a unique vertex $\iota(v) \in P_4$ such that $\phi\iota(v) = v$. For the remaining vertices we make the following choices:

$$(x_1 x_2)(x_3(x_4 x_5)) \mapsto 4312 \qquad (x_1 x_2)((x_3 x_4) x_5) \mapsto 3412 \qquad ((x_1 x_2)(x_3 x_4)) x_5 \mapsto 3124$$
$$((x_1 x_2) x_3)(x_4 x_5) \mapsto 1243 \qquad (x_1(x_2 x_3))(x_4 x_5) \mapsto 2143 \qquad x_1((x_2 x_3)(x_4 x_5)) \mapsto 2431$$

We check the coplanarity of the vertices $\{\iota(v) : v$ a vertex of $F\}$ for the faces F of the associahedron. For vertices v of the pentagon $F = (x_1 x_2) x_3 x_4 x_5$ we note that the $\iota(v)$ all lie in the plane $\lambda_1 + 1 = \lambda_2$ (and of course $\sum \lambda_i = 10$) in $\mathbb{R}^4 = \{(\lambda_1, \lambda_2, \lambda_3, \lambda_4)\}$. Similarly ι maps the vertices of $F = x_1 x_2 x_3(x_4 x_5)$ to the plane $\lambda_4 + 1 = \lambda_3$. For the remaining pentagonal and square faces the vertices are mapped to vertices of original faces of the permutohedron. In fact we have

PROPOSITION 2.1. The associahedron K_5 may be defined as the convex hull of subset $\{\iota(v) : v$ a vertex of $K_5\}$ of the usual vertices of the permutohedron P_4 in $\mathbb{R}^4$. Furthermore K_5 may be obtained from P_4 by intersection with the region $\lambda_1 + 1 \geqslant \lambda_2$, $\lambda_4 + 1 \geqslant \lambda_3$.

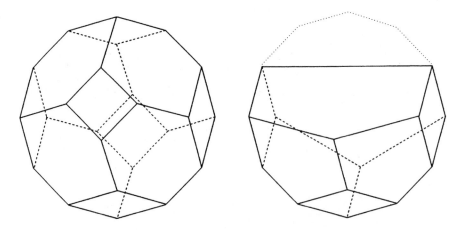

FIGURE 3. K_5 obtained from P_4 by two perpendicular cuts.

REMARK 2.2. There is no corresponding result for K_6 and P_5. The vertices of the faces $x_1(x_2x_3)x_4x_5x_6$ and $x_1x_2x_3(x_4x_5)x_6$ of K_6 would have to be mapped to the hyperplanes $\lambda_2 = 1$ and $\lambda_4 = 1$ respectively. But then ι must map the vertices of the intersection $x_1(x_2x_3)(x_4x_5)x_6$ to points with $\lambda_2 = \lambda_4 = 1$, which is clearly not the case for any vertices of P_5.

References

[1] H.-J. BAUES. *Geometry of loop spaces and the cobar construction.* Memoirs of the AMS **230** (1980).

[2] G. CARLSSON and R.J. MILGRAM. Stable homotopy and iterated loop spaces. Handbook of Algebraic Topology, edited I. M. James (North-Holland, Amsterdam, 1995), 505–583.

[3] I.M. GELFAND, M.M. KAPRANOV and A.V. ZELEVINSKY. Newton polytopes of principal A-determinants. *Soviet Math. Dokl.* **40** (1990), 278–281.

[4] M.M. KAPRANOV. The permutoassociahedron, MacLane's coherence theorem and asymptotic zones for the KZ equation. *J. Pure Appl. Algebra* **85** (1993), 119–142.

[5] A. LASCOUX and M.-P. SCHÜTZENBERGER. Symmetry and flag manifolds. Lecture Notes in Mathematics **996** (Springer, Berlin, 1983), 118–144.

[6] C. LEE. The associahedron and the triangulations of the n-gon. *European J. Combin.* **10** (1989), 551–560.

[7] J.-L. LODAY. Diassociativity. *In preparation.*

[8] R.J. MILGRAM. Iterated loop spaces. *Ann. of Math.* **84** (1966), 386–403.

[9] J.D. STASHEFF. Homotopy associativity of H-spaces I. *Trans. Amer. Math. Soc.* **108** (1963), 275–292.

[10] J.D. STASHEFF. *H-spaces from a homotopy point of view.* Lecture Notes in Mathematics **161** (Springer, Berlin, 1970).

MAX PLANCK INSTITUTE FOR MATHEMATICS, GOTTFRIED-CLAREN-STRASSE 26, 53225 BONN. GERMANY.

Contemporary Mathematics
Volume **202**, 1997

Combinatorial models for
real configuration spaces and E_n-operads

CLEMENS BERGER

ABSTRACT. We define several partially ordered sets with the equivariant homotopy type of real configuration spaces $F(\mathbb{R}^n, p)$. The main tool is a general method for constructing E_n-suboperads of a given E_∞-operad by appropriate cellular subdivision.

Introduction

The configuration space $F(\mathbb{R}^\infty, p)$ of p-tuples of pairwise distinct points of $\mathbb{R}^\infty$ can serve as universal $\mathfrak{S}_p$-bundle, the symmetric group acting freely by permutation of the p points. The main result of this paper is a combinatorial construction of the natural filtration of $F(\mathbb{R}^\infty, p)$ induced by the finite-dimensional configuration spaces.

More generally, an E_∞-operad with some extra cell structure has a combinatorially defined filtration by E_n-suboperads. As a byproduct, we obtain several partially ordered sets with the equivariant homotopy type of $F(\mathbb{R}^n, p)$. In particular, we rediscover the Smith-filtration [19] of Barratt-Eccles' Γ-functor [4] and also Milgram's permutohedral models of $F(\mathbb{R}^n, p)$ [17], [3].

We have tried to concentrate here on the combinatorial aspects of E_n-operads and to trace connections to other similar developments (cf. [1], [11]) when we were aware of them. We completely left out the application of E_n-operads to n-fold iterated loop spaces and refer the interested reader to [15], [8], [5].

The combinatorial aspects of the theory of E_n-operads have perhaps been underestimated for some time. This is quite surprising, if one considers F. Cohen's already classical computation [8] of the homology and cohomology of $F(\mathbb{R}^{n+1}, p)$ which among others identifies (in modern language) the cohomology ring with the Orlik-Solomon n-algebra of the complete graph on p vertices and the homology with the multilinear part of the free Poisson n-algebra on p generators (cf. [11]). It would be nice to have a purely combinatorial proof of this result (possibly along these lines) relating it to some surprising combinatorial work (cf. [2]).

1991 *Mathematics Subject Classification.* 18B35 06A07 20B30.
Key words and phrases. Configuration spaces, cellular E_n-operads, symmetric groups, poset-models, permutohedra.

We have divided our exposition into two parts :

Part One introduces the language of cellular E_∞-(pre)operads, defines their combinatorial filtration and relates this filtration to the equivariant homotopy type of the configuration spaces $F(\mathbb{R}^n, p)$.

Part Two discusses three basic examples : the configuration preoperad F itself, the simplicial operad Γ and the permutohedral operad J.

I would like to thank Takuji Kashiwabara for helpful remarks and suggestions and Zig Fiedorowicz for his careful and prompt reading of the first draft, which led to an important comparison theorem.

1. Homotopy uniqueness of cellular E_n-operads.

Following Cohen, May and Taylor [9] we neglect at first the multiplicative structure of an *operad* and retain only the "functorial" part of the structure which will be sufficient to determine the homotopy types we are interested in.

DEFINITION 1.1. Define Λ to be the category whose objects are the finite (non empty) sets $\mathbf{p} = \{1, 2, \ldots, p\}$ and whose morphisms are the injective maps.

A *preoperad* with values in the category $\mathcal{C}$ is a contravariant functor $\mathcal{O} : \Lambda \to \mathcal{C}$, written $(\mathcal{O}_p)_{p>0}$ on objects and $\phi^* : \mathcal{O}_q \to \mathcal{O}_p$ on morphisms $\phi \in \Lambda(\mathbf{p}, \mathbf{q})$.

A *map of preoperads* is a natural transformation of functors. If there is a notion of (weak) equivalence in $\mathcal{C}$, we shall call *(weak) Λ-equivalence* any map of preoperads $f : \mathcal{O} \to \mathcal{O}'$ such that for each $p > 0$, the induced map $f_p : \mathcal{O}_p \to \mathcal{O}'_p$ is a (weak) equivalence. Two preoperads will then be called (weakly) Λ-equivalent if they can be joined by a chain of not necessarily composable (weak) Λ-equivalences.

These notions apply in particular to partially ordered, simplicial and topological preoperads. The *nerve* functor $\mathcal{N} : \mathbf{Poset} \to \mathbf{Set}^\Delta$ transforms by composition partially ordered preoperads into simplicial preoperads and the *realization* functor $|-| : \mathbf{Set}^\Delta \to \mathbf{Top}$ transforms simplicial into topological preoperads. Both transformations preserve weak equivalences.

NOTATION 1.2. Each morphism of the category Λ decomposes uniquely in a *bijection* followed by an *increasing* map. For $\phi \in \Lambda(\mathbf{p}, \mathbf{q})$, we write

$$\phi = \phi^{inc} \circ \phi^\natural,$$

with $\phi^\natural \in \Lambda(\mathbf{p}, \mathbf{p})$ and $\phi^{inc} \in \Lambda^{inc}(\mathbf{p}, \mathbf{q}) = \{\phi \in \Lambda(\mathbf{p}, \mathbf{q}) | \phi(i) < \phi(j) \text{ for } i < j\}$.

For p distinct integers $i_1, \ldots, i_p$ in $\mathbf{q}$ we shall denote

$$\phi_{i_1, \ldots, i_p} : \mathbf{p} \to \mathbf{q}$$

the morphism which maps $(1, \ldots, p)$ onto $(i_1, \ldots, i_p)$.

EXAMPLES 1.3. *(a) The symmetric groups and their universal bundles.*
The collection of symmetric groups $\mathfrak{S}_p = \Lambda(\mathbf{p}, \mathbf{p})$ defines a set-valued preoperad $\mathfrak{S} : \Lambda \to \mathbf{Set}$ by setting for $\phi \in \Lambda(\mathbf{p}, \mathbf{q})$:

$$\begin{aligned} \phi^* : \mathfrak{S}_q &\to \mathfrak{S}_p \\ \sigma &\mapsto (\sigma \circ \phi)^\natural. \end{aligned}$$

Composing $\mathfrak{S}$ with the universal bundle construction $W : \mathbf{Set} \to \mathbf{Set}^\Delta$ one gets a simplicial preoperad $\Gamma = W \circ \mathfrak{S}$ whose rich combinatorial structure has been studied by Barratt-Eccles [4] and Smith [19].

(b) The configuration preoperad.

The collection of configuration spaces $F_p = F(\mathbb{R}^\infty, p)$ defines a topological preoperad $F : \Lambda \to \mathbf{Top}$ by setting for $\phi \in \Lambda(\mathbf{p}, \mathbf{q})$:

$$\phi^* : F_q \quad \to \quad F_p$$
$$(x_1, \ldots, x_q) \quad \mapsto \quad (x_{\phi(1)}, \ldots, x_{\phi(p)}).$$

Like Γ_p each F_p is a universal $\mathfrak{S}_p$-bundle. This suggests some relationship between the two. Indeed Smith [19] constructed a filtration $\Gamma_p^{(n)}$ of the simplicial set Γ_p which was shown by Kashiwabara [14] to be homotopy equivalent to the geometric filtration $F_p^{(n)} = F(\mathbb{R}^n, p) \times \mathbb{R}^\infty$ of the configuration preoperad F. This result was the starting point of our investigation, and we shall see below that this filtered homotopy equivalence is based on some functorially constructed cell decompositions of both preoperads.

(c) The complete graph preoperad.

Let $\mathbb{N}^{\binom{p}{2}}$ denote the cartesian product of $\binom{p}{2}$ copies of the set $\mathbb{N}$ of natural numbers. An element $\mu \in \mathbb{N}^{\binom{p}{2}}$ will be written with a *double index* : $\mu = (\mu_{ij})_{1 \leq i < j \leq p}$. Such an element is most naturally interpreted as an *edge-labeling* (by natural numbers) of the complete graph on p vertices. The collection of these labeling sets $\mathbb{N}^{\binom{p}{2}}$ extends to a (set-valued) preoperad. For $\phi \in \Lambda(\mathbf{p}, \mathbf{q})$ the induced map $\phi^* : \mathbb{N}^{\binom{q}{2}} \to \mathbb{N}^{\binom{p}{2}}$ is given by the evident formula :

$$\phi^*(\mu)_{ij} = \begin{cases} \mu_{\phi(i), \phi(j)} & \text{if } \phi(i) < \phi(j); \\ \mu_{\phi(j), \phi(i)} & \text{if } \phi(j) < \phi(i). \end{cases}$$

DEFINITION 1.4. Let $\mathcal{K} : \Lambda \to \mathbf{Poset}$ be the partially ordered preoperad defined by $\mathcal{K}_p = \mathbb{N}^{\binom{p}{2}} \times \mathfrak{S}_p$ and for $\phi \in \Lambda(\mathbf{p}, \mathbf{q})$:

$$\phi^* : \mathcal{K}_q \quad \to \quad \mathcal{K}_p$$
$$(\mu, \sigma) \quad \mapsto \quad (\phi^*(\mu), \phi^*(\sigma)),$$

where the partial order on $\mathcal{K}_p$ is given by

$$(\mu, \sigma) \leq (\nu, \tau) \Leftrightarrow \forall i < j \text{ either } \phi_{ij}^*(\mu, \sigma) = \phi_{ij}^*(\nu, \tau) \text{ or } \mu_{ij} < \nu_{ij}.$$

REMARK 1.5. The universal $\mathfrak{S}_2$-bundle can be realized as the unit-sphere S^∞ in $\mathbb{R}^\infty$, the non trivial element of $\mathfrak{S}_2$ acting as antipodal map. The minimal CW-structure of S^∞ compatible with this action is given by the hemispheres of each dimension. The set of these cells, ordered by inclusion, is canonically isomorphic to $\mathcal{K}_2$. The partially ordered sets $\mathcal{K}_p$ serve to define analogous cell decompositions of the universal $\mathfrak{S}_p$-bundles. Observe in particular that the partial order on $\mathcal{K}_p$ is the least fine partial order such that all maps $\phi^* : \mathcal{K}_p \to \mathcal{K}_2$ are monotone. Formally,

$$(\mu, \sigma) \leq (\nu, \tau) \Leftrightarrow \phi_{ij}^*(\mu, \sigma) \leq \phi_{ij}^*(\nu, \tau) \text{ for all } i, j.$$

DEFINITION 1.6. Let A be a partially ordered set and X a topological space. A collection $(c_\alpha)_{\alpha \in A}$ of closed contractible subspaces (the *"cells"*) of X will be called a *cellular A-decomposition* of X if the following three conditions hold :

1. $c_\alpha \subseteq c_\beta \Leftrightarrow \alpha \leq \beta$;
2. the cell inclusions are (closed) cofibrations ;
3. $X = \varinjlim_A c_\alpha$, so X equals the union of its cells and has the weak topology with respect to its cells.

LEMMA 1.7. *If a topological space X admits a cellular A-decomposition, then there is a cellular homotopy equivalence from X to $|\mathcal{N}A|$.*

PROOF. – Since cell inclusions are closed cofibrations, the homotopy colimit h-$\varinjlim_A c_\alpha$ contains the ordinary colimit $\varinjlim_A c_\alpha$ as a deformation retract. On the other hand, contracting the cells to a point defines a homotopy equivalence from the homotopy colimit to the realization of the nerve $|\mathcal{N}A| = \text{h-}\varinjlim_A | * |$.   □

REMARK 1.8. Condition (1) of a cellular A-decomposition can be replaced by a weaker condition without losing property (1.7). To this purpose let us formally define the *cell-interior* $\check{c}_\alpha$ to be the difference

$$\check{c}_\alpha = c_\alpha \setminus (\bigcup_{\beta < \alpha} c_\beta).$$

Suppose now that instead of (1) we have only

(1′) $\alpha \leq \beta \Rightarrow c_\alpha \subseteq c_\beta$, while equivalence holds if $\check{c}_\alpha \neq \emptyset$.

By (1′) and (3), each cell is the union of cells with nonempty interior. Thus, X is the colimit over A as well as the colimit over the partially ordered set A' of cells with nonempty interior. But for A', condition (1) holds, so by Lemma 1.7, there is a homotopy equivalence from X to $|\mathcal{N}A'|$.

On the other hand, by Quillen's Theorem A, the poset inclusion of A' into A induces a homotopy equivalence $|\mathcal{N}A'| \xrightarrow{\sim} |\mathcal{N}A|$ since the "homotopy fibers" $i_\alpha = \{\beta \in A' \,|\, \beta \leq \alpha\}$ are all contractible, again by Lemma 1.7 : $c_\alpha \xrightarrow{\sim} |\mathcal{N}i_\alpha|, \alpha \in A$. Hence, Lemma 1.7 remains valid for cellular A-decompositions which satisfy only (1′), (2), (3).

To facilitate language we shall call cells with nonempty cell-interior *proper* cells and those with empty cell-interior *improper* cells. What we have shown reads as follows : only the poset of proper cells forms a cell-decomposition in the strict sense, but adjoining improper cells does not modify the homotopy type of the poset as long as the improper cells are contractible.

DEFINITION 1.9. A topological preoperad $\mathcal{O}$ is called a *cellular E_∞-preoperad* if the $\mathfrak{S}_2$-space $\mathcal{O}_2$ admits a cellular $\mathcal{K}_2$-decomposition $(\mathcal{O}_2^{(\alpha)})_{\alpha \in \mathcal{K}_2}$, compatible with the action of $\mathfrak{S}_2$, such that

1. for each $p > 0$ and each $\alpha \in \mathcal{K}_p$ the formally defined "cell"

$$\mathcal{O}_p^{(\alpha)} = \bigcap_{1 \leq i < j \leq p} (\phi_{ij}^*)^{-1}(\mathcal{O}_2^{\phi_{ij}^\bullet(\alpha)})$$

is contractible, and for each $\alpha, \beta \in \mathcal{K}_p$ with $\alpha \leq \beta$ the natural "cell-inclusion" $\mathcal{O}_p^{(\alpha)} \subseteq \mathcal{O}_p^{(\beta)}$ is a cofibration ;

2. each $\mathfrak{S}_p$-orbit of $\mathcal{O}_p$ contains an *ordered* point, i.e. a point $x \in \mathcal{O}_p$ whose projections $\phi_{ij}^*(x)$ belong to cell-interiors of the form $\check{\mathcal{O}}_2^{(\mu, id_2)}$.

NOTATION 1.10. A cellular E_∞-preoperad $\mathcal{O}$ induces the partially ordered preoperad $\mathcal{K}(\mathcal{O})$ consisting of those elements of the complete graph preoperad $\mathcal{K}$ which index the *proper* cells of $\mathcal{O}$ (cf. 1.8). The formally defined cell-interior $\check{\mathcal{O}}_p^{(\alpha)}$ can also be defined as an intersection of inverse images of cell-interiors :

$$\check{\mathcal{O}}_p^{(\alpha)} = \bigcap_{1 \leq i < j \leq p} (\phi_{ij}^*)^{-1}(\check{\mathcal{O}}_2^{\phi_{ij}^\bullet(\alpha)}).$$

This shows that the partially ordered sets $\mathcal{K}(\mathcal{O})_p = \{\alpha \in \mathcal{K}_p \mid \breve{\mathcal{O}}_p^{(\alpha)} \neq \emptyset\}$ are indeed subposets of $\mathcal{K}_p$ closed under the operations of the category Λ.

On the other hand, the natural filtration of the complete graph preoperad $\mathcal{K}$ by $\mathcal{K}^{(n)} = \{(\mu, \sigma) \in \mathcal{K} \mid \mu_{ij} < n \text{ for } i < j\}$ induces a filtration $\mathcal{O}^{(n)}$ of the cellular E_∞-preoperad $\mathcal{O}$; explicitly :

$$\mathcal{O}_p^{(n)} = \bigcup_{\alpha \in \mathcal{K}_p^{(n)}} \mathcal{O}_p^{(\alpha)}.$$

It follows at once that for each p, the filtration of $\mathcal{O}_p$ is induced from the canonical filtration of $\mathcal{O}_2$ through the projections ϕ_{ij}^*.

Note the dimensional shift : $\mathcal{O}_2^{(n)}$ has the equivariant homotopy type of an $(n-1)$-dimensional sphere.

Topological preoperads of the form $\mathcal{O}^{(n)}$ (for a cellular E_∞-preoperad $\mathcal{O}$) will be called *cellular E_n-preoperads*.

THEOREM 1.11. *Let $\mathcal{O}$ be a cellular E_∞-preoperad.*

(a) For each p, the set of proper cells $(\mathcal{O}_p^{(\alpha)})_{\alpha \in \mathcal{K}(\mathcal{O})_p}$ defines a cellular $\mathcal{K}(\mathcal{O})_p$-decomposition of $\mathcal{O}_p$.

(b) The inclusion of $\mathcal{K}(\mathcal{O})$ in $\mathcal{K}$ is a filtered Λ-equivalence of partially ordered preoperads. In particular, there is a Λ-equivalence $\mathcal{O}^{(n)} \xrightarrow{\sim} |\mathcal{N}\mathcal{K}^{(n)}|$. Hence, any two cellular E_n-preoperads are Λ-equivalent ($1 \leq n \leq \infty$).

PROOF. – In view of Remark 1.8, it remains to show that the cells $\mathcal{O}_p^{(\alpha)}, \alpha \in \mathcal{K}_p$, satisfy the weak form of a cell-decomposition, i.e. conditions (1′), (2), (3) of (1.6-1.8). For this, let $\mathcal{O}_p^{(\alpha)}$ be a *proper* cell such that $\mathcal{O}_p^{(\alpha)} \subseteq \mathcal{O}_p^{(\beta)}$ and suppose (by contraposition) that α does not precede β in $\mathcal{K}_p$. Then there is a map $\phi \in \Lambda(\mathbf{2}, \mathbf{p})$ such that $\phi^*(\alpha)$ does not precede $\phi^*(\beta)$ in $\mathcal{K}_2$; we thus have empty intersections $\breve{\mathcal{O}}_2^{\phi^*(\alpha)} \cap \mathcal{O}_2^{\phi^*(\beta)}$ and $\breve{\mathcal{O}}_p^{(\alpha)} \cap \mathcal{O}_p^{(\beta)}$, in contradiction with the hypothesis.

Furthermore, as each cell is the union of proper cells, the colimit condition (3) is equivalent to the statement that the space $\mathcal{O}_p$ decomposes into a disjoint union of cell-interiors $\breve{\mathcal{O}}_p^{(\alpha)}, \alpha \in \mathcal{K}(\mathcal{O})_p$. Now, given a point $x \in \mathcal{O}_p$, there are unique indices $(\mu_{ij}^x, \sigma_{ij}^x) \in \mathcal{K}_2$ such that $\phi_{ij}^*(x)$ belongs to the cell-interior $\breve{\mathcal{O}}_2^{(\mu_{ij}^x, \sigma_{ij}^x)}$. By condition (1.9.2) there is also a permutation $\sigma \in \mathfrak{S}_p$ such that $(\sigma^{-1})^*(x)$ is an *ordered* point, which is unique because of the relation $\sigma_{ij}^x = \phi_{ij}^*(\sigma)$ for all i, j. The cell-interior $\breve{\mathcal{O}}_p^{(\alpha)}$ containing x has thus index $\alpha = (\mu, \sigma)$ with $\mu = (\mu_{ij}^x)_{1 \leq i < j \leq p}$ and $\sigma = (\sigma_{ij}^x)_{1 \leq i < j \leq p}$.   □

REMARK 1.12. The preceding theorem suggests a slight generalization of the concept of a cellular E_∞-preoperad $\mathcal{O}$. All we need for the comparison with the complete graph preoperad is the contractibility of the *proper* cells and the filtered Λ-equivalence $\mathcal{K}(\mathcal{O}) \xrightarrow{\sim} \mathcal{K}$. The contractibility of the improper cells is not the only way of obtaining the latter equivalence, cf. Quillen's Theorem B [18].

The configuration preoperad F is actually a cellular preoperad with contractible proper cells but some noncontractible improper cells, yet the inclusion of $\mathcal{K}(F)$ into $\mathcal{K}$ is a filtered Λ-equivalence. Moreover, the cell-structure of F gives $\mathfrak{S}_p$-equivariant homeomorphisms $F_p^{(n)} \cong F(\mathbb{R}^n, p) \times \mathbb{R}^\infty$ relating very naturally the filtration of the complete graph preoperad to the finite-dimensional configuration spaces $F(\mathbb{R}^n, p)$.

On the other hand, the *Smith-filtration* [19] of $\Gamma_p = W\mathfrak{S}_p$ coincides with the filtration $\Gamma_p^{(n)}$ formally derived from the cellular E_∞-structure of Γ. The comparison theorem thus gives the following corollary which was conjectured by Smith and proved by Kashiwabara [14] :

COROLLARY 1.13. *The realization of the simplicial set $\Gamma_p^{(n)}$ has the same $\mathfrak{S}_p$-equivariant homotopy type as the real configuration space $F(\mathbb{R}^n, p)$.*

The above corollary is true for each cellular E_n-preoperad $\mathcal{O}^{(n)}$. In the next chapter we shall examine several cellular E_n-preoperads from a combinatorial viewpoint. Often, they come equipped with a combinatorially defined multiplication

$$m_{i_1\ldots i_p}^{\mathcal{O}^{(n)}} : \mathcal{O}_p^{(n)} \times \mathcal{O}_{i_1}^{(n)} \times \cdots \times \mathcal{O}_{i_p}^{(n)} \quad \to \quad \mathcal{O}_{i_1+\cdots+i_p}^{(n)}$$
$$(z, z_1, \ldots, z_p) \quad \mapsto \quad z(z_1, \ldots, z_p)$$

turning them into an E_n-operad. We refer the reader to [16], [15] or [5] for the exact definition of an operad and more specifically for the relationship between E_n-operads and n-fold iterated loop spaces. The *degeneracy* operators of a (unital) operad define the actions by increasing maps of the category Λ so that each (unital) operad has an underlying preoperad structure, see [16], Variant 4(iii).

DEFINITION 1.14. An operad $\mathcal{O}$ is called a *cellular E_∞-operad* if the underlying preoperad is a cellular E_∞-preoperad such that the multiplication $m_{i_1\ldots i_p}^{\mathcal{O}}$ preserves the cellular structure in the sense specified below (1.15b).
We then call the suboperads $\mathcal{O}^{(n)}$ *cellular E_n-operads.*

EXAMPLES 1.15. (a) *The permutation operad.*
The set-valued preoperad $\mathfrak{S}$ is in fact an operad with the obvious unit $1 \in \mathfrak{S}_1$ and multiplication given by

$$m_{i_1\ldots i_p}^{\mathfrak{S}} : \mathfrak{S}_p \times \mathfrak{S}_{i_1} \times \cdots \times \mathfrak{S}_{i_p} \quad \to \quad \mathfrak{S}_{i_1+\cdots+i_p}$$
$$(\sigma; \sigma_1, \ldots, \sigma_p) \quad \mapsto \quad \sigma(i_1, \ldots, i_p) \circ (\sigma_1 \oplus \cdots \oplus \sigma_p),$$

where $\sigma(i_1, \ldots, i_p)$ permutes the p subsets $\mathbf{i_k} \hookrightarrow \mathbf{i_1} + \cdots + \mathbf{i_p}$ according to σ.

(b) *The complete graph operad $\mathcal{K}$.*
The preoperad $\mathcal{K}$ is an operad with obvious unit $1 \in \mathcal{K}_1$ and multiplication

$$m_{i_1\ldots i_p}^{\mathcal{K}} : \mathcal{K}_p \times \mathcal{K}_{i_1} \times \cdots \times \mathcal{K}_{i_p} \quad \to \quad \mathcal{K}_{i_1+\cdots+i_p}$$
$$((\mu, \sigma); (\mu_1, \sigma_1), \ldots, (\mu_p, \sigma_p)) \quad \mapsto \quad (\mu(\mu_1, \ldots, \mu_p), \sigma(\sigma_1, \ldots, \sigma_p)),$$

where $\sigma(\sigma_1, \ldots, \sigma_p) = \sigma(i_1, \ldots, i_p) \circ (\sigma_1 \oplus \cdots \oplus \sigma_p)$ as above, and where the edge-labeling $\mu(\mu_1, \ldots, \mu_p)$ of the complete graph on $i_1 + \cdots + i_p$ vertices is defined by the following formula (for sake of precision $\psi_r : \mathbf{i_r} \hookrightarrow \mathbf{i_1} + \cdots + \mathbf{i_p}$ denotes the canonical inclusion) :

$$\mu(\mu_1, \ldots, \mu_p)_{jk} = \begin{cases} (\mu_r)_{\psi_r^{-1}(j), \psi_r^{-1}(k)} & \text{if } j, k \in \psi_r(\mathbf{i_r}), \\ \mu_{rs} & \text{if } j \in \psi_r(\mathbf{i_r}) \text{ and } k \in \psi_s(\mathbf{i_s}), r < s. \end{cases}$$

In other words, on edges of the complete subgraph spanned by $\psi_r(\mathbf{i_r})$ the labeling $\mu(\mu_1, \ldots, \mu_p)$ coincides with μ_r, whereas on edges joining vertices of different subsets $\psi_r(\mathbf{i_r})$ and $\psi_s(\mathbf{i_s})$ the labeling $\mu(\mu_1, \ldots, \mu_p)$ is induced by μ.
The complete graph operad $\mathcal{K}$ is thus a cellular E_∞-operad, naturally filtered by suboperads $\mathcal{K}^{(n)}$. They will serve as *universal models* for cellular E_n-operads. Indeed, given an arbitrary cellular E_∞-operad $\mathcal{O}$, we assume that the multiplication $m_{i_1\ldots i_p}^{\mathcal{O}}$ sends each cell-product into the cell prescribed by the complete graph

operad:

$$m^{\mathcal{O}}_{i_1\ldots i_p}(\mathcal{O}^{(\mu,\sigma)}_p \times \mathcal{O}^{(\mu_1,\sigma_1)}_{i_1} \times \cdots \times \mathcal{O}^{(\mu_p,\sigma_p)}_{i_p}) \subseteq \mathcal{O}^{(\mu(\mu_1,\ldots,\mu_p),\sigma(\sigma_1,\ldots,\sigma_p))}_{i_1+\cdots+i_p}$$

This implies that multiplication is filtration-preserving, making our definition of a cellular E_n-operad meaningful. Both parts of the following theorem are due to Zig Fiedorowicz and improve considerably an earlier "up to homotopy" version. In particular, cellular E_n-operads are actually E_n-operads in May's [15] sense endowed with some extra cell-structure.

THEOREM 1.16. *(Fiedorowicz). Any two cellular E_n-operads are multiplicatively Λ-equivalent (i.e. equivalent as operads). Moreover, the little n-cubes operad of Boardman-Vogt has the structure of a cellular E_n-operad.* $(1 \leq n \leq \infty)$

PROOF. – The main lemma of Section 5 of [1] shows that the collection of homotopy colimits $(\text{h-}\varinjlim_{\alpha \in \mathcal{K}_p} \mathcal{O}^{(\alpha)}_p)_{p>0}$ defines a topological operad. This operad retracts by multiplicative Λ-equivalences onto the given E_∞-operad $\mathcal{O}$ as well as onto the complete graph operad $|\mathcal{N}\mathcal{K}|$, cf. 1.7-1.8. It follows that any cellular E_n-operad is multiplicatively Λ-equivalent to the n-th filtration of the complete graph operad, whence the first part of the theorem.

Let $\mathcal{C}([0,1]^n, p)$ denote the space of p-fold configurations of (open) "little n-cubes" (cf. [7], [15]) and $\mathcal{C}_p$ the inductive limit $\varinjlim_n \mathcal{C}([0,1]^n, p)$, where $\mathcal{C}([0,1]^n, p)$ is embedded in $\mathcal{C}([0,1]^{n+1}, p)$ as the space of those little $(n+1)$-cubes having last coordinate equal to $id_{]0,1[}$.

For $(c_1, c_2) \in \mathcal{C}_2$ we write $c_1 \square_\mu c_2$ if c_1 and c_2 are separated by a hyperplane H_i perpendicular to the i-th coordinate axis for some $i \leq \mu + 1$ such that, whenever there is no separating hyperplane H_i for $i < \mu + 1$, the left cube c_1 lies on the negative side of $H_{\mu+1}$ and the right cube c_2 on the positive side of $H_{\mu+1}$. In the latter case we write more precisely $c_1 \; \square_\mu \; c_2$.

For $(\mu, \sigma) \in \mathcal{K}_p$ we then define the associated cell by

$$\mathcal{C}^{(\mu,\sigma)}_p = \{(c_1, c_2, \ldots, c_p) \in \mathcal{C}_p \mid c_i \underset{\mu_{ij}}{\square} c_j \text{ if } \sigma(i) < \sigma(j), \text{ and } c_j \underset{\mu_{ij}}{\square} c_i \text{ if } \sigma(j) < \sigma(i)\}.$$

These cells endow the little cubes operad $(\mathcal{C}_p)_{p>0}$ with the structure of a cellular E_∞-operad. In particular, the little cubes multiplication plainly preserves this cellular structure in the aforementioned sense (1.15b). The only subtle point is the *contractibility* of the cells; we shall sketch a proof.

Suppose first that the cell $\mathcal{C}^{(\mu,\sigma)}_p$ contains an interior point $(c_1, \ldots, c_p) \in \breve{\mathcal{C}}^{(\mu,\sigma)}_p$. By the definition of the cell-interior (1.8), this means that $c_i \square_{\mu_{ij}} c_j$ if $\sigma(i) < \sigma(j)$, and $c_j \square_{\mu_{ij}} c_i$ if $\sigma(j) < \sigma(i)$. Furthermore, if $(\mu, \sigma) \in \mathcal{K}^{(n)}_p$, the cell $\mathcal{C}^{(\mu,\sigma)}_p$ projects onto $\mathcal{C}^{(\mu,\sigma)}_p \cap \mathcal{C}([0,1]^n, p)$ by a fibration with contractible fibers; it will thus be sufficient to show the base is contractible. Indeed, the projected cell contracts to the projected configuration $(\bar{c}_1, \ldots, \bar{c}_p) \in \breve{\mathcal{C}}^{(\mu,\sigma)}_p \cap \mathcal{C}([0,1]^n, p)$ by an n-step *affine* contraction which deforms the little n-cubes *coordinatewise*, beginning with the last coordinate and ending with the first. The descending order guarantees that the contraction stays within $\mathcal{C}^{(\mu,\sigma)}_p \cap \mathcal{C}([0,1]^n, p)$ since the appropriate separating hyperplanes exist at each moment of the deformation.

It remains to show that *improper* cells are contractible. In our case, even more is true ; each cell $\mathcal{C}^{(\mu,\sigma)}_p$ contains a unique maximal proper cell $\mathcal{C}^{(\hat{\mu},\hat{\sigma})}_p$ with which it can be identified, or equivalently : the inclusion of $\mathcal{K}(\mathcal{C})_p$ into $\mathcal{K}_p$ has a right adjoint

$(\mu, \sigma) \mapsto (\hat{\mu}, \hat{\sigma})$. The explicit formula for this right adjoint is given by $\hat{\sigma} = \sigma$ and $\hat{\alpha}_{ij} = \min\{a \mid \alpha_{i,i_1} = \alpha_{i_1,i_2} = \cdots = \alpha_{i_s,j} = a$ for $i < i_1 < \cdots < i_s < j\}$, where we have used the indexing 2.2 for $\alpha = (\mu, \sigma)$ and $\hat{\alpha} = (\hat{\mu}, \hat{\sigma})$.

The suboperad $\mathcal{C}^{(n)}$ of the little cubes operad is thus a cellular E_n-operad which projects, as above, onto the little n-cubes operad by a multiplicative Λ-equivalence. □

2. On the combinatorial structure of cellular E_n-operads.

As Theorem 1.11 suggests there is some fruitful interplay between the combinatorial structure and the geometry of a cellular E_n-operad $\mathcal{O}^{(n)}$ because the subposet $\mathcal{K}(\mathcal{O})_p^{(n)}$ of $\mathcal{K}_p^{(n)}$ defined by the proper cells represents in its own right a combinatorial model of the equivariant homotopy type of the configuration space $F(\mathbb{R}^n, p)$. So, there might be cellular E_n-operads which are "combinatorially smaller" than others.

We shall compare here three cellular E_∞-(pre)operads : the *configuration* preoperad F with its natural filtration by "dimension", the *simplicial* operad Γ of Barratt-Eccles with its Smith-filtration and the *permutohedral* operad J which is based on Milgram's combinatorial models of iterated loop spaces.

EXAMPLE 2.1. *The cell structure of the configuration preoperad F.*
The points of $\mathbb{R}^\infty$ will be written as real number series $(x^{(i)})_{i \geq 0}$ satisfying $x^{(i)} = 0$ for large i. For $x, y \in \mathbb{R}^\infty$ we introduce different semi-order relations by

$$x \underset{i}{\leq} y \text{ iff } x^{(i)} \leq y^{(i)} \text{ and } x^{(k)} = y^{(k)} \text{ for } k > i,$$

and similarly for $x \underset{i}{<} y$.

By the very definition of a cellular E_∞-preoperad, the cell decomposition of F_p relies on the cell decomposition of F_2. It is often possible and convenient to give explicit cell-decompositions for each p and to verify a posteriori that these cell-decompositions satisfy the necessary properties and relations.

For $(\mu, \sigma) \in \mathcal{K}_p$ we find :

$$F_p^{(\mu, \sigma)} = \{(x_1, \ldots, x_p) \in F_p \mid x_i \underset{\mu_{ij}}{\leq} x_j \text{ if } \sigma(i) < \sigma(j), \text{ and } x_j \underset{\mu_{ij}}{\leq} x_i \text{ if } \sigma(j) < \sigma(i)\}.$$

It is easy to check that for $p = 2$ this defines a cellular $\mathcal{K}_2$-decomposition of F_2 which retracts by a $\mathfrak{S}_2$-deformation onto the canonical $\mathcal{K}_2$-decomposition of the unit sphere in $\mathbb{R}^\infty$. Note that the (formal) cell-interior $\check{F}_p^{(\sigma, \mu)}$ is obtained by replacing everywhere $\leq$ by $<$; in fact, this is true for $p = 2$ by direct verification and follows for the general case from the definition of the cell-interiors (cf. 1.10).

In view of 2.3c and Remark 1.12, it is sufficient to show that *proper* cells are contractible; there are actually improper cells with several components, for example $\mu_{12} = \mu_{13} = 0$, $\mu_{23} = 1$ defines an improper cell $F_3^{(\mu, id_3)}$ with two components. Now, using 2.3b and the affine structure of F_p, each proper cell contracts conically to any of its interior points. Moreover, proper cells are defined by inequalities of coordinates, so they are (up to a factor $\mathbb{R}^\infty$) polyhedral cones in some finite dimensional euclidean space whence cell-inclusions are cofibrations. Finally, condition 1.9.2 is also satisfied, since a point $(x_1, x_2, \ldots, x_p) \in F_p$ is ordered iff $x_1 \underset{\mu_{12}}{<} x_2 \underset{\mu_{23}}{<} \cdots \underset{\mu_{p-1,p}}{<} x_p$ and 2.3b shows that each point of F_p is of this form up to permutation.

NOTATION 2.2. For $\alpha = (\mu, \sigma) \in \mathcal{K}_p$ it is often convenient to use the following indexing : $\alpha_{ij} = (\sigma^{-1})^*(\mu)_{ij}$, which is equivalent to $\alpha = \sigma^*((\alpha_{ij})_{1 \leq i < j \leq p}, id_{\mathbf{p}})$.

PROPOSITION 2.3. *Let F be the configuration preoperad.*
(a) The posets of proper cells are given by

$$\mathcal{K}(F)_p = \{\alpha \in \mathcal{K}_p \,|\, \alpha_{ik} = \max(\alpha_{ij}, \alpha_{jk}) \text{ for } i < j < k\}$$
$$= \{\alpha \in \mathcal{K}_p \,|\, \alpha_{ij} = \max_{i \leq k < j} \alpha_{kk+1}\}.$$

(b) The cell-interior $\breve{F}_p^{(\alpha)}$ for $\alpha = (\mu, \sigma) \in \mathcal{K}(F)_p$ is given by :

$$\breve{F}_p^{(\alpha)} = \{(x_1, \ldots, x_p) \in F_p \,|\, x_{\sigma^{-1}(1)} \underset{\alpha_{12}}{<} x_{\sigma^{-1}(2)} \underset{\alpha_{23}}{<} \cdots \underset{\alpha_{p-1,p}}{<} x_{\sigma^{-1}(p)}\}$$

(c) The inclusion of $\mathcal{K}(F)$ into $\mathcal{K}$ is a filtered Λ-equivalence of partially ordered preoperads. Moreover, each intermediate preoperad $\mathcal{K}'$ breaks the latter inclusion into two filtered Λ-equivalences $\mathcal{K}(F) \overset{\sim}{\hookrightarrow} \mathcal{K}' \overset{\sim}{\hookrightarrow} \mathcal{K}$.

PROOF. – The existence of an interior point in $F_p^{(\mu,\sigma)}$ implies that for each triple index (ijk) such that $\phi_{ij}^*(\sigma) = \phi_{jk}^*(\sigma)$ we have either

$$x_i \underset{\mu_{ij}}{<} x_j \underset{\mu_{jk}}{<} x_k \text{ and } x_i \underset{\mu_{ik}}{<} x_k$$

or the opposite inequalities, thus in both cases $\mu_{ik} = \max(\mu_{ij}, \mu_{jk})$. This shows that proper cells $F_p^{(\alpha)}$ satisfy $\alpha_{ik} = \max(\alpha_{ij}, \alpha_{jk})$ for all $i < j < k$. Conversely, if the latter property holds for an index $\alpha \in \mathcal{K}_p$ then the cell-interior $\breve{F}_p^{(\alpha)}$ contains all points of the form $x_{\sigma^{-1}(1)} \underset{\alpha_{12}}{<} x_{\sigma^{-1}(2)} \underset{\alpha_{23}}{<} \cdots \underset{\alpha_{p-1,p}}{<} x_{\sigma^{-1}(p)}$, so it is clearly nonempty, proving (a) and (b).

For (c), it will be sufficient to show that the induced inclusion of quotient categories $\mathcal{K}(F)_p/\mathfrak{S}_p \to \mathcal{K}_p/\mathfrak{S}_p$ admits a filtration-preserving right adjoint. Indeed, since the action of the symmetric group is free, the nerve of the quotient map is a Kan fibration, whence (by the five lemma) the inclusion of $\mathcal{K}(F)_p$ into $\mathcal{K}_p$ is an equivalence iff the quotient inclusion is. Furthermore, restriction of the right adjoint to $\mathcal{K}'_p/\mathfrak{S}_p$ yields the second part of (c).

The $\mathfrak{S}_p$-invariance of the indexing $(\alpha_{ij})_{1 \leq i < j \leq p}$ for $\alpha = (\mu, \sigma) \in \mathcal{K}_p$ defines a canonical bijection between "labelings" and $\mathfrak{S}_p$-orbits. Under this bijection the morphisms $\alpha \overset{\rho}{\to} \beta$ of the quotient category $\mathcal{K}_p/\mathfrak{S}_p$ correspond to permutations $\rho \in \mathfrak{S}_p$ such that $(\alpha, id_{\mathbf{p}}) \leq \rho^*(\beta, id_{\mathbf{p}})$ in $\mathcal{K}_p$.

We now define the right adjoint $\mathcal{K}_p/\mathfrak{S}_p \to \mathcal{K}(F)_p/\mathfrak{S}_p : \alpha \mapsto \hat{\alpha}$ by the components of the counit of the adjunction, i.e. by a family of universal maps

$$\hat{\alpha} \overset{id}{\to} \alpha, \text{ where } \hat{\alpha}_{ij} = \max_{i \leq k < j} \min_{r \leq k < s} \alpha_{rs}.$$

It follows from (a) that $\hat{\alpha}$ belongs to $\mathcal{K}(F)_p/\mathfrak{S}_p$ and that for each $\beta \overset{\rho}{\to} \alpha$ such that $\beta \in \mathcal{K}(F)_p/\mathfrak{S}_p$ we get $(\beta, id_{\mathbf{p}}) \leq \rho^*(\hat{\alpha}, id_{\mathbf{p}}) \leq \rho^*(\alpha, id_{\mathbf{p}})$ which gives the desired universal property. $\square$

REMARK 2.4. The poset $\mathcal{K}(F)_p^{(n)}$ can be found in Getzler-Jones' article [11] under the name *lexicographical cell-decomposition* of $F(\mathbb{R}^n, p)$. As their description is sligthly different from ours we shall recall it here, especially since this second description will reappear quite naturally when dealing with the permutohedral operad.

DEFINITION 2.5. An *ordered partition* of an integer $p > 0$ is an ordered decomposition $p = i_1 + \cdots + i_r$ into a sum of integers $i_k > 0$.

We associate to an ordered partition three combinatorially equivalent objects:

1. the *direct sum decomposition* $\mathbf{i_1} + \cdots + \mathbf{i_r} \cong \mathbf{p}$;
2. the *subgroup* $\mathfrak{S}_{(i_1,\ldots,i_r)}$ of $\mathfrak{S}_p$ consisting of all permutations of the form $m^{\mathfrak{S}}_{i_1,\ldots,i_r}(1; \sigma_1, \ldots, \sigma_r)$, which is canonically isomorphic to $\mathfrak{S}_{i_1} \times \cdots \times \mathfrak{S}_{i_r}$;
3. the *bar code* $(\epsilon_i)_{1 \leq i < p} \in [1]^{p-1}$ where ϵ_i is 1 (resp. 0) iff the ordered partition separates (resp. does not separate) the integers i and $i+1$.

We partially order the set of ordered partitions by *refinement*. This order is opposite to *subgroup-inclusion* but equals the *product order* on the bar codes $(\epsilon_i)_{1 \leq i < p} \in [1]^{p-1}$, where $[1] = \{0 < 1\}$. So we have

$$(i_1, \ldots, i_r) \preceq (j_1, \ldots, j_s)$$

iff one of the following three equivalent conditions is satisfied :

1. there is an ordered partition $s = k_1 + \cdots + k_r$ such that

$$i_1 = j_1 + \cdots + j_{k_1},$$
$$i_2 = j_{k_1+1} + \cdots + j_{k_1+k_2},$$
$$\cdots$$
$$i_r = j_{k_1+\cdots+k_{r-1}+1} + \cdots + j_{k_1+\cdots+k_r};$$

2. $\mathfrak{S}_{(i_1,\ldots,i_r)} \supseteq \mathfrak{S}_{(j_1,\ldots,j_s)}$
3. the associated bar codes $(\epsilon_i)_{1 \leq i < p}, (\zeta_i)_{1 \leq i < p}$ verify $\epsilon_i \leq \zeta_i$ for all i.

The correspondence between ordered partitions and bar codes extends naturally to a correspondence

$$\{\text{ascending chains of ordered partitions}\} \leftrightarrow \{\text{multiple bar codes}\}$$

$$part_1 \preceq part_2 \preceq \cdots \preceq part_l \leftrightarrow (\epsilon_i)_{1 \leq i < p} \in \mathbb{N}^{p-1}$$

where the *multiple bar code* is simply obtained by summing up the bar codes of the chain-elements; conversely, $part_k$ is the least fine ordered partition separating all couples $i, i+1$ such that $\epsilon_i > l - k$.

PROPOSITION 2.6. *The natural bijection between* $\mathcal{K}(F)^{(n)}_p$ *and* $[n-1]^{p-1} \times \mathfrak{S}_p$ *identifies* $\mathcal{K}(F)^{(n)}_p$ *with the covering category of a* $\mathfrak{S}_p$-*valued "shuffle" functor* $Sh^{(n)}_p : [n-1]^{p-1} \to \mathbf{Set}$.

PROOF. – By 2.3a, each index $\alpha = (\mu, \sigma) \in \mathcal{K}(F)^{(n)}_p$ is uniquely determined by the permutation σ and the integer-family $(\alpha_{k,k+1})_{1 \leq k < p} \in [n-1]^{p-1}$. It remains to determine the order relation induced by $\mathcal{K}(F)^{(n)}_p$. The partial order on $\mathcal{K}^{(n)}_p$ can be characterized as follows, cf. 2.3c :

$$\alpha = (\mu, \sigma) \leq \beta = (\nu, \tau) \Leftrightarrow \begin{cases} \alpha_{ij} \leq \beta_{\rho(i),\rho(j)} & \text{if } \rho(i) < \rho(j), \\ \alpha_{ij} < \beta_{\rho(j),\rho(i)} & \text{if } \rho(j) < \rho(i), \end{cases}$$

where the permutation ρ is determined by $\tau = \rho \circ \sigma$.

This leads to the following category structure on $[n-1]^{p-1}$:

$$[n-1]^{p-1}((\epsilon_k)_{1\leq k<p}, (\zeta_k)_{1\leq k<p}) =$$

$$\{\rho \in \mathfrak{S}_p \mid \max_{i\leq k<j} \epsilon_k \leq \max_{\rho(i)\leq k<\rho(j)} \zeta_k \text{ if } \rho(i) < \rho(j), \text{ and}$$

$$\max_{i\leq k<j} \epsilon_k < \max_{\rho(j)\leq k<\rho(i)} \zeta_k \text{ if } \rho(j) < \rho(i)\}$$

Indeed, the above characterization of the order relation identifies $\mathcal{K}(F)_p^{(n)}$ with the covering category (see [18]) of the functor

$$Sh_p^{(n)} : \frac{[n-1]^{p-1}}{(\epsilon_k)_{1\leq k<p}} \xrightarrow{} \frac{\mathbf{Set}}{\mathfrak{S}_p}$$
$$\rho \mapsto \rho_*.$$

$\square$

REMARK 2.7. The name "shuffle" functor is chosen because, for $n=2$, *adjacent* elements of the poset $[1]^{p-1}$ define a *morphism-set* in $[1]^{p-1}$ containing only shuffle-orderings, i.e. inverses of shuffle-permutations.

The categories $[n-1]^{p-1}$ are models for the quotient spaces $F(\mathbb{R}^n, p)/\mathfrak{S}_p$. In particular, for $n=2$, the nerve of $[1]^{p-1}$ is a classifying space for the braid group B_p on p strands; this model appears already in a paper of Greenberg [12], see also Fox and Neuwirth's combinatorial deduction of Artin's presentation of the braid groups B_p [10].

As an illustration, let us have a look at the category $[1]^2$, whose nerve is thus a classifying space for B_3 (we use the bar code for the objects) :

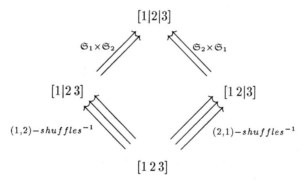

Balteanu, Fiedorowicz, Schwänzl and Vogt [1] embed the covering category $\mathcal{K}(F)_p^{(n)}$ in the "multilinear part" $\mathcal{M}_n(p)$ of the *free n-fold monoidal category* generated by p objects. The collection of categories $(\mathcal{M}_n(p))_{p>0}$ defines a cellular E_n-operad. In particular, nerves of connected n-fold monoidal categories are n-fold iterated loop spaces. A central role in their proof is played by the so-called *Coherence Theorem*, which roughly states that the category $\mathcal{M}_n(p)$ underlies a poset. Fiedorowicz pointed out that there is a natural poset-inclusion of $\mathcal{M}_n(p)$ into $\mathcal{K}_p^{(n)}$ compatible with the operad structure and generalizing the $K_2^{(n)}$-decomposition of the "octahedral" $(n-1)$-sphere $\mathcal{M}_n(2)$. There is actually a chain of Λ-equivariant poset-inclusions

$$\mathcal{K}(F)_p^{(n)} \subset \mathcal{M}_n(p) \subset \mathcal{K}(\mathcal{C})_p^{(n)} \subset \mathcal{K}_p^{(n)}$$

proving geometrically (1.16, [1]) as well as combinatorially (2.3c) that the n-fold monoidal operad $(\mathcal{M}_n(p))_{p>0}$ is a cellular E_n-operad.

Getzler-Jones [11] also embed the poset $\mathcal{K}(F)_p^{(n)}$ in a larger poset which corresponds to the cell structure of Fulton-MacPherson's compactification of $F(\mathbb{R}^n, p)$. It turns out that the latter cell structure is in some precise sense "freely generated" by the former via the formalism of planary trees (see also [7]). The underlying combinatorics are intimately related to Stasheff's *associahedra*.

There is a similar relationship between $\mathcal{K}(F)_p^{(n)}$ and $\mathcal{M}_n(p)$ in the form of a canonical surjective map $\mathcal{K}(F)_p^{(n)} \times \mathcal{A}_p \twoheadrightarrow \mathcal{M}_n(p)$, where $\mathcal{A}_p$ denotes the set of all bracketings of a p-element set. The inclusion of $\mathcal{K}(F)_p^{(n)}$ into $\mathcal{M}_n(p)$ endows each element $(\mu, id_p) \in \mathcal{K}(F)_p^{(n)}$ with a natural bracketing such as

$$((1 \underset{\mu_{12}}{\square} 2) \underset{\mu_{23}}{\square} (3 \underset{\mu_{34}}{\square} \cdots (\cdots \underset{\mu_{p-2,p-1}}{\square} p-1) \underset{\mu_{p-1,p}}{\square} p)).$$

In this setting, Property 2.3a becomes equivalent to the condition that the indices of the composition laws increase "from inside to outside" in the bracketing. If all $p-1$ composition indices are distinct, the permutation of the composition indices defines a map $\mathfrak{S}_{p-1} \to \mathcal{A}_p$ studied in Tonk's paper in this volume [20].

EXAMPLE 2.8. *The cell structure of the simplicial operad Γ.*
As the universal bundle functor W commutes with cartesian products, unit and multiplication of the permutation operad (1.15a) induce a unit and a multiplication of the composite functor $\Gamma = W\mathfrak{S}$, turning it into a cellular E_∞-operad, as we shall see.

The cells of $\Gamma_p = W\mathfrak{S}_p$ are realized by certain simplicial subsets $\Gamma_p^{(\mu,\sigma)}$ of Γ_p. We recall that a k-simplex of Γ_p is written as a $(k+1)$-tuple of elements of $\mathfrak{S}_p$. We shall write σ_x for the *last* component of the simplex $x \in \Gamma_p$. The n-skeleton of Γ_p will be denoted by $sk_n\Gamma_p$. For $(\mu, \sigma) \in \mathcal{K}_p$ we find :

$$\Gamma_p^{(\mu,\sigma)} = \{x \in \Gamma_p \,|\, \phi_{ij}^*(x) \in sk_{\mu_{ij}}\Gamma_2 \text{ and } \phi_{ij}^*(\sigma_x) = \phi_{ij}^*(\sigma) \text{ if } \phi_{ij}^*(x) \notin sk_{\mu_{ij}-1}\Gamma_2\}$$

Condition (1.9.1) is satisfied, since there are simplicial contractions

$$\begin{aligned} \gamma : \Gamma_p^{(\mu,\sigma)} &\rightarrow \Gamma_p^{(\mu,\sigma)} \\ (\sigma_0, \ldots, \sigma_k) &\mapsto (\sigma_0, \ldots, \sigma_k, \sigma) \end{aligned}$$

and geometric realization transforms cell-inclusions into closed cofibrations. Condition (1.9.2) is also satisfied since a point in $|\Gamma_p|$ is ordered iff it is contained in the interior of a simplex of Γ_p whose last component is the neutral element of $\mathfrak{S}_p$.

The multiplication $m_{i_1 \ldots i_p}^{\Gamma} : \Gamma_p \times \Gamma_{i_1} \times \cdots \times \Gamma_{i_p} \to \Gamma_{i_1 + \cdots + i_p}$ preserves the cellular structure, since we deduce from $m_{i_1 \ldots i_p}^{\Gamma} = W m_{i_1 \ldots i_p}^{\mathfrak{S}}$ the relations (cf. 1.15):

$$\phi_{\psi_r(i),\psi_r(j)}^*(m_{i_1 \ldots i_p}^{\Gamma}(x; x_1, \ldots, x_p)) = \phi_{ij}^*(x_r) \text{ for } i, j \in \mathbf{i_r}, \text{ and}$$

$$\phi_{\psi_r(i),\psi_s(j)}^*(m_{i_1 \ldots i_p}^{\Gamma}(x; x_1, \ldots, x_p)) = \phi_{rs}^*(x) \text{ for } i \in \mathbf{i_r}, j \in \mathbf{i_s}, r < s.$$

The cell-interior of $|\Gamma_p^{(\mu,\sigma)}|$ is spanned by the interiors of the simplices of the following subset (which is not a simplicial subset) :

$$\check{\Gamma}_p^{(\mu,\sigma)} = \{x \in \Gamma_p \,|\, \sigma_x = \sigma \text{ and } \phi_{ij}^*(x) \in sk_{\mu_{ij}}\Gamma_2 \backslash sk_{\mu_{ij}-1}\Gamma_2\}.$$

So, a cell $|\Gamma_p^{(\mu,\sigma)}|$ is proper iff there exists a simplex $(\sigma_0, \sigma_1, \ldots, \sigma_k) \in \Gamma_p$ such that $\sigma = \sigma_k$ and such that for each $i < j$, the image-sequence

$$(\phi_{ij}^*(\sigma_0), \ldots, \phi_{ij}^*(\sigma_k)) \in \Gamma_2$$

contains exactly μ_{ij} changes. This leads to the following proposition, for which a proof can be found in [**5**].

PROPOSITION 2.9. *Let Γ be the simplicial E_∞-operad of Barratt-Eccles. Then the posets of proper cells are given by :*

$$\mathcal{K}(\Gamma)_p = \{\alpha \in \mathcal{K}_p \,|\, \text{there exists a descending chain of labelings } (\alpha_{ij}^{(r)}) \in \mathbb{N}^{\binom{p}{2}}$$

$$\text{beginning with } (\alpha_{ij}^{(0)}) = (\alpha_{ij}) \text{ and ending at } (0) \text{ such that}$$

$$\alpha_{ij}^{(r)} \equiv \alpha_{jk}^{(r)} \bmod 2 \text{ implies } \alpha_{ij}^{(r)} \equiv \alpha_{jk}^{(r)} \equiv \alpha_{ik}^{(r)} \bmod 2 \text{ and}$$

$$0 \le \alpha_{ij}^{(r)} - \alpha_{ij}^{(r+1)} \le 1 \text{ for all } i, j, r\}.$$

REMARK 2.10. The posets $\mathcal{K}(F)_p$ and $\mathcal{K}(\Gamma)_p$ are quite different, which explains the difficulty in showing directly that $F_p^{(n)}$ and $|\Gamma_p^{(n)}|$ are homotopy equivalent (cf. [**14**]). The importance of the Γ-construction comes from the fact that the Smith-filtration

$$\Gamma_p^{(n)} = \{x \in \Gamma_p \,|\, \phi_{ij}^*(x) \in sk_{n-1}\Gamma_2 \text{ for all } i < j\}$$

defines a family of cellular E_n-operads in the category of simplicial sets so that, by Fiedorowicz's comparison theorem, May's entire theory of E_n-operads can be applied in this simplicial context, including the approximation and detection theorems, cf. [**15**], [**19**] or [**5**]. The case $n = \infty$ was the initial motivation of Barratt-Eccles [**4**]. Moreover, the operad structure of Γ is purely group-theoretic, so that the cellular E_n-operads $\Gamma^{(n)}$ relate the homology of the symmetric groups rather directly to universal phenomena occuring in the theory of n-fold iterated loop spaces.

NOTATION 2.11. As a last example of a cellular E_∞-operad we present here Milgram's permutohedral models [**17**]. Like Kapranov [**13**] we shall write P_n for the permutohedron embedded in $\mathbb{R}^n$, i.e. for the convex hull of the point set

$$\{(\sigma(1), \ldots, \sigma(n)) \in \mathbb{R}^n \,|\, \sigma \in \mathfrak{S}_n\}.$$

The permutohedron P_n is a *convex polytope* whose *face-poset* is canonically isomorphic to the poset $\mathcal{P}(\mathfrak{S}_n)$ formed by all (right) cosets

$$\mathfrak{S}_{(i_1,\ldots,i_r)}\sigma \in \mathfrak{S}_{(i_1,\ldots,i_r)}\backslash\mathfrak{S}_n,$$

with respect to subgroups of $\mathfrak{S}_n$ of the form $\mathfrak{S}_{(i_1,\ldots,i_r)}$, where $(i_1,\ldots,i_r)$ is an *ordered partition* of n, cf. 2.5. Indeed, each coset $\mathfrak{S}_{(i_1,\ldots,i_r)}\sigma$ corresponds to the convex hull of the *vertices* $(\tau(1),\ldots,\tau(n)) \in \mathbb{R}^n$, as τ runs through $\mathfrak{S}_{(i_1,\ldots,i_r)}\sigma$.

In the literature ([**17**], [**3**], [**13**]), the faces of the permutohedra are labelled by *left* cosets instead of right cosets and the symmetric group acts by left multiplication instead of right multiplication. Our convention follows the definition of the category Λ where the permutations act on themselves by right multiplication. There is however an easy way to switch between the two conventions by means of the involution $inv : \sigma \mapsto \sigma^{-1}$ which associates to a permutation its *"ordering"* and vice versa (cf. [**19**], [**14**], [**13**]). In particular, our multiplication 1.15a of the permutation operad also reflects this notational convention, so that a reader who prefers left notation, has to change $m^{\mathfrak{S}}_{i_1\ldots i_n}$ into $inv \circ m^{\mathfrak{S}}_{i_1\ldots i_n} \circ (inv \times \cdots \times inv)$ which corresponds to "place-permutation" rather than "element-permutation".

The definition of the Milgram operads relies formally on the existence of an operad structure on the collection of permutohedra $(P_n)_{n>0}$ which

(a) restricts to the permutation operad on vertices, and

(b) satisfies the *boundary condition*, i.e. the multiplication $m^P_{i_1\ldots i_n}$ sends the boundary of $P_n \times P_{i_1} \times \cdots \times P_{i_n}$ to the boundary of $P_{i_1+\cdots+i_n}$.

The affine extension of the permutation operad does *not* satisfy the boundary condition, so that some additional combinatorial properties of the permutohedra have to be used. I am indebted to Fiedorowicz for insisting on this point and for sending me some helpful pictures.

There is actually a natural *cubical* subdivision of the permutohedron P_n induced by simplicial stars with respect to the barycentric subdivision of P_n. This cubical subdivision admits the following geometric description : Each vertex $\sigma \in P_n$ carries a natural $(n-1)$-*frame* f_σ defined by the union of all σ-incident edges in the barycentrically subdivided 1-skeleton of P_n. The *simplicial hull* of f_σ (i.e. the cocell of f_σ in the nerve of $\mathcal{P}(\mathfrak{S}_n)$) yields a standard simplicial $(n-1)$-cube bipointed by σ and the barycenter of P_n. The cubical decomposition of P_n thus corresponds to a *frame-decomposition* of the barycentrically subdivided 1-skeleton of P_n.

We shall show below that the permutation operad is naturally "framed": for each vertex $(\sigma; \sigma_1, \ldots \sigma_n) \in P_n \times P_{i_1} \times \cdots \times P_{i_n}$ the image of the product-frame $f_\sigma \times f_{\sigma_1} \times \cdots \times f_{\sigma_n}$ under $m^{\mathfrak{S}}_{i_1\ldots i_n}$ defines a frame $f_{(\sigma;\sigma_1,\ldots,\sigma_n)}$ in $P_{i_1+\cdots+i_n}$ whose simplicial hull is a $(i_1+\cdots+i_n-1)$-cube. We then define the *permutohedral operad* to be the *cubical extension of this "framed" permutation operad*. The underlying *preoperad* coincides with the affine extension of the permutation preoperad, since the image-frame induced by a Λ-action is the natural one.

The definition of the image-frames uses an alternative description of the face-poset $\mathcal{P}(\mathfrak{S}_n)$ based on the beautiful theorem of Blind and Mani [6] that the face-poset of a *simple* polytope is uniquely determined by its 1-skeleton. The permuto-hedron is simple and its (oriented) 1-skeleton coincides with the (left) *weak Bruhat order* on the symmetric group $\mathfrak{S}_n$. To be more precise : each cell of the permutohedron P_n has a canonical *initial* (resp. *final*) vertex given by the unique permutation $\tau \in \mathfrak{S}_{(i_1,\ldots,i_r)}\sigma$ such that τ^{-1} is increasing (resp. decreasing) on subsets $\mathbf{i_k}$ of $\mathbf{i_1} + \cdots + \mathbf{i_r}$, cf. [3]. This orientation of the 1-skeleton of P_n defines precisely the weak Bruhat order on $\mathfrak{S}_n$, and each coset in $\mathcal{P}(\mathfrak{S}_n)$ is an *interval* for the weak Bruhat order. In other words, cells of the permutohedron correspond bijectively to "admissible" intervals $[\tau_1, \tau_2]$ of the weak Bruhat order on $\mathfrak{S}_n$, where admissible means that the lower and upper bounds are the initial and final vertices of some coset in $\mathcal{P}(\mathfrak{S}_n)$.

The permutation operad *preserves* the weak Bruhat order, i.e. the multiplica-tion $m^{\mathfrak{S}}_{i_1\ldots i_n}$ embeds $\mathfrak{S}_n \times \mathfrak{S}_{i_1} \times \cdots \times \mathfrak{S}_{i_n}$ in $\mathfrak{S}_{i_1+\cdots+i_n}$ as a *subposet*. Furthermore, we define the *geodesic* between two comparable vertices of the permutohedron P_n to be the barycenter of all oriented edge-paths between them, and the *barycenter* of an arbitrary interval $[\tau_1, \tau_2]$ to be the middle of the geodesic between τ_1 and τ_2. The barycenter of an admissible interval coincides with the barycenter of the associated cell. The extremal vertices of the $(n-1)$-frame f_σ at $\sigma \in P_n$ are now precisely the barycenters of the 2-element intervals of $\mathfrak{S}_n$ containing σ. More generally, the extremal vertices of the product-frame $f_\sigma \times f_{\sigma_1} \times \cdots \times f_{\sigma_n}$ are precisely the barycenters of the 2-element intervals of $\mathfrak{S}_n \times \mathfrak{S}_{i_1} \times \cdots \times \mathfrak{S}_{i_n}$ containing $(\sigma; \sigma_1, \ldots, \sigma_n)$. The multiplication $m^{\mathfrak{S}}_{i_1 \ldots i_n}$ sends these intervals to well defined intervals of $\mathfrak{S}_{i_1 + \cdots + i_n}$ containing $\sigma(\sigma_1, \ldots, \sigma_n)$. The *image-frame* $f_{(\sigma; \sigma_1, \ldots, \sigma_n)}$ is then by definition the union of the geodesics between $\sigma(\sigma_1, \ldots, \sigma_n)$ and the barycenters of these image-intervals. The simplicial hull of $f_{(\sigma; \sigma_1, \ldots, \sigma_n)}$ is an $(i_1 + \cdots + i_n - 1)$-cube bipointed by $\sigma(\sigma_1, \ldots, \sigma_n)$ and the barycenter of $P_{i_1 + \cdots + i_n}$, actually isomorphic to a well defined subdivision of the standard simplicial $(i_1 + \cdots + i_n - 1)$-cube, see [5] for more details.

Finally we need the *convex projectors*

$$D_{(i_1, \ldots, i_r)} = \psi_1^* \times \cdots \times \psi_r^* : P_n \to P_{i_1} \times \cdots \times P_{i_r} \cong c_{(i_1, \ldots, i_r)},$$

where ψ_k is the canonical inclusion of $\mathbf{i_k}$ in $\mathbf{n} = \mathbf{i_1} + \cdots + \mathbf{i_r}$, and where the convex hull $c_{(i_1, \ldots, i_r)}$ of the subgroup $\mathfrak{S}_{(i_1, \ldots, i_r)}$ is identified with the cartesian product of the corresponding permutohedra by affine extension of the canonical isomorphism

$$\mathfrak{S}_{i_1} \times \cdots \times \mathfrak{S}_{i_r} \cong \mathfrak{S}_{(i_1, \ldots, i_r)}.$$

For each coset $\mathfrak{S}_{(i_1, \ldots, i_r)} \tau \in \mathcal{P}(\mathfrak{S}_n)$ such that τ is initial, this defines a convex projector $D^\tau_{(i_1, \ldots, i_r)} = \tau^* D_{(i_1, \ldots, i_r)} (\tau^*)^{-1}$ onto the corresponding cell of the permutohedron. The map which associates to a coset its convex projector "transforms" the permutation operad into the permutohedral operad.

DEFINITION 2.12. Milgram's E_k-operads are defined as quotient spaces

$$J_n^{(k)} = (P_n)^{k-1} \times \mathfrak{S}_n / \sim$$

where the equivalence relation identifies certain boundary cells of the cartesian product. Explicitly, for each point $(\tau^*(x_1), \ldots, \tau^*(x_{k-1}); \sigma) \in (P_n)^{k-1} \times \mathfrak{S}_n$ such that x_s belongs to the convex hull of a proper subgroup $\mathfrak{S}_{(i_1, \ldots, i_r)}$ of $\mathfrak{S}_n$ and such that τ is the initial vertex of the coset $\mathfrak{S}_{(i_1, \ldots, i_r)} \tau$, we have the relation

$$(\tau^*(x_1), \ldots, \tau^*(x_{k-1}); \sigma) \sim$$
$$(x_1, \ldots, x_s, D_{(i_1, \ldots, i_r)}(x_{s+1}), \ldots, D_{(i_1, \ldots, i_r)}(x_{k-1}); \tau\sigma).$$

The action of $\phi \in \Lambda(\mathbf{m}, \mathbf{n})$ is induced by

$$\phi^* : (P_n)^{k-1} \times \mathfrak{S}_n \quad \to \quad (P_m)^{k-1} \times \mathfrak{S}_m$$
$$(x_1, \ldots, x_{k-1}; \tau) \quad \mapsto \quad (((\tau\phi)^{inc})^*(x_1), \ldots, ((\tau\phi)^{inc})^*(x_{k-1}); (\tau\phi)^\natural).$$

The space $J_n^{(k)}$ embeds in $J_n^{(k+1)}$ by identifying $(P_n)^{k-1}$ with the subset of $(P_n)^k$ formed by the points whose first component is the barycenter of the permutohedron, i.e. the fixed point under the action of $\mathfrak{S}_n$.

The previously defined Λ-structure as well as the diagonal multiplication on $(P_n)^{k-1} \times \mathfrak{S}_n$ are compatible with the equivalence relation and induce thus a natural operad structure on the spaces $(J_n^{(k)})_{n>0}$. The boundary condition of the permutohedral operad is crucial at this point, since it implies that the gluing of the cells is preserved under multiplication. The associated monad can be identified with Milgram's construction J_k which models for connected CW-spaces the functor $\Omega^k S^k$ [17]. The cellular E_k-structure of $(J_n^{(k)})_{n>0}$ is based on the following lemma, which relates our equivalence relation to that of Baues [3], [5] :

LEMMA 2.13. *Each point $x \in J_n^{(k)}$ has a unique representative*

$$x_{can} = (x_1, \ldots, x_{k-1}; \sigma) \in (P_n)^{k-1} \times \mathfrak{S}_n,$$

such that the minimal cells c_i whose cell-interiors contain x_i are convex hulls of a decreasing chain of subgroups of $\mathfrak{S}_n$, in particular $c_1 \supset c_2 \supset \cdots \supset c_{k-1}$.

The collection of these cells $c_1 \times \cdots \times c_{k-1} \times \sigma$ induces a cell-decomposition of $J_n^{(k)}$ whose poset is antiisomorphic *to $\mathcal{K}(F)_n^{(k)}$, cf. 2.5.*

THEOREM 2.14. *Milgram's operads $J^{(k)}$ form the natural filtration of a cellular E_∞-operad. In particular, the space $J_n^{(k)}$ has the $\mathfrak{S}_n$-equivariant homotopy type of the real configuration space $F(\mathbb{R}^k, n)$. The posets of proper cells are given by*
$$\mathcal{K}(J)_n^{(k)} = \mathcal{K}(F)_n^{(k)}.$$

PROOF. – By the comparison theorem it remains to define the underlying cell structure beginning with a cellular $\mathcal{K}_2$-decomposition of the inductive limit $J_2 = \varinjlim_k J_2^{(k)}$. But the previous lemma shows that we have only to *dualize* the cell-decompositions defined by the cartesian products, which is possible because of the compactness of the $J_2^{(k)}$ and the underlying affine structure. This dualization process works Λ-equivariantly for all n and gives thus the asserted posets of proper cells. $\square$

References

[1] C. Balteanu, Z. Fiedorowicz, R. Schwänzl and R. Vogt – *Iterated monoidal categories*, preprint (1995).

[2] H. Barcelo and N. Bergeron – *The Orlik-Solomon Algebra on the Partition Lattice and the Free Lie Algebra*, J. of Comb. Theory (A) **55** (1990), 80-92.

[3] H.-J. Baues – *Geometry of loop spaces and the cobar-construction*, Mem. AMS **230** (1980).

[4] M. G. Barratt and P. J. Eccles – *Γ^+-Structures*, Topology **13** (1974), 25-45, 113-126, 199-207.

[5] C. Berger – *Opérades cellulaires et espaces de lacets itérés*, Ann. de l'Inst. Fourier **46** (1996).

[6] R. Blind and P. Mani – *On puzzles and polytope isomorphism*, Aequationes Math. **34** (1987), 287-297.

[7] J. M. Boardman and R. M. Vogt – *Homotopy invariant algebraic structures on topological spaces*, Lecture Notes in Math. **347**, Springer Verlag (1973).

[8] F. R. Cohen – *The homology of C_{n+1}-spaces*, Lecture Notes in Math. **533**, Springer Verlag (1976), 207-351.

[9] F. R. Cohen, J. P. May and L. R. Taylor – *Splitting of certain spaces CX*, Math. Proc. Camb. Phil. Soc. **84** (1978), 465-496.

[10] R. Fox and L. Neuwirth – *The braid groups*, Math. Scand. **10** (1962), 119-126.

[11] E. Getzler and J. D. S. Jones – *Operads, homotopy algebra, and iterated integrals for double loop spaces*, preprint (1995).

[12] P. Greenberg – *Triangulating groups ; two examples*, preprint (1992).

[13] M. M. Kapranov – *The permutoassociahedron, MacLane's coherence theorem and asymptotic zones for KZ equation*, J. of Pure and Applied Algebra **85** (1993), 119-142.

[14] T. Kashiwabara – *On the Homotopy Type of Configuration Complexes*, Contemp. Math. **146** (1993), 159-170.

[15] J. P. May – *The Geometry of iterated loop spaces*, Lecture Notes in Math. 277, Springer Verlag (1972).

[16] J. P. May – *Definitions : Operads, algebras and modules*, this volume.

[17] R. J. Milgram – *Iterated loop spaces*, Annals of Math. **84** (1966), 386-403.

[18] D. Quillen – *Higher algebraic K-theory I*, Lecture Notes in Math. **341**, Springer Verlag (1973), 85-147.

[19] J. H. Smith – *Simplicial Group Models for $\Omega^n S^n X$*, Israel J. of Math. **66** (1989), 330-350.

[20] A. Tonk – *Relating the permutohedra and associahedra*, this volume.

UNIVERSITÉ DE NICE-SOPHIA ANTIPOLIS, LABORATOIRE J. A. DIEUDONNÉ, URA 168, PARC VALROSE, F-06108 NICE CEDEX 2, FRANCE.

E-mail address: `cberger@math.unice.fr`

Contemporary Mathematics
Volume **202**, 1997

From operads to 'physically' inspired theories

Jim Stasheff

An operad-chik looks at configuration spaces, moduli spaces
and mathematical physics

1. Introduction

As evidenced by these conferences (Hartford and Luminy), operads have had a renaissance in recent years for a variety of reasons.
Originally studied entirely as a tool in homotopy theory, operads have
recently received new inspirations from homological algebra, category
theory, algebraic geometry and mathematical physics. I'll try to provide a transition from the foundations to the frontier with mathematical physics.

For me, the transition occurred in two stages. First, there is the generalization of Lie algebra cohomology known as BRST (Becchi-Rouet-
Stora-Tyutin) cohomology, which turned out to be very closely related
to strong homotopy Lie (L_∞) algebras, which I will describe later in
homological algebraic terms - along the lines of Balavoine's talk at this
conference [**5**]. That description makes no use of operads, but the relevance of operads appeared later in the work of Hinich and Schechtman
[**25**].

Operads revealed themselves as more directly involved in String
Field Theories (SFTs) and Vertex Operator Algebras (VOAs), both of
which in turn draw on Conformal Field Theories (CFTs). Central to
CFTs are algebraic structures parameterized by moduli spaces $\mathcal{M}_{g,n}$ of
n-punctured Riemann surfaces of genus g. Operads are almost visible

1991 *Mathematics Subject Classification*. Primary 55P99, 18G99, 81R99,
81T70, Secondary 57T99, 18G55, 81R50.

Research supported in part by the NSF throughout most of my career, currently
under grant DMS-9504871.

in $\mathcal{M}_{0,n+1}$ if we choose one point "∞" as outgoing and the other n as incoming.

2. Background

But perhaps I should back up and recall the essentials of operads. At the front of this volume appears the formal, official definition due to Peter May [42] as well as the alternate, assuming there is an 'identity' in $\mathcal{C}(1)$, in terms of $\circ_i$ operations, cf. Gerstenhaber's comp algebras [19]. (For a full discussion of the appropriate definitions and relations in the absence of an 'identity' in $\mathcal{C}(1)$, see [40].) In my talk at Luminy on the pre-history of operads [48], I emphasized planar rooted trees. For operads proper, rooted trees form the 'mother of all operads'.

2.1. The tree operad. Let $\mathcal{T}(n)$ be the set of trees with 1 root and n leaves labelled 1 through n. The collection $\{\mathcal{T}(n)\}$ of tree spaces forms a non-symmetric operad (or comp algebra): given trees $f \in \mathcal{T}(k), g \in \mathcal{T}(j)$, for each i, let $f \circ_i g$ be the tree obtained by grafting the root of g to the leaf of f labelled i. More generally, given trees $f \in \mathcal{T}(k), g_i \in \mathcal{T}(n_i)$, let $\gamma(f; g_1, \ldots, g_k) \in \mathcal{T}(\Sigma n_i)$ be the tree obtained by grafting the root of g_i to the leaf of f labelled i.

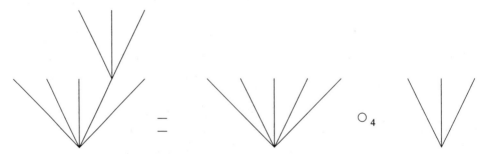

This and the basic examples that appear in the introduction to this volume [43]and in [48] are discrete or topological operads. To move to (vector space) algebra from discrete operads, just take the vector spaces spanned by the discrete sets. To move to differential graded algebra from topological operads, suitable chain functors serve well. For any topological operad $\mathcal{C}$, the singular chains $C_*(\mathcal{C}(n))$ form a dg (differential graded) operad. Since the associahedra are presented as cell complexes and the $\circ_i$ operations are cellular, the cellular chains $\mathcal{A}(n) = CC_*(K_n)$ form a dg operad $\mathcal{A}$.

2.2. A_∞-algebras. An algebra over $\mathcal{A}$ is known as an A_∞-algebra (or strongly homotopy associative algebra) and consists of a graded module V with maps

$$m_n : V^{\otimes n} \to V \text{ of degree } n - 2$$

satisfying suitable compatibility conditions. In particular,
$m_1 = d$ is a differential,
$m = m_2 : V \otimes V \to V$ is a chain map, that is, d is a derivation with respect to $m = m_2$,
$m_3 : V^{\otimes 3} \to V$ is a chain homotopy for associativity, i.e.

$$dm_3 + m_3 d = m(m \otimes 1) - m(1 \otimes m),$$

m_4 is a 'higher homotopy' such that $dm_4 - m_4 d$ equals a sum of five terms, the 'pentagonal' relation which would be precisely zero in Drinfel'd's quasi-Hopf algebras [17]. (The homotopy interpretation exists, for example, in the classifying space of the category of representations of a quasi-Hopf algebra [27].)

The original and motivating example of an A_∞-algebra was provided by the singular chains on a based loop space ΩX.

Warning! When $d = 0$ (for example, after passing to (co)homology from (co)chains), the conditions give a strictly associative algebra **but** there may still be non-trivial maps m_n. Here are two natural examples:

EXAMPLE 2.1. Consider the Borromean rings

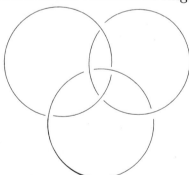

Borromean rings

consisting of three circles which are pairwise unlinked but all together are linked. (The name 'Borromean' derives from their appearance in the coat of arms of the House of Borromeo in Italy.) If we regard them as situated in S^3, then the cohomology ring of the complement is a trivial algebra, but m_3 is non-zero in cohomology, being represented by Massey products and detecting the simultaneous linking of all three circles [41].

The relevant Massey product is defined as follows: Let u, v, w be cocyles which are Alexander dual to the fundamental homology classes of the three circles. Because the circles are pairwise unlinked, the corresponding cocycles in the complement have cup products which are cohomologically trivial, e.g. $u \cup v = \delta a$ and $v \cup w = \delta b$. Massey shows that the triple product $< u, v, w >$ represented by $ub + aw$ generates H^1 of the complement.

EXAMPLE 2.2. Consider the space $\mathrm{Sp}(5)/\mathrm{SU}(5)$ where $\mathrm{Sp}(5)$ denotes the 'orthogonal' group in 5 quaternionic dimensional space. According to Borel's calculations [13], the cohomology algebra has a single generator in each of dimensions 6, 10, 21 and 25 with only the pairing to the fundamental class in dimension 31 being non-trivial. This is isomorphic to the cohomology algebra of the connected sum of $S^6 \times S^{25}$ and $S^{10} \times S^{21}$, but these spaces have distinct homotopy types, even rationally, since triple Massey products are trivial in the connected sum but non-trivial in the homogeneous space. Borel calculates the cohomology via the Leray-Serre spectral sequence of the fibration

$$\mathrm{Sp}(5) \to \mathrm{Sp}(5)/\mathrm{SU}(5) \to \mathrm{BSU}(5)$$

and the Massey product calculation is manifest, though not named as such.

3. Configuration spaces

Now let's return briefly to the moduli space $\mathcal{M}_{0,n+1}$. 'Genus 0' is called in physspeak 'tree level', for reasons which should become apparent. The modular equivalence relation allows us to choose representatives in which the first two points are $0, 1 \in \boldsymbol{C}P(1)$ and the last is '∞', thus identifying $\mathcal{M}_{0,n+1}$ with the configuration space $F(R^2 - \{0, 1\}, n - 2)$. Configuration spaces do not form an operad, but there are two ways to get an operad from the configuration spaces:

3.1. The little disks operad. First, decorate the chosen points with non-overlapping 'little' disks. Let D be the unit disk in the plane and let $\mathcal{D}(n)$ be the space of all maps f from $\coprod_{i=1}^n D \to D$ such that f, when restricted to each disk, is the composition of translation and multiplication by a positive real number and images of the components of f are disjoint.

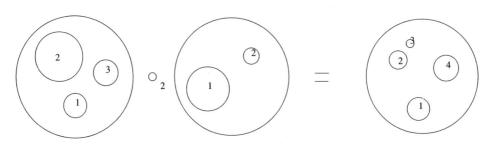

Of course, this generalizes to little n-balls for any n. For certain purposes, little n-cubes [12] are more useful [42].

3.2. Compactifications. Instead of adding decorations to the configurations, we can obtain an operad by suitable compactification of configuration space. This is the approach relevant to many of the applications to physics and to knot invariants.

In my Hartford talk, I tried to span 162 years of mathematics, from Gauss' linking number to some recent developments involving the concept of operad. That sketch of the historical development appears as Appendix A to this paper. Here I will go directly to the 'cyclohedra', the cousins of the associahedra which are relevant to knot invariants. (The portion of my Hartford talk which summarized the paper of Bott and Taubes [14] is omitted in deference to the original. Some updates due to Labastida and to Dylan Thurston are included in Appendix A here.)

To generalize Gauss' approach for links to knots, we can try the configuration space of pairs of distinct points on a single circle, but then the space is not compact. One way around this difficulty, due to Calugareanu in 1959 [15] and in an improved version to Pohl in 1968 [45], uses a 'nearby' parallel K', that is a framing of K. The new approach due to Kontsevich [34] involves the compactification of the configuration space of n distinct points on a circle as a manifold-with-corners, keeping track of how points on S^1 approach coincidence.

The way to retain that information is well known in algebraic geometry under the rubric of 'blowing up' and has been worked out in great detail for configurations on non-singular algebraic varieties by Fulton and MacPherson [18]. The real, as opposed to complex, analog has been studied by Axelrod and Singer [4]; not only do we want to work over the real instead of the complex field but we also do not want to projectivize the information. The essential idea is to compactify by adding a boundary where two (or more) points collide. For example, in $\mathbb{R}^3$, where two points collide, adjoin the unit sphere bundle of this subspace of $(\mathbb{R}^3)^n$.

Bott and Taubes looked at the details in the case of a circle, saw the result as the product of a circle S^1 with a polytope and drew pictures of the polytopes for 2, 3 and 4 points, although for more points they were content with the description in terms of blow-ups. In his lecture at MSRI, Bott drew the pictures and mentioned that the polytopes resembled the ones I had constructed for studying loop spaces some 34 years before. On closer examination I realized that my polytopes, since dubbed the associahedra, appeared as faces of the Bott-Taubes polytopes W_n, which I will call cyclohedra. (I have reindexed the polytopes by the number of points in the relevant configurations, rather than by their dimensions.) Further analysis provided a simple description of the compactifications as truncations of the simplex. This led me to reconsider the associahedra and describe their realization also as truncated simplices - something I had not noticed for 35 years! In discussing these matters with Shnider and Sternberg, I found that not only do they describe the associahedra as truncated simplices in their book, but the truncation appears (somewhat obscurely - at least to me) on the cover! (Further details of this truncation are in Appendix B, which is joint work with Shnider.)

4. Cyclohedra

The polytopes W_n which realize the compactification of $F(S^1, n)$ as $S^1 \times W_n$ I suggest calling **cyclohedra**. (Bott and Taubes use the notation $C(n, S^1)$ for $F(S^1, n)$ - see Appendix A.) Parameterize the circle as usual by arclength normalized to have circumference 1 (or 2π if you insist). Number the points $x_0, \ldots, x_{n-1}$ in order of increasing parameter. Let θ be the parameter for x_0 and let t_i be the arclength from x_i to x_{i+1} with t_{n-1} the arclength from x_{n-1} to x_0. Noticing that $\Sigma t_i = 1$, the space $F(S^1, n)$ is seen to be homeomorphic to $S^1 \times \overset{\circ}{\Delta}{}^{n-1}$, the product of the circle with the open $(n-1)$-simplex. The cyclohedra W_n are the 'blow-up' compactifications of $\overset{\circ}{\Delta}{}^{n-1}$ very much as the associahedra K_n are for the interval, except that x_n is interpreted as x_0. Thus they can be viewed also as truncations of simplices - in fact the only change is that since x_0 can run into x_{n-1}, the truncations include additionally the same formulas but with the indices interpreted *mod n*. (See Appendix B where the indices are subintervals $I = (i, \ldots, j)$ of $(1, \ldots, n-1)$, to be interpreted cyclically for the cyclohedron, e.g. $(n-1, 1, 2)$.)

Just as the associahedron K_n can be described as a convex polytope with one vertex for each way of inserting parentheses in a meaningful way in a word of n letters, so the cyclohedron W_n has vertices which are given by all meaningful ways of inserting parentheses in a string of n symbols arranged on a circle.

12 ——————————— 21

$$W_2$$

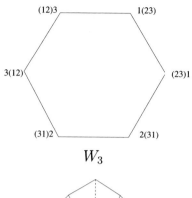

$$W_3$$

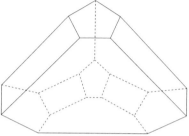

$$W_4$$

The facets of W_n are of the following forms:

$$W_{n-1}$$

$$K_n$$

$$W_r \times K_s \quad \text{with} \quad r + s - 1 = n$$

Faces of lower dimension are of the form

$$W_k \times K_{n_1} \times \cdots \times K_{n_k} \quad \text{with} \quad \Sigma\, n_i = n.$$

That is, we have various inclusions:

$$W_k \times K_{n_1} \times \cdots \times K_{n_k} \hookrightarrow \partial W_n,$$

just as we had

$$K_k \times K_{n_1} \times \cdots \times K_{n_k} \hookrightarrow \partial K_n$$

for the associahera.

In my work with Kimura and Voronov on string and conformal field theories [**31**, **32**], we were concerned with the (Deligne-)Knudsen(-Mumford) compactification of moduli spaces of punctured Riemann surfaces and their way of visualing points on the compactification divisor in terms of 'n-punctured stable curves'. We found an analogous representation of points on the boundary of a real compactification of moduli spaces as stable Riemann surfaces with pairs of tangent directions at double points mod diagonal rotations [**31**]. Then Kimura and I found analogous pictures for the associahedra and the cyclohedra. A schematic picture of a point on the boundary of $\overline{F}(S^1, n)$ is given by a 'tree' of circles attached at double points with punctures distinct from the double points.

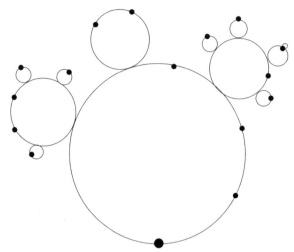

Observe that (the compactification of) the moduli space of n distinct points on the real line, modulo translation and dilation which can fix the first and last points as 0 and 1, is diffeomorphic to (the compactification of) the configuration space $F(\overset{\circ}{I}, n-2)$. The biggest circle is to be interpreted as an element of $F(S^1, k)$ while the smaller circles are to be interpreted as elements of $F(\overset{\circ}{I}, n_i - 2)$ with the end points 0 and 1 identified with a puncture on the adjacent circle. For the special case in which only one n_i is different from 2, we have a representation of a facet of the form

$$W_r \times K_s \quad \text{with} \quad r + s - 1 = n.$$

The inclusions

$$K_k \times K_{n_1} \times \cdots \times K_{n_k} \hookrightarrow \partial K_n$$

form the structure maps of an operad, while the cyclohedra have facets of the form

$$W_k \times K_{n_1} \times \cdots \times K_{n_k} \hookrightarrow \partial W_n$$

and so do not form an operad, but rather what Markl [**40**] has recently described and dubbed a *module over an operad* (which is not the same as a module over an algebra over an operad). That is, we can regard the facet inclusions for both K_n and W_n as $\circ_i$ operations and then compose them appropriately. We have compatibility in the sense of commutativity of the following diagram:

$$W_p \times K_q \times K_r \longrightarrow W_p \times K_{q+r-1}$$
$$\downarrow \qquad\qquad\qquad\qquad \downarrow$$
$$W_{p+q-1} \times K_r \longrightarrow W_{p+q+r-2}$$

where the arrows are given by suitable $\circ_i$ operations.

Although the cyclohedra first arose in the context of compactified configuration spaces, their structure as a module over the operad of associahedra suggests other questions, perhaps the most intriguing of which are the subjects of the next section.

4.1. Approximations. The operads originally invented by Peter May for studying iterated loop spaces were used in a variety of ways. For the associahedra, we have the following construction:

Just as we have degeneracy maps $s_i : \Delta_n \to \Delta_{n-1}$ for $i = 0, 1, \ldots, n-1$, so I defined (with more effort) degeneracy maps $s_i : K_n \to K_{n-1}$ [49]. For a space X with distinguished point $*$, form the space

$$KX := \coprod K_n \times X^n / \sim$$

where $(s_i(t), x_1, \ldots, x_{n-1}) \sim (t, x_1, \ldots, x_{i-1}, *, x_{i+1}, \ldots, x_{n-1})$ for $t \in K_n$.

[Corollary to a Theorem of May [42]]: For a connected CW complex X, the space KX has the homotopy type of $\Omega\Sigma X$, the based loop space on the suspension of X.

What about WX, similarly defined using the cyclohedra instead of the associahedra? That is, we can define (again with effort) degeneracy maps $s_i : W_n \to W_{n-1}$ and form

$$WX := \coprod W_n \times X^n / \sim$$

where $(s_i(t), x_1, \ldots, x_{n-1}) \sim (t, , x_1, \ldots, x_{i-1}, *, x_{i+1}, \ldots, x_{n-1})$ for $t \in W_n$.

The module structure of the W_n over the K_n as operad means that we can define maps $WKX \to WX$ and further that we have commutativity of the diagram

$$WKKX \longrightarrow WKX$$
$$\downarrow \qquad\qquad\qquad \downarrow$$
$$WKX \longrightarrow WX.$$

The space KX is an approximation to $\Omega\Sigma X$, but what does WX approximate?

Added in 1996: Markl has observed that there are also 'degeneracies' $W_n \to K_n$ which are deformation retractions compatible with the facet structure, thus WX is also an approximation to $\Omega\Sigma X$. This makes the following the more interesting question.

Since the cyclic symmetry of $F(S^1, n)$ extends to an action of the cyclic group $\mathbf{Z}/n$ on W_n and we can define the degeneracy s_0 first and then use the action of $\mathbf{Z}/n$ to define the other s_i's, we could also construct

$$W_{\mathbf{Z}}X := \coprod W_n \times_{\mathbf{Z}/n} X^n / \sim$$

or yet again

$$CWX := \coprod \overline{F}(S^1, n) \times_{\mathbf{Z}/n} X^n / \sim .$$

It is unknown what the analog of May's theorem should be for these constructions; we have approximations to as yet unknown functors.

5. L_∞-algebras and BRST cohomology

Although A_∞-spaces arose first and led to A_∞-algebras via their chain algebras, the Lie analogs occurred first as algebras [**46**] and the relevant topology much later.

5.1. L_∞-algebras or (Strong) Homotopy Lie algebras.

DEFINITION 5.1 (L_∞ or (Strong) Homotopy Lie algebras). An L_∞-*algebra* is a graded vector space $V = \sum_{i \in \mathbf{Z}} V_i$ with a differential Q of degree 1 (with $Q^2 = 0$) and a collection of n-ary brackets:

$$[v_1, \dots, v_n] \in V, \qquad v_1, \dots, v_n \in V, \ n \geq 2,$$

which are homogeneous of degree $3 - 2n$ and super (or graded) symmetric:

$$[v_1, \dots, v_i, v_{i+1}, \dots, v_n] = (-1)^{|v_i||v_{i+1}|}[v_1, \dots, v_{i+1}, v_i, \dots, v_n],$$

$|v|$ denoting the degree of $v \in V$, and satisfy the relations

$$Q[v_1, \dots, v_n] + \sum_{i=1}^n \epsilon(i)[v_1, \dots, Qv_i, \dots, v_n]$$

$$= \sum_{k+l=n+1} \sum_{\text{unshuffles } \sigma} \epsilon(\sigma)[[v_{i_1}, \dots, v_{i_k}], v_{j_1}, \dots, v_{j_{l-1}}],$$

where $\epsilon(i) = (-1)^{|v_1|+\cdots+|v_{i-1}|}$ is the sign picked up by taking Q through $v_1, \dots, v_{i-1}$ and, for the unshuffle

$$\sigma : \{1, 2, \dots, n\} \mapsto \{i_1, \dots, i_k, j_1, \dots, j_{l-1}\},$$

the sign $\epsilon(\sigma)$ is the sign picked up by the elements v_i passing through the v_j's during the unshuffle of $v_1, \dots, v_n$, as usual in superalgebra.

REMARK 5.1. Here we follow the physics grading and sign conventions in our definition of a homotopy Lie algebra [**52, 53**]. These are equivalent to but different from those in the existing mathematics literature, cf. Lada and Stasheff [**36**], in which the n-ary bracket has degree $2 - n$. With those mathematical conventions, homotopy Lie algebras occur naturally as deformations of Lie algebras. If L is a Lie algebra and V is a complex with a homotopy equivalence to the trivial complex $0 \to L \to 0$, then V is naturally a homotopy Lie algebra; see Schlessinger and Stasheff [**46**], also Getzler and Jones [**21**]. Similarly, with the physics conventions, L_∞-algebras can occur naturally as deformations of "graded Lie algebras with a bracket of degree -1", which are equivalent to ordinary graded Lie algebras after a shift of grading and redefining the bracket by a sign, see [**53**, Section 4.1]. (For topologists, the physics conventions correspond to the algebra of homotopy groups of a space with respect to Whitehead product while the math conventions correspond to the algebra of homotopy groups of a loop space with respect to Samelson product. Indeed, L_∞-algebras first appeared implicitly in Sullivan's minimal models of rational homotopy types, as evidenced in the following alternative.)

DEFINITION 5.2. An L_∞-algebra is equivalent to the following data: A graded vector space $W = \oplus W_i$.

The **free graded commutative coalgebra** generated by sW, which is W with the grading shifted up by 1, denoted

$$\Lambda sW = \bigoplus \Lambda^n sW.$$

A linear map

$$D = d_0 + d_1 + d_2 + \cdots : \Lambda sW \to \Lambda sW$$

where each d_i is a **co**-derivation which lowers n by i such that $D^2 = 0$. We say that D is a coderivation to summarize the several conditions:

$$d_i(x_1 \wedge \cdots \wedge x_n) = \Sigma \pm d_i(x_{j_1} \wedge \cdots \wedge x_{j_{i+1}}) \wedge x_{j_{i+2}} \wedge \cdots \wedge x_{j_n}$$

where the sum is over all *unshuffles* of $\{1, \ldots, n\}$, that is, all permutations that keep each of the subsets $1, \ldots, i$ and $i+1, \ldots, n$ in the same relative order. Thus, d_i is determined by $d_i|_{W^{\otimes i+1}} : W^{\otimes i+1} \to W$.

In accordance with the physics conventions, the V of Definition 5.1 is the sW of Definition 5.2.

Even ordinary Lie algebras came late to being defined operadically [**25**], although an appropriate operad was already there implicitly in the work of Fred Cohen [**16**].

DEFINITION 5.3. The Lie operad $\mathcal{L} = \{\mathcal{L}(n)\}$ is defined by $\{\mathcal{L}(n) = H_{n-1}(F(\mathbb{R}^2, n))$ for $n \geq 1\}$, where $F(\mathbb{R}^2, n)$ denotes the configuration space of ordered n-tuples of distinct points in $\mathbb{R}^2$.

The operad structure is best seen via the deformation retractions $\mathcal{D}(n) \to F(\mathbb{R}^2, n)$, where $\mathcal{D}$ is the little disks operad.

REMARK 5.2. The L_∞-operad is more subtle. The first description is due to Hinich and Schechtman [25]. According to their theorem, L_∞-algebras can be described as algebras over a certain linear operad built from trees, which can be identified with an operad constructed from one row of the E^1 term of the spectral sequence derived from the filtration by strata of the real compactification of the moduli spaces of punctured Riemann spheres [31]. (An operad structure on the stratification spectral sequence of the DKM-compactification was first observed by Beilinson and Ginzburg [11].)

A beautiful extension of these ideas can be found in Ginzburg and Kapranov [22].

5.2. BRST cohomology and L_∞-algebras.

BRS refers to Becchi, Rouet and Stora, who in 1975 [10] called attention to the "so-called Slavnov identities which express an invariance of the Fadde'ev-Popov Lagrangian". The T refers to Tyutin who, at about the same time [51], had a preprint on the same subject - the symmetry of gauge transformations.

The acronym BRST has come to be applied in mathematical physics very widely; it seems most justified in the following context. From the point of view of differential homological algebra, BRST cohomology is a generalization of Lie algebra cohomology (with coefficients in a module); the coboundary operator typically written in terms of a basis contains a term of the same form as that of Chevalley-Eilenberg. The generalization involves a differential graded vector space L with a bracket and an action of L on a differential graded vector space M (usually a Koszul-Tate resolution), but both the Jacobi identity and the representation property are no longer satisfied precisely but only up to homotopy, in the strong sense that the BRST differential corresponds to an L_∞-structure on $L \oplus M$.

In the Lagrangian formalism, this L_∞-algebra appears as a deformation of an abelian graded Lie algebra. The physicists' 'master equation' [9] is precisely the integrability equation [20] of formal algebraic deformation theory [8].

6. Gerstenhaber algebras and BV-algebras

The full homologies $H_*(F(\mathbb{R}^2, n))$ also form an operad in the category of graded vector spaces. Since $H_*(F(\mathbb{R}^2, 2))$ is 0 except for $H_0 \approx H_1 \approx k$, we have two basic operations: a graded commutative product (denoted $a \otimes b \mapsto ab$) and a Whitehead bracket $[\ ,\ ]$ of degree 1 (corresponding to a graded Lie bracket after suspension). The defining relation between the two operations, a Leibniz identity with suitable signs:

$$[a, bc] = [a, b]c + (-1)^{|a+1||b|}b[a, c]$$

is that of a Gerstenhaber algebra [19]. Gerstenhaber defined his bracket in the context of Hochschild cohomology of an algebra with coefficients in itself precisely by constructing the $\circ_i$-operations of a comp algebra structure on Hochschild's cochain complex. However, various of the tri-linear relations on the cohomology held only up to homotopy on the cochain level. A similar story holds in certain phyically inspired examples [37], [33]. Which set of homotopies are to be the basis for a definition of "homotopy Gerstenhaber algebra" should be dictated by applications. Those we now are beginning to see in physics and physically inspired mathematics lead to the various operads discussed in [33], especially the one denoted G_∞.

Still more complicated is what is now known as a BV-algebra (Batalin-Vilkovisky) [9], again motivated by structures inspired by mathematical physics, which can simplistically be described as a Gerstenhaber algebra with an additional operator Δ which is a derivation with respect to the bracket such that $\Delta^2 = 0$ but

$$[a, b] = \Delta(ab) - \Delta(a)b - (-1)^{|a|}a\Delta(b);$$

that is, the failure of Δ to be a derivation of the graded commutative product is given by the bracket. Since Δ is a derivation of the bracket, it can be described as a 'differential operator of order 2'. This has motivated Akman [1] to study 'differential operators of order r' in the context of algebras A without any assumption of commutativity - or even of associativity! (Her definition agrees with Koszul's [35] if the algebra is commutative, but differs from that of Grothendieck [23] in the most general case.) For any linear map $\Delta : A \to A$, Akman defines inductively a sequence of obstructions Φ^{r+1} which are $(r + 1)$-linear forms with values in A. The operator Δ is **differential of order** $\leq r$ if $\Phi^{r+1} = 0$. Remarkably, if A is (graded) commutative and $\Delta^2 = 0$, then with $\Phi^0 = \Delta$, the sequence of Φ^r's forms an L_∞-algebra. If A is not (graded) commutative but $\Delta^2 = 0$, Akman's relations among the Φ^r's provides a good definition of a (strong homotopy) *Leibniz$_\infty$*-algebra.

7. Operads and Conformal Field Theories

As mentioned in the introduction, operads revealed themselves in the physical context of String Field Theories (SFTs) and Vertex Operator Algebras (VOAs), both of which in turn draw on Conformal Field Theories (CFTs).

7.1. The moduli space operad of Riemann spheres with coordinatized punctures.
Instead of creating an operad from the moduli spaces of punctured Riemann surfaces by compactifying, one can also succeed by decorating the punctures with local coordinates. We can 'sew' two Riemann surfaces together unambiguously (up to the modular equivalence) if we have suitable local coordinates at the punctures. The term 'sewing' occurred in the physics literature for some time before Huang, in the contex of vertex opertor algebras, gave it an operadic formulation [26].

Let $\mathcal{P}_n$ be the moduli space of nondegenerate Riemann spheres Σ with n *labelled* punctures and non-overlapping holomorphic disks at each puncture (holomorphic embeddings of the standard disk $|z| < 1$ to Σ centered at the puncture). The spaces $\mathcal{P}_{n+1}$, $n \geq 1$, form an operad under sewing Riemann spheres at punctures (cutting out the disks $|z| \leq r$ and $|w| \leq r$ for some $r = 1 - \epsilon$ at sewn punctures and identifying the annuli $r < |z| < 1/r$ and $r < |w| < 1/r$ via $w = 1/z$). The symmetric group interchanges punctures along with the holomorphic disks, as usual.

The essence of a CFT can now be described as follows [32]:

Consider the Virasoro algebra Vir, which is the algebra of complex-valued vector fields on the circle in this text. Vir is generated by the elements $L_m = z^{m+1}\partial/\partial z$, $m \in \mathbb{Z}$, with the commutators given by the formula $[L_m, L_n] = (n - m)L_{m+n}$. By V we will denote the complexification of this algebra, $V := \mathrm{Vir} \otimes_{\mathbb{R}} \mathbb{C} = \mathrm{Vir} \oplus \overline{\mathrm{Vir}}$.

A *CFT* (*at the tree level*) consists of the following *data*:

1. A topological vector space $\mathcal{H}$ (a *state space*).
2. An action $T : V \otimes \mathcal{H} \to \mathcal{H}$ of the complexified Virasoro algebra V on $\mathcal{H}$.
3. A vector $|\Sigma\rangle \in \mathrm{Hom}(\mathcal{H}^{\otimes n}, \mathcal{H})$ for each $\Sigma \in \mathcal{P}_{n+1}$ depending smoothly on Σ.

These data must satisfy the following compatibility *axioms*:

4. $T(\mathbf{v})|\Sigma\rangle = |\delta(\mathbf{v})\Sigma\rangle$, where $\mathbf{v} = (v_1, \ldots, v_{n+1}) \in V$ and δ is the natural action of V^{n+1} on $\mathcal{P}_{n+1}$ by infinitesimal reparameterizations at punctures. In particular, $T(\mathbf{v})|\Sigma\rangle = T(\bar{\mathbf{v}})|\Sigma\rangle = 0$, whenever $\mathbf{v}$ can be extended to a holomorphic vector field on Σ outside of the disks.

5. The correspondence $\Sigma \mapsto |\Sigma\rangle$ defines the structure of an algebra over the operad $\mathcal{P}_{n+1}$ on the space of states $\mathcal{H}$.

No physicist would express the last axiom in those terms, and certainly not the more succinct: a CFT is an algebra over the operad $\mathcal{P}_{n+1}$, implying an action of V^{n+1} on the states $|\Sigma\rangle$.

7.2. L_∞-algebras and Closed String Field Theory. A CSFT is built on a special kind of CFT. A *string background (at the tree level)* is a CFT based on a vector space $\mathcal{H}$ with the following additional *data*:

1. A $\mathbb{Z}$-grading $\mathcal{H} = \bigoplus_{i \in \mathbb{Z}} \mathcal{H}_i$ on the state space.
2. An action of the Clifford algebra $C(V \oplus V^*)$, which is denoted usually by $b : V \otimes \mathcal{H} \to \mathcal{H}$ and $c : V^* \otimes \mathcal{H} \to \mathcal{H}$ for generators of the Clifford algebra, the degree of b is -1, and the degree of c is 1.
3. A differential $Q : \mathcal{H} \to \mathcal{H}$, $Q^2 = 0$, of degree 1, called a *BRST operator*, such that
4. $Qb + bQ = T$.

The graded space $\mathcal{H}$ with the operator Q is called a *BRST complex*. For $\psi \in \mathcal{H}_i$, the degree $\mathrm{gh}\,\psi := i$ is called the *ghost number*. The BRST complex here refers to that for the Virasoro algebra.

One of the nicest implications of a string background is the construction of a morphism of complexes $\Omega_{n+1} : \mathrm{Hom}(\mathcal{H}, \mathcal{H}^{\otimes n}) \to \Omega^\bullet(\mathcal{P}_{n+1})$, $n \geq 1$, from the complex of linear mappings between tensor powers of the BRST complex $\mathcal{H}$ to the de Rham complex of the space $\mathcal{P}_{n+1}$.

The physicists' notion of a CSFT can be rephrased in our terms.

DEFINITION 7.1. A CSFT (Closed String Field Theory) is a string background together with a morphism of operads $\underline{M} \to \mathcal{P}$.

A CSFT defines the structure of a homotopy Lie algebra on the space $\mathcal{H}_{\mathrm{rel}}$ of relative states. The brackets defining this structure are given by the formula:

$$[\cdot, \ldots, \cdot] = \int_{\underline{\mathcal{M}}_{n+1}} \omega_{n+1} \in \mathrm{Hom}((\mathcal{H}_{\mathrm{rel}})^n, \mathcal{H}_{\mathrm{rel}})$$

where $\underline{\mathcal{M}}_{n+1}$ is yet another compactified and decorated moduli space of dimension $2n - 4$. The form ω_{n+1} is related to Ω_{n+1} by a pull-back.

The homotopy Lie algebra structure in CSFT was first constructed by Zwiebach [54], but it was in [31] that the moduli spaces $\underline{\mathcal{M}}_{n+1}$ were defined and the homotopy Lie structure was explained as coming just from geometry of the $\underline{\mathcal{M}}_{n+1}$'s.

8. Homotopy commutativity and C_∞-algebras

One issue noticeably missing in the above discussion but prominent in the earliest use of operads is that of homotopy commutativity. From the point of view of Koszul duality [22], in which Lie algebras and commutative algebras are dual, the concept Koszul dual to an L_∞-algebra is that of a C_∞-algebra:

DEFINITION 8.1. A C_∞-algebra is equivalent to the following data:

A graded vector space $W = \oplus W_i$.

The **free graded Lie coalgebra** generated by sW, which is W with the grading shifted up by 1, denoted

$$\mathcal{L}sW = \bigoplus \mathcal{L}^n sW.$$

A linear map

$$D = d_0 + d_1 + d_2 + \cdots : \mathcal{L}sW \to \mathcal{L}sW$$

where each d_i is a **co**-derivation which lowers n by i such that $D^2 = 0$. Thus, d_i is determined by d_i taking i-fold brackets to elements of W.

Notice that a C_∞-algebra is an A_∞-algebra with strict commutativity of the bilinear product, not homotopy commutativity, and suitable symmetry of the higher order products. In the context of (strong) homotopy algebras as given by multilinear operations on W, the simplest definition of a C_∞-algebra first appeared in work of Kadeishvili [29, 28] and then in that of Smirnov [47] (both of whom called them commutative A_∞-algebras). Such algebras appeared also in [39] under the name of 'balanced A_∞-algebras'.

DEFINITION 8.2. A C_∞-*algebra* is an A_∞-algebra $(A, \{m_n\})$ such that each map $m_n : A^{\otimes n} \to A$ is a Harrison cochain, i.e. m_n vanishes on the sum of all (p, q)-shuffles for $p + q = n$, the sign of the shuffle coming from the grading of A shifted by 1.

In characteristic 0, when confronted with a bilinear product which is only homotopy commutative, we can always symmetrize to obtain strict (graded) commutativity. The price we pay if the original bilinear product were associative is that the symmetrized one no longer is. It therefore is in organizing the higher homotopies that we may be led to choose one alternative over the other. This issue also arises for L_∞-algebras and even more dramatically for homotopy Gerstenhaber or BV-algebras [**37, 33**].

The relevance of C_∞-algebras to moduli spaces and physics ('filtered topological gravity') is pointed out in [**32**] where we consider the operad of Deligne-Knudsen-Mumford compactifications of the moduli spaces of punctured Riemann spheres and the corresponding operad $\mathcal{S}$ of singular chains.

Let V be an algebra over $\mathcal{S}$ such that the structure morphism $\mu :$ $\mathcal{S}(n) \to \text{Hom}(V^{\otimes n}, V)$ vanishes on all elements in $\mathcal{S}(n)$ of chain degree $p > n - 2$; then V has the structure of a C_∞-algebra.

If we choose not to symmetrize but have strict associativity, then the permutahedra P_n [**44, 30**] are relevant. If we have both homotopy commutativity and homotopy associativity, then we confront the permutassociahedra KP_n [**30**] which are relevant to Mac Lane's coherence conditions [**38**] or the full panoply of E_∞-operads as they were originally intended.

Notice that we have *different* hexagons in these various settings, most easily distinguished by noting the vertex labellings.

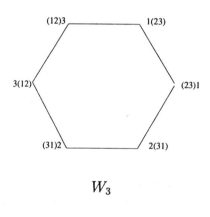

$$W_3$$

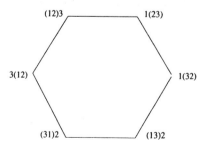

KP_3 Mac Lane's hexagon

JIM STASHEFF

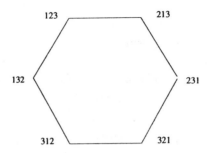

P_3

Finally , if we have strict commutativity and homotopy associativity, we can consider the hexagon

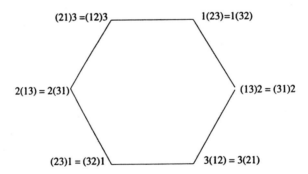

which together with the pentagon K_4 can give rise to the 'alternate universe soccer ball', the vertices being given by all meaningful bracketings of 4 variables subject to the commutativity $xy = yx$.

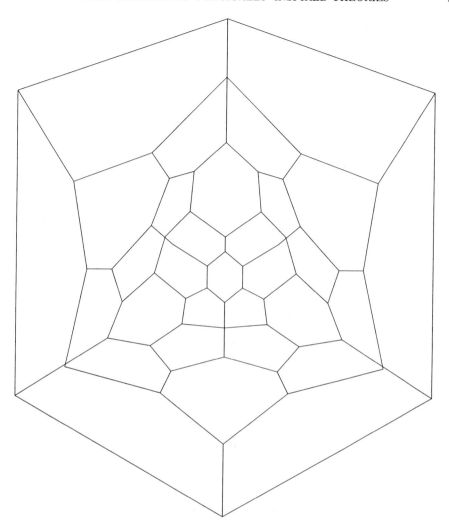

9. Appendix A: Knot invariants and Cyclohedra

It all began with Gauss in 1833. He defined the linking number of two disjoint closed or infinite curves:

[4.]

Von der *Geometria Situs*, die LEIBNITZ ahnte und in die nur einem Paar Geometern (EULER und VANDERMONDE) einen schwachen Blick zu thun vergönnt war, wissen und haben wir nach anderthalbhundert Jahren noch nicht viel mehr wie nichts.

Eine Hauptaufgabe aus dem *Grenzgebiet* der *Geometria Situs* und der *Geometria Magnitudinis* wird die sein, die Umschlingungen zweier geschlossener oder unendlicher Linien zu zählen.

Es seien die Coordinaten eines unbestimmten Punkts der ersten Linie x, y, z; der zweiten x', y', z' und

$$\iint \frac{(x'-x)(dy\,dz'-dz\,dy')+(y'-y)(dz\,dx'-dx\,dz')+(z-z')(dx\,dy'-dy\,dx')}{[(x'-x)^2+(y'-y)^2+(z'-z)^2]^{\frac{3}{2}}} = V$$

dann ist dies Integral durch beide Linien ausgedehnt

$$= 4\,m\pi$$

und m die Anzahl der Umschlingungen.

Der Werth ist gegenseitig, d. i. er bleibt derselbe, wenn beide Linien gegen einander umgetauscht werden. 1833. Jan. 22.

Of *Geometria Situs*, that Leibnitz guessed and of which only a pair of geometers (Euler and Vandermonde) were privilieged to have had a weak sight, we know not much more that nothing after a century and a half.

A major task from the boundary of *Geometria Situs* and *Geometria Magnitudinus* would be to count the linking number of two closed or infinite curves.

Let the coordinates of an arbitrary point on the first curve be x, y, z; on the second x', y', z' and

$$\iint \frac{(x'-x)(dy\,dz'-dz\,dy')+(y'-y)(dz\,dx'-dx\,dz')+(z-z')(dx\,dy'-dy\,dx')}{[(x'-x)^2+(y'-y)^2+(z'-z)^2]^{\frac{3}{2}}} = V$$

then the integral extended over both curves

$$= 4\pi m$$

and m is the linking number.

The value is symmetric, i.e. it remains the same, if the two curves are interchanged. 1833, Jan. 22.

Today with vector and differential form notation, we can write this more succinctly as follows:

$$\phi : S^1 \times S^1 \longrightarrow \mathbb{R}^3 - 0 \;\simeq\; S^2$$

$$(x, y) \longmapsto \frac{K_1(x) - K_2(y)}{|K_1(x) - K_2(y)|}$$

$$Link(K_1, K_2) = 1/4\pi \int_{S^1 \times S^1} \phi^*(d\,\mathrm{vol}_{S^2}).$$

At a lecture at MSRI, Bott presented some of the work which became his paper with Taubes: 'On the self-linking of knots' [14]. Their purpose was 'to present a "topological" account of the self-linking invariant of a knot $K \subset \mathbb{R}^3$, discovered independently by Guadagnini, Martellini and Mintchev [24] and Bar-Natan [7, 6]'. They were highly motivated by a generalization due to Kontsevich [34] which can be regarded as being in the Gaussian tradition, once we realize Gauss' integral is over the configuration space of pairs of distinct points, one on each of the two curves.

Bott and Taubes use the notation $C(n, \mathbb{R}^3)$ for configurations of n ordered points in $\mathbb{R}^3$, also known as $F(\mathbb{R}^3, n)$ above, and $\overline{C}(n, \mathbb{R}^3)$ for the compactification. We will stay with the notation F and $\overline{F}$ as in the body of this paper.

Recall the map ϕ used in defining the Gauss linking number, or rather, the retraction $\mathbb{R}^3 - 0 \to S^2$. Similarly, for $F(n, \mathbb{R}^3)$ we have several maps $\phi_{ij} : F(n, \mathbb{R}^3) \to S^2$ given by

$$(x_1, \ldots, x_n) \mapsto \frac{x_i - x_j}{|x_i - x_j|} \text{ for } i \neq j.$$

(Bott-Taubes) The maps ϕ_{ij} extend smoothly to the compactifications $\overline{F}(n, \mathbb{R}^3)$. We denote by θ_{ij} the pullback under ϕ_{ij} of the (normalized) volume form on S^2. The forms θ_{ij} generate $H^*(F(n, \mathbb{R}^3))$ [16, 3], though their relation to $H^*(\overline{C}(n, S^3))$ is less clear.

Since a knot K is an imbedding of S^1 in $\mathbb{R}^3$, we have an induced map

$$\overline{F}(n, K) : \overline{F}(n, S^1) \to \overline{F}(n, \mathbb{R}^3)$$

and can further pull back the forms θ_{ij} to $\overline{F}(n, S^1)$. Recall that $\overline{F}(n, S^1)$ can be identified with $S^1 \times W_n$ as we have reindexed the W_n by the number of points, rather than its dimension as a topological space. To get a numerical potential knot invariant, we can integrate a polynomial in the θ_{ij} over $\overline{F}(n, S^1)$ if the polynomial has degree equal to one half the dimension of $\overline{F}(n, S^1)$. (It is for purposes of such integration that Kontsevich introduced the compactifications.)

To verify if the resulting number is a knot invariant, we must consider its behavior under an isotopy of the knot. For this purpose, let $\mathcal{K}$ denote the space of knots, i.e. of smooth embeddings of S^1 in $\mathbb{R}^3$ and consider the map

$$\mathcal{K} \times \overline{C}(n, S^1) \to \overline{C}(n, \mathbb{R}^3)$$

$$(K, x) \mapsto \overline{C}(n, K)(x).$$

Now we can fibre integrate a polynomial in the (pullbacks of the) θ_{ij} over $\overline{F}(n, S^1)$ to obtain a numerical function on $\mathcal{K}$. It will be a knot invariant if it is constant on each path component of $\mathcal{K}$, i.e. if it is locally constant on $\mathcal{K}$ or, equivalently, has exterior derivative equal to zero. Computing the exterior derivative, we see that 'boundary terms' intrude as integrals over the facets of W_n. To obtain a knot invariant, the trick is to add counter terms involving configurations of $n+t$ points of which n are on the knot and the remaining t anywhere else in $\mathbb{R}^3$. The first example, v_2, of such an invariant occurred in the physics literature in a paper by Guadagnini, Martellini and Mintchev [24] and was worked out independently by Bar-Natan [6].

To simplify the notation in the integrals, we denote $\overline{F}(n, \mathbb{R}^3)$ by F_n and the compactification of configurations of $n+t$ points with n on the knot by $F_{n,t}$. We can then write this invariant as

$$v_2 = 1/4 \int_{F_4} \theta_{13}\theta_{24} - 1/3 \int_{F_{3,1}} \theta_{14}\theta_{24}\theta_{34}.$$

The next example appears in a paper of Alvarez and Labastida [2]; as transcribed with some correction of signs by Dylan Thurston [50], $6v_3$ can be written as:

$$3 \int_{F_6} \theta_{14}\theta_{26}\theta_{35} - 2 \int_{F_6} \theta_{14}\theta_{25}\theta_{36} + 6 \int_{F_{5,1}} \theta_{16}\theta_{25}\theta_{36}\theta_{46} + 3 \int_{F_{4,2}} \theta_{15}\theta_{25}\theta_{56}\theta_{36}\theta_{46}.$$

Remarkably, in almost all cases, it can be shown that only the 'principal' faces of the form W_{n-1} require the addition of counterterms. At the present writing, a few troublesome and very special cases remain to be verified and a conceptual approach to this aspect of the need for counterterms is still missing.

10. Appendix B: Associahedra and Cyclohedra as truncated simplices (with Steven Shnider)

We will construct a convex polyhedron L_n isomorphic to the associahedron K_n by truncation of a simplex and then a simple modification of the construction will produce a realization of the cyclohedron W_n.

We use the following realization of the simplex as a subset of $\mathbf{R}^{n-1}$,

$$\Delta^{n-2} = \{t = (t_1, \dots, t_{n-1}) | t_i \geq 0, \sum_{k=1}^{n-1} t_i = c_{n-1}\},$$

and denote the vertices of Δ^{n-2} by e_k for $1 \leq k \leq n-1$. In order to keep track of the truncation procedure, it will be helpful to let the value of the constant c_k depend on k. The truncation procedure will be by hyperplanes parallel to some, but not all, faces of Δ^{n-2}. The relevant ones will correspond to sets of vertices indexed by intervals $I = (i, \dots, j)$ of natural numbers, so we consider the class of functions from intervals of natural numbers to the positive reals satisfying the following conditions:

1. $c(I) > 0$ for all intervals I
2. $c(I_1) + c(I_2) < c(I_1 \cup I_2)$ if $I_1 \cup I_2$ properly contains both I_1 and I_2.

We call these functions "suitable". For example, let $c(I) = 3^{\#I}$. Having chosen such a function, we set $c_k = c((1, \cdots, k))$.

For any interval $I \subset (1, \dots, n-1)$, the face Δ_I of Δ^{n-2} is the one that *excludes* the vertices e_k for $k \in I$ and thus is defined by

$$\Delta_I = \{t \in \Delta^{n-2} | \sum_{k \in I} t_k = 0\}.$$

For any function $c(I)$ satisfying the conditions above we define a family of truncating hyperplanes,

$$P_I^c = \{t | \sum_{k \in I} t_k = c(I)\}.$$

Define an $n-2$ dimensional convex polytope in $\mathbf{R}^{n-1}$ by

$$L_n^c = \{t | \sum_{k=1}^{n-1} t_k = c_{n-1}, \quad \sum_{k \in I} t_k \geq c(I) \quad \text{for each} \quad I \subset (1, \dots, n-1)\}.$$

We shall prove For any suitable function c, there is an isomorphism of cell complexes between L_n^c and the cell complex K_n, where the k cells of L_n^c are its k-dimensional faces.

The proof is based on a lemma, which uses the concept of compatible intervals of indices. Two intervals of natural numbers, I and J, will be called *compatible* if $I \cup J$ is not an interval properly containing both I and J, that is, either

1. $J \subset I$, or
2. $I \subset J$, or
3. $I \cup J$ is not an interval.

For the purpose of stating the lemma, a single interval will be said to be compatible with itself.

LEMMA 10.1. The polytope L_n^c has nonempty interior in the $n-2$ dimensional hyperplane $\{t | \sum t_j = c_{n-1}\}$ for all suitable functions c. The intersection over p hyperplanes, $L_n^c \cap \bigcap P_{I_a}^c$, defines a nonempty $(n-p-2)$-face if and only if the intervals I_a are pairwise compatible.

First we show that the intersection $P_I \cap P_J \cap L_c^n$ is empty if the intervals I and J are not compatible. Indeed, the conditions on the function c imply that when $I \cup J$ is an interval properly containing both I and J, then for $t \in P_I^c \cap P_J^c$ and $t_i \geq 0$, we have

$$(10.1) \qquad \sum_{i \in I \cup J} t_i \leq \sum_{i \in I} t_i + \sum_{i \in J} t_i = c(I) + c(J) < c(I \cup J),$$

violating the inequality for the interval $I \cup J$ which appears in the definition of L_n.

The rest of the proof of the lemma is by induction on n using the following reduction procedure. Suppose the values of t_j for $k \leq j \leq k+s-1$ have been fixed so that the sum is $c((k, \ldots, k+s-1))$, for some pair (k, s) where $1 \leq k \leq n-1$ and $1 \leq s \leq n-k$. Define an associated modification, $c_{k,s}$, of the function c. For any interval $I = (i, \cdots, j)$, let I_{+s} be the interval $(i, \ldots, j+s)$. Define $c_{k,s}$ by

1. $c_{k,s}(I) = c(I_{+s}) - c((k, \cdots, k+s-1))$ if $k \in I$,
2. $c_{k,s}(I) = c(I)$ if $k \notin I$.

One verifies immediately that $c_{k,s}$ is suitable if c was. Define new variables t_j',

1. $t_j' = t_j$ for $j < k$,
2. $t_j' = t_{j+s}$ for $k \leq j \leq n-s-1$.

Once the values of t_j in the given interval $(k, \ldots, k+s-1)$ are fixed the inequalities for the the remaining t_j outside this interval are the same as for the polytope $L_{n-s}^{c_{k,s}}$ in the $n-s-1$ space with variables t_j'.

To prove that L_n^c has nonempty interior in the $n-2$ hyperplane, we first show that L_c^n is nonempty by an easy induction, using the construction just given, which shows that the intersection of the affine hyperplane $t_1 = c((1))$ with the polytope L_n^c is a polytope $L_{n-1}^{c'}$. Beginning with L_2^c, which is a point for any c, we proceed inductively to show that L_n^c is nonempty. Next for a given suitable function and a fixed n, there exists an $\epsilon > 0$ such that setting $c_{n-1}' = c_{n-1}$ and $c'(I) = (1 + \epsilon)c(I)$ for any proper subinterval of $(1, \ldots, n-1)$ defines a suitable function on the subintervals of $(1, \ldots, n-1)$. A point of $L_n^{c'}$ is an interior point of L_n^c as a set in $n-2$ space.

Finally, we use induction on n to prove that the intersection $L_n^c \cap \bigcap P_{I_a}^c$ is a nonempty $(n-p-2)$-dimensional set if the p intervals, I_a are compatible. Compatibility implies that the system of linear equations defining $\cap P_{I_a}$ has rank p, so if the intersection is not empty, it has dimension $n - p - 2$. Label the intervals so that I_1 does not contain any other interval I_a. Assume $I_1 = (k, \ldots, k + s - 1)$. Since L_{s+1}^c is nonempty, there is a point $(t_k, \ldots, t_{k+s-1})$ satisfying all the required conditions which involve only these coordinates. Consider the polytope $L_{n-s}^{c_{k,s}}$ in the coordinates t_j' defined above. The intervals of indices in the t_j' coordinates are compatible and so by the induction assumption, the intersection of the corresponding hyperplanes $P_{I_a'}'$ for $a > 1$ with the polytope $L_{n-s}^{c_{k,s}}$ is nonempty. To complete the proof we must show that returning to the original coordinates, $(t_1, \ldots, t_k, \ldots, t_{k+s-1}, \ldots, t_{n-1})$ is in the required intersection. We need only verify the inequalities in the definition of L_n^c which involve intervals which are not compatible with I_1, namely those intervals J which intersect I_1 but do not contain I_1 and such that $I_1 \cup J$ properly contains I_1, since all others involve either the full sum $t_k + \ldots + t_{k+s-1}$ as a summand or involve only the coordinates $\{t_k, \ldots, t_{k+s-1}\}$. Let $J' = I_1 \cap J$ and $J'' = J - J'$, then

$$\sum_{j \in J} t_j = \sum_{j \in J'} t_j + \sum_{j \in J''} t_j \geq c(J') + c(I_1 \cup J) - c(I_1) \geq c(J).$$

This concludes the proof of the lemma.

In K_n the p-dimensional cells are labelled by bracketings of n •'s with $n - 2 - p$ pairs of parentheses and two cells intersect if and only if the two bracketing labels are consistent with a single bracketing, that is, there is a single big bracketing such that each of the given ones consists of a subset of the parentheses in the big bracketing.

We can use the same labelling for the faces of L_n^c. For $I = (k, \ldots, k + s - 1)$, let $b(I)$ be the bracketing which is represented by a row of n •'s with parentheses around the $s + 1$ •'s in positions k to $k + s$. For the singleton intervals (k), $b((k))$ is a sequence of n •'s with parentheses around the kth and $k + 1^{st}$. With this convention, for a set of p compatible intervals $\{I_1, \ldots, I_p\}$ the corresponding bracketings $b(I_a)$ are consistent with a single bracketing with p parentheses, $b(I_1, \ldots, I_p)$, which will label the intersection. Note that if I_j and I_k are compatible, then the bracketings $b(I_j)$ and $b(I_k)$ are either nested or disjoint. Adjacency for the intervals I_j, I_k would imply that the bracketings in the $\bullet$'s overlap. This gives the following lemma, which finishes the proof of the proposition.

LEMMA 10.2. *The map* $P_I \mapsto b(I)$ *induces an incidence isomorphism between (the poset of) the cells of* L_n *and those of* K_n.

To truncate the simplex Δ^{n-1} to realize W_n, we need only interpret the interval $I = (i, \ldots, j) \subset (1, \ldots, n-1)$ as consisting of integers $mod\ n-1$ and similarly for anything indexed by such an interval.

Acknowledgments. This work is partly a result of my collaboration with Kimura and Voronov. It has also benefited from interactions with F. Akman, G. Barnich, R. Bott, C. Taubes, M. Kontsevich, M. Markl, S. Shnider and D. Thurston.

References

1. F. Akman, *On some generalizations of Batalin-Vilkovisky algebras*, Preprint, Cornell University, 1995, to appear in JPAA q-alg/9506027.
2. M. Alvarez and J.M.F. Labastida, *Vassiliev invariants for torus knots*, Tech. report, CTP-MIT, 1995, to appear in Journal of Knot Theory and its Ramifications, q-alg/9506009.
3. V. I. Arnold, *The cohomology ring of the colored braid group*, Mat. Zametki (1969), 227–231.
4. S. Axelrod and I. M. Singer, *Chern-Simons perturbation theory II*, J. Diff. Geom. **39** (1994), 173–213, hep-th/9304087.
5. D. Balavoine, *Deformation of algebras over a quadratic operad*, Operads: Proceedings of Renaissance Conferences (J.-L. Loday, J. Stasheff, and A. A. Voronov, eds.), Amer. Math. Soc., 1996, in this volume.
6. D. Bar-Natan, *On Vassiliev knot invariants*, Topology **34 No. 2** (1995), 423 – 472.
7. _____, *Perturbative aspects of the Chern-Simons topological quantum field theory*, J. Knot Theory and its Ram. **4** (1995), 503–548, a mildly updated version of his Princeton University Ph.D. thesis, 1991.
8. G. Barnich and M. Henneaux, *Consistent couplings between fields with a guage freedom and deformations of the master equation*, Phys. Lett. **B 311** (1993), 123–129.
9. I.A. Batalin and G.S. Vilkovisky, *Gauge algebra and quantization*, Phys. Lett. **102 B** (1981), 27–31.

10. C. Becchi, A. Rouet, and R. Stora, *Renormalization of the abelian Higgs-Kibble model*, Commun. Math. Phys. **42** (1975), 127–162.
11. A. Beilinson and V. Ginzburg, *Infinitesimal structure of moduli spaces of G-bundles*, Internat. Math. Research Notices **4** (1992), 63–74.
12. J. M. Boardman and R. M. Vogt, *Homotopy invariant algebraic structures on topological spaces*, Lecture Notes in Math., vol. 347, Springer-Verlag, 1973.
13. A. Borel, *Sur la cohomologie des espaces fibrés principaux et des espaces homogenes de groupes de Lie compacts*, Annals of Math. **57** (1953), 115–207.
14. R. Bott and C. Taubes, *On the self-linking of knots*, J. Math. Phys. **35** (1994), 5247–5287.
15. G. Calugareanu, *L'integral de Gauss et l'analyse de noueds tridimensionnels*, Rev. Math. Pures Appl **4** (1959), 5–20.
16. F. R. Cohen, *Artin's braid groups, classical homotopy theory and sundry other curiosities*, Contemp. Math. **78** (1988), 167–206.
17. V.G. Drinfel'd, *Quasi-Hopf algebras*, Leningrad Math J. **1** (1990), 1419–1457.
18. W. Fulton and R. MacPherson, *A compactification of configuration spaces*, Ann. Math. **139** (1994), 183–225.
19. M. Gerstenhaber, *The cohomology structure of an associative ring*, Ann. Math. **78** (1962), 267–288.
20. M. Gerstenhaber, *On the deformation of rings and algebras*, Ann. of Math. **79** (1964), 59–103.
21. E. Getzler and J.D.S. Jones, *n-algebras and Batalin-Vilkovisky algebras*, preprint, 1993.
22. V. Ginzburg and M. Kapranov, *Koszul duality for operads*, Duke Math. J. **76** (1994), 203–272.
23. A. Grothendieck, *Elements de Geometrie Algebrique IV, Études Locals des Schemas et des Morphismes de Schemas*, Pub. Math. IHES **32** (1967), Proposition 16.8.8 on p.42.
24. E. Guadagnini, M. Martellini, and M. Mintchev, *Wilson lines in Chern-Simons theory and link invariants*, Nucl. Phys. **B330** (1990), 575–607.
25. V. Hinich and V. Schechtman, *Homotopy Lie algebras*, Adv. Studies Sov. Math. **16** (1993), 1–18.
26. Y.-Z. Huang, *Operadic formulation of topological vertex algebras and Gerstenhaber or Batalin-Vilkovisky algebras*, Comm. Math. Phys. (1994), 105–144, hep-th/9306021.
27. Stasheff J.D., *Differential graded lie algebras, quasi-Hopf algebras and higher homotopy algebras*, Proceedings of the Workshop on Quantum Groups, Deformation Theory, and Representation Theory, Euler International Mathematical Institute, Leningrad, October 1990, vol. 1510, 1992, pp. 120–137.
28. T. Kadeishvili, *The category of differential coalgebras and the category of A(∞)-algebras*, Proc. Tbilisi Math. Inst. **77** (1985), 50–70.
29. ———, *A(∞)-algebra structure in cohomology and the rational homotopy type*, preprint 37, Forschungsschwerpunkt Geometrie, Universität Heidelberg, Mathematisches Institut, 1988.
30. M. M. Kapranov, *The permutoassociahedron, Mac Lane's coherence theorem and asymptotic zones for the KZ equation*, J. Pure and Appl. Alg. **85** (1993), 119–142.

31. T. Kimura, J. Stasheff, and A. A. Voronov, *On operad structures of moduli spaces and string theory*, Commun. Math. Phys. **171** (1995), 1–25, hep-th/9307114.

32. T. Kimura, J. Stasheff, and A. A. Voronov, *Homology of moduli spaces of curves and commutative homotopy algebras*, The Gelfand Mathematics Seminars, 1993–1994 (J. Lepowsky and M. M. Smirnov, eds.), Birkhäuser, 1996.

33. T. Kimura, A.A. Voronov, and G. Zuckerman, *Homotopy Gerstenhaber algebras and topological field theory*, Operads: Proceedings of Renaissance Conferences (J.-L. Loday, J. Stasheff, and A. A. Voronov, eds.), Amer. Math. Soc., 1996, in this volume.

34. M. Kontsevich, *Feynman diagrams and low-dimensional topology*, First European Congress of Mathematics, Vol. II (Paris, 1992) (Basel), Progr. Math., vol. 120, Birkhäuser, 1994, pp. 97–121.

35. J.-L. Koszul, *Crochet de Schouten-Nijenhuis et cohomologie*, Astérisque (1985), 257–271.

36. T. Lada and J.D. Stasheff, *Introduction to sh Lie algebras for physicists*, Intern'l J. Theor. Phys. **32** (1993), 1087–1103.

37. B. H. Lian and G. J. Zuckerman, *New perspectives on the BRST-algebraic structure of string theory*, Commun. Math. Phys. **154** (1993), 613–646, hep-th/9211072.

38. S. Mac Lane, *Natural associativity and commutativity*, Rice Univ. Studies **49** (1963), 28–46.

39. M. Markl, *A cohomology theory for $A(m)$-algebras and applications*, JPAA (1992), 141–175.

40. _____ , *Models for operads*, Comm. in Algebra **24** (1996), 1471–1500.

41. W.S Massey, *Higher order linking numbers*, Conference on Algebraic Topology, 1968, pp. 174–205.

42. J. P. May, *The geometry of iterated loop spaces*, Lecture Notes in Math., vol. 271, Springer-Verlag, 1972.

43. _____ , *Definitions: Operads, algebras and modules*, Operads: Proceedings of Renaissance Conferences (J.-L. Loday, J. Stasheff, and A. A. Voronov, eds.), Amer. Math. Soc., 1996, in this volume.

44. R.J. Milgram, *Iterated loop spaces*, Annals of Math. **84** (1966), 386–403.

45. W.F. Pohl, *The self linking number of a closed space curve*, Jour. Math. and Mech. **17** (1968), 975–986.

46. M. Schlessinger and J. D. Stasheff, *The Lie algebra structure of tangent cohomology and deformation theory*, J. of Pure and Appl. Alg. **38** (1985), 313–322.

47. V. A. Smirnov, *On the cochain complex of topological spaces*, Math. USSR Sbornik **43** (1992), 133–144.

48. J. Stasheff, *The pre-history of operads*, Operads: Proceedings of Renaissance Conferences (J.-L. Loday, J. Stasheff, and A. A. Voronov, eds.), Amer. Math. Soc., 1996, in this volume.

49. J. D. Stasheff, *On the homotopy associativity of H-spaces, I*, Trans. Amer. Math. Soc. **108** (1963), 275–292.

50. D. Thurston, *Integral expressions for the Vassiliev knot invariants*, Ph.D. thesis, Harvard Univ., 1995, senior thesis.

51. I. V. Tyutin, *Gauge invariance in field theory and statistical physics in operator formulation (in Russian)*, Tech. report, Lebedev Physics Inst., 1975.

52. E. Witten and B. Zwiebach, *Algebraic structures and differential geometry in two-dimensional string theory*, Nucl. Phys. B **377** (1992), 55–112.

53. B. Zwiebach, *Closed string field theory: Quantum action and the Batalin-Vilkovisky master equation*, Nucl. Phys. B **390** (1993), 33–152.

54. ———, *Closed string field theory: Quantum action and the Batalin-Vilkovisky master equation*, Nucl. Phys. B **390** (1993), 33–152.

DEPARTMENT OF MATHEMATICS, UNIVERSITY OF NORTH CAROLINA, CHAPEL HILL, NC 27599-3250, USA

E-mail address: jds@@math.unc.edu

Contemporary Mathematics
Volume **202**, 1997

OPÉRADES DES ALGÈBRES $(k + 1)$-AIRES

Allahtan Victor GNEDBAYE

0.—Introduction[1].

V. Ginzburg et M. Kapranov viennent de donner un nouveau souffle à la théorie des *opérades* (*cf.* [2]). Cependant, en développant la notion d'opérades quadratiques, ils abordent essentiellement le cas des algèbres binaires, c'est-à-dire les opérades engendrées par des $\mathbb{S}$-modules concentrés en degré 2. Or, lorsqu'on regarde une opérade comme une algèbre associative dans la catégorie monoïdale des $\mathbb{S}$-modules, on s'aperçoit que la *quadraticité* porte en fait sur les relations et non sur les générateurs.

Le but de ce travail est précisément d'étendre la *théorie de dualité* de Ginzburg et Kapranov aux opérades engendrées par des $\mathbb{S}$-modules concentrés en degré $(k + 1)$ où $k \geq 1$ est un entier quelconque. Nous travaillons donc avec des *algèbres* $(k + 1)$-*aires*, c'est-à-dire dont la multiplication porte sur $(k+1)$ variables. Le cas classique correspond à la valeur $k = 1$.

Dans le chapitre I, nous définissons les notions générales d'algèbres $(k + 1)$-aires et leurs divers types. Les deux manières d'écrire l'*asso-*

---

[1]Mathematics Subject Classification : 17A30, 17A42, 17D99, 18G60, 18G99.

ciativité classique i.e.,

$$(ab)c = a(bc) \qquad \text{ou} \qquad (ab)c - a(bc) = 0$$

donnent respectivement naissance à l'*associativité totale* (k relations)

$$(a_0 \ldots a_{i-1}(a_i \ldots a_{i+k})a_{i+k+1} \ldots a_{2k}) = ((a_0 \ldots a_k)a_{k+1} \ldots a_{2k})$$

où $i = 1, \cdots, k$, et à l'*associativité partielle* (une relation)

$$\sum_{i=0}^{k} (-1)^{ik}(a_0 \ldots a_{i-1}(a_i \ldots a_{i+k})a_{i+k+1} \ldots a_{2k}) = 0 .$$

De même nous généralisons les notions de *commutativité*, de *symétrie*, et la *relation de Jacobi* définissant les *algèbres de Lie* $(k+1)$-*aires*. Notons que ces dernières, déjà considérées par Ph. Hanlon et M. Wachs (*cf.* [4]), interviennent dans l'étude des problèmes combinatoires liés au *réseau de partition*.

Nous donnons des exemples d'algèbres $(k + 1)$-aires provenant du cas classique, et nous construisons certaines algèbres $(k+1)$-aires libres utiles pour la suite.

Le chapitre II est consacré à la notion d'*opérades quadratiques* $(k+1)$-*aires* et à leurs opérades duales : les générateurs sont placés en degré $(k+1)$ et les relations en degré $(2k+1)$. Nous appliquons cette généralisation en déterminant les opérades de nos exemples d'algèbres $(k + 1)$-aires. Nous montrons que les deux types d'algèbres $(k + 1)$-aires associatives ont des opérades quadratiques $(k + 1)$-aires duales l'une de l'autre; puis que l'opérade des algèbres de Lie $(k+1)$-aires est en dualité avec celle des algèbres $(k + 1)$-aires totalement associatives et symétriques.

Dans le chapitre III, nous construisons des théories d'homologie pour les algèbres $(k+1)$-aires : l'homologie de Hochschild des algèbres $(k + 1)$-aires partiellement associatives, et l'analogue Chevalley-Eilenberg pour les algèbres de Lie $(k + 1)$-aires construite par Hanlon et Wachs. Nous montrons que l'homologie de Hochschild de toute algèbre $(k + 1)$-aire partiellement associative libre est triviale. Nous donnons une décomposition du complexe de Hanlon-Wachs en somme directe de sous-complexes et nous en déduisons le «bon» complexe pour l'homologie des algèbres de Lie $(k + 1)$-aires.

Notations.— Dans tout ce qui suit le symbole $\mathbb{K}$ désigne un corps commutatif de caractéristique nulle (bien que la majeure partie de cet article soit valable sur un anneau quelconque); sauf mention expresse du contraire, les espaces vectoriels, les applications linéaires, les produits tensoriels seront relatifs à $\mathbb{K}$.

Pour tout couple d'entiers (p, q) tel que $p \le q$, nous désignons par $|_p^q$ l'ensemble $\{n \in \mathbb{N} : p \le n \le q\}$. Pour tout entier $n \ge 0$, nous notons S_n (*resp.* $\operatorname{sgn}(\sigma)$) le groupe symétrique opérant sur l'ensemble $|_0^{n-1}$ (*resp.* la signature de la permutation σ de S_n). Enfin, nous notons $\operatorname{Sh}_{p,q}$ l'ensemble des (p, q)-shuffles *i.e.*, les permutations σ de S_{p+q} telles que

$$\sigma(0) < \sigma(1) < \cdots < \sigma(p-1) \quad \text{et} \quad \sigma(p) < \sigma(p+1) < \cdots < \sigma(p+q-1).$$

Nous fixons une fois pour toutes un entier $k \ge 1$.

I.— ALGÈBRES $(k+1)$-AIRES.

Nous commençons par introduire la notion d'algèbre $(k+1)$-aire et quelques unes de ses sous-structures : sous-algèbre, idéal, *etc.* Nous donnons ensuite des exemples qui serviront de références, et des outils pour en fabriquer : les algèbres $(k+1)$-aires libres.

1.— Définitions générales.

Une *algèbre $(k+1)$-aire* est la donnée d'un espace vectoriel A muni d'une application multilinéaire $(\ )\ :\ A^{\otimes k+1} \longrightarrow A$ appelée *multiopération*. Le résultat $(x_0 \dots x_k)$ de la multiopération est dit *multiproduit des facteurs* $x_0, \dots, x_k$.

Par *cas classique*, nous entendons le cas d'une opération binaire, qui correspond à la valeur $k = 1$.

Une *sous-algèbre* d'une algèbre $(k+1)$-aire A est un sous-espace vectoriel B de A stable pour la multiopération *i.e.*, tout multiproduit $(x_0 \dots x_k)$ est dans B lorsque les éléments $x_0, \dots, x_k$ sont dans B.

Un *idéal multilatère* d'une algèbre $(k+1)$-aire A est un sous-espace vectoriel $\mathcal{I}$ de A tel que tout multiproduit $(x_0 \dots x_k)$ soit dans $\mathcal{I}$ dès que l'un des facteurs x_i est dans $\mathcal{I}$. Il est clair que tout quotient d'une algèbre $(k+1)$-aire par un idéal multilatère hérite d'une structure d'algèbre $(k+1)$-aire.

De manière évidente, toute intersection d'une famille de sous-algèbres (*resp.* d'idéaux multilatères) d'une algèbre $(k+1)$-aire est une sous-algèbre (*resp.* un idéal multilatère).

Soit A une algèbre $(k+1)$-aire et soit X une partie quelconque de A. L'intersection de toutes les sous-algèbres (*resp.* de tous les idéaux multilatères) de A contenant X est appelée *la sous-algèbre* (*resp. l'idéal multilatère*) *de A engendrée par X.*

Une *unité* dans une algèbre $(k+1)$-aire A est un élément e tel que l'on ait les relations

$$(\underbrace{e \ldots e}_{j \; facteurs} xe \ldots e) = x, \quad \forall\, x \in A, \quad \forall\, j \in |_0^k.$$

On observera qu'une unité n'est plus nécessairement unique dès que l'entier k est supérieur ou égal à 2.

Par morphisme d'algèbres $(k+1)$-aires, nous entendons toute application linéaire commutant avec les multiopérations. Il est clair que les algèbres $(k+1)$-aires et leurs morphismes forment une catégorie.

2.— Types d'algèbres $(k+1)$-aires.

Comme dans le cas classique, la multiopération peut être amenée à satisfaire à certaines compatibilités que nous abordons maintenant.

2.1.— Algèbres $(k+1)$-aires associatives.

Il y a deux notions d'associativité pour les algèbres $(k+1)$-aires : *l'associativité totale* caractérisée par les relations

$$((x_0 \ldots x_k)x_{k+1} \ldots x_{2k}) = (x_0 \ldots x_{i-1}(x_i \ldots x_{i+k})x_{i+k+1} \ldots x_{2k})$$

pour tous $i \in |_1^k$; *l'associativité partielle* caractérisée par la relation

$$\sum_{i=0}^{k} (-1)^{ik}(x_0 \ldots x_{i-1}(x_i \ldots x_{i+k})x_{i+k+1} \ldots x_{2k}) = 0 \, .$$

Evidemment, pour $k = 1$, ces deux notions coïncident et nous retrouvons les algèbres associatives classiques. Remarquons aussi que, si l'entier k est impair, toute algèbre $(k+1)$-aire totalement associative est partiellement associative; nous donnerons ultérieurement un exemple d'algèbre $(k+1)$-aire partiellement associative qui n'est pas totalement associative. D'autres exemples d'algèbres $(k+1)$-aires associatives s'obtiennent par la proposition suivante.

Proposition 1. *Soit $(A, \bullet)$ une algèbre associative au sens classique. Alors la multiopération, définie par*

$$(a_0 \dots a_k) := a_0 \bullet \dots \bullet a_k \quad où \quad a_0, \dots, a_k \in A,$$

confère à l'espace vectoriel A une structure d'algèbre $(k+1)$-aire totalement associative. Cette structure est donc aussi partiellement associative si l'entier k est impair.

Démonstration.— L'associativité totale (*resp.* partielle) découle immédiatement de celle de l'algèbre A (*resp.* et de la parité de k). ∎

2.2.— Algèbres $(k+1)$-aires commutatives. De même, il y a deux notions possibles de commutativité pour les algèbres $(k+1)$-aires :

la symétrie caractérisée par les relations

(2.2.1) $$(x_{\sigma(0)} \dots x_{\sigma(k)}) = (x_0 \dots x_k), \quad \forall \, \sigma \in S_{k+1} \; ;$$

la commutativité caractérisée par la relation

(2.2.2) $$\sum_{\sigma \in S_{k+1}} \mathrm{sgn}(\sigma)(x_{\sigma(0)} \dots x_{\sigma(k)}) = 0 \; .$$

Puisqu'il y a autant de permutations paires qu'impaires, il est clair qu'une algèbre $(k+1)$-aire symétrique est aussi commutative. Pour toute algèbre associative et commutative au sens classique, l'algèbre $(k+1)$-aire totalement associative associée (*cf.* Proposition 1) est symétrique et donc commutative.

Proposition 2 (Amitsur-Levitzki). *Soit R une algèbre associative et commutative au sens classique. Alors, pour tout entier $n \geq 1$, l'algèbre classique des matrices carrées $\mathcal{M}_n(R)$ est une algèbre $(2n)$-aire commutative, totalement et partiellement associative.*

Démonstration.— Comme $\mathcal{M}_n(R)$ est une algèbre associative classique, la relation d'associativité totale (resp. partielle) de la multiopération découle de la Proposition 1 où $k = 2n - 1$. La commutativité résulte de la formule d'Amitsur-Levitzki (*cf.* [1], [6]) :

$$\sum_{\sigma \in S_{2n}} \mathrm{sgn}(\sigma)(M_{\sigma(0)} \dots M_{\sigma(2n-1)}) = 0$$

pour tous $M_0, \ldots, M_{2n-1} \in \mathcal{M}_n(R)$. ∎

Remarque 3.— On observera qu'il faut prendre le même entier n pour avoir la commutativité de l'algèbre $(2n)$-aire $\mathcal{M}_n(R)$. En effet, $\mathcal{M}_{n+1}(R)$ est une algèbre $(2n)$-aire partiellement et totalement associative mais pas commutative (prendre par exemple $n = 1$).

2.3.— **Algèbres de Lie $(k+1)$-aires.** Introduites par Ph. Hanlon et M. Wachs (*cf.* [4]), les algèbres de Lie $(k+1)$-aires sont les algèbres $(k+1)$-aires dont la multiopération, appelée *crochet* et notée $[x_0 \ldots x_k]$, est assujettie aux axiomes suivants

l'antisymétrie caractérisée par les relations

$$(2.3.1) \qquad [x_{\sigma(0)} \ldots x_{\sigma(k)}] = \text{sgn}(\sigma)[x_0 \ldots x_k] \, , \quad \forall \, \sigma \in S_{k+1}$$

et *la relation de Jacobi généralisée* suivante

$$(2.3.2) \quad J(x_0, \ldots, x_{2k}) := \sum_{\sigma \in S_{2k+1}} \text{sgn}(\sigma)[[x_{\sigma(0)} \ldots x_{\sigma(k)}] \, x_{\sigma(k+1)} \ldots x_{\sigma(2k)}] = 0.$$

Lorsque $k = 1$, nous retrouvons la notion classique d'algèbres de Lie. Notons qu'en présence de l'antisymétrie, la relation (2.3.2) (qui possède $(2k+1)!$ termes) se ramène à une relation ayant moins de termes :

$$(2.3.3) \quad J'(x_0, \ldots, x_{2k}) := \sum_{\sigma \in Sh_{k+1,k}} \text{sgn}(\sigma)[[x_{\sigma(0)} \ldots x_{\sigma(k)}] \, x_{\sigma(k+1)} \ldots x_{\sigma(2k)}] = 0,$$

et nous avons $J(x_0, \ldots, x_{2k}) = k!(k+1)!J'(x_0, \ldots, x_{2k})$.

Proposition 4. *Soit $(A, (\ ))$ une algèbre $(k+1)$-aire partiellement associative. Alors le crochet, défini par la formule*

$$(2.3.4) \qquad [a_0 \ldots a_k] := \sum_{\sigma \in S_{k+1}} \text{sgn}(\sigma)(a_{\sigma(0)} \ldots a_{\sigma(k)}),$$

confère à l'espace vectoriel A une structure d'algèbre de Lie $(k+1)-$ aire. C'est l'algèbre de Lie $(k+1)-$aire universelle associée à A et on la note $(A_L, [\ ])$.

Démonstration.— L'antisymétrie est évidente par définition du crochet.

Pour vérifier la relation de Jacobi, notons que $J(a_0, \ldots, a_{2k})$ est une combinaison linéaire sur $\mathbb{K}$ des éléments

$$\boldsymbol{a}_{\sigma,i} := \left(a_{\sigma(0)} \ldots a_{\sigma(i-1)}(a_{\sigma(i)} \ldots a_{\sigma(i+k)})a_{\sigma(i+k+1)} \ldots a_{\sigma(2k)}\right)$$

où $\sigma \in S_{2k+1}$ et $i \in |_0^k$. Le coefficient devant $\boldsymbol{a}_{\sigma,i}$ valant $\mathrm{sgn}(\sigma)$ fois celui devant $\boldsymbol{a}_{id,i}$, nous pouvons donc les regrouper et nous contenter de rechercher les $\boldsymbol{a}_{id,i}$ impliqués dans $J'(a_0, \ldots, a_{2k})$.

L'élément $\boldsymbol{a}_{id,i}$ apparaît une fois, et une seule, dans $J'(a_0, \ldots, a_{2k})$ et provient du crochet itéré $[[a_i \ldots a_{i+k}]a_0 \ldots a_{i-1}a_{i+k+1} \ldots a_{2k}]$. Ce dernier a pour signe $(-1)^{i(k+1)}$. Pour obtenir $\boldsymbol{a}_{id,i}$ à partir de ce crochet, il faut faire sauter un élément au-dessus de i autres, d'où le nouveau signe $(-1)^i$. En multipliant, nous obtenons $(-1)^{ik}$. Ainsi, nous avons montré que

$$J'(a_0, \ldots, a_{2k}) = \sum_{\sigma \in Sh_{k+1,k}} \mathrm{sgn}(\sigma) \sum_{i=0}^{k} (-1)^{ik} \boldsymbol{a}_{\sigma,i} \ .$$

La relation de Jacobi résulte alors de la condition d'associativité partielle.

Soit $f : (A, (\ )) \to (L, [\ ])$ une application linéaire de $(A, (\ ))$ dans une algèbre de Lie $(k+1)$-aire $(L, [\ ])$, commutant avec les multiopérations. On vérifie aisément que l'application linéaire $f_L := f : A_L \to L$ devient un morphisme d'algèbres de Lie $(k+1)$-aires. C'est clairement l'unique qui étende la structure initiale. Ce qui démontre l'universalité de l'algèbre de Lie $(k+1)$-aire $(A_L, [\ ])$. ∎

Remarques 5.— Cette proposition permet d'obtenir d'autres exemples d'algèbres de Lie $(k+1)$-aires : si l'entier k est impair, toute algèbre $(k+1)$-aire totalement associative donne naissance à une algèbre de Lie $(k+1)$-aire, et nous en connaissons par la Proposition 1. Nous donnerons ultérieurement la construction d'enveloppes universelles d'algèbres de Lie $(k+1)$-aires.

D'autre part, si $h : A \to B$ est un morphisme d'algèbres $(k+1)$-aires partiellement associatives, alors il devient un morphisme d'algèbres de Lie $(k+1)$-aires que nous notons aussi $h_L := h : A_L \to B_L$.

2.4.— Algèbres $(k+1)$-aires et produit tensoriel. Soient A et B des algèbres $(k+1)$-aires. Munissons le produit tensoriel $A \otimes B$ de la *multiopération diagonale* donnée par

$$(2.4.1) \qquad (a_0 \otimes b_0 \ldots a_k \otimes b_k) := (a_0 \ldots a_k) \otimes (b_0 \ldots b_k)$$

où les éléments $a_0, \cdots, a_k$ (*resp.* $b_0, \cdots, b_k$) parcourent A (*resp.* B).
On démontre facilement que

Lemme 6. *Si l'algèbre* $(k+1)$-*aire* A (*resp. B*) *est totalement (resp. partiellement) associative alors l'algèbre* $(k+1)$-*aire* $A \otimes B$ *est partiellement associative.* ∎

De même nous avons

Lemme 7. *Si* A *est une algèbre* $(k+1)$-*aire totalement associative et symétrique, et si* B *est une algèbre de Lie* $(k+1)$-*aire alors* $A \otimes B$ *est une algèbre de Lie* $(k+1)$-*aire.* ∎

3.— Algèbres $(k+1)$-aires libres et enveloppes universelles.

Nous nous proposons de construire, pour tout espace vectoriel V, l'objet libre dans la catégorie des algèbres $(k+1)$-aires du type #. Il s'agit d'une algèbre $(k+1)$-aire $\#\mathcal{A}^{<k>}(V)$ du type #, munie d'une application canonique $\iota : V \hookrightarrow \#\mathcal{A}^{<k>}(V)$, et ayant la propriété universelle suivante.

Pour toute application linéaire f *de* V *dans une algèbre* $(k+1)$-*aire* A *du type* #, *il existe un unique morphisme d' algèbres* $\tilde{f} :$ $\#\mathcal{A}^{<k>}(V) \to A$, *tel que l'on ait* $\tilde{f} \circ \iota = f$.

Comme à l'accoutumée, lorsqu'une solution $(\#\mathcal{A}^{<k>}(V), \iota)$ existe, elle est unique à (un unique) isomorphisme d'algèbres près.

Nous commençons par construire l'algèbre $(k+1)$-aire libre sur un espace vectoriel donné V. Ce qui permet, par passage au quotient par les relations qu'il faut, d'en déduire les autres types d'algèbres $(k+1)$-aires libres sur V. Nous donnons les constructions explicites de l'algèbre $(k+1)$-aire libre sur V pour les types "totalement associative", "symétrique et totalement associative", "partiellement associative" et de "Lie" (l'entier k étant supposé impair pour les deux derniers types). Parallèlement, nous construisons aussi *l'enveloppe universelle* des algèbres de Lie $(k+1)$-aires dans la catégorie des algèbres $(k+1)$-aires partiellement (resp. totalement) associatives.

3.1.— Algèbre $(k+1)$-aire libre. Soit V un espace vectoriel et soit $(\mathcal{A}_n^{<k>}(V))_{n \geq 0}$ la suite d'espaces vectoriels définie par la relation de récurrence

$$\mathcal{A}_0^{<k>} = V \quad \text{et} \quad \mathcal{A}_n^{<k>} = \bigoplus_{\substack{0 \leq n_0, \ldots, n_k \leq n-1 \\ n_0 + \cdots + n_k = n-1}} \mathcal{A}_{n_0}^{<k>} \otimes \ldots \otimes \mathcal{A}_{n_k}^{<k>}.$$

Munissons l'espace vectoriel $\mathcal{A}^{<k>}(V) := \bigoplus_{n \geq 0} \mathcal{A}_n^{<k>}(V)$ de la multiopération induite par la $(k+1)$-concaténation : pour tous $w_i \in \mathcal{A}_{n_i}^{<k>}(V)$ où $i \in |_0^k$ et $n_i \in \mathbb{N}$, nous posons

$$(w_0 \ldots w_k) := w_0 \otimes \ldots \otimes w_k \in \mathcal{A}_n^{<k>}(V) \quad \text{où} \quad n = n_0 + \cdots + n_k + 1 .$$

Il est clair que nous avons défini sur $\mathcal{A}^{<k>}(V)$ une structure d'algèbre $(k+1)$-aire.

Proposition 8. *L'algèbre $(k+1)$-aire $\mathcal{A}^{<k>}(V)$, munie de l'inclusion canonique $\iota : V = \mathcal{A}_0^{<k>} \hookrightarrow \mathcal{A}^{<k>}(V)$, est l'algèbre $(k+1)$-aire libre sur l'espace vectoriel V .*

Démonstration.— Il nous reste à vérifier l'universalité pour les applications linéaires de V dans une algèbre $(k+1)$-aire. Soit $f : V \to A$ une telle application. Par définition de la multiopération sur $\mathcal{A}^{<k>}(V)$, les restrictions $\tilde{f}_j := \tilde{f}_{|\mathcal{A}_j^{<k>}}$ de tout morphisme d'algèbres $\tilde{f} : \mathcal{A}^{<k>}(V) \to A$ doivent satisfaire à la relation de récurrence

$$\tilde{f}_n(w_0 \otimes \ldots \otimes w_k) = (\tilde{f}_{n_0}(w_0) \ldots \tilde{f}_{n_k}(w_k)) \quad \text{où} \quad n_0 + \cdots + n_k = n-1 .$$

Ceci assure l'unicité d'un morphisme d'algèbres $(k+1)$-aires $\tilde{f} : \mathcal{A}^{<k>}(V) \to A$, tel que $\tilde{f} \circ \iota = f$, puisque $\tilde{f}_0 = \tilde{f} \circ \iota$.

Par construction, il est clair que l'application linéaire $\tilde{f}$, définie par la relation de récurrence ci-dessus et par $\tilde{f}_0 = f$, est un morphisme d'algèbres qui prolonge f. ■

9.— **Série de Poincaré de $\mathcal{A}^{<k>}(V)$.** On observera que l'espace vectoriel $\mathcal{A}_n^{<k>}(V)$ est isomorphe à un nombre de copies de $V^{\otimes nk+1}$ égal à $a_n^{<k>}$ où la suite $(a_n^{<k>})_{n \geq 0}$ est définie par la relation de récurrence

$$a_0^{<k>} = 1 \quad \text{et} \quad a_n^{<k>} = \sum_{\substack{0 \leq n_0, \ldots, n_k \leq n-1 \\ n_0 + \cdots + n_k = n-1}} a_{n_0}^{<k>} \cdots a_{n_k}^{<k>} \quad \text{pour} \quad n \geq 1.$$

Par exemple, on a

$$a_1^{<k>} = 1, \quad a_2^{<k>} = k+1, \quad a_3^{<k>} = \frac{(k+1)(3k+2)}{2}.$$

Ainsi, la série formelle $Y = \sum_{n \geq 0} a_n^{<k>} X^{nk+1}$ vérifie l'équation

$$Y^{k+1} = Y - X.$$

3.2.— Algèbre $(k+1)$-aire totalement associative libre. Soit V un espace vectoriel et soit $T^{<k>}(V)$ l'espace vectoriel suivant :

$$T^{<k>}(V) := V \oplus V^{\otimes k+1} \oplus \cdots \oplus V^{\otimes nk+1} \oplus \cdots \quad .$$

Considérons la multiopération induite par la $(k+1)$-concaténation :

$$(3.2.1) \qquad (w_0 \ldots w_k) := x_0^0 \otimes \ldots \otimes x_{n_0 k}^0 \otimes x_0^1 \otimes \ldots \otimes x_{n_k k}^k$$

pour tous $w_i = x_0^i \otimes \ldots \otimes x_{n_i k}^i \in V^{\otimes n_i k+1}$. Comme nous avons l'égalité

$$(n_0 k + 1) + \cdots + (n_k k + 1) = (n_0 + \cdots + n_k + 1)k + 1 \;,$$

il est clair que nous avons ainsi défini sur l'espace $T^{<k>}(V)$ une structure d'algèbre $(k+1)$-aire totalement associative. De plus, l'élément générique $w = x_0 \otimes \ldots \otimes x_{nk}$ de $V^{\otimes nk+1}$ s'obtient comme produit itéré

$$(3.2.2) \qquad w = (\ldots((x_0 \ldots x_k)x_{k+1} \ldots x_{2k}) \ldots x_{nk}) \;.$$

Proposition 10. *L'algèbre* $tAs^{<k>}(V) := (T^{<k>}(V), (3.2.1))$, *munie de l'inclusion canonique* $V \hookrightarrow T^{<k>}(V)$, *est l'algèbre* $(k+1)$-*aire totalement associative libre sur l'espace vectoriel* V.

Démonstration.— Vérifions l'universalité pour les applications linéaires de V dans une algèbre $(k+1)$-aire totalement associative. Soit $f : V \to A$ une telle application. Il est clair que l'espace vectoriel $T^{<k>}(V)$ est engendré par les éléments $w = x_0 \otimes \ldots \otimes x_{nk}$ de $V^{\otimes nk+1}$ où $n \in \mathbb{N}$. En vertu de la relation (3.2.2), tout morphisme d'algèbres $\tilde{f} : T^{<k>}(V) \to A$ prolongeant f est nécessairement donné par la formule

$$(3.2.3) \quad \tilde{f}(w) := (\ldots((f(x_0) \ldots f(x_k))f(x_{k+1}) \ldots f(x_{2k})) \ldots f(x_{nk})).$$

D'où l'unicité d'un tel morphisme. On vérifie aisément que la formule (3.2.3) définit bien un morphisme d'algèbres $(k+1)$-aires. ∎

3.3.— Algèbre $(k+1)$-aire symétrique et totalement associative libre. Soit V un espace vectoriel et soit $T^{<k>}(V)_{sym}$ le quotient de l'algèbre $(k+1)$-aire $tAs^{<k>}(V)$ par l'idéal multilatère $\mathcal{I}_{sym}$ engendré par les éléments de la forme

$$(x_{\sigma(0)} \ldots x_{\sigma(k)}) - (x_0 \ldots x_k) \quad \text{où} \quad x_i \in V \quad \text{et} \quad \sigma \in \mathrm{S}_{k+1} .$$

En tant qu'espace vectoriel nous avons

$$T^{<k>}(V)_{sym} = V \oplus (V^{\otimes k+1})_{S_{k+1}} \oplus \cdots \oplus (V^{\otimes nk+1})_{S_{nk+1}} \oplus \cdots$$

où $(V^{\otimes nk+1})_{S_{nk+1}}$ désigne l'espace des coinvariants pour l'action naturelle du groupe S_{nk+1} sur l'espace vectoriel $V^{\otimes nk+1}$.

Il est clair que la multiopération (3.2.2) induit sur l'espace vectoriel $T^{<k>}(V)_{sym}$ une structure d'algèbre $(k+1)$-aire symétrique et totalement associative. Notons-la $stAs^{<k>}(V)$; comme conséquence immédiate de la Proposition 10, nous avons le

Corollaire 11. *L'algèbre $(k+1)$-aire $stAs^{<k>}(V)$, munie de l'inclusion canonique $V \hookrightarrow T^{<k>}(V)_{sym}$, est l'algèbre $(k+1)$-aire symétrique et totalement associative libre sur l'espace vectoriel V .* ∎

3.4.— Algèbre $(k+1)$-aire partiellement associative libre. Soit V un espace vectoriel et soit $(pAs_n^{<k>}(V))_{n \in \mathbb{N}}$ la suite d'espaces vectoriels définie par la relation de récurrence

$$pAs_0^{<k>} = V, pAs_n^{<k>} = \bigoplus_{\substack{0 \le n_1,\ldots,n_k \le n-1 \\ n_1+\cdots+n_k=n-1}} V \otimes pAs_{n_1}^{<k>} \otimes \ldots \otimes pAs_{n_k}^{<k>}.$$

L'espace vectoriel $pAs_n^{<k>}(V)$ est isomorphe à un nombre de copies de $V^{\otimes nk+1}$ égal à $p_n^{<k>}$ où la suite d'entiers $(p_n^{<k>})_{n \in \mathbb{N}}$ vérifie la relation de récurrence

$$p_0^{<k>} = 1 \quad \text{et} \quad p_n^{<k>} = \sum_{\substack{0 \le n_1,\ldots,n_k \le n-1 \\ n_1+\cdots+n_k=n-1}} p_{n_1}^{<k>} \cdots p_{n_k}^{<k>} \quad \text{pour} \quad n \ge 1.$$

Un calcul immédiat nous donne les premiers coefficients

$$p_1^{<k>} = 1, \; p_2^{<k>} = k, \; p_3^{<k>} = \frac{k(3k-1)}{2}.$$

On munit l'espace vectoriel $pAs^{<k>}(V) := \bigoplus_{n\geq 0} pAs_n^{<k>}(V)$ de la multiopération définie par récurrence sur *le degré relatif* du premier facteur : pour tous $W_i \in pAs_{n_i}^{<k>}(V)$ où $i \in |_0^k$ et $n_i \in \mathbb{N}$, on pose

$$(3.4.1) \quad (W_0 \ldots W_k) := W_0 \otimes \ldots \otimes W_k \quad \text{si} \quad n_0 = 0 \quad i.e. \ W_0 \in V,$$

et, si $n_0 \geq 1$ et pour $W_0 = w_0^0 \otimes \ldots \otimes w_k^0$ avec $w_0^0 \in V$, on pose

$$(3.4.2) \quad (W_0 \ldots W_k) := -\sum_{i=1}^{k} (-1)^{ik} w_0^0 \otimes \ldots \otimes w_{i-1}^0 \otimes (w_i^0 \ldots w_k^0 W_1 \ldots W_i) \otimes W_{i+1} \otimes \ldots \otimes W_k.$$

Proposition 12. *Supposons que l'entier k soit impair. Alors l'espace vectoriel $pAs^{<k>}(V)$, muni de la multiopération définie par les formules (3.4.1) et (3.4.2) et de l'inclusion canonique*

$$\iota : V = pAs_0^{<k>}(V) \hookrightarrow pAs^{<k>}(V),$$

est l'algèbre $(k+1)$-aire partiellement associative libre sur l'espace vectoriel V.

Démonstration.— Nous allons vérifier l'associativité partielle par récurrence sur le degré relatif du premier facteur. Considérons des éléments $W_i \in pAs_{n_i}^{<k>}(V)$ où $i \in |_0^{2k}$:

Si $n_0 = 0$, c'est-à-dire que $W_0 \in V$, alors nous avons

$$\begin{aligned}
&((W_0 \ldots W_k)W_{k+1} \ldots W_{2k}) \\
=\ &((W_0 \otimes \ldots \otimes W_k)W_{k+1} \ldots W_{2k}) \\
=\ &-\sum_{i=1}^{k} (-1)^{ik} W_0 \otimes \ldots \otimes W_{i-1} \otimes (W_i \ldots W_{k+i}) \otimes \ldots \otimes W_{2k} \\
=\ &-\sum_{i=1}^{k} (-1)^{ik} (W_0 \ldots W_{i-1}(W_i \ldots W_{k+i})W_{k+i+1} \ldots W_{2k}),
\end{aligned}$$

d'où la relation d'associativité partielle lorsque $n_0 = 0$.

Supposons que $W_0 = w_0^0 \otimes \ldots \otimes w_k^0$ avec $w_0^0 \in V$ et $n_0 \geq 1$, puis posons

$$Z_i := (W_0 \ldots W_{i-1}(W_i \ldots W_{k+i})W_{k+i+1} \ldots W_{2k}) \quad \text{où } i \in |_0^k.$$

Alors, par les relations (3.4.1) et (3.4.2), nous avons

$$Z_0 = \sum_{1 \leq p, q \leq k} (-1)^{k(p+q)} Z_0^{p,q}$$

où $Z_0^{p,q}$ est égal à

$$(w_0^0 \ldots w_{q-1}^0 (w_q^0 \ldots w_{p-1}^0 (w_p^0 \ldots w_k^0 W_1 \ldots W_p) \ldots W_{k+q}) \ldots W_{2k})$$

si $q \leq p$, ou égal à

$$(w_0^0 \ldots w_{p-1}^0 (w_p^0 \ldots w_k^0 W_1 \ldots W_p) \ldots (W_q \ldots W_{k+q}) \ldots W_{2k})$$

si $p < q$. De même, pour tout $i \in |_1^k$, nous obtenons

$$Z_i = -\sum_{j=1}^k (-1)^{jk} Z_i^j$$

où Z_i^j est égal à

$$(w_0^0 \ldots w_{j-1}^0 (w_j^0 \ldots w_k^0 W_1 \ldots W_j) \ldots (W_i \ldots W_{k+i}) \ldots W_{2k})$$

si $j < i$, ou égal à

$$(w_0^0 \ldots w_{j-1}^0 (w_j^0 \ldots w_k^0 W_1 \ldots W_{i-1} (W_i \ldots W_{k+i}) \ldots W_{k+j}) \ldots W_{2k})$$

si $i \geq j$. Puisque $Z_0^{p,q} = Z_q^p$ pour tout couple d'entiers (p,q) tels que $p+1 \leq q$, il vient

$$\sum_{i=0}^k (-1)^{ik} Z_i = \sum_{1 \leq q \leq p \leq k} (-1)^{k(p+q)} Z_0^{p,q} - \sum_{1 \leq i \leq j \leq k} (-1)^{k(i+j)} Z_i^j$$

$$= \sum_{q=1}^k \left(\sum_{p=q}^k (-1)^{k(p+q)} Z_0^{p,q} - \sum_{i=1}^q (-1)^{k(i+q)} Z_i^q \right).$$

Soit $q \in |_1^k$; l'hypothèse de récurrence appliquée à la famille d'éléments $(w_q^0, \ldots, w_k^0, W_1, \ldots, W_{k+q})$ se traduit par la relation

$$Y_q := \sum_{p=q}^k (-1)^{k(p+q)} (w_q^0 \ldots w_{p-1}^0 (w_p^0 \ldots w_k^0 W_1 \ldots W_p) W_{p+1} \ldots W_{k+q})$$

$$+ (-1)^{k^2} \sum_{i=1}^q (-1)^{k(i+q)} (w_q^0 \ldots w_k^0 W_1 \ldots W_{i-1} (W_i \ldots W_{k+i}) \ldots W_{k+q})$$

$$= 0 \, .$$

Comme l'entier k est impair, nous en déduisons que

$$\sum_{p=q}^{k}(-1)^{k(p+q)}Z_0^{p,q} - \sum_{i=1}^{q}(-1)^{k(i+q)}Z_i^q$$

$$= (w_0^0 \ldots w_{q-1}^0 Y_q W_{k+q+1} \ldots W_{2k}) = 0 .$$

Ainsi nous avons montré que l'algèbre $(k+1)$-aire $pAs^{<k>}(V)$ est partiellement associative.

Soit f une application linéaire de V dans une algèbre $(k+1)$-aire partiellement associative A. Comme l'espace vectoriel $pAs_n^{<k>}(V)$ est engendré par les éléments

$$w_0 \otimes \ldots \otimes w_k = (w_0 \ldots w_k) \quad \text{où} \quad w_0 \in V, \ w_i \in pAs_{n_i}^{<k>}(V), \ i \in |_1^k,$$

les restrictions $\tilde{f}_n := \tilde{f}_{|pAs_n^{<k>}}$ de tout morphisme d'algèbres $\tilde{f}$: $pAs^{<k>}(V) \to A$, doivent satisfaire à la relation

$$(3.4.3) \qquad \tilde{f}_n(w_0 \otimes \ldots \otimes w_k) = (\tilde{f}_0(w_0)\tilde{f}_{n_1}(w_1)\ldots\tilde{f}_{n_k}(w_k)) ;$$

d'où l'unicité d'un morphisme d'algèbres $(k+1)$-aires $\tilde{f}$ tel que $\tilde{f}_0 = \tilde{f} \circ \iota = f$.

Montrons, par récurrence sur le degré relatif du premier facteur, que l'application linéaire $\tilde{f}$, dont les restrictions sont données par (3.4.3) et par $\tilde{f}_0 = f$, est un morphisme d'algèbres $(k+1)$-aires. Soient $W_i \in pAs_{n_i}^{<k>}(V)$ où $i \in |_0^{2k}$:

Si $n_0 = 0$, c'est-à-dire que $W_0 \in V$, alors par définition nous avons

$$\tilde{f}(W_0 \ldots W_k) = \tilde{f}(W_0 \otimes \ldots \otimes W_k)$$
$$= (\tilde{f}_0(W_0)\ldots\tilde{f}_{n_k}(W_k)) = (\tilde{f}(W_0)\ldots\tilde{f}(W_k)) .$$

Supposons que $n_0 \geq 1$ et écrivons

$$W_0 = w_0^0 \otimes \ldots \otimes w_k^0 \quad \text{avec} \quad w_0^0 \in V \quad \text{et} \quad w_i \in pAs_{n_i^0}^{<k>}(V), \ i \in |_1^k.$$

Par les relations (3.4.2) et (3.4.3), nous avons

$$\tilde{f}(W_0 \ldots W_k)$$
$$= -\sum_{i=1}^{k}(-1)^{ik}\tilde{f}(w_0^0 \otimes \ldots \otimes (w_i^0 \ldots w_k^0 W_1 \ldots W_i) \otimes \ldots \otimes W_k)$$
$$= -\sum_{i=1}^{k}(-1)^{ik}(\tilde{f}_0(w_0^0)\ldots\tilde{f}(w_i^0 \ldots w_k^0 W_1 \ldots W_i)\ldots\tilde{f}_{n_k}(W_k)) .$$

Par hypothèse de récurrence, nous avons

$$\tilde{f}(w_i^0 \ldots w_k^0 W_1 \ldots W_i) = (\tilde{f}_{n_i^0}(w_i^0) \ldots \tilde{f}_{n_k^0}(w_k^0) \tilde{f}_{n_1}(W_1) \ldots \tilde{f}_{n_i}(W_i)) \ .$$

Comme l'algèbre $(k+1)$-aire A est partiellement associative, il vient

$$\begin{aligned}
\tilde{f}(W_0 \ldots W_k) &= ((\tilde{f}_0(w_0^0) \ldots \tilde{f}_{n_k^0}(w_k^0)) \tilde{f}_{n_1}(W_1) \ldots \tilde{f}_{n_k}(W_k)) \\
&= (\tilde{f}_{n_0}(W_0) \ldots \tilde{f}_{n_k}(W_k)) = (\tilde{f}(W_0) \ldots \tilde{f}(W_k)) \ .
\end{aligned}$$

Ainsi l'application linéaire $\tilde{f} : p\mathcal{A}s^{<k>}(V) \to A$ est bien un morphisme d'algèbres $(k+1)$-aires; ce qui achève de démontrer l'universalité de l'algèbre $(k+1)$-aire $p\mathcal{A}s^{<k>}(V)$. ∎

3.5.— Enveloppe universelle d'une algèbre de Lie $(k+1)$-aire. *Dans cette partie nous supposons que l'entier k est impair.* Soit $(\mathcal{L}, [\ ])$ une algèbre de Lie $(k+1)$-aire. Nous cherchons une algèbre $(k+1)$-aire partiellement associative $pU^{<k>}(\mathcal{L})$, munie d'un morphisme d'algèbres de Lie $(k+1)$-aires $\varphi : \mathcal{L} \to pU^{<k>}(\mathcal{L})_L$, et ayant la propriété universelle suivante.

Pour toute algèbre $(k+1)$-aire partiellement associative A et pour tout morphisme d'algèbres de Lie $(k+1)$-aires $g : \mathcal{L} \to A_L$, il existe un unique morphisme d'algèbres $(k+1)$-aires $\tilde{g} : pU^{<k>}(\mathcal{L}) \to A$ tel que l'on ait $\tilde{g}_L \circ \varphi = g$

Nous allons construire la solution de ce problème universel. Soit $p\mathcal{A}s^{<k>}(\mathcal{L})$ l'algèbre $(k+1)$-aire partiellement associative libre sur l'espace vectoriel $\mathcal{L}$ et soit $\mathcal{I}(\mathcal{L})$ l'idéal multilatère de $p\mathcal{A}s^{<k>}(\mathcal{L})$ engendré par les éléments de la forme
(3.5.1)
$$[x_0 \ldots x_k] - \sum_{\sigma \in S_{k+1}} \mathrm{sgn}(\sigma)(x_{\sigma(0)} \ldots x_{\sigma(k)}) \quad \text{où} \quad x_0, \ldots x_k \in \mathcal{L} \ .$$

Ensuite posons $pU^{<k>}(\mathcal{L}) := p\mathcal{A}s^{<k>}(\mathcal{L})/\mathcal{I}(\mathcal{L})$ l'algèbre $(k+1)$-aire partiellement associative quotient, et notons $\varphi : \mathcal{L} \to pU^{<k>}(\mathcal{L})_L$ l'inclusion naturelle d'espaces vectoriels. Alors, pour tous $x_0, \ldots, x_k \in \mathcal{L}$, nous avons

$$\begin{aligned}
\varphi([x_0 \ldots x_k]) &= [x_0 \ldots x_k] \\
&= \sum_{\sigma \in S_{k+1}} \mathrm{sgn}(\sigma)(x_{\sigma(0)} \ldots x_{\sigma(k)}) \quad \text{dans } pU^{<k>}(\mathcal{L}) \\
&= [x_0 \ldots x_k] = [\varphi(x_0) \ldots \varphi(x_k)] \quad \text{dans } pU^{<k>}(\mathcal{L})_L.
\end{aligned}$$

Ainsi, l'application φ est un morphisme d'algèbres de Lie $(k+1)$-aires.

Proposition 13. *Soit k un entier impair. L'algèbre $(k+1)$-aire $pU^{<k>}(\mathcal{L})$, munie du morphisme d'algèbres $\varphi : \mathcal{L} \hookrightarrow pU^{<k>}(\mathcal{L})_L$, est l'enveloppe universelle de l'algèbre de Lie $(k+1)$-aire $\mathcal{L}$ dans la catégorie des algèbres $(k+1)$-aires partiellement associatives.*

Démonstration.— Soit A une algèbre $(k+1)$-aire partiellement associative et soit $g : \mathcal{L} \to A_L$ un morphisme d'algèbres $(k+1)$-aires. Comme l'algèbre $(k+1)$-aire $pU^{<k>}(\mathcal{L})$ est engendrée par l'espace vectoriel $\mathcal{L} = \varphi(\mathcal{L})$, tout morphisme d'algèbres $(k+1)$-aires $\tilde{g} : pU^{<k>}(\mathcal{L}) \to A$, tel que $\tilde{g}_L \circ \varphi = g$, est déterminé de manière unique.

Par ailleurs, l'universalité de $p\mathcal{A}s^{<k>}(\mathcal{L})$ assure l'existence d'un morphisme d'algèbres $(k+1)$-aires $\tilde{g} : p\mathcal{A}s^{<k>}(\mathcal{L}) \to A$ qui prolonge l'application linéaire $g : \mathcal{L} \to A_L = A$. Pour tous $x_0, \ldots, x_k \in \mathcal{L}$ nous avons

$$
\begin{aligned}
\tilde{g}([x_0 \ldots x_k]) &= g([x_0 \ldots x_k]) && \text{car } [x_0 \ldots x_k] \in \mathcal{L} \\
&= [g(x_0) \ldots g(x_k)] && \text{dans } A_L \\
&= \sum_{\sigma \in S_{k+1}} \text{sgn}(\sigma)(g(x_{\sigma(0)}) \ldots g(x_{\sigma(k)})) \\
&= \tilde{g}(\sum_{\sigma \in S_{k+1}} \text{sgn}(\sigma)(x_{\sigma(0)} \ldots x_{\sigma(k)})).
\end{aligned}
$$

Par conséquent, le morphisme $\tilde{g}$ s'annule sur l'idéal $\mathcal{I}(\mathcal{L})$. Il induit donc par passage au quotient un morphisme d'algèbres $(k+1)$-aires (encore noté) $\tilde{g} : pU^{<k>}(\mathcal{L}) \to A$. L'égalité $\tilde{g}_L \circ \varphi = g$ est évidente par construction. ∎

Remarque 14.— De manière analogue, nous avons une notion d'enveloppe universelle dans la catégorie des algèbres $(k+1)$-aires totalement associatives : il suffit de reprendre ce qui précède en remplaçant le mot «partiellement» (resp. l'algèbre $p\mathcal{A}s^{<k>}(\mathcal{L})$) par le mot «totalement» (resp. l'algèbre $t\mathcal{A}s^{<k>}(\mathcal{L})$). Nous noterons $tU^{<k>}(\mathcal{L})$ l'algèbre $(k+1)$-aire totalement associative ainsi obtenue.

3.6.— Algèbre de Lie $(k+1)$-aire libre. Puisque nous ne connaissons l'algèbre $(k+1)$-aire partiellement associative libre que lorsque l'entier k est impair, nous ne construisons l'algèbre de Lie $(k+1)$-aire libre sur un espace vectoriel V que pour cette parité. L'on trouvera dans l'article de Ph. Hanlon et M. Wachs (*cf.* [4]) une description en termes d'arbres indépendante de la parité de k.

Soit V un espace vectoriel et soit $p\mathcal{A}s^{<k>}(V)$ l'algèbre $(k+1)$-aire partiellement associative libre sur V. Considérons la structure d'algèbre de Lie $(k+1)$-aire $(p\mathcal{A}s^{<k>}(V)_L, [\;\;])$ canoniquement associée. Soit $\mathcal{L}^{<k>}(V)$ la sous-algèbre de $p\mathcal{A}s^{<k>}(V)_L$ engendrée par l'espace vectoriel V. Alors nous avons

Proposition 15. *Supposons que k soit impair. L'algèbre $(k+1)$-aire $\mathcal{L}^{<k>}(V)$, munie de l'inclusion canonique $\iota : V \hookrightarrow \mathcal{L}^{<k>}(V)$, est l'algèbre de Lie $(k+1)$-aire libre sur l'espace vectoriel V.*

Démonstration.— Soit $\mathcal{L}$ une algèbre de Lie $(k+1)$-aire et soit $h : V \to \mathcal{L}$ une application linéaire.

Comme l'algèbre de Lie $(k+1)$-aire $\mathcal{L}^{<k>}(V)$ est engendrée par V, tout morphisme d'algèbres défini sur $\mathcal{L}^{<k>}(V)$ est déterminé par ses valeurs sur $V = \iota(V)$. Par conséquent, s'il existe un morphisme d'algèbres $\tilde{h} : \mathcal{L}^{<k>}(V) \to \mathcal{L}$ tel que $\tilde{h} \circ \iota = h$, il est unique.

Considérons l'application linéaire $f := \varphi \circ h : V \to Up^{<k>}(\mathcal{L})$ où φ désigne l'inclusion de $\mathcal{L}$ dans son enveloppe universelle $pU^{<k>}(\mathcal{L})$. Il existe un morphisme d'algèbres $\tilde{f} : p\mathcal{A}s^{<k>}(V) \to pU^{<k>}(\mathcal{L})$ qui coïncide avec f sur V. Soit $\tilde{f}_L : p\mathcal{A}s^{<k>}(V)_L \to pU^{<k>}(\mathcal{L})_L$ le morphisme d'algèbres de Lie $(k+1)$-aires induit par $\tilde{f}$. Pour tous $x_0, \dots, x_k \in V$, nous avons

$$\tilde{f}_L([x_0 \dots x_k]) = [\tilde{f}(x_0) \dots \tilde{f}(x_k)] = [f(x_0) \dots f(x_k)] \in \mathcal{L}.$$

Par conséquent la restriction $\tilde{h}$ de $\tilde{f}_L$ sur l'algèbre $\mathcal{L}^{<k>}(V)$ est à valeurs dans $\mathcal{L} \subset pU^{<k>}(\mathcal{L})_L$, et prolonge de manière évidente l'application linéaire h. ∎

4.— Cas pathologiques.

Nous montrons qu'il existe des exemples d'algèbres $(k+1)$-aires partiellement associatives dans le cas particulier où l'entier k est pair; nous en déduisons l'existence d'algèbres de Lie $(k+1)$-aires pour cette même parité.

4.1.— Soit $B^{<k>}$ un espace vectoriel de dimension $(k+2)$ et soit $\{e_\infty, e_0, \cdots, e_k\}$ une base de $B^{<k>}$. Munissons $B^{<k>}$ de la multiopération telle que tout multiproduit soit nul à l'exception des suivants

$$(e_\infty \ldots e_\infty) = e_0 \ , \quad (e_0 e_\infty \cdots e_\infty) = -\sum_{i=1}^{k} (-1)^{ik} e_i \ ,$$

$$(\underbrace{e_\infty \ldots e_\infty}_{j \ facteurs} e_0 e_\infty \ldots e_\infty) = e_j \ , \quad j = 1, \cdots, k.$$

Par construction, il est clair que $B^{<k>}$ est une algèbre $(k+1)$-aire partiellement et non totalement associative, et ce pour tout entier k.

4.2.— Soit V un espace vectoriel de dimension $(k+1)$ muni d'une base $\{f_0, \cdots, f_k\}$. L'algèbre $(k+1)$-aire $t\mathcal{A}s^{<k>}(V) \otimes B^{<k>}$ est partiellement associative (Lemme 6). Nous pouvons donc considérer l'algèbre de Lie $(k+1)$-aire canoniquement associée $(t\mathcal{A}s^{<k>}(V) \otimes B^{<k>})_L$. Son crochet n'est pas identiquement nul puisque nous avons :

$$[f_0 \otimes e_\infty \ldots f_k \otimes e_\infty] = (\sum_{\sigma \in S_{k+1}} \text{sgn}(\sigma)(f_{\sigma(0)} \ldots f_{\sigma(k)})) \otimes e_0 \neq 0.$$

Ainsi nous avons construit une algèbre de Lie $(k+1)$-aire non abélienne pour tout entier k.

Remarques 16.— Nous avons dû «grossir» l'algèbre $B^{<k>}$ parce que son algèbre de Lie associée $(B^{<k>})_L$ est abélienne.

Par ailleurs, la plus petite algèbre de Lie $(k+1)$-aire non abélienne a pour dimension $(k+1)$. On peut la décrire en prenant une base $\{x_0, \cdots, x_k\}$ et en décrétant que tous les crochets sont nuls exceptés

$$[x_{\sigma(0)} \ldots x_{\sigma(k)}] := \text{sgn}(\sigma)x_0 \ , \quad \forall \ \sigma \in S_{k+1} \ .$$

II.— OPÉRADES QUADRATIQUES $(k+1)$-AIRES.

Nous nous proposons de généraliser les constructions de Ginzburg et Kapranov afin d'obtenir des opérades pour les algèbres $(k+1)$-aires. Nous appliquons cette généralisation en déterminant les opérades des algèbres $(k+1)$-aires rencontrées précédemment. Puisque nous travaillons avec des opérades non nécessairement unitaires, nous suivrons les notations de Loday (*cf.* [5]).

1.— Définitions générales.

Soit E un espace vectoriel de dimension finie, muni d'une action du groupe symétrique S_{k+1}. Considérons l'opérade libre $\mathcal{T}(E)$ construite sur la famille d'espaces vectoriels $\{E(n) = 0,\ n \neq k+1;\ E(k+1) = E\}$. Puisque cette famille est concentrée en degré $(k+1)$, seules les classes d'isomorphisme d'arbres $(k+1)$-aires (c'est-à-dire dont chaque sommet possède exactement $(k+1)$ arêtes rentrantes) apportent une contribution non triviale dans $\mathcal{T}(E)$. En d'autres termes, nous avons

$$\mathcal{T}(E)(n) = 0 \text{ si } n \not\equiv 1 \bmod k, \mathcal{T}(E)(mk+1) = \bigoplus_{\substack{(mk+1)-arbres \\ (k+1)-aires\ T}} E(T).$$

Lorsque $R \subset \mathcal{T}(E)(2k+1)$ est un sous-espace vectoriel stable sous l'action de S_{2k+1}, nous désignons par (R) l'idéal de $\mathcal{T}(E)$ engendré par R, et nous notons $\mathcal{P}^{<k>}(\mathbb{K}, E, R) := \mathcal{T}(E)/(R)$ l'opérade quotient. Nous appelons *opérade quadratique $(k+1)$-aire*, toute opérade de la forme $\mathcal{P}^{<k>}(\mathbb{K}, E, R)$.

Notons $E^\vee := \mathrm{Hom}_{\mathbb{K}}(E, \mathbb{K}) \otimes (\mathbf{sgn})$ le dual linéaire tensorisé par la représentation signature de S_{k+1}, et $R^\perp$ l'orthogonal de R dans $\mathcal{T}(E^\vee)(2k+1) \cong \mathcal{T}(E)(2k+1)^\vee$. Par définition, *l'opérade duale* de l'opérade quadratique $(k+1)$-aire $\mathcal{P}^{<k>}(\mathbb{K}, E, R)$ est l'opérade

$$\mathcal{P}^{<k>}(\mathbb{K}, E, R)^! := \mathcal{P}^{<k>}(\mathbb{K}, E^\vee, R^\perp).$$

Il est clair que l'on retrouve la même chose lorsque l'on prend deux fois la duale d'une opérade quadratique $(k+1)$-aire :

$$(\mathcal{P}^{<k>}(\mathbb{K}, E, R)^!)^! = \mathcal{P}^{<k>}(\mathbb{K}, E, R).$$

Soient $\mathcal{P}^{<k>}(\mathbb{K}, E, R)$ et $\mathcal{P}^{<k>}(\mathbb{K}, E', R')$ des opérades quadratiques $(k+1)$-aires et soit $\phi : E \to E'$ un morphisme S_{k+1}-équivariant. Alors le morphisme ϕ induit fonctoriellement un morphisme d'opérades libres $\Phi : \mathcal{T}(E) \to \mathcal{T}(E')$. On appelle *morphisme* (resp. *isomorphisme*) *d'opérades quadratiques $(k+1)$-aires*, tout morphisme (resp. isomorphisme) S_{k+1}-équivariant $\phi : E \to E'$ tel que $\Phi(R) \subset R'$ (resp. $\Phi(R) = R'$).

2.— Description du S_{2k+1}-module $\mathcal{T}(E)(2k+1)$.

Nous allons caractériser la structure de S_{2k+1}-module de l'espace $\mathcal{T}(E)(2k+1)$:

Proposition 1. *On a un isomrphisme de S_{2k+1}-modules*

$$\mathcal{T}(E)(2k+1) \cong \mathrm{Ind}_{S_{k+1} \times S_k}^{S_{2k+1}}(E \otimes E).$$

L'espace vectoriel $\mathcal{T}(E)(2k+1)$ est donc isomorphe à un nombre de copies de $E \otimes E$ égal au coefficient binômial $\binom{2k+1}{k+1}$.

Nous allons démontrer ce résultat en deux étapes. Nous donnons d'abord une présentation générale du produit $S_p \times S_q$ comme sous-groupe de S_{p+q} et les classes à gauche. Ensuite nous montrons que l'espace $\mathcal{T}(E)(2k+1)$ se réalise ainsi.

2.1.— Soient p et q deux entiers au moins égaux à 1. Nous considérons le groupe produit $S_p \times S_q$ comme un sous-groupe de S_{p+q}, S_p opérant sur les p premiers éléments et S_q sur les q derniers. Plus précisément, on définit un homomorphisme injectif de groupes en associant à tout couple (σ, τ) de $S_p \times S_q$, l'élément de S_{p+q} donné par

$$(\sigma, \tau) : j \longmapsto \sigma(j) \text{ si } j \in \left|_0^{p-1}, \tau(j-p)+p \text{ si } j \in \right|_p^{p+q-1}.$$

Lemme 2. *Les classes à gauche de $S_p \times S_q$ dans S_{p+q} sont déterminées par les (p,q)-shuffles i.e.,*

$$S_{p+q} = \coprod_{s \in Sh_{p,q}} s.(S_p \times S_q) .$$

Démonstration.— Remarquons pour commencer que l'indice de $S_p \times S_q$ dans S_{p+q} est égal à $(p+q)!/p!q!$, c'est-à-dire le cardinal de l'ensemble $Sh_{p,q}$. Il suffit donc de montrer que deux (p,q)-shuffles distincts ne peuvent être dans la même classe à gauche.

En effet, supposons que l'on ait $s' = s \circ (\sigma, \tau)$ où $s', s \in Sh_{p,q}$ et $(\sigma, \tau) \in S_p \times S_q$. Alors nous avons

$$\sigma^{-1}(j) \in \left|_0^{p-1} \text{ et } s'(\sigma^{-1}(j)) = s \circ (\sigma, \tau)(\sigma^{-1}(j)) = s(j), \quad \forall j \in \right|_0^{p-1}.$$

Par conséquent nous obtenons $s(j) \in \{s'(0), \ldots, s'(p-1)\}$ pour tout $j \in |_0^{p-1}$. Et puisque les permutations sont des injections, les deux ensembles

$$\{s(0), \ldots, s(p-1)\} \quad \text{et} \quad \{s'(0), \ldots, s'(p-1)\},$$

qui ont même cardinal, sont égaux. Nous en déduisons l'égalité des (p, q)-shuffles s et s'. Ce qui achève de démontrer le Lemme. ∎

Remarque 3.— Dans la pratique, on procède de la manière suivante. Etant donné un élément λ de S_{p+q}, on range dans l'ordre croissant les parties $\{\lambda(0), \ldots, \lambda(p-1)\}$ et $\{\lambda(p), \ldots, \lambda(p+q-1)\}$; cet ordre caractérise un unique (p, q)-shuffle qui détermine la classe de λ.

2.2.— Rappelons que par définition nous avons

$$\mathcal{T}(E)(2k+1) = \bigoplus_{\substack{(2k+1)-arbres \\ (k+1)-aires \ T}} E(T) \text{ et } E(T) = \bigotimes_{\substack{sommets \\ v \in T}} E(In(v)) \, ,$$

le groupe S_{2k+1} opérant sur $\mathcal{T}(E)(2k+1)$ par permutation des feuilles des arbres.

Tout $(2k+1)$-arbre $(k+1)$-aire T admet exactement deux sommets, et sa classe d'isomorphisme détermine un unique $(k+1, k)$-shuffle s caractérisé par

$$\{s(0), \ldots, s(k)\} := \{\text{numéros des arêtes rentrantes}$$
$$\text{dans le premier sommet de } T\}$$

Munissons l'espace vectoriel $E \otimes E$ de l'action diagonale du produit $S_{k+1} \times S_k$ déduite de celle de S_{k+1} sur E (S_k étant vu comme sous-groupe de S_{k+1} laissant invariant un élément préalablement fixé dans $\{0, \ldots, k\}$). Comme les $(k+1, k)$-shuffles constituent un système de représentants des classes à gauche de $S_{k+1} \times S_k$ dans S_{2k+1} (Lemme 2), nous avons

$$\text{Ind}_{S_{k+1} \times S_k}^{S_{2k+1}} (E \otimes E) \cong \bigoplus_{s \in Sh_{k+1,k}} s.(E \otimes E) \, .$$

Pour tout $(k+1, k)$-shuffle s, la composante $s.(E \otimes E)$ correspond précisément à $E(T_s)$ où T_s est un représentant de la classe d'isomorphisme d'arbres déterminée par s. D'où la Proposition 1. ∎

3.— Exemples d'opérades des algèbres $(k+1)$-aires.

Rappelons que lorsque $\mathcal{P}$ représente l'opérade des algèbres d'un certain type, $\mathcal{P}(n)$ s'identifie à la partie n-multilinéaire (*i.e.*, engendrée par les monômes contenant chaque x_i une fois et une seule) de l'algèbre du même type libre sur l'espace vectoriel de base $\{x_1, \ldots, x_n\}$.

Nous désignerons par $\mathbf{tAs}^{<k>}$(resp. $\mathbf{pAs}^{<k>}$, resp. $\mathbf{stAs}^{<k>}$, resp. $\mathbf{Lie}^{<k>}$) l'opérade des algèbres $(k+1)$-aires totalement associatives (resp. partiellement associatives, resp. symétriques et totalement associatives, resp. de Lie).

3.1.— L'opérade $\mathbf{tAs}^{<k>}$. Nous avons vu que l'espace vectoriel sous-jacent à l'algèbre $(k+1)$-aire totalement associative libre sur un espace V est donné par (*cf.* I-3.2):

$$V \oplus V^{\otimes k+1} \oplus \cdots \oplus V^{\otimes mk+1} \oplus \ldots \quad .$$

Nous en déduisons que l'opérade $\mathbf{tAs}^{<k>}$est telle que $\mathbf{tAs}^{<k>}(n) = 0$ si $n \not\equiv 1 \bmod k$ et

$$(3.1.1) \quad \mathbf{tAs}^{<k>}(mk+1) = \mathbb{K}[S_{mk+1}] \quad \text{(représentation régulière)}.$$

L'espace vectoriel $E = \mathbf{tAs}^{<k>}(k+1)$ est alors de dimension $(k+1)!$ et est engendré par les éléments

$$(3.1.2) \quad . \qquad \boldsymbol{x}_\sigma := (x_{\sigma(0)} \ldots x_{\sigma(k)}) \ , \quad \forall\, \sigma \in S_{k+1} \ .$$

L'espace vectoriel $\mathcal{T}(\mathbf{tAs}^{<k>}(k+1))(2k+1) \cong \binom{2k+1}{k+1}(E \otimes E)$ est donc de dimension $(2k+1)!(k+1)$ et est engendré par les éléments :

$$(3.1.3)\ \boldsymbol{x}_{\sigma,i} := (x_{\sigma(0)} \ldots x_{\sigma(i-1)}(x_{\sigma(i)} \ldots x_{\sigma(i+k)})x_{\sigma(i+k+1)} \ldots x_{\sigma(2k)})$$

où $\sigma \in S_{2k+1}$ et $i \in |_0^k$. Le sous-espace $R_{\mathbf{tAs}^{<k>}}$ est engendré par les $(2k+1)!k$ associateurs totaux

$$(3.1.4) \qquad \boldsymbol{y}_{\sigma,i} := \boldsymbol{x}_{\sigma,i} - \boldsymbol{x}_{\sigma,0} \ , \ \forall\, \sigma \in S_{2k+1} \ , \ \forall\, i \in |_1^k \ .$$

Ainsi, nous avons

$$\mathbf{tAs}^{<k>} = \mathcal{P}^{<k>}(\mathbb{K}, \mathbb{K}[S_{k+1}], R_{\mathbf{tAs}^{<k>}}).$$

3.2.— L'opérade pAs$^{<k>}$. Lorsque l'entier k est impair, on sait que l'espace vectoriel sous-jacent à l'algèbre $(k+1)$-aire partiellement associative libre sur un espace vectoriel V est donné par (*cf.* I-3.4):

$$p_0^{<k>}.V \ \oplus \ p_1^{<k>}.V^{\otimes k+1} \ \oplus \cdots \oplus \ p_m^{<k>}.V^{\otimes mk+1} \ \oplus \cdots$$

où $p_m^{<k>}.V^{\otimes mk+1}$ signifie $p_m^{<k>}$ copies de $V^{\otimes mk+1}$. Ainsi, l'opérade **pAs**$^{<k>}$ est telle que **pAs**$^{<k>}(n) = 0$ si $n \not\equiv 1 \ mod \ k$ et

$$(3.2.1) \qquad \mathbf{pAs}^{<k>}(mk+1) = p_m^{<k>}.\mathbb{K}[\mathrm{S}_{mk+1}].$$

Puisque les espaces **tAs**$^{<k>}(k+1)$ et **pAs**$^{<k>}(k+1)$ sont identiques, il en est de même pour les opérades libres $\mathcal{T}(\mathbf{tAs}^{<k>}(k+1))$ et $\mathcal{T}(\mathbf{pAs}^{<k>}(k+1))$. Le sous-espace $R_{\mathbf{pAs}^{<k>}}$ est engendré par les $(2k+1)!$ associateurs partiels

$$(3.2.2) \qquad \boldsymbol{z}_\sigma := \sum_{i=0}^{k} (-1)^{ik} \boldsymbol{x}_{\sigma,i} \ , \quad \forall \ \sigma \in \mathrm{S}_{2k+1} \ .$$

Ainsi, nous avons

$$\mathbf{pAs}^{<k>} = \mathcal{P}^{<k>}(\mathbb{K}, \mathbb{K}[\mathrm{S}_{k+1}], R_{\mathbf{pAs}^{<k>}}).$$

3.3.— L'opérade stAs$^{<k>}$. Nous avons vu que l'espace vectoriel sous-jacent à l'algèbre $(k+1)$-aire symétrique et totalement associative libre sur un espace vectoriel V est donné par (*cf.* I-3.3):

$$V \ \oplus \ (V^{\otimes k+1})_{S_{k+1}} \ \oplus \cdots \oplus \ (V^{\otimes nk+1})_{S_{nk+1}} \ \oplus \cdots \quad .$$

Donc l'opérade **stAs**$^{<k>}$ est telle que **stAs**$^{<k>}(n) = 0$ si $n \not\equiv 1 \ mod \ k$ et

$$(3.3.1) \qquad \mathbf{stAs}^{<k>}(mk+1) = \mathbf{1} \quad \text{(représentation triviale)}.$$

Ainsi, l'espace $E = \mathbf{stAs}^{<k>}(k+1)$ est engendré par l'élément $(x_0 \ldots x_k)$. L'espace $\mathcal{T}(\mathbf{stAs}^{<k>}(k+1))(2k+1) \cong \binom{2k+1}{k+1}(E \otimes E)$ est de dimension $\binom{2k+1}{k+1}$ et engendré par les éléments :

$$(3.3.2) \quad \boldsymbol{s}_\sigma := ((x_{\sigma(0)} \ldots x_{\sigma(k)})x_{\sigma(k+1)} \ldots x_{\sigma(2k)}) \ , \quad \forall \ \sigma \in \mathrm{Sh}_{k+1,k} \ .$$

Le sous-espace $R_{\mathbf{stAs}^{<k>}}$ est engendré par les $(\binom{2k+1}{k+1}-1)$ associateurs totaux symétriques

$$(3.3.3) \qquad \mathbf{st}_\sigma := \boldsymbol{s}_\sigma - \boldsymbol{s}_{id} \ , \quad \forall \ \sigma \in \mathrm{Sh}_{k+1,k} \setminus \{id\} \ .$$

Ainsi, nous avons

$$\mathbf{stAs}^{<k>} = \mathcal{P}^{<k>}(\mathbb{K}, \mathbf{1}, R_{\mathbf{stAs}^{<k>}}).$$

3.4.— L'opérade Lie$^{<k>}$. Lorsque l'entier k est impair, nous avons donné une construction de l'algèbre de Lie $(k+1)$-aire libre sur V, mais nous ne connaissons pas parfaitement la structure d'espace vectoriel sous-jacent. Toutefois, *l'antisymétrie* du crochet permet d'affirmer que l'espace vectoriel $E = \mathbf{Lie}^{<k>}(k+1)$ est engendré par l'élément $[x_0 \ldots x_k]$ et est la représentation signature (**sgn**) de S_{k+1}. L'espace $\mathcal{T}(\mathbf{Lie}^{<k>}(k+1))(2k+1) \cong \binom{2k+1}{k+1}.(E \otimes E)$ est donc de dimension $\binom{2k+1}{k+1}$ et est engendré par les éléments :

$$(3.4.1) \quad \boldsymbol{l}_\sigma := [[x_{\sigma(0)} \ldots x_{\sigma(k)}]x_{\sigma(k+1)} \ldots x_{\sigma(2k)}] , \quad \forall\, \sigma \in \mathrm{Sh}_{k+1,k} .$$

Le sous-espace S_{2k+1}-invariant $R_{\mathbf{Lie}^{<k>}}$ est la représentation signature et est engendré par le relateur de Jacobi $J'(x_0, \ldots, x_{2k})$. Et nous avons

$$\mathbf{Lie}^{<k>} = \mathcal{P}^{<k>}(\mathbb{K}, (\mathbf{sgn}), (\mathbf{sgn})).$$

4.— Dualités entre les opérades des algèbres $(k+1)$-aires.

Théorème 4. *Soit k un entier impair. Les opérades quadratiques $(k+1)$-aires $\mathbf{tAs}^{<k>}$ et $\mathbf{pAs}^{<k>}$ sont duales l'une de l'autre.*

Démonstration.— Comme dans [3, 2.1.11], introduisons sur l'espace vectoriel $\mathcal{T}(\mathbf{tAs}^{<k>}(k+1))(2k+1)$ le produit scalaire caractérisé par l'orthogonalité de la famille $(\boldsymbol{x}_{\sigma,i})$ et les relations

$$< \boldsymbol{x}_{\sigma,i}, \boldsymbol{x}_{\sigma,i} > := (-1)^{k(i+1)}\mathrm{sgn}(\sigma) .$$

Montrons que l'annihilateur de $R_{\mathbf{tAs}^{<k>}}$ est $R_{\mathbf{pAs}^{<k>}}$. En effet, pour tout $z = \sum \alpha_{\tau,j}\boldsymbol{x}_{\tau,j}$ de $\mathcal{T}(\mathbf{tAs}^{<k>}(k+1))(2k+1)$, nous avons

$$< \boldsymbol{y}_{\sigma,i}, z > = (-1)^k \mathrm{sgn}(\sigma)(\alpha_{\sigma,0} - (-1)^{ik}\alpha_{\sigma,i}) .$$

Par conséquent, l'élément z appartient à l'annihilateur de l'espace $R_{\mathbf{tAs}^{<k>}}$ si, et seulement si, $\alpha_{\sigma,i} = (-1)^{ik}\alpha_{\sigma,0}$ pour tout $\sigma \in S_{2k+1}$ et tout $i = 1, \ldots, k$; c'est-à-dire que z est de la forme

$$z = \sum_{\tau \in S_{2k+1}} \alpha_{\tau,0} \sum_{i=0}^{k} (-1)^{ik}\boldsymbol{x}_{\tau,i} = \sum_{\tau \in S_{2k+1}} \alpha_{\tau,0}\boldsymbol{z}_\tau .$$

Ainsi, l'annihilateur de l'espace $R_{\mathbf{tAs}^{<k>}}$ est le sous-espace engendré par les éléments $\boldsymbol{z}_\sigma$, c'est-à-dire l'espace $R_{\mathbf{pAs}^{<k>}}$. La dualité entre les opérades $\mathbf{tAs}^{<k>}$ et $\mathbf{pAs}^{<k>}$ résulte alors du

Lemme 5. *Les représentations* $\mathbb{K}[S_{k+1}]^{\vee}$ *et* $\mathbb{K}[S_{k+1}]$ *sont isomorphes.*

Démonstration.— Soit $\{\sigma^* : \sigma \in S_{k+1}\}$ la base duale de l'espace vectoriel $\mathrm{Hom}_{\mathbb{K}}(\mathbb{K}[S_{k+1}], \mathbb{K})$ et soit φ l'application linéaire

$$\varphi : \mathrm{Hom}_{\mathbb{K}}(\mathbb{K}[S_{k+1}], \mathbb{K}) \otimes (\mathbf{sgn}) \to \mathbb{K}[S_{k+1}], \quad \sigma^* \otimes 1 \mapsto \mathrm{sgn}(\sigma)\sigma.$$

Il est clair que φ est un isomorphisme d'espaces vectoriels. De plus, pour tous éléments σ, τ de S_{k+1}, nous avons

$$\varphi(\tau.(\sigma^* \otimes 1)) = \mathrm{sgn}(\tau)\varphi((\tau\sigma)^* \otimes 1) = \mathrm{sgn}(\tau)\mathrm{sgn}(\tau\sigma)\tau\sigma = \tau.\varphi(\sigma^* \otimes 1).$$

Donc l'application φ est un isomorphisme de représentations. ∎

De même, nous avons

Théorème 6. *Soit k un entier impair. Les opérades quadratiques $(k+1)$-aires* $\mathbf{stAs}^{<k>}$ *et* $\mathbf{Lie}^{<k>}$ *sont duales l'une de l'autre.*

Démonstration.— Considérons la restriction du produit scalaire défini ci-dessus au sous-espace $\mathcal{T}(\mathbf{stAs}^{<k>}(k+1))(2k+1)$ de l'espace $\mathcal{T}(\mathbf{tAs}^{<k>}(k+1))(2k+1)$. La famille $(\boldsymbol{s}_\sigma)$ est orthogonale et nous avons

$$< \boldsymbol{s}_\sigma, \boldsymbol{s}_\sigma > = (-1)^k \mathrm{sgn}(\sigma) .$$

Nous allons déterminer l'annihilateur de l'espace $R_{\mathbf{stAs}^{<k>}}$. Si $s = \sum \beta_\tau \boldsymbol{s}_\tau$ est un élément de $\mathcal{T}(\mathbf{stAs}^{<k>}(k+1))(2k+1)$, alors nous avons

$$< \mathbf{st}_\sigma, s > = (-1)^k (\beta_\sigma \mathrm{sgn}(\sigma) - \beta_{id}) .$$

Par conséquent, l'élément s appartient à l'annihilateur de l'espace $R_{\mathbf{stAs}^{<k>}}$ si, et seulement si, $\beta_\sigma = \mathrm{sgn}(\sigma)\beta_{id}$, pour tout $\sigma \in \mathrm{Sh}_{k+1,k}$. Ainsi l'annihilateur de l'espace $R_{\mathbf{stAs}^{<k>}}$ est le sous-espace engendré par l'élément

$$\boldsymbol{s} := \sum_{\sigma \in Sh_{k+1,k}} \mathrm{sgn}(\sigma)\boldsymbol{s}_\sigma .$$

Comme $\mathbf{stAs}^{<k>}(k+1)$ est la représentation unité $\mathbf{1}$, son dual tordu $\mathbf{stAs}^{<k>}(k+1)^{\vee}$ n'est autre que la représentation signature $(\mathbf{sgn})$. On peut donc associer au générateur $(x_0 \ldots x_k)$ de $\mathbf{stAs}^{<k>}(k+1)$, le générateur $[x_0 \ldots x_k]$ de $\mathbf{Lie}^{<k>}(k+1) \cong \mathbf{stAs}^{<k>}(k+1)^{\vee}$. Et dans cette correspondance, le générateur $\boldsymbol{s}$ de l'annihilateur de $R_{\mathbf{stAs}^{<k>}}$

est associé au relateur de Jacobi $J'(x_0, \ldots, x_{2k})$. D'où la dualité entre les opérades $\mathbf{stAs}^{<k>}$ et $\mathbf{Lie}^{<k>}$. ∎

Remarques 7.— Dans les Théorèmes 4 et 6, nous avons supposé que l'entier k est impair parce que nous avons décrit les opérades $\mathbf{pAs}^{<k>}$ et $\mathbf{Lie}^{<k>}$ uniquement pour cette parité. Toutefois, on observera que la démonstration de ces théorèmes permet de dire que pour les entiers pairs k, on a

$$\mathbf{tAs}^{<k>^!} \cong \mathcal{P}^{<k>}(\mathbb{K}, \mathbb{K}[S_{k+1}], R_{\mathbf{pAs}^{<k>}})$$

et

$$\mathbf{stAs}^{<k>^!} \cong \mathcal{P}^{<k>}(\mathbb{K}, (\mathbf{sgn}), (\mathbf{sgn})).$$

III.— HOMOLOGIE DES ALGÈBRES $(k+1)$-AIRES.

Nous construisons une théorie d'homologie pour les algèbres $(k+1)$-aires partiellement associatives et nous montrons que l'homologie de toute algèbre $(k+1)$-aire partiellement associative libre est triviale. Ensuite nous montrons que le complexe de Hanlon-Wachs se scinde en somme directe de sous-complexes et nous donnons le complexe prédit par la théorie des opérades pour l'homologie des algèbres de Lie $(k+1)$-aires.

1.— Homologie de Hochschild des algèbres $(k+1)$-aires partiellement associatives.

Dans toute cette partie nous supposons que l'entier k est impair; nous conservons délibérément les signes $(-1)^{ik}$, même si les entiers i et ik ont la même parité.

Soit $(A, (\ ))$ une algèbre $(k+1)$-aire partiellement associative. Adoptons les notations $C_n^{<k>}(A) := A^{\otimes nk+1}$ et $(a_0, \ldots, a_{nk}) := a_0 \otimes \ldots \otimes a_{nk} \in C_n^{<k>}(A)$. Ensuite considérons les applications linéaires

$$b_0^{<k>} :\equiv 0 \quad \text{et} \quad b_n^{<k>} := \sum_{i=0}^{k(n-1)} (-1)^{ik} d_i : C_n^{<k>}(A) \to C_{n-1}^{<k>}(A)$$

où les opérateurs (d_i) sont définis par la formule

$$d_i(a_0, \ldots, a_{nk}) := (a_0, \ldots, a_{i-1}, (a_i \ldots a_{i+k}), a_{i+k+1}, \ldots, a_{nk})$$

et $(a_i \ldots a_{i+k})$ désigne le multiproduit dans A.

Lemme 1. *a) Pour tout couple d'entiers (i,j) tels que $0 \leq i < j - k$, nous avons*

$$d_i d_j = d_{j-k} d_i.$$

b) Pour tout entier $n \geq 2$ et tout entier $i \in |_0^{k(n-2)}$, nous avons

$$\sum_{j=i}^{k+i} (-1)^{k(i+j)} d_i d_j = 0 .$$

Démonstration. — En effet, soient $\{a_0, \dots, a_{nk}\}$ $(nk+1)$ éléments de A.

Si $i + k < j$ alors nous avons

$$
\begin{aligned}
d_i d_j(a_0, \dots, a_{nk}) &= d_i(a_0, \dots, a_{j-1}, (a_j \dots a_{j+k}), \dots, a_{nk}) \\
&= (a_0, \dots, a_{i-1}, (a_i \dots a_{i+k}), \dots, a_{j-1}, (a_j \dots a_{j+k}), \dots, a_{nk}) \\
&= d_{j-k}(a_0, \dots, a_{i-1}, (a_i \dots a_{i+k}), a_{i+k+1}, \dots, a_{nk}) \\
&= d_{j-k} d_i(a_0, \dots, a_{nk}) ,
\end{aligned}
$$

d'où l'assertion a) du Lemme.

Si $i \leq j \leq k + i$ alors nous avons

$$d_i d_j(a_0, \dots, a_{nk}) =$$
$$(a_0, \dots, a_{i-1}, (a_i \dots a_{j-1}(a_j \dots a_{j+k}) \dots a_{2k+i}), \dots, a_{nk}).$$

L'associativité partielle appliquée au $(2k+1)$-uplet $\{a_i, \dots, a_{2k+i}\}$ nous donne

$$y_i := \sum_{j=i}^{k+i} (-1)^{k(j-i)} (a_i \dots a_{j-1}(a_j \dots a_{j+k}) \dots a_{2k+i}) = 0.$$

Par conséquent, il vient

$$\sum_{j=i}^{k+i} (-1)^{k(i+j)} d_i d_j(a_0, \dots, a_{nk}) =$$
$$(a_0, \dots, a_{i-1}, y_i, a_{2k+i+1}, \dots, a_{nk}) = 0,$$

d'où l'assertion b) du Lemme. ∎

Corollaire 2. *Pour tout entier $n \geq 1$, nous avons $b_{n-1}^{<k>} b_n^{<k>} = 0$.*

Démonstration.— Puisque $b_0^{<k>} \equiv 0$, la propriété est évidente pour $n = 1$. Si $n \geq 2$, alors nous avons

$$b_{n-1}^{<k>} b_n^{<k>} = \sum_{\substack{0 \leq i \leq k(n-2) \\ 0 \leq j \leq k(n-1)}} (-1)^{k(i+j)} d_i d_j$$

$$= \sum_{0 \leq i \leq j-k-1} (-1)^{k(i+j)} d_{j-k} d_i + \sum_{0 \leq j \leq i+k} (-1)^{k(i+j)} d_i d_j$$

par   a) Lemme 1

$$= (-1)^{k^2} \sum_{0 \leq j \leq i-1 \leq k(n-2)-1} (-1)^{k(i+j)} d_i d_j +$$

$$\sum_{0 \leq j \leq i+k \leq k(n-1)} (-1)^{k(i+j)} d_i d_j$$

$$= - \sum_{i=1}^{k(n-2)} \sum_{j=0}^{i-1} (-1)^{k(i+j)} d_i d_j + \sum_{i=0}^{k(n-2)} \sum_{j=0}^{k+i} (-1)^{k(i+j)} d_i d_j$$

$$= \sum_{i=0}^{k(n-2)} \sum_{j=i}^{k+i} (-1)^{k(i+j)} d_i d_j = 0 \quad \text{par} \quad \text{b) Lemme 1 .} \quad \blacksquare$$

Ainsi nous avons défini un complexe de chaînes $(C_*^{<k>}(A), b_*^{<k>})$ dont l'homologie est appelée *l'homologie de Hochschild de l'algèbre $(k+1)$-aire partiellement associative A* et est notée

$$pH_n^{<k>}(A) := H_n(C_*^{<k>}(A), b_*^{<k>}), \quad n \in \mathbb{N} .$$

Théorème 3. *Soit V un espace vectoriel et soit $p\mathcal{A}s^{<k>}(V)$ l'algèbre $(k+1)$-aire partiellement associative libre sur V. Alors l'homologie de Hochschild de $p\mathcal{A}s^{<k>}(V)$ est donnée par*

$$pH_0^{<k>}(p\mathcal{A}s^{<k>}(V)) \cong V \ \text{ et } \ pH_n^{<k>}(p\mathcal{A}s^{<k>}(V)) \cong 0 \quad si \quad n > 0.$$

Démonstration.— Nous allons construire une homotopie entre les applications *identité* et *nulle* sur $(C_*^{<k>}(p\mathcal{A}s^{<k>}(V)), b_*^{<k>})$. Considérons les applications linéaires

$$h_n : C_n^{<k>}(p\mathcal{A}s^{<k>}(V)) \to C_{n+1}^{<k>}(p\mathcal{A}s^{<k>}(V)) \quad \text{où} \quad n \in \mathbb{N}$$

définies pour tous $W_i \in p\mathcal{A}s_{n_i}^{<k>}(V)$, $i \in |_0^{nk}$, par $h_n(W_0, \dots, W_{nk}) := 0$ si $n_0 = 0$ et

$$h_n(W_0, \dots, W_{nk}) := (w_0^0, \dots, w_k^0, W_1, \dots, W_{nk})$$

si $n_0 > 0$ et pour $W_0 = w_0^0 \otimes \dots \otimes w_k^0$.

Lemme 4. *Nous avons* $h_{n-1}b_n^{<k>} + b_{n+1}^{<k>}h_n = \mathrm{id}_n, \forall n \geq 1.$

Nous en déduisons que $pH_n^{<k>}(p\mathcal{A}s^{<k>}(V)) \cong 0$ pour tout entier $n \geq 1$. D'autre part, pour tout entier $n \geq 1$, l'espace $p\mathcal{A}s_n^{<k>}(V)$ est engendré par les éléments de la forme

$$w_0 \otimes \ldots \otimes w_k = (w_0 \ldots w_k) = b_1^{<k>}(w_0, \ldots, w_k) \, .$$

Par conséquent, les seuls $b_0^{<k>}$-cycles de $p\mathcal{A}s^{<k>}(V)$ qui ne sont pas des $b_1^{<k>}$-bords sont les éléments de $p\mathcal{A}s_0^{<k>}(V) = V$. Ainsi nous obtenons $pH_0^{<k>}(p\mathcal{A}s^{<k>}(V)) \cong V$; ce qui achève de démontrer le Théorème. ∎

Preuve du Lemme 4.— Soient $W_i \in p\mathcal{A}s_{n_i}^{<k>}(V)$ où $i \in |_0^{nk}$.

Si $n_0 = 0$, alors nous avons $b_{n+1}^{<k>}h_n(W_0, \ldots, W_{nk}) = 0$ et

$$
\begin{aligned}
&h_{n-1}b_n^{<k>}(W_0, \ldots, W_{nk})\\
=\ & \sum_{i=0}^{k(n-1)} (-1)^{ik} h_{n-1}(W_0, \ldots, W_{i-1}, (W_i \ldots W_{i+k}), \ldots, W_{nk})\\
=\ & h_{n-1}((W_0 \ldots W_k), \ldots, W_{nk})\\
=\ & (W_0, \ldots, W_k, W_{k+1}, \ldots, W_{nk}) \, ,
\end{aligned}
$$

d'où la relation $h_{n-1}b_n^{<k>} + b_{n+1}^{<k>}h_n = \mathrm{id}_n$ pour $n_0 = 0$.

Supposons que $n_0 > 0$; alors nous avons

Lemme 5. *Pour tous entiers* $n \geq 1$ *et* $j \geq 1$, *on a les relations*

$$d_0 h_n = \mathrm{id}_n, \quad h_{n-1}d_j = d_{j+k}h_n, \quad h_{n-1}d_0 = -\sum_{i=1}^{k} (-1)^{ik} d_i h_n$$

sur les éléments $(W_0, \ldots, W_{nk})$ *tels que* $n_0 > 0$.

Il en résulte immédiatement que

$$h_{n-1}b_n^{<k>} = \sum_{j=0}^{k(n-1)} (-1)^{jk} h_{n-1} d_j$$

$$= -\sum_{i=1}^{k} (-1)^{ik} d_i h_n + \sum_{j=1}^{k(n-1)} (-1)^{jk} d_{j+k} h_n$$

$$= -\sum_{i=1}^{k} (-1)^{ik} d_i h_n + (-1)^{k^2} \sum_{i=k+1}^{nk} (-1)^{ik} d_i h_n$$

$$= -\left(\sum_{i=1}^{nk} (-1)^{ik} d_i\right) h_n = -(b_{n+1}^{<k>} - d_0) h_n.$$

Ainsi nous avons la relation $h_{n-1}b_n^{<k>} + b_{n+1}^{<k>} h_n = d_0 h_n = \mathrm{id}_n$, pour $n_0 > 0$. Ce qui démontre le Lemme 4. ■

Preuve du Lemme 5.— Soient $W_0 = w_0^0 \otimes \ldots \otimes w_k^0 \in pAs_{n_0}^{<k>}(V)$ avec $n_0 > 0$ et $W_i \in pAs_{n_i}^{<k>}(V)$ où $i \in |_1^{nk}$; alors nous avons

$$d_0 h_n(W_0, \ldots, W_{nk}) = d_0(w_0^0, \ldots, w_k^0, W_1, \ldots, W_{nk})$$
$$= ((w_0^0 \ldots w_k^0), W_1, \ldots, W_{nk})$$
$$= (W_0, W_1, \ldots, W_{nk}).$$

Si l'entier $j > 0$, on a

$$h_{n-1} d_j(W_0, \ldots, W_{nk})$$
$$= h_{n-1}(W_0, \ldots, W_{j-1}, (W_j \ldots W_{j+k}), W_{j+k+1}, \ldots, W_{nk})$$
$$= (w_0^0, \ldots, w_k^0, W_1, \ldots, W_{j-1}, (W_j \ldots W_{j+k}), W_{j+k+1}, \ldots, W_{nk})$$
$$= d_{j+k}(w_0^0, \ldots, w_k^0, W_1, \ldots, W_{nk}) = d_{j+k} h_n(W_0, \ldots, W_{nk}) .$$

Enfin, par la formule (3.4.2) du Chapitre I, nous obtenons

$$h_{n-1} d_0(W_0, \ldots, W_{nk}) = h_{n-1}((W_0 \ldots W_k), W_{k+1}, \ldots, W_{nk})$$

$$= -\sum_{i=1}^{k} (-1)^{ik} h_{n-1}(w_0^0 \otimes \ldots \otimes (w_i^0 \ldots w_k^0 W_1 \ldots W_i), \ldots, W_{nk})$$

$$= -\sum_{i=1}^{k} (-1)^{ik} d_i(w_0^0, \ldots, w_k^0, W_1, \ldots, W_{nk})$$

$$= -\sum_{i=1}^{k} (-1)^{ik} d_i h_n(W_0, \ldots, W_{nk}) .$$

Ce qui achève de prouver le Lemme 5. ■

2.— Homologie des algèbres de Lie $(k+1)$-aires.

Ph. Hanlon et M. Wachs ont construit une théorie d'homologie pour les algèbres de Lie $(k+1)$-aires. Il s'agit de l'homologie du complexe $(\Lambda^*(L), \partial)$ où la différentielle $\partial : \Lambda^r(L) \to \Lambda^{r-k}(L)$ est donnée par $\partial \equiv 0$ si $r \leq k$ et, pour $r > k$, on a :

$$\partial(x_1 \wedge \cdots \wedge x_r) :=$$

$$\sum_{\sigma \in Sh_{r-k-1,k+1}} \mathrm{sgn}(\sigma) x_{\sigma(1)} \wedge \cdots \wedge x_{\sigma(r-k-1)} \wedge [x_{\sigma(r-k)} \ldots x_{\sigma(r)}].$$

Ensuite ils montrent que l'homologie de toute algèbre de Lie $(k+1)$-aire libre est triviale.

Comme ∂ envoie $\Lambda^r(L)$ sur $\Lambda^{r-k}(L)$, il est clair que le complexe $(\Lambda^*(L), \partial)$ se scinde en somme directe de k sous-complexes

$$\cdots \to \Lambda^{nk+i}(L) \to \Lambda^{(n-1)k+i}(L) \to \cdots \to \Lambda^{k+i}(L) \to \Lambda^i(L) \to 0$$

où $i = 1, \cdots, k$. Le «bon» complexe prédit par la théorie des opérades est le sous-complexe correspondant à la valeur $i = 1$ que nous notons

$$(D_*^{<k>}(L) = \Lambda^{*k+1}(L), \partial).$$

REFERENCES

[1] Amitsur (S.A.), Levitzki (J.), *Minimal identities for algebras*, Proc. Amer. Math. Soc., 1950, t. 1, 449-463.

[2] Ginzburg (V.), Kapranov (M.M.), *Koszul duality for operads*, Duke Math. J., 1994, t. 76, 203-272.

[3] Gnedbaye (A.V.), *Les algèbres k-aires et leurs opérades*, C. R. Acad. Sci. Paris, Série I, 1995, t. 321, 147-152.

[4] Hanlon (Ph.), Wachs (M.), *On Lie k-algebras*, Advances in Math., 1995, t. 113, 206-236.

[5] Loday (J.-L.), *La renaissance des opérades*, Exposé 792 Séminaire Bourbaki, novembre 1994, Astérisque à paraître.

[6] Rosset (S.), *A new proof of the Amitsur-Levitzki identity*, Israel J. Math., 1976, t. 23, 187-188.

I.R.M.A.
Université Louis Pasteur et C.N.R.S.
7, rue René Descartes
67084 Strasbourg Cedex, FRANCE
e-mail : gnedbaye@math.u-strasbg.fr

Contemporary Mathematics
Volume **202**, 1997

Coproduct and Cogroups in the Category of Graded Dual Leibniz Algebras

Jean-Michel OUDOM

Introduction.

Throughout this paper, we will consider k a commutative field of characteristic different from 2, except in section 4 in which k is of characteristic 0. All vector spaces and tensor products are taken over k.

First, let us come back to the origins of Leibniz algebras. In [L4], J.-L. Loday has shown that the classical differential map d of the Chevalley Eilenberg complex

$$
\begin{array}{ccc}
\mathfrak{g}^{\otimes n} & \xrightarrow{\widetilde{d}} & \mathfrak{g}^{\otimes n-1} \\
\downarrow & & \downarrow \\
\Lambda^n \mathfrak{g} & \xrightarrow{d} & \Lambda^{n-1}\mathfrak{g}
\end{array}
$$

can be lifted to a differential map $\widetilde{d}$ on the tensor algebra. The map $\widetilde{d}$ is given by:

$$\widetilde{d}(x_1 \otimes \cdots \otimes x_n) = \sum_{1 \leq i < j \leq n} (-1)^{j+1} x_1 \otimes \cdots \otimes [x_i, x_j] \otimes \cdots \otimes \hat{x}_j \otimes \cdots$$

for $x_1, \cdots, x_n \in \mathfrak{g}$. Moreover, for a vector space $\mathfrak{g}$ given with a bilinear bracket $[\,,\,]$, one obtains the following equivalence:

$$\widetilde{d}^2 = 0 \iff [x, [y, z]] = [[x, y], z] - [[x, z], y] \text{ for } x, y, z \in \mathfrak{g}.$$

We thus have a new kind of algebra, namely the Leibniz algebras, which are defined as the vector spaces given with a bilinear bracket which satisfies the above identity. The class of Lie algebras is obviously contained in the class of Leibniz algebras and furthermore, one has a homological theory defined on this latter class of algebras, namely the Leibniz homology.

Therefore, one has two homological theories for Lie algebras: the Lie homology and the Leibniz homology. The latter is defined on a larger class of algebras, the Leibniz algebras. For these two theories, one has a Künneth formula theorem:

Theorem ([K]): *For any Lie algebras $\mathfrak{g}'$ and $\mathfrak{g}''$, there is an isomorphism:*

$$H_*(\mathfrak{g}' \times \mathfrak{g}'') \underset{\text{coalgebras}}{\simeq} H_*(\mathfrak{g}') \otimes H_*(\mathfrak{g}'').$$

Theorem ([L1]): *For any Leibniz algebras $\mathfrak{g}'$ and $\mathfrak{g}''$, there is an isomorphism:*

$$HL_*(\mathfrak{g}' \times \mathfrak{g}'') \underset{\text{vect.}}{\simeq} HL_*(\mathfrak{g}') * HL_*(\mathfrak{g}'').$$

1991 *Mathematics Subject Classification.* 17A30, 17B55, 17B56, 18D35.

A Christine, pour son soutien quotidien et sa précieuse collaboration à la rédaction de ce travail en anglais !

Now, let A be an associative algebra with unit, and let us consider $\mathfrak{gl}(A)$, the Lie algebra of matrices with coefficients in A. As $\mathfrak{gl}(A)$ is a Lie algebra, it is a Leibniz algebra. So, one can apply either Lie homology or Leibniz homology on $\mathfrak{gl}(A)$. When k is a characteristic 0 field, these two homological functors behave on $\mathfrak{gl}(A)$ in the following two parallel ways:

Theorem ([L-Q], [T]):

$$H_*(\mathfrak{gl}(A)) \underset{\text{Hopf algebras}}{\simeq} \Lambda HC_{*-1}(A).$$

Theorem ([C],[L3]):

$$HL_*(\mathfrak{gl}(A)) \underset{\text{vect.}}{\simeq} THH_{*-1}(A).$$

In these two examples of parallel developments of Lie homology and Leibniz homology, it is worth emphasizing that one has isomorphisms of coalgebras and Hopf algebras on the left hand side. However, one only has graded vector space isomorphisms on the right hand side.

The aim of this study is twofold. It is first to have a better understanding of the above parallelism between Lie homology and Leibniz homology. It is also to exhibit algebraic structures for which the isomorphisms on the right hand side would be isomorphisms of coalgebras or algebras.

To a given algebraic structure, the Koszul duality for operads ([G-K]) associates another algebraic structure, namely the dual structure, which plays an important role in the theory of cohomology of algebras over the first structure. Therefore, we will base our study on the introduction of dual Leibniz algebras ([L2]).

We first study the relationship between free dual Leibniz algebras and the Leibniz cohomology. In the second part, we state that the free product $*$ is the coproduct in the category of dual Leibniz algebras and we show that the Künneth formula for the cohomology of Leibniz algebras: $HL_*(\mathfrak{g}' \times \mathfrak{g}'') \simeq HL_*(\mathfrak{g}') * HL_*(\mathfrak{g}'')$ is an isomorphism of graded dual Leibniz algebras. The third part is devoted to an investigation of the cogroups in the category of connected graded dual Leibniz algebras: we construct a full and faithful functor from the category of such cogroups to the category of graded Leibniz coalgebras. Finally, the fourth part is dedicated to a new proof of the Cuvier-Loday theorem computing $HL_*(\mathfrak{gl}(A))$ and we give an interpretation of $HL^*(\mathfrak{gl}(A))$ in terms of cogroups in the category of graded dual Leibniz algebras.

1. Free dual Leibniz algebras and the cohomology of Leibniz algebras

NOTATION 1.1. For any graded vector space R and for any homogeneous vector x of R, the degree of x is denoted by $|x|$.

For any graded bilinear product μ defined on R, let $*$ denote the graded bilinear product defined as follows:

$$x * y = \mu(x,y) + (-1)^{|x||y|}\mu(y,x) \text{ for } x,y \in R.$$

DEFINITION 1.2 ([L2]). A graded dual Leibniz algebra is a vector space R given with a graded bilinear product satisfying the dual Leibniz identity:

$$(xy)z = x(yz) + (-1)^{|y||z|}x(zy) \text{ for } x,y,z \in R.$$

NOTATION 1.3. For any graded vector space V, let TV denote the graded vector space defined by $TV = \bigoplus_{n \geq 1} V^{\otimes^n}$.

THEOREM 1.4 ([L2]). *Let V be a graded vector space. There is a unique graded dual Leibniz product μ defined on TV such that $\mu(v,\omega) = v \otimes \omega$ for $v, \omega \in V \times TV$. Furthermore (TV, μ) is the free graded dual Leibniz algebra generated by V.*

(For any graded dual Leibniz algebra R and any graded map $f : V \to R$, there is a unique map of graded dual Leibniz algebra $\overline{f} : TV \to R$ such that:

$$
\begin{array}{ccc}
V & = & V \\
\downarrow & c & \downarrow f \\
TV & \xrightarrow[\overline{f}]{} & R
\end{array}
\quad)
$$

PROOF. (Proposition 1.8 of [L2]). □

DEFINITION 1.5. Let R be a graded vector space given with a graded bilinear product and let n be an integer. A n-degree derivation of R is a linear map δ of degree n satisfying the following identity: $\delta(rs) = \delta(r)s + (-1)^{n|r|}r\delta(s)$ for $r, s \in R$.

PROPOSITION 1.6. *Let d and δ be two derivations of degree p and q respectively. Then $d \circ \delta - (-1)^{pq}\delta \circ d$ is a $(p+q)$-derivation.*

PROOF. Straightforward. □

PROPOSITION 1.7. *Let V be a graded vector space and let n be an integer. We consider the free graded dual Leibniz algebra (TV, μ) generated by V. Then, for any linear map of degree n, $f : V \to TV$, there is a unique n-derivation δ of TV which extends f on V.*

PROOF. For any $x, \omega \in V \times TV$, we necessarily have $\delta(x \otimes \omega) = \mu(f(x), \omega) + (-1)^{n|x|}x \otimes \delta(\omega)$. This identity gives us an inductive definition of δ. Then, it is easy to check that δ is an n-derivation of TV. □

We can now define the cohomology of a Leibniz algebra. From now on and until the end of the first part, all the graded vector spaces that we will consider will be of finite dimension in each degree.

NOTATION 1.8. For any graded vector spaces M and N, let $\tau : M \otimes N \simeq N \otimes M$ denote the isomorphism of graded vector spaces defined by:
$$\tau(m \otimes n) = (-1)^{pq}n \otimes m \text{ for } m, n \in M_p \times N_q.$$
The graded vector space dual to M is denoted by M^*, and we use the usual identification of $(M \otimes N)^*$ with $M^* \otimes N^*$ given by:
$$(f \otimes g)(m \otimes n) = (-1)^{pq}f(m)g(n) \text{ for } f, g \in M_p^* \times N_q^* \text{ and } m, n \in M_p \times N_q.$$
Let sM denote the graded vector space defined by $sM_n = M_{n-1}$ for $n \in \mathbb{N}$.

A graded Leibniz algebra is a graded vector space $\mathfrak{g}$ given with a graded bilinear bracket [,] satisfying the Leibniz identity:
$$[\,,\,](1 \otimes [\,,\,]) = [\,,\,]([\,,\,] \otimes 1) - [\,,\,]([\,,\,] \otimes 1)(1 \otimes \tau).$$
By dualization, one gets a coproduct Δ on $\mathfrak{g}^*$, which satisfies the following identity:
$$(1 \otimes \Delta)\Delta = (\Delta \otimes 1)\Delta - (1 \otimes \tau)(\Delta \otimes 1)\Delta.$$
For any $f \in \mathfrak{g}^*$ we put: $\Delta(f) = \sum\limits_{(1),(2)} f_{(1)} \otimes f_{(2)}$.

PROPOSITION 1.9. *Let $\mathfrak{g}$ be a graded Leibniz algebra. Let us consider the free graded dual Leibniz algebra $T(s\mathfrak{g}^*)$ generated by $s\mathfrak{g}^*$ and δ the unique 1-derivation which extends $\phi : s\mathfrak{g}^* \longrightarrow s\mathfrak{g}^* \otimes s\mathfrak{g}^*$ defined by:*
$$\phi(f) = \sum_{(1),(2)} (-1)^{|f_{(1)}|}f_{(1)} \otimes f_{(2)} \text{ for } f \in s\mathfrak{g}^*.$$
Then $(T(s\mathfrak{g}^*), \delta)$ is a differential graded dual Leibniz algebra.

By definition, its homology is the cohomology of $\mathfrak{g}$ with coefficients in k. We will write $(C^*(\mathfrak{g}), \delta)$ for the above differential graded dual Leibniz algebra and $HL^*(\mathfrak{g})$ for its homology.

PROOF.

$$\delta^2(f) = \delta\Big(\sum_{(1),(2)} (-1)^{|f_{(1)}|} f_{(1)} \otimes f_{(2)} \Big)$$

$$= \sum_{(1),(2)} (-1)^{|f_{(1)}|} \delta(f_{(1)}) f_{(2)} - \sum_{(1),(2)} f_{(1)} \otimes \delta(f_{(2)})$$

$$= \sum_{(1),(2)} \sum_{(11),(12)} (-1)^{|f_{(12)}|} f_{(11)} \otimes f_{(12)} \otimes f_{(2)}$$

$$+ (-1)^{(|f_{(12)}|+1)(|f_{(2)}|+1)+|f_{(12)}|} f_{(11)} \otimes f_{(2)} \otimes f_{(12)}$$

$$- \sum_{(1),(2)} \sum_{(21),(22)} (-1)^{|f_{(21)}|} f_{(1)} \otimes f_{(21)} \otimes f_{(22)}$$

By composing δ^2 with the isomorphism of graded vector spaces which consists in multiplying by $(-1)^p$, where p is the degree of the second variable, we get the following identity:

$$\sum_{(1),(2)} \sum_{(11),(12)} f_{(11)} \otimes f_{(12)} \otimes f_{(2)} - (-1)^{|f_{(12)}||f_{(2)}|} f_{(11)} \otimes f_{(2)} \otimes f_{(12)}$$

$$- \sum_{(1),(2)} \sum_{(21),(22)} f_{(1)} \otimes f_{(21)} \otimes f_{(22)} = \big((\Delta \otimes 1)\Delta - (1 \otimes \tau)(\Delta \otimes 1)\Delta - (1 \otimes \Delta)\Delta\big)(f) = 0.$$

So, $\delta^2(f) = 0$ for $f \in s\mathfrak{g}^*$. From (1.6), $2\delta^2$ is a 2-derivation on $T(s\mathfrak{g}^*)$. Therefore, according to (1.7), it is entirely determined by its restriction to $s\mathfrak{g}^*$. $\square$

REMARKS 1.10. 1. The cohomology of the Leibniz algebra $\mathfrak{g}$, defined as above, is equal to the one defined in [L-P]. Indeed, the underlying graded vector spaces of the two complexes are obviously equal. Moreover, note that the differentials are equal in degree 1. Now, as the differential defined in [L-P] is a 1-derivation for the dual Leibniz product ([L2]), it follows from (1.7) that it is the same as the one defined in (1.9).

2. If we work in the category of graded dual Leibniz coalgebras, we can go through the homological framework. In this case, we can avoid the condition of dimension finiteness of $\mathfrak{g}$, but on the other hand, we have to deal with dual Leibniz coproducts and suppose that $\mathfrak{g}$ is positively graded.

2. Coproduct of dual Leibniz algebras and the Künneth formula

The aim of this part is to give an interpretation of the Künneth formula for the cohomology of Leibniz algebras in terms of graded dual Leibniz algebras. First, we investigate coproducts in the category of graded dual Leibniz algebras. To begin with, let us review some notations.

NOTATION 2.1 ([B]). Let $\{',''\}$ be a set of two symbols, let i be in $\{','' \}$, and let R' and R'' be two graded vector spaces. We will use the following notations:

$\mathcal{I}(',''')$ is the set of sequences of finite length in $'$ and $''$,

$\mathcal{M}(',''')$ is the set of mixed sequences of finite length in $'$ and $''$,

   (i.e. where $'$ and $''$ appear at least once),

$\mathcal{R}(',''')$ is the set of alternating sequences of finite length in $'$ and $''$,

$R' * R'' = \bigoplus_{I \in \mathcal{R}(',''')} R^I$, where $R^I = R^{i_1} \otimes \cdots \otimes R^{i_n}$ for $I = (i_1, \cdots, i_n)$.

Three more notations are needed to state the next proposition:

$\mathcal{C}^i(',')$ is the set of $I \in \mathcal{R}(',')$ such that $i_1 = i$,

$$(R' * R'')^i = \bigoplus_{I \in \mathcal{C}^i(',')} R^I,$$

$$(R' * R'')^{\leq n} = \bigoplus_{\substack{I \in \mathcal{R}(',') \\ |I| \leq n}} R^I, \text{ where } |I| \text{ stands for the length } I.$$

PROPOSITION 2.2. *Let R' and R'' be two graded dual Leibniz algebras. Then there is a unique graded dual Leibniz product μ defined on $R' * R''$ such that:*
1. *$\mu(x^i, y^i) = x^i y^i$ for $i \in \{','\}$ and $x^i, y^i \in R^i$,*
2. *$\mu(x^i, y^J) = x^i \otimes y^J$ for $i \in \{','\}$, $J \in \mathcal{R}(',')$ such that $j_1 \neq i$, and for $x^i \in R^i$ and $x^J \in R^J$.*

PROOF. We define μ and check by induction on the length of the sequences of $\mathcal{R}(',')$ that it is a dual Leibniz product. The inductive hypothesis at the stage $n \geq 3$ is as follows:

For any $I, J, K \in \mathcal{R}(',')$, $x^I, y^J, z^K \in R^I \times R^J \times R^K$,
  (i)- $\mu(x^I, y^J)$ is defined for $|I| + |J| \leq n$ and $\mu(x^I, y^J) \in (R' * R'')^{\leq |I| + |J|}$,
  (ii)- if $I \in \mathcal{C}^i(',')$ and $|I| + |J| \leq n$, then $\mu(x^I, y^J) \in (R' * R'')^i$,
  (iii)- $\mu(\mu(x^I, y^J), z^K) = \mu(x^I, y^J * z^K)$ for $|I| + |J| + |K| \leq n$.

The construction of $\mu(x^I, y^J)$ for $|I| + |J| \leq 3$ is straightforward. Moreover, we can easily check (i), (ii) and (iii) at the stage 3. Now, let us assume the inductive hypothesis at the stage n.
First, we define $\mu(x^i, y^J)$ for $x^i \in R^i, y^J \in R^J$ with $J \in \mathcal{R}(',')$ such that $|J| \leq n$.
  - when $j_1 \neq i$, then we necessarily have $\mu(x^i, y^J) = x^i \otimes y^J$.
  - if $j_1 = i$, it yields:
$$\mu(x^i, y^J) = \mu(x^i, \mu(y^i, y^{J \setminus j_1})) \quad \text{and necessarily,}$$

$$= \mu(\mu(x^i, y^i), y^{J \setminus j_1}) - (-1)^{|y^i||y^{J \setminus j_1}|} \mu(x^i, \mu(y^{J \setminus j_1}, y^i))$$

$$= x^i y^i \otimes y^{J \setminus j_1} - (-1)^{|y^i||y^{J \setminus j_1}|} x^i \otimes \mu(y^{J \setminus j_1}, y^i).$$

Then, we can define $\mu(x^I, y^J)$ for $x^I \in R^I, y^J \in R^J$, where $I, J \in \mathcal{R}(',')$ with $|I| + |J| \leq n + 1$.

$$\mu(x^I, y^J) = \mu(\mu(x^{i_1}, x^{I \setminus i_1}), y^J)$$

$$= \mu(x^{i_1}, x^{I \setminus i_1} * y^J) \quad \text{by necessary condition.}$$

We easily check that μ defined as such satisfies (i) and (ii) for $|I| + |J| \leq n + 1$. Therefore, it suffices to show that it verifies (iii) for $|I| + |J| + |K| \leq n + 1$.

First, we show (iii) in the case when $|I| = 1$, $|J| = 1$ and $|K| = n - 1$.
Let $i, j \in \{','\}$, $K \in \mathcal{R}_{n-1}(',')$, $x^i, y^j, z^K \in R^i \times R^j \times R^K$. Then
  - when $i \neq j$, (iii) directly follows from the definition.
  - when $i = j \neq k_1$, it is an obvious consequence of (ii).
  - if $i = j = k_1$, we obtain:

$$\mu(\mu(x^i, y^i), z^K) = \mu(x^i y^i, z^K)$$

$$= (x^i y^i) z^i \otimes z^{K \setminus k_1} - (-1)^{|z^i||z^{K \setminus k_1}|} x^i y^i \otimes \mu(z^{K \setminus k_1}, z^i) \quad \text{by definition.}$$

$$\mu(x^i, \mu(y^i, z^K)) = \mu(x^i, y^i z^i \otimes z^{K \setminus k_1} - (-1)^{|z^i||z^{K \setminus k_1}|} y^i \otimes \mu(z^{K \setminus k_1}, z^i))$$

$$= x^i (y^i z^i) \otimes z^{K \setminus k_1} - (-1)^{|z^{K \setminus k_1}||y^i z^i|} x^i \otimes \mu(z^{K \setminus k_1}, y^i z^i)$$

$$- (-1)^{|z^{K \setminus k_1}||z^i|} x^i y^i \otimes \mu(z^{K \setminus k_1}, z^i)$$

$$+ (-1)^{|z^{K \backslash k_1}||z^i| + |\mu(z^{K \backslash k_1}, z^i)||y^i|} x^i \otimes \mu(\mu(z^{K \backslash k_1}, z^i), y^i)$$

$$= x^i(y^i z^i) \otimes z^{K \backslash k_1} - (-1)^{|z^{K \backslash k_1}||y^i z^i|} x^i \otimes \mu(z^{K \backslash k_1}, y^i z^i)$$

$$- (-1)^{|z^{K \backslash k_1}||z^i|} x^i y^i \otimes \mu(z^{K \backslash k_1}, z^i)$$

$$+ (-1)^{|z^{K \backslash k_1}||z^i| + |z^{K \backslash k_1}||y^i| + |y^i||z^i|} x^i \otimes \mu(z^{K \backslash k_1}, z^i y^i)$$

$$+ (-1)^{|z^{K \backslash k_1}||y^i z^i|} x^i \otimes \mu(z^{K \backslash k_1}, y^i z^i) \quad \text{by induction,}$$

$$= x^i(y^i z^i) \otimes z^{K \backslash k_1} - (-1)^{|z^{K \backslash k_1}||z^i|} x^i y^i \otimes \mu(z^{K \backslash k_1}, z^i)$$

$$+ (-1)^{|z^{K \backslash k_1}||z^i| + |z^{K \backslash k_1}||y^i| + |y^i||z^i|} x^i \otimes \mu(z^{K \backslash k_1}, z^i y^i).$$

$$\mu(x^i, \mu(z^K, y^i)) = (-1)^{|y^i||z^{K \backslash k_1}|} \mu(x^i, z^i y^i \otimes z^{K \backslash k_1})$$

$$= (-1)^{|y^i||z^{K \backslash k_1}|} x^i(z^i y^i) \otimes z^{K \backslash k_1} - (-1)^{|z^i||z^{K \backslash k_1}|} x^i \otimes \mu(z^{K \backslash k_1}, z^i y^i).$$

This identity yields:

(1) $\mu(\mu(x^I, y^J), z^K) = \mu(x^I, y^J * z^K)$ for $|I| = |J| = 1$ and $|K| \leq n - 1$.

In a second step, we check (iii) for $|I| = 1$ and when $|J| + |K| \leq n$.
Let $i \in \{', ''\}$, $J, K \in \mathcal{R}(', '')$ such that $|J| + |K| \leq n$, $x^i, y^J, z^K \in R^i \times R^J \times R^K$.
Then we have:

- if $i \neq j_1$, (iii) follows directly from the definition.
- if $i = j_1$, we get:

$$\mu(\mu(x^i, y^J), z^K) = \mu\big(x^i y^i \otimes y^{J \backslash j_1} - (-1)^{|y^{J \backslash j_1}||y^i|} x^i \otimes \mu(y^{J \backslash j^1}, y^i), z^K\big)$$

$$= \mu(x^i y^i, y^{J \backslash j_1} * z^K) - (-1)^{|y^{J \backslash j_1}||y^i|} \mu(x^i, \mu(y^{J \backslash j_1}, y^i) * z^K)$$

$$= \mu(x^i y^i, y^{J \backslash j_1} * z^K) - (-1)^{|y^{J \backslash j_1}||y^i|} \mu(x^i, \mu(y^{J \backslash j_1}, y^i * z^K))$$

$$- (-1)^{|y^{J \backslash j_1}||y^i| + |z^K||y^J|} \mu(x^i, \mu(z^K, \mu(y^{J \backslash j_1}, y^i))) \quad \text{by induction,}$$

$$\mu(x^i, \mu(y^J, z^K)) = \mu(x^i, \mu(y^i, y^{J \backslash j_1} * z^K))$$

$$= \mu(x^i y^i, y^{J \backslash j_1} * z^K) - (-1)^{|y^i|(|y^{J \backslash j_1}| + |z^K|)} \mu(x^i, \mu(y^{J \backslash j_1} * z^K, y^i))$$

from (1),

$$= \mu(x^i y^i, y^{J \backslash j_1} * z^K) - (-1)^{|y^i|(|y^{J \backslash j_1}| + |z^K|)} \mu(x^i, \mu(\mu(y^{J \backslash j_1}, z^K), y^i))$$

$$- (-1)^{|y^i|(|y^{J \backslash j_1}| + |z^K|) + |y^{J \backslash j_1}||z^K|} \mu(x^i, \mu(\mu(z^K, y^{J \backslash j_1}), y^i))$$

$$= \mu(x^i y^i, y^{J \backslash j_1} * z^K) - (-1)^{|y^i|(|y^{J \backslash j_1}| + |z^K|)} \mu(x^i, \mu(y^{J \backslash j_1}, z^K * y^i))$$

$$- (-1)^{|y^i|(|y^{J \backslash j_1}| + |z^K|) + |y^{J \backslash j_1}||z^K|} \mu(x^i, \mu(z^K, \mu(y^{J \backslash j_1}, y^i)))$$

$$- (-1)^{|y^J||z^K|} \mu(x^i, \mu(z^K, y^J)) \quad \text{by induction,}$$

which yields:

(2) $\mu(\mu(x^i, y^J), z^K) = \mu(x^i, y^J * z^K)$ for $|J| + |K| \leq n$.

Now, we can state (iii) in the general case: let $I, J, K \in \mathcal{R}(', '')$ such that $|I| + |J| + |K| \leq n + 1$, and let $x^I, y^J, z^K \in R^I \times R^J \times R^K$. Then we have:

$$\mu(\mu(x^I, y^J), z^K) = \mu(\mu(x^{i_1}, x^{I \backslash i_1} * y^J), z^K)$$

$$= \mu(x^{i_1}, (x^{I \backslash i_1} * y^J) * z^K) \quad \text{from (2) and}$$

$$\mu(x^I, \ y^J * z^K) = \mu(\mu(x^{i_1}, x^{I \backslash i_1}), y^J * z^K)$$

$$= \mu(x^{i_1}, x^{I \backslash i_1} * (y^J * z^K)) \quad \text{from (2) as well.}$$

Then the following lemma ends the proof by induction. $\square$

LEMMA ([L2]). *Let R be a graded vector space given with a graded bilinear product μ and let $x_1, x_2, x_3 \in R$. If μ satisfies the dual Leibniz identity on all triples $(x_{\sigma(1)}, x_{\sigma(2)}, x_{\sigma(3)})$ for $\sigma \in \Sigma_3$, then:*

$$(x_1 * x_2) * x_3 = x_1 * (x_2 * x_3).$$

PROOF. (proposition 1.5 of [L2]). □

THEOREM 2.3. *Let R' and R'' be two graded dual Leibniz algebras. Then $(R' * R'', \mu)$ where μ is defined as in (2.2), is the coproduct of R' and R'' in the category of graded dual Leibniz algebras.*

PROOF. Let Q be a graded dual Leibniz algebra, let $f' : R' \to Q$ and $f'' : R'' \to Q$ be two maps of graded dual Leibniz algebras. The canonical inclusions of R' and R'' in $R' * R''$ are respectively denoted by $i' : R' \longrightarrow R' * R''$ and $i'' : R'' \longrightarrow R' * R''$. Our purpose is to define a map of dual Leibniz algebras $f : R' * R'' \longrightarrow Q$ such that $f' = f \circ i'$ and $f'' = f \circ i''$.

For any $I = (i_1, \cdots, i_n) \in \mathcal{R}(',\,'')$, $x^I \in R^I$, we necessarily have:

$$f(x^I) = f^{i_1}(x^{i_1}(f^{i_2}(x^{i_2}(\cdots (f^{i_{n-1}}(x^{i_{n-1}}f^{i_n}(x^{i_n})\cdots).$$

Let us show that f is a map of graded dual Leibniz algebras. This is done by induction on $|I| + |J|$ for $I, J \in \mathcal{R}(',\,'')$. The induction obviously starts with $|I| + |J| = 2$.

Let us assume that if $I, J \in \mathcal{R}(',\,'')$ such that $|I| + |J| \leq n$,

$$f(\mu(x^I, y^J)) = f(x^I)f(y^J) \text{ for } x^I, y^J \in R^I \times R^J.$$

Let $i \in \{',\,''\}$, $J \in \mathcal{R}(',\,'')$ such that $|J| \leq n$, $x^i, y^J \in R^i \times R^J$,

- when $i \neq j_1$, then from the definition of f, we get:

$$f(\mu(x^i, y^J)) = f(x^i \otimes y^J) = f(x^i)f(y^J),$$

- if $i = j_1$, then:

$$f(\mu(x^i, y^J)) = f(x^i y^{j_1} \otimes y^{J \setminus j_1} - (-1)^{|y^{J \setminus j_1}||y^i|} x^i \otimes \mu(y^{J \setminus j_1}, y^i)) \text{ by definition of } \mu,$$

$$= f(x^i y^{j_1})f(y^{J \setminus j_1}) - (-1)^{|y^{J \setminus j_1}||y^i|} f(x^i)f(\mu(y^{J \setminus j_1}, y^i)) \text{ by definition of } f,$$

$$= (f(x^i)f(y^i))f(y^{J \setminus j_1}) - (-1)^{|y^{J \setminus j_1}||y^i|} f(x^i)(f(y^{J \setminus j_1})f(y^i)) \text{ by induction,}$$

$$= f(x^i)(f(y^i)f(y^J)) \text{ since } Q \text{ is a dual Leibniz algebra,}$$

$$= f(x^i)f(y^J) \text{ by induction.}$$

Let $I, J \in \mathcal{R}(',\,'')$ such that $|I| + |J| \leq n + 1$, and let $x^I, y^J \in R^I \times R^J$. We have:

$$f(\mu(x^I, y^J)) = f(\mu(x^{i_1}, x^{I \setminus i_1} * y^J)$$

$$= f(x^{i_1})f(x^{I \setminus i_1} * y^J)$$

$$= f(x^{i_1})(f(x^{I \setminus i_1}) * f(y^J)) \text{ by induction,}$$

$$= (f(x^{i_1})f(x^{I \setminus i_1}))f(y^J) \text{ since } Q \text{ is a dual Leibniz algebra,}$$

$$= f(x^I)f(y^J). □$$

PROPOSITION 2.4. *Let R' et R'' be two graded dual Leibniz algebras, and let δ' and δ'' be two n-derivations of R' and R'' respectively. There is a unique n-derivation δ of $(R' * R'', \mu)$ which extends δ' and δ''.*

PROOF. Let $I \in \mathcal{R}(',\,'')$; we necessarily have

$\delta(x^I) = \delta^{i_1}(x^{i_1}) \otimes x^{I \setminus i_1} + (-1)^{n|x^{i_1}|} x^{i_1} \otimes \delta(x^{I \setminus i_1})$ for $x^I \in R^I$. Let us show that δ is a n-degree derivation. We proceed by induction on the length of the involved sequences I and J. The step $|I| + |J| = 2$ is obvious and we perform the induction in the same way as in (2.3): we first treat the case when $|I| = 1$, and then we extend it to the general case. $\square$

REMARKS 2.5. 1. With the same data as in (2.4), let $I = (i_1, \cdots, i_p) \in \mathcal{R}(',\!'')$ and let $x^I \in R^I$. Then δ is defined by:

$$\delta(x^I) = \sum_{k=1}^{p} (-1)^{n(|x^{i_1}| + \cdots + |x^{i_{k-1}}|)} x^{i_1} \otimes \cdots \otimes x^{i_{k-1}} \otimes \delta^{i_k}(x^{i_k}) \otimes x^{i_{k+1}} \otimes \cdots \otimes x^{i_p}.$$

2. All this framework can be done in the category of connected graded dual Leibniz coalgebras if we replace *free algebra* by *cofree coalgebra*, *coproduct* by *product* and *derivation* by *coderivation*.

From now and until the end of this part, let $\mathfrak{g}'$ and $\mathfrak{g}''$ be two graded Leibniz algebras of finite dimension in each degree.

PROPOSITION 2.6. *Let $(\mathfrak{g}' \times \mathfrak{g}'', [,])$ be the product of $\mathfrak{g}'$ and $\mathfrak{g}''$ as graded vector spaces, given with the following bracket:*
$$[(x', x''), (y', y'')] = ([x', y'], [x'', y'']) \text{ for } (x', x''), (y', y'') \in \mathfrak{g}' \times \mathfrak{g}''.$$
This bracket induces a structure of graded Leibniz algebra on $\mathfrak{g}' \times \mathfrak{g}''$, which is the product of $\mathfrak{g}'$ and $\mathfrak{g}''$ in the category of graded Leibniz algebras.

PROOF. Straightforward. $\square$

PROPOSITION 2.7. *Let V' and V'' be two graded vector spaces, and let $T(V')$, $T(V'')$ and $T(V' \oplus V'')$ be the free graded dual Leibniz algebras generated by V', V'' and $V' \oplus V''$ respectively. Let ψ be the unique morphism of dual Leibniz algebras from $T(V') * T(V'')$ to $T(V' \oplus V'')$ which extends the canonical inclusions i' and i'' of $T(V')$ and $T(V'')$ in $T(V' \oplus V'')$. Then ψ is an isomorphism of graded dual Leibniz algebras.*

PROOF. It directly follows from the universal properties that free graded dual Leibniz algebras and the coproduct of two graded dual Leibniz algebras, respectively, satisfy. $\square$

REMARK 2.8. The isomorphism ψ is defined as follows:
$$\psi(\omega^I) = \omega_{i_1}(\omega_{i_2}(\cdots(\omega_{i_{n-2}}(\omega_{i_{n-1}}\omega_{i_n}))\cdots)) \text{ for } I \in \mathcal{R}_n(',\!'') \text{ and } \omega^I \in TV^I,$$
where the products are made in $T(V' \oplus V'')$. For instance, for $x', y' \in V'$ and $z'' \in V''$, we obtain $\psi(x' \otimes y' \otimes z'') = x' \otimes y' \otimes z'' + (-1)^{|y'||z''|} x' \otimes z'' \otimes y'$.

THEOREM 2.9. *Let us consider the two following differential graded vector spaces: - $(C^*(\mathfrak{g}') * C^*(\mathfrak{g}''), d)$ where d is the differential defined from δ' and δ'' by using (2.4),*

- $(C^*(\mathfrak{g}' \times \mathfrak{g}''), \delta)$, the differential graded dual Leibniz algebra which calculates the cohomology $HL^*(\mathfrak{g}' \times \mathfrak{g}'')$.
Then the isomorphism of graded dual Leibniz algebras
$$\psi : C^*(\mathfrak{g}') * C^*(\mathfrak{g}'') \longrightarrow C^*(\mathfrak{g}' \times \mathfrak{g}'')$$
defined as above, is an isomorphism of differential graded vector spaces.

PROOF. As ψ is an isomorphism, it remains to see that (i) $\psi d \psi^{-1} = \delta$. Since ψ is a morphism of graded dual Leibniz algebras, it follows that $\psi d \psi^{-1}$ is a 1-derivation of $C^*(\mathfrak{g}' \times \mathfrak{g}'')$. Thus, following (1.7) we only need to state (i) on $s(\mathfrak{g}' \times \mathfrak{g}'')^* = s(\mathfrak{g}')^* \oplus s(\mathfrak{g}'')^*$.

Let Δ', Δ'' and Δ be the dual maps of the brackets of $\mathfrak{g}'$, $\mathfrak{g}''$ and $\mathfrak{g}' \times \mathfrak{g}''$ respectively. Let f be a graded linear form on $\mathfrak{g}' \times \mathfrak{g}''$, and let f' and f'' be its restrictions to $\mathfrak{g}'$ and $\mathfrak{g}''$ respectively. Then, it is easy to check that
$$\Delta(f) = \Delta'(f') + \Delta''(f'').$$
Since $\psi|_{C^*(\mathfrak{g}')} = 1_{C^*(\mathfrak{g}')}$ and $\psi|_{C^*(\mathfrak{g}'')} = 1_{C^*(\mathfrak{g}'')}$, it follows that $\psi d\psi^{-1}$ and δ are equal on $s(\mathfrak{g}')^* \oplus s(\mathfrak{g}'')^*$. $\square$

COROLLARY 2.10. *Let $\mathfrak{g}'$ and $\mathfrak{g}''$ be two graded Leibniz algebras. Then there is a canonical isomorphism of graded dual Leibniz algebras:*
$$HL^*(\mathfrak{g}' \times \mathfrak{g}'') \simeq HL^*(\mathfrak{g}') * HL^*(\mathfrak{g}''). \quad \square$$

REMARKS 2.11. 1. The above isomorphism ψ is precisely the dual map of the isomorphism of the Künneth formula described in [L1]. By using the same arguments as in (1.10.2), we can state a Künneth formula for the homology of Leibniz algebras which would be precisely the one described in [L1].

2. The diagonal map $\mathfrak{g} \to \mathfrak{g} \times \mathfrak{g}$ gives the codiagonal map in the category of graded dual Leibniz algebras on $HL^*(\mathfrak{g})$. More precisely,
$$HL^*(\mathfrak{g}) * HL^*(\mathfrak{g}) \simeq HL^*(\mathfrak{g} \times \mathfrak{g}) \longrightarrow HL^*(\mathfrak{g})$$
is the unique map of graded dual Leibniz algebras which extends the identity map on each factor $HL^*(\mathfrak{g})$ of $HL^*(\mathfrak{g}) * HL^*(\mathfrak{g})$. Up to transposition, this gives all the results stated in [O].

3. Cogroups in the category of connected graded dual Leibniz algebras

First, let us recall what a cogroup in a category is. Let $\mathcal{C}$ be a category. The set of morphisms between two objects of $\mathcal{C}$ is denoted by $Hom_{\mathcal{C}}(-, -)$.

DEFINITION 3.1. An object A of $\mathcal{C}$ is called a comultiplicative object (resp. associative comultiplicative object, resp. cogroup, resp. abelian cogroup) in $\mathcal{C}$ if $Hom_{\mathcal{C}}(A, -)$ is a functor from $\mathcal{C}$ to the category of monoids with unit (resp. associative monoids with unit, resp. groups, resp. abelian groups).

A category $\mathcal{C}$ is called with coproducts and zero maps if:
- the coproduct of any two objects of $\mathcal{C}$ exists,
- for any two objects A, B of $\mathcal{C}$, there is a map $0_{AB} \in Hom_{\mathcal{C}}(A, B)$ which is absorbent for the composition of maps in $\mathcal{C}$.

For any two objects (A, B) of $\mathcal{C}$, we will use the following notations:
- $A * B$ is the coproduct of A and B in $\mathcal{C}$,
- $< f, g > \in Hom_{\mathcal{C}}(A * B, C)$ is the unique map defined by f and g for any object C of $\mathcal{C}$ and any two maps $(f, g) \in Hom_{\mathcal{C}}(A, C) \times Hom_{\mathcal{C}}(B, C)$,
- $T : A * B \to B * A$ is the isomorphism $T_{A,B} = < i_A, i_B >$ where i_A and i_B are the respective imbeddings of A and B in $B * A$.

THEOREM 3.2 [E-H]. *Let $\mathcal{C}$ be a category with coproducts and zero maps and let A be an object in $\mathcal{C}$. A structure of comultiplicative object in $\mathcal{C}$ on A is given by a morphism $\Phi : A \longrightarrow A * A$ such that $< 0_A, 1_A > \Phi = < 1_A, 0_A > \Phi = 1_A$.*

Moreover, A is an associative comultiplicative object if
$$(\Phi * 1)\, \Phi = (1 * \Phi)\, \Phi,$$
A is a cogroup if there is a map $\nu \in Hom_{\mathcal{C}}(A, A)$ such that
$$< 1, \nu > \Phi = < \nu, 1 > \Phi = 0_A,$$
and A is an abelian cogroup if furthermore $T\Phi = \Phi$. $\square$

124   JEAN-MICHEL OUDOM

We saw in (2.3) that the category of graded dual Leibniz algebras is a category with coproducts (for any algebraic operad $\mathcal{P}$, the category of $\mathcal{P}$-algebras is a category with coproducts). Moreover, there is an obvious final and initial object: the zero dual Leibniz algebra.

THEOREM 3.3. *Let R be a comultiplicative object in the category of connected graded dual Leibniz algebras. There is a natural graded sub-vector space $N \subset R$ such that the unique map of graded dual Leibniz algebras $\eta : TN \to R$ induced by the inclusion is an isomorphism of graded dual Leibniz algebras.*

PROOF. The construction of N is made by induction on the degree. The induction hypothesis is the following:

$N_k \subset R_k$ is constructed for $k \leq p$, and for $N^p = \bigoplus_{k \leq p} N_k$, the map of graded dual Leibniz algebras $\eta : TN^p \to R$ is an isomorphism in degree $\leq p$.

Note that for $N_1 = R_1$, the hypothesis is obviously true. To begin with, let us review some notations and definitions:

NOTATION 3.4. Let N be a graded sub-vector space of R such that $N_k = 0$ for $k > p + 1$, and such that $\eta : TN \to R$ is an isomorphism in degree $\leq p$. Then $\Phi : R \to R * R$ restricted to N yields a graded linear map from N to $R * R$. This map can be lifted to a graded linear map from $N \to TN * TN$. Let Φ_T be the unique map of graded dual Leibniz algebras which extends the preceding map on TN. Then we get the following commutative diagram:

$$\begin{array}{ccc} TN & \xrightarrow{\Phi_T} & TN * TN \\ \eta \downarrow & & \downarrow \eta*\eta \\ R & \xrightarrow{\Phi} & R * R \end{array}$$

Let $\widetilde{\Phi}$ denote the map of graded dual Leibniz algebras given by:
$$\widetilde{\Phi} : TN \xrightarrow{\Phi_T} TN * TN \xrightarrow{\Psi} T(N + N).$$
As a graded vector space, $T(N+N) \simeq \bigoplus_{I \in \mathcal{I}(',\,'')} N^I$. For $I \in \mathcal{I}(',\,'')$, the projection of $\widetilde{\Phi}$ on N^I is denoted by $\widetilde{\Phi}_I$.

For any integer $r \geq 2$, let I_r denote the unique alternating sequence in $\{',\,''\}$ of length r which begins with $''$, and let us write $\widetilde{\Phi}_r$ for $\widetilde{\Phi}_{I_r}$.

The graded dual Leibniz algebras $T(N)$ and $T(N + N)$ are bigraded by the underlying gradation and the length of the words. Let $F^{p,q}TN$ (resp. $F^{p,q}T(N + N)$) denote the vector spaces generated by the words of length p and of degree q in TN (resp. $T(N + N)$).

Let $\nabla : T(N + N) \to TN$ be the unique map of graded dual Leibniz algebras which extends the linear map $N + N \longrightarrow N$
$$(x, y) \longmapsto x + y.$$
Notice that ∇ is a bigraded map.

LEMMA 3.5.
1. *For any $I \in \mathcal{I}(',\,'')$ of length n, $\nabla|_{N^I} : N^I \to N^{\otimes^n}$ is equal to the identity map.*
2. $\widetilde{\Phi}(F^{p,k}TN) \subset \sum_{q \geq p} F^{q,k}T(N + N)$ *for all $p, k \in \mathbb{N}^*$.*
3. $\widetilde{\Phi}_r(TN) \subset \sum_{k \geq 1} F^{r,k}T(N + N)$ *for $r \geq 2$.*
4. $\widetilde{\Phi}_r(F^{p,*}TN) = 0$ *for $r \geq 2$ and $p > r$.*

5. $\nabla\widetilde{\Phi}_r(a) = a$ for $a \in F^{r,*}TN$ where $r \geq 2$.

6. $\widetilde{\Phi}_r(F^{*,k}TN) = 0$ for $k \in \mathbb{N}^*$ and $r > k$.

PROOF.

1. It follows from the definition of the dual Leibniz product of TN.

2. Let $n_1 \otimes \cdots \otimes n_p \in F^{p,*}TN$. As $\widetilde{\Phi}$ is a map of graded dual Leibniz algebras, we have $\widetilde{\Phi}(n_1 \otimes \cdots \otimes n_p) = \widetilde{\Phi}(n_1)\big(\widetilde{\Phi}(n_2)\big(\cdots\big(\widetilde{\Phi}(n_{p-1})\widetilde{\Phi}(n_p)\big)\cdots\big)$.

Since the dual Leibniz product of $T(N + N)$ is bigraded, (2) follows.

3. It follows from the definition of $\widetilde{\Phi}_r$.

4. This is a direct consequence of (2) and (3).

5. Let $n_1 \otimes \cdots \otimes n_r \in F^{r,*}TN$, then: $\widetilde{\Phi}_r(n_1 \otimes \cdots \otimes n_r) =$
$$\pi_{I_r}\big((n'_1 + n''_1 + \overline{\widetilde{\Phi}}(n_1))\big(\cdots\big((n'_{r-1} + n''_{r-1} + \overline{\widetilde{\Phi}}(n_{r-1}))(n'_r + n''_r + \overline{\widetilde{\Phi}}(n_r))\big)\cdots\big).$$

Since the length of $\overline{\widetilde{\Phi}}(n_i)$ is > 1, and since the dual Leibniz product of $T(N + N)$ is bigraded, it follows that

$$\widetilde{\Phi}_r(n_1 \otimes \cdots \otimes n_r)$$
$$= \pi_{I_r}\big((n'_1 + n''_1)\big(\cdots\big((n'_{r-1} + n''_{r-1})(n'_r + n''_r)\big)\big)\cdots\big)$$
$$= \pi_{I_r}\big(\sum_{I \in \mathcal{I}_r(',\,'')} n^I\big)$$
$$= n^{I_r} \qquad \text{which, from (1), yields (5).}$$

6. This is a direct consequence of the fact that $F^{p,q}TN = 0$ for $p > q$. $\quad\square$

LEMMA 3.6. *Let N be a graded sub-vector space of R such that $N_k = 0$ for $k \geq p+1$ and such that the map of graded dual Leibniz algebras $\eta : TN \to R$, which extends the inclusion, is an isomorphism in degree $\leq p$. Then η is one to one in degree $p + 1$.*

PROOF. The fact that N_{p+1} is a sub-vector space of R_{p+1} implies that η is one to one on $F^{1,p+1}TN = N_{p+1}$. Let $r \geq 2$, $a \in F^{r,p+1}TN$. Then by using (3.5.5), it follows that $\nabla\widetilde{\Phi}_r(a) = a$. It follows from the definition of Φ_T that $(\eta * \eta)\Phi_T(a) = \Phi\eta(a)$. But $\eta * \eta$ is one to one on $F^{r,p+1}(TN * TN)$ and ψ is a bigraded map, which gives rise to the following sequence of implications:
$$\eta(a) = 0 \Rightarrow \Phi\eta(a) = 0 \Rightarrow (\eta * \eta)\Phi_T(a) = 0 \Rightarrow \overline{\Phi}_T(a) = 0 \Rightarrow \psi\overline{\Phi}_T(a) = 0$$
$$\Rightarrow \widetilde{\Phi}_r(a) = 0$$
$$\Rightarrow a = 0 \quad \text{from (6.2.5).} \quad\square$$

PROOF OF (3.4). Let $a \in R_{p+1}$. Then $\Phi(a) = a' + a'' + \overline{\Phi}(a)$, where
$$\overline{\Phi}(a) : R_{p+1} \longrightarrow \bigoplus_{\substack{J \in \mathcal{R}(',\,'')\\ |J| \geq 2}} \big(R^I\big)_{p+1}.$$

As η is an isomorphism in degree $\leq p$, for any $I \in \mathcal{R}(',\,'')$ such that $|I| \geq 2$, $\eta * \eta$ is an isomorphism from $\big(TN^I\big)_{p+1}$ to $\big(R^I\big)_{p+1}$. Therefore, $\overline{\Phi}$ lifts to a map
$$\overline{\widetilde{\Phi}} : R_{p+1} \longrightarrow TN * TN \overset{\psi}{\simeq} T(N + N).$$

We define $\widetilde{\Phi}_r$ by $\widetilde{\Phi}_r = \pi_{I_r}\overline{\widetilde{\Phi}}$. Let us define N_{p+1} as follows:
$$N_{p+1} = \{a \in r_{p+1} \,;\, \forall r \geq 2, \widetilde{\Phi}_r(a) = 0\}.$$

Then $\eta : TN^{p+1} \longrightarrow R$ is an isomorphism in degree $\leq p$ and, in view of (3.6), it is one to one in degree $p + 1$.

Now, let us show that it is onto in degree $p + 1$. Let $a \in R_{p+1}$, and let us consider the following sequence in R_{p+1}:

$$a_1 = a, k \geq 2, \quad a_k = a_{k-1} - \eta \nabla \widetilde{\Phi}_k(a_{k-1}) \text{ for } k \geq 2.$$

Then, from $p + 1$, the sequence $(a_k)_{k \in \mathbb{N}}$ is stationary. Indeed from (3.5.6),

$$a_{p+2} = a_{p+1} - \eta \nabla \widetilde{\Phi}_{p+2}(a_{p+1}) = a_{p+1}.$$

Let us show by induction that $\widetilde{\Phi}_r(a_k) = 0$ for $r \leq k$. The first step of our induction is given by:

$$\nabla \widetilde{\Phi}_2(a_2) = \nabla \widetilde{\Phi}_2(a_1) - \nabla \widetilde{\Phi}_2 \nabla \widetilde{\Phi}_2(a_1) \quad \text{from (3.5.5)},$$
$$= \nabla \widetilde{\Phi}_2(a_1) - \nabla \widetilde{\Phi}_2(a_1)$$
$$= 0,$$

which combined with (3.5.1), yields $\widetilde{\Phi}_2(a_2) = 0$. Now, assume that $\widetilde{\Phi}_r(a_k) = 0$ for $r \leq k$. Then

$$\widetilde{\Phi}_r(a_{k+1}) = \widetilde{\Phi}_r(a_k) - \widetilde{\Phi}_r \nabla \widetilde{\Phi}_{k+1}(a_k)$$
$$= \widetilde{\Phi}_r(a_k) \quad \text{from (3.5.4)},$$
$$= 0.$$

$$\nabla \widetilde{\Phi}_{k+1}(a_{k+1}) = \nabla \widetilde{\Phi}_{k+1}(a_k) - \nabla \widetilde{\Phi}_{k+1} \nabla \widetilde{\Phi}_{k+1}(a_k)$$
$$= \nabla \widetilde{\Phi}_{k+1}(a_k) - \nabla \widetilde{\Phi}_{k+1}(a_k) \quad \text{from (3.5.5)},$$
$$= 0.$$

So, from (3.5.1), $\widetilde{\Phi}_{k+1}(a_{k+1}) = 0$. It follows from (3.5.4) that $\widetilde{\Phi}_r(a_{p+1}) = 0$ for all $r \geq 2$, that is, $a_{p+1} \in N_{p+1}$. But by construction, $a_{k+1} - a_k \in Im(\eta)$ for any k. $\square$

REMARK 3.7. A similar result as (3.3) for the category of connected graded associative algebras stems from the work of I. Berstein [B]. The proof of (3.3) that we developed is an adaptation to the case of dual Leibniz algebras of the proof given by Berstein in [B]. In fact his proof can be adapted to the category of connected graded $\mathcal{P}$-algebras for any operad $\mathcal{P}$ such that for any $n \in \mathbb{N}^*$, $\mathcal{P}(n)$ is a free $k[S_n]$-module.

PROPOSITION 3.8. *Let* (R, Φ) *be an associative comultiplicative object in the category of connected graded dual Leibniz algebras, and let S be the graded sub-vector space of R defined as follows:*

$$S = \{a \in R \; ; \Phi(a) = a' + a'' + \sum_{(1),(2)} a'_{(1)} \otimes a''_{(2)} \}.$$

Then the map of graded dual Leibniz algebras $\eta : TS \to R$ *is an isomorphism.*

PROOF. It follows from (3.3) that $R \simeq TN$. Let us show that $N = S$. We get the inclusion $S \subset N$ by definition of N and S. The associativity of Φ is equivalent to the commutativity of the following diagram:

$$
\begin{array}{ccc}
TN & \xrightarrow{\widetilde{\Phi}} & T(N' + N'') \\
\widetilde{\Phi} \downarrow & & \downarrow (\widetilde{\Phi},1) \\
T(N' + N'') & \xrightarrow[(1,\widetilde{\Phi})]{} & T(N' + N'' + N''')
\end{array}
$$

where $(\widetilde{\Phi}, 1)$ (resp. $(1, \widetilde{\Phi})$) is the unique map of graded dual Leibniz algebras which extends $\widetilde{\Phi}$ (resp. 1) on N' and 1 (resp. $\widetilde{\Phi}$) on N''. Let $\mathcal{I}(','',''')$ (resp. $\mathcal{M}(','',''')$)

denote the set of the sequences of finite length in $'$, $''$ and $'''$ (resp. mixed sequences). Then, as graded vector spaces: $T(N' + N'' + N''') \simeq \bigoplus_{I \in \mathcal{I}(',''',''')} N^I$

For any $I \in \mathcal{I}(',''',''')$ and for any linear map φ to $T(N'+N''+N''')$, the projection of φ on N^I is denoted by φ_I.

LEMMA 3.9. *Let $n \geq 2$. Let us consider the following notations:*
$$\mathcal{N}_n(',''',''') = \{I \in \mathcal{M}(',''','''); \ |I| = n \ and \ i_1 \neq '\}$$
$$\mathcal{N}_n(',''') = \{J \in \mathcal{M}(','''); \ |J| = n \ and \ J \neq (','',''',\cdots,'')\}.$$
Let α (resp. β) be the map from $\mathcal{N}_n(',''',''')$ to $\mathcal{N}_n(',''')$ which consists in transforming $'$ and $''$ to $'$, and $'''$ to $''$ (resp. $'$ to $'$, and $''$ and $'''$ to $''$).

Let $\sim$ be the smallest equivalence relation of $\mathcal{N}_n(',''')$ which contains the binary relation $\mathcal{R}$ defined as follows: let $J, K \in \mathcal{N}_n(',''')$,
$$J \mathcal{R} K \iff \exists I \in \mathcal{N}_n(',''','''); \ \alpha(I) = J \ and \ \beta(I) = K.$$
Then all elements of $\mathcal{N}_n(',''')$ are equivalent under $\sim$.

PROOF. Let us show that every $J \in \mathcal{N}_n(',''')$ belongs to the equivalence class of the unique alternating sequence in $'$ and $''$ of length n that we denote by L. We consider the following order defined on $\{',''','''\}$: $' <'' <'''$.

For $J = (j_1, \cdots, j_n) \neq K = (k_1, \cdots, k_n) \in \mathcal{N}_n(',''')$ such that $j_1 = k_1 =''$ and $j_\gamma \leq k_\gamma$ for $\gamma > 1$, let $I = (i_1, \cdots, i_n) \in \mathcal{I}_n(',''',''')$ be the sequence defined by:
$$i_\gamma =' \ \text{if} \ j_\gamma = k_\gamma =',$$
$$i_\gamma ='' \ \text{if} \ j_\gamma =' \ \text{and} \ k_\gamma ='',$$
$$i_\gamma =''' \ \text{if} \ j_\gamma = k_\gamma ='' .$$
As $J \neq K$ and since for any γ, $j_\gamma \leq k_\gamma$, I belongs to $\mathcal{M}(',''',''')$. Since $j_1 = k_1 =''$, it follows that $I \in \mathcal{N}_n(',''',''')$. Moreover, it is easy to check that $\alpha(I) = J$ and $\beta(I) = K$.

Let $J \in \mathcal{N}_n(',''')$ such that $j_1 ='$. Then $K = (''', i_2, \cdots, i_n) \in \mathcal{N}_n(',''')$. By taking I defined as above, we easily check that $I \in \mathcal{N}_n(',''',''')$, $\alpha(I) = J$ and $\beta(I) = K$.

Now, let $K = (k_1, \cdots, k_n) \neq L \in \mathcal{N}_n(',''')$ such that $j_1 =''$. We define $J \in \mathcal{I}(',''')$ by: $j_\gamma = \inf(k_\gamma, l_\gamma)$. Then $J \in \mathcal{N}_n(',''')$ and from the above relations we get that $J \mathcal{R} K$ and $J \mathcal{R} L$. □

PROOF OF (3.8). Let $n \in N$. Then one has: $\widetilde{\Phi}(n) = n' + n'' + \sum_{\substack{J \in \mathcal{R}(',''') \\ |J| \geq 2}} \widetilde{\Phi}_J(n).$

Let us show that $\widetilde{\Phi}_J(n) = 0$ for $p \geq 2$ and $J \in \mathcal{N}_p(',''')$. This will give the inclusion $N \subset S$. We proceed by induction on p. The first step ($p = 2$) is given by the definition of N itself. By the same arguments as in (3.5.5), we check that:

for any $I \in \mathcal{I}(',''',''')$, and any $J \in \mathcal{I}(',''')$ such that $|I| = |J|$,

(i)
$$(\widetilde{\Phi}, 1)_I N^J = 0 \ \text{if} \ J \neq \alpha(I) \quad \text{and} \quad (\widetilde{\Phi}, 1)_I : N^{\alpha(I)} \xrightarrow{\sim} N^I$$
$$(1, \widetilde{\Phi})_I N^J = 0 \ \text{if} \ J \neq \beta(I) \quad \text{and} \quad (1, \widetilde{\Phi})_I : N^{\beta(I)} \xrightarrow{\sim} N^I.$$

Let J_k be the unique sequence of length k of the type $(','',\cdots,'')$ and let $\Delta_k = \widetilde{\Phi}_{J_k}$. Then, from the induction hypothesis, we get:
$$\widetilde{\Phi}(n) = n' + n'' + \sum_{2 \leq k \leq p} \Delta_k(n) + \sum_{\substack{J \in \mathcal{R}(',''') \\ |J| \geq p+1}} \widetilde{\Phi}_J(n).$$

Thanks to the definition of the dual Leibniz product on $T(N' + N'' + N''')$, we easily check that for $I \in \mathcal{N}_{p+1}(',\,'',\,''')$,

$$(\widetilde{\Phi}, 1)_I \big(n' + n'' + \sum_{2 \le k \le p} \Delta_k(n)\big) = (1, \widetilde{\Phi})_I \big(n' + n'' + \sum_{2 \le k \le p} \Delta_k(n)\big) = 0.$$

Therefore, for $I \in \mathcal{N}_{p+1}(',\,'',\,''')$,

$$(\widetilde{\Phi}, 1)_I \widetilde{\Phi}(n) = (\widetilde{\Phi}, 1)_I \widetilde{\Phi}_{\alpha(I)}(n)$$

$$= (\widetilde{\Phi}, 1)_I \widetilde{\Phi}(n) = (1, \widetilde{\Phi})_I \widetilde{\Phi}_{\beta(I)}(n).$$

Now let us consider the equivalence relation on $\mathcal{N}_{p+1}(',\,'')$ defined as follows: let $J, K \in \mathcal{N}_{p+1}(',\,'')$; $J \simeq K \iff Ker(\widetilde{\Phi}_J) \cap N = Ker(\widetilde{\Phi}_K) \cap N$. Then, from (i), $\simeq$ contains $\mathcal{R}$ and thus contains $\sim$. So, we get from (3.9) that $J \simeq K$ for all $J, K \in \mathcal{N}_{p+1}(',\,'')$. But by definition of N, one has $Ker(\widetilde{\Phi}_L) \cap N = N$. $\square$

DEFINITION 3.10. A graded Leibniz coalgebra is a graded vector space C, given with a cobracket $\Delta : C \to C \otimes C$ which satisfies the following identity:
$$(1 \otimes \Delta)\Delta = (\Delta \otimes 1)\Delta - (1 \otimes \tau)(\Delta \otimes 1)\Delta.$$

PROPOSITION 3.11. *Let R be an associative comultiplicative object in the category of connected graded dual Leibniz algebras. Let $S \subset R$ be the graded sub-vector space of (3.8) and let $\Delta : S \longrightarrow S' \otimes S''$ be defined by $\Delta = \widetilde{\Phi}_{(',\,'')}$. Then (S, Δ) is a graded Leibniz coalgebra.*

PROOF. From the associativity of Φ, one gets for any $a \in S$,

$$(\widetilde{\Phi}, 1)_{(',\,'',\,''')} \widetilde{\Phi}(a) = (1, \widetilde{\Phi})_{(',\,'',\,''')} \widetilde{\Phi}(a) \text{ and } (\widetilde{\Phi}, 1)_{(',\,''',\,'')} \widetilde{\Phi}(a) = (1, \widetilde{\Phi})_{(',\,''',\,'')} \widetilde{\Phi}(a).$$

Let $a \in S$; we use the following notations:

$$\Delta(a) = \sum_{(1),(2)} a'_{(1)} \otimes a''_{(2)},$$

$$\widetilde{\Phi}_{(',\,'',\,'')}(a) = \sum_{(\alpha),(\beta),(\gamma)} a'_{(\alpha)} \otimes a''_{(\beta)} \otimes a''_{(\gamma)}.$$

Then, for any $a \in S$,

$$(\widetilde{\Phi}, 1)_{(',\,'',\,''')} \widetilde{\Phi}(a) = \sum_{(1),(2)} \sum_{(11),(12)} a'_{(11)} \otimes a''_{(12)} \otimes a'''_{(2)}$$

$$= (1, \widetilde{\Phi})_{(',\,'',\,''')} \widetilde{\Phi}(a)$$

$$= \sum_{(1),(2)} \sum_{(21),(22)} a'_{(1)} \otimes a''_{(21)} \otimes a'''_{(22)} + \sum_{(\alpha),(\beta),(\gamma)} a'_{(\alpha)} \otimes a''_{(\beta)} \otimes a'''_{(\gamma)}.$$

$$(\widetilde{\Phi}, 1)_{(',\,''',\,'')} \widetilde{\Phi}(a) = \sum_{(1),(2)} \sum_{(11),(12)} (-1)^{|a_{(12)}||a_{(2)}|} a'_{(11)} \otimes a'''_{(2)} \otimes a''_{(12)}$$

$$= (1, \widetilde{\Phi})_{(',\,''',\,'')} \widetilde{\Phi}(a) = \sum_{(\alpha),(\beta),(\gamma)} a'_{(\alpha)} \otimes a'''_{(\beta)} \otimes a''_{(\gamma)}.$$

From the last equality, it follows that $\widetilde{\Phi}_{(',\,'',\,'')}(a) = (1 \otimes \tau)(\Delta \otimes 1)\Delta(a)$. Hence, the first one gives $(\Delta \otimes 1)\Delta = (1 \otimes \Delta)\Delta + (1 \otimes \tau)(\Delta \otimes 1)\Delta$. $\square$

Let $CGDL$-$cogrp$ be the category of cogroups in the category of connected graded dual Leibniz algebras, and let CGL-$coalg$ be the category of connected graded Leibniz coalgebras.

PROPOSITION 3.12. *The functor* $\quad CGDL\text{-}cogrp \longrightarrow CGL\text{-}coalg$

$$(R, \Phi) \quad \longrightarrow \quad (S, \Delta)$$

is full and faithful.

PROOF. Given (R, Φ) a cogroup in the category $CGDL\text{-}alg$, we need to prove that the Leibniz coproduct Δ gives rise to a complete determination of Φ on TS. To this end, we only need to show that Φ is completely determined by Δ on S, as Φ is a map of dual Leibniz algebras.

By definition of S, for $a \in S$, $\widetilde{\Phi}(a) = a' + a'' + \sum\limits_{2 \leq k} \Delta_k(a)$,

with Δ_k as it is defined in the proof of (3.8). Let K_r be the unique sequence of length r such that $k_1 = '$, $k_2 = '''$ and $k_l = ''$ for $l > 2$. As $'''$-marked factors of $(\widetilde{\Phi}, 1)\, \widetilde{\Phi}(a)$ come from the $''$-marked factors of $\widetilde{\Phi}(a)$, and since only one $'''$ is involved in K_r, the following identity holds:

$$(\widetilde{\Phi}, 1)_{K_r} \widetilde{\Phi}(a) = \big(\prod_{1 \leq i \leq r-2} 1^{\otimes^i} \otimes T \otimes 1^{\otimes^{r-2-i}} \big)(\Delta_{r-1} \otimes 1)\Delta.$$

As for any $x, y \in V' \oplus V'' \oplus V'''$, and for any $\omega_1, \omega_2 \in T(V' \oplus V'' \oplus V''')$,

$$x\big((y \otimes \omega_1)\omega_2\big) = x \otimes y \otimes (\omega_1 * \omega_2),$$

and since only one $'''$ is involved in K_r, we get from the above expression of $\widetilde{\Phi}(a)$ the following identity:

$$(1, \widetilde{\Phi})_{K_r} \widetilde{\Phi}(a) = \Delta_r \ ', \ '' \text{ and } ''' \text{ being placed accordingly.}$$

Thus, the associativity of Φ gives us an inductive expression of Δ_r using Δ. Therefore Δ completely determines the cogroup structure of R. This gives us the faithfulness of our functor. In order to get the fullness, we need to see that any morphism between two (S, Δ) is provided by a morphism between the initial cogroups, which is now straightforward. $\square$

COROLLARY 3.13. *Let $CGDL\text{-}ab$ be the category of abelian cogroups in $CGDL$ and let $CGVect$ be the category of connected graded vector spaces. Then the functor*

$$CGDL\text{-}ab \longrightarrow CGVect$$
$$(R, \Phi) \longrightarrow S$$

is an equivalence of categories.

PROOF. As the twist map $T : R * R \to R * R$ consists in changing $'$ to $''$ and vice versa, one easily checks that the functor $\quad CGVect \longrightarrow CGDL\text{-}ab$
$$V \longrightarrow (TV, \Phi)$$
is a quasi-inverse of the previous functor, where $\Phi : TV \longrightarrow TV * TV$ is the unique dual Leibniz algebras map which extends the map $\quad V \longrightarrow TV * TV$
$$v \longmapsto v' + v''$$
to TV. $\square$

CONJECTURE 3.14: ON COGROUPS IN THE CATEGORY OF $\mathcal{P}$-ALGEBRAS.

Let k be a characteristic zero field, let $\mathcal{P}$ be a quadratic operad over k and let $\mathcal{P}^!$ be its Koszul dual ([G-K]). Let $\mathcal{P}CG\text{-}cogrp$ denote the category of cogroups in the category of connected graded $\mathcal{P}$-algebras, and let $\mathcal{P}^!CG\text{-}coalg$ denote the category of connected graded $\mathcal{P}^!$-coalgebras. We expect that an equivalence of categories is furnished between: $\qquad \mathcal{P}CG\text{-}cogrp \simeq \mathcal{P}^!CG\text{-}coalg.$

REMARKS 3.15: ON THE CONJECTURE.

1- Berstein has proved this statement for $\mathcal{P} = \mathcal{A}ss$. In this paper, we have constructed a full and faithful functor in the case $\mathcal{P} = \mathcal{L}eib^!$, where $\mathcal{L}eib$ stands for the operad of Leibniz algebras. We expect it to be an equivalence of categories.

Finally, the Cartan-Milnor-Moore theorem, [M-M], states an equivalence of categories between the category of connected graded cocommutative Hopf algebras and the category of connected graded Lie algebras. Since cocommutative Hopf algebras are exactly the group-objects in the category cocommutative coalgebras, the Cartan- Milnor-Moore theorem states an equivalence between the category of group-objects in the category of connected graded cocommutative coalgebras and the category of connected graded Lie algebras. Thus, the Cartan-Milnor-Moore theorem is exactly the dual statement of what we count on for $\mathcal{P} = \mathcal{C}omm$.

2- As explained in (3.7), we got partial results in the case of an operad which is free as a $k[S_n]$-module in each degree. Precisely, we know that a comultiplicative object is a free object.

4. Leibniz homology and cohomology of the Lie algebra of matrices

This part is devoted to a new proof of the Cuvier-Loday theorem and an interpretation of $HL^*(\mathfrak{gl}(A))$ in terms of cogroups in the category of graded dual Leibniz algebras. From now on, k will be of characteristic zero.

THEOREM 4.1 ([C],[L3]). *Let k be a characteristic zero field, and A an associative k-algebra with unit. Then there is an isomorphism of graded vector spaces:*
$$HL_*(\mathfrak{gl}(A)) \simeq THH_{*-1}(A).$$

SKETCH OF PROOF. The complex which calculates $HL_*(\mathfrak{gl}(A))$ is the following:
$$\cdots \longrightarrow \mathfrak{gl}(A)^{\otimes n} \longrightarrow \mathfrak{gl}(A)^{\otimes n-1} \longrightarrow \cdots \longrightarrow \mathfrak{gl}(A)$$
where the differential map was given earlier in the introduction. For convenience, this differential map will be denoted by d.

One can easily check that d commutes with the left action of $\mathfrak{gl}(A)$ and moreover, this action is homotopic to zero ([C],[L3]). Now for any integer r, $\mathfrak{gl}_r(k)$ is a reductive subalgebra of $\mathfrak{gl}_r(A)$. Therefore by taking the coinvariants under $\mathfrak{gl}_r(k)$, one gets a quasi-isomorphic complex ([L3]):

$$
\begin{array}{ccccccc}
\cdots \longrightarrow & \mathfrak{gl}_r(A)^{\otimes n} & \xrightarrow{d} & \mathfrak{gl}_r(A)^{\otimes n-1} & \longrightarrow \cdots \longrightarrow & \mathfrak{gl}_r(A) \\
& \downarrow & & \downarrow & & \downarrow \\
\cdots \longrightarrow & \left(\mathfrak{gl}_r(A)^{\otimes n}\right)_{\mathfrak{gl}_r} & \xrightarrow{d} & \left(\mathfrak{gl}_r(A)^{\otimes n-1}\right)_{\mathfrak{gl}_r} & \longrightarrow \cdots \longrightarrow & \left(\mathfrak{gl}_r(A)\right)_{\mathfrak{gl}_r}
\end{array}
$$

By taking the limit and by using the invariant theory ([L3]), one obtains the following quasi-isomorphism of complexes:

$$
\begin{array}{ccccccc}
\cdots \longrightarrow & \mathfrak{gl}(A)^{\otimes n} & \xrightarrow{d} & \mathfrak{gl}(A)^{\otimes n-1} & \longrightarrow \cdots \longrightarrow & \mathfrak{gl}(A) \\
& \downarrow & & \downarrow & & \downarrow \\
\cdots \longrightarrow & k[S_n] \otimes A^{\otimes n} & \xrightarrow{d} & k[S_{n-1}] \otimes A^{\otimes n-1} & \longrightarrow \cdots \longrightarrow & A
\end{array}
$$

Let us denote by L_* the last complex and T the quasi-isomorphism. The following lemma gives a description of T using elementary matrices.

LEMMA 4.2 ([L3]). *For any $\sigma \in S_n$ and $a_1, \cdots, a_n \in A$,*
$$T(E_{1,\sigma(1)}^{a_1} \otimes \cdots \otimes E_{n,\sigma(n)}^{a_n}) = \sigma \otimes (a_1 \otimes \cdots \otimes a_n). \quad \square$$

Our approach will now differ from Cuvier's and Loday's as regards the proof of (4.1). It consists in gathering their steps together by defining a filtration on L_*. To this end, we propose the following procedure: we will first define the degree of the filtration and then show that it is a filtration on the complex L_*. The result will then be established by examining the spectral sequence associated to the filtration.

NOTATION 4.3. Let $\omega \in S_n$. We define $deg(\omega) = \#\{i < j \mid \omega(i) > \omega(j)\}$. Let $\sigma \in S_n$. We consider the following decomposition of σ into products of cycles:

$$\sigma = (\omega_1, \omega_2, \cdots, \omega_{\alpha(2)-1})(\omega_{\alpha(2)}, \cdots, \omega_{\alpha(3)-1}) \cdots\cdots (\omega_{\alpha(r)}, \cdots, \omega_n)$$

where $1 = \omega_1 < \omega_{\alpha(2)} < \cdots < \omega_{\alpha(r)}$
and $\omega_{\alpha(i)} < \omega_k$ when $\alpha(i) \leq k \leq \alpha(i+1) - 1$.

We thus define a permutation $\omega(\sigma) \in S_n$ by $\omega(\sigma)(i) = \omega_i$ for $1 \leq i \leq n$, such that $\sigma = \omega(\sigma)\tau_\sigma \omega(\sigma)^{-1}$, where $\tau_\sigma = (1, \cdots, \alpha(2) - 1) \cdots\cdots (\alpha(r), \cdots, n)$.

DEFINITION 4.4. Let $\sigma \in S_n$; we define $|\sigma| \in \mathbb{N}$ by: $|\sigma| = deg(\omega(\sigma))$.

This degree $|\ |$ is extended to L_* by $|\sigma \otimes (a_1 \otimes \cdots \otimes a_n)| = |\sigma|$. Our goal is to show that it defines a filtration on the complex L_*.

DEFINITION 4.5. Let $\sigma \in S_n$ and $1 \leq i \leq n$ such that $\sigma(i) \neq i$. We define $d_i \sigma \in S_{n-1}$ by:

$$
\begin{aligned}
d_i\sigma(j) &= \sigma(j) & \text{when } j < \sigma(i) \text{ and } \sigma(j) < \sigma(i) \\
&= \sigma(j) - 1 & \text{when } j < \sigma(i) \text{ and } \sigma(j) > \sigma(i) \\
d_i\sigma(i) &= \sigma^2(i) & \text{when } i < \sigma(i) \text{ and } \sigma^2(i) < \sigma(i) \\
&= \sigma^2(i) - 1 & \text{when } i < \sigma(i) \text{ and } \sigma^2(i) > \sigma(i) \\
d_i\sigma(j) &= \sigma(j+1) & \text{when } j > \sigma(i) \text{ and } \sigma(j+1) < \sigma(i) \\
&= \sigma(j+1) - 1 & \text{when } j > \sigma(i) \text{ and } \sigma(j+1) > \sigma(i) \\
d_i\sigma(i-1) &= \sigma^2(i) & \text{when } i > \sigma(i) \text{ and } \sigma^2(i) < \sigma(i) \\
&= \sigma^2(i) - 1 & \text{when } i > \sigma(i) \text{ and } \sigma^2(i) > \sigma(i) .
\end{aligned}
$$

Let $\sigma \in S_n$ and $1 \leq i \leq n$ such that $i \neq \sigma^{-1}(i)$. We define $\overline{d}_i \sigma \in S_{n-1}$ by:

$$
\begin{aligned}
\overline{d}_i\sigma(j) &= \sigma(j) & \text{when } j < \sigma^{-1}(i) & \text{ and } \sigma(j) < \sigma^{-1}(i) \\
&= \sigma(j) - 1 & \text{when } j < \sigma^{-1}(i) & \text{ and } \sigma(j) > \sigma^{-1}(i) \\
\overline{d}_i\sigma(j) &= i & \text{when } \sigma(j) = \sigma^{-1}(i) & \text{ and } i < \sigma^{-1}(i) \\
&= i - 1 & \text{when } \sigma(j) = \sigma^{-1}(i) & \text{ and } i > \sigma^{-1}(i) \\
\overline{d}_i\sigma(j) &= \sigma(j+1) & \text{when } j > \sigma^{-1}(i) & \text{ and } \sigma(j+1) < \sigma^{-1}(i) \\
&= \sigma(j+1) - 1 & \text{when } j > \sigma^{-1}(i) & \text{ and } \sigma(j+1) > \sigma^{-1}(i) \\
\overline{d}_i\sigma(j) &= i & \text{when } \sigma(j+1) = \sigma^{-1}(i) & \text{ and } i < \sigma^{-1}(i) \\
&= i - 1 & \text{when } \sigma(j+1) = \sigma^{-1}(i) & \text{ and } i > \sigma^{-1}(i) .
\end{aligned}
$$

PROPOSITION 4.6. Let $\sigma \in S_n$ and let $a_1, \cdots, a_n \in A$. Then the following equality holds:

$$
\begin{aligned}
d(\sigma \otimes (a_1 \otimes \cdots \otimes a_n)) &= \sum_{1 \leq i < \sigma(i) \leq n} (-1)^{\sigma(i)+1} d_i\sigma \otimes (\cdots \otimes a_i a_{\sigma(i)} \otimes \cdots \hat{a}_{\sigma(i)} \cdots) \\
&+ \sum_{1 \leq i < \sigma^{-1}(i) \leq n} (-1)^{\sigma^{-1}(i)+1} \overline{d}_i\sigma \otimes (\cdots \otimes a_{\sigma^{-1}(i)} a_i \otimes \cdots \hat{a}_{\sigma^{-1}(i)} \cdots).
\end{aligned}
$$

PROOF. Using elementary properties of the elementary matrices, one easily gets that

$$
\begin{aligned}
d(E^{a_1}_{1,\sigma(1)} \otimes \cdots \otimes E^{a_n}_{n,\sigma(n)}) &= \sum_{1 \leq i < \sigma(i) \leq n} (-1)^{\sigma(i)+1} \cdots \otimes E^{a_i a_{\sigma(i)}}_{i,\sigma^2(i)} \otimes \cdots \hat{E}^{a_{\sigma(i)}}_{\sigma(i),\sigma^2(i)} \cdots \\
&+ \sum_{1 \leq i < \sigma-1(i) \leq n} (-1)^{\sigma^{-1}(i)+1} \cdots \otimes E^{a_{\sigma^{-1}(i)} a_i}_{\sigma^{-1}(i),\sigma(i)} \otimes \cdots \hat{E}^{a_{\sigma^{-1}(i)}}_{\sigma^{-1}(i),i} \cdots
\end{aligned}
$$

which gives us the result by taking the coinvariants. $\square$

LEMMA 4.7. *1- Let $\sigma \in S_n$, $|d_i\sigma| \leq |\sigma|$ for all $i \neq \sigma(i)$.*
          2- Let $\sigma \in S_n$, $|\overline{d}_i\sigma| \leq |\sigma|$ for all $i \neq \sigma^{-1}(i)$.

PROOF.

1- We easily check that $\omega(d_i\sigma) = \delta_i\omega$, where $\delta_i\omega$ is defined as follows:

$$\delta_i\omega(j) = \omega(j) \qquad \text{for } j < \omega^{-1}\sigma(i) \text{ such that } \omega(j) < \sigma(i)$$
$$= \omega(j) - 1 \qquad \text{for } j < \omega^{-1}\sigma(i) \text{ such that } \omega(j) > \sigma(i)$$
$$\delta_i\omega(j) = \omega(j+1) \qquad \text{for } j \geq \omega^{-1}\sigma(i) \text{ such that } \omega(j+1) < \sigma(i)$$
$$= \omega(j+1) - 1 \text{ for } j \geq \omega^{-1}\sigma(i) \text{ such that } \omega(j+1) > \sigma(i) .$$

But
$$\{k < l \mid \delta_i\omega(k) > \delta_i\omega(l)\}$$
$$= \{k < l \mid k < \omega^{-1}\sigma(i), l < \omega^{-1}\sigma(i), \omega(k) > \omega(l)\}$$
$$\cup \{k < l \mid k < \omega^{-1}\sigma(i), l \geq \omega^{-1}\sigma(i), \omega(k) > \omega(l+1)\}$$
$$\cup \{k < l \mid k \geq \omega^{-1}\sigma(i), l \geq \omega^{-1}\sigma(i), \omega(k+1) > \omega(l+1)\}$$
$$= \{k < l \mid \omega(k) > \omega(l)\} \setminus (\{k < \omega^{-1}\sigma(i) \mid \omega(k) > \sigma(i)\}$$
$$\cup \{l > \omega^{-1}\sigma(i) \mid \omega(l) < \sigma(i)\})$$

2- We easily check that $\omega(\overline{d}_i\sigma) = \overline{\delta}_i\omega$ where $\overline{\delta}_i\omega$ is defined as follows:

$$\overline{\delta}_i\omega(j) = \omega(j) \qquad \text{for } j < \omega^{-1}\sigma^{-1}(i) \text{ such that } \omega(j) < \sigma^{-1}(i)$$
$$= \omega(j) - 1 \qquad \text{for } j < \omega^{-1}\sigma^{-1}(i) \text{ such that } \omega(j) > \sigma^{-1}(i)$$
$$\overline{\delta}_i\omega(j) = \omega(j+1) \qquad \text{for } j \geq \omega^{-1}\sigma^{-1}(i) \text{ such that } \omega(j+1) < \sigma^{-1}(i)$$
$$= \omega(j+1) - 1 \text{ for } j \geq \omega^{-1}\sigma^{-1}(i) \text{ such that } \omega(j+1) > \sigma^{-1}(i) .$$

Furthermore,
$$\{k < l \mid \overline{\delta}_i\omega(k) > \overline{\delta}_i\omega(l)\} = \{k < l \mid k < \omega^{-1}\sigma^{-1}(i), l < \omega^{-1}\sigma^{-1}(i), \omega(k) > \omega(l)\}$$
$$\cup \{k < l \mid k < \omega^{-1}\sigma^{-1}(i), l \geq \omega^{-1}\sigma^{-1}(i), \omega(k) > \omega(l+1)\}$$
$$\cup \{k < l \mid k \geq \omega^{-1}\sigma^{-1}(i), l \geq \omega^{-1}\sigma^{-1}(i), \omega(k+1) > \omega(l+1)\}$$
$$= \{k < l \mid \omega(k) > \omega(l)\} \setminus (\{k < \omega^{-1}\sigma^{-1}(i) \mid \omega(k) > \sigma^{-1}(i)\}$$
$$\cup \{l > \omega^{-1}\sigma^{-1}(i) \mid \omega(l) < \sigma^{-1}(i)\}). \quad \square$$

Thus we have a filtration on the complex L_*, which is denoted by F_p.

DEFINITION 4.8. Let $\sigma \in S_n$ such that $|\sigma| \geq 1$, and let i be the first integer such that $\sigma(i) > i + 1$. We define $s(\sigma) \in S_{n+1}$ by:

$$s(\sigma)(j) = \sigma(j) \quad \text{for } j < i$$
$$s(\sigma)(i) = i + 1$$
$$s(\sigma)(j) = \sigma(j-1) \quad \text{when } j < i \text{ and } \sigma(j-1) \leq i$$
$$s(\sigma)(j) = \sigma(j-1) + 1 \quad \text{when } j < i \text{ and } \sigma(j-1) > i.$$

LEMMA 4.9. *With the same data as in the above definition,*
  1- $d_j s(\sigma) = s(d_j\sigma)$ for $j < \sigma(j) \leq i$.
  2- $d_i s(\sigma) = \sigma$.
  3- $d_j s(\sigma) = s(d_{j-1}\sigma)$ for $i < j < \sigma(j-1) + 1$.
  4- $|d_{i+1}s(\sigma)| < |\sigma|$ and $|d_i\sigma| < |\sigma|$.
  5- $\overline{d}_j s(\sigma) = s(\overline{d}_j\sigma)$ for $j < \sigma^{-1}(j)$ and $j \leq i$.

6- $\bar{d}_j s(\sigma) = s(\bar{d}_{j-1}\sigma)$ for $i < j < \sigma^{-1}(j-1) + 1$.

PROOF.

1- We check that:

$d_j s(\sigma)(k) = s(d_j\sigma)(k)$

$= \sigma(k)$ for $k < j$,

$= \sigma(j+1)$ for $k = j < i-1$ and $\sigma(j+1) < \sigma(j) = j+1$,

$= \sigma(j+1) - 1$ for $k = j < i-1$ and $\sigma(j+1) = j+2 > \sigma(j) = j+1$,

$= \sigma(k+1)$ for $j < k < i-1$ and $\sigma(k+1) < \sigma(j) = j+1$,

$= \sigma(k+1) - 1$ for $j < k < i-1$ and $\sigma(k+1) > \sigma(j) = j+1$,

$= \sigma(k)$ for $i \leq k$ and $\sigma(k) < \sigma(j) = j+1$,

$= \sigma(k) - 1$ for $i \leq k$ and $\sigma(j) = j+1 \leq \sigma(k) \leq i$,

$= \sigma(k)$ for $i \leq k$ and $\sigma(k) > i$.

The assertions 2-, 3-,5- and 6-, are proved by similar verifications.

4- We just have to state that for any $\sigma \in S_n$ such that $|\sigma| \geq 1$ and when i is the lowest index such that $\sigma(i) > i+1$, then we have $|d_i\sigma| < |\sigma|$.

We saw above in the proof of (4.7) that

$\{k < l \,|\, \omega(d_i\sigma)(k) > \omega(d_i\sigma)(l)\}$

$= \{k < l \,|\, \omega(k) > \omega(l)\} \setminus \big(\{k < \omega^{-1}\sigma(i) \,|\, \omega(k) > \sigma(i)\}$

$\cup \{l > \omega^{-1}\sigma(i) \,|\, \omega(l) < \sigma(i)\}\big)$

$= \{k < l \,|\, \omega(k) > \omega(l)\} \setminus \{l > \omega^{-1}\sigma(i) \,|\, \omega(l) < \sigma(i)\}$.

Moreover, we easily get that $\omega^{-1}(i+1) \in \{l > \omega^{-1}\sigma(i) \,|\, \omega(l) < \sigma(i)\}$. Thus this previous set is not empty. □

PROOF OF (4.1). As L_* is a filtered complex and since its filtration is complete and bounded below, there is a spectral sequence:

$E^0_{p,q} = \big(F_p L_*/F_{p-1}L_*\big)_{p+q} \implies HL_*(\mathfrak{gl}(A))$

where the differential of $E^0_{p,q}$ is induced by the differential of L_*. Now, for any $p \geq 1$, the complex $(F_p L_*/F_{p-1}L_*, d)$ is acyclic. Indeed, let us define

$s : (F_p L_*/F_{p-1}L_*)_n \longrightarrow (F_p L_*/F_{p-1}L_*)_{n+1}$

as follows: for any $\sigma \in S_n$ such that $|\sigma| = p$ and for $a_1, \cdots, a_n \in A$,

$s\big(\sigma \otimes (a_1 \otimes \cdots \otimes a_n)\big) = (-1)^{i+1} s(\sigma) \otimes (a_1 \otimes \cdots \otimes a_i \otimes 1 \otimes a_{i+1} \otimes \cdots \otimes a_n)$,

where i is the smallest integer such that $\sigma(i) > i+1$. Then by using (4.9) and (4.6), the following identity holds in $F_p L_*/F_{p-1}L_*$:

$ds\big(\sigma \otimes (a_1 \otimes \cdots \otimes a_n)\big) + sd\big(\sigma \otimes (a_1 \otimes \cdots \otimes a_n)\big) = \sigma \otimes (a_1 \otimes \cdots \otimes a_n)$.

Therefore, s is a homotopy from the identity to zero on $F_p L_*/F_{p-1}L_*$, which implies that $E^1_{p,q} \simeq 0$ for $p > 0$ and $q \geq 0$. This leads us to the following isomorphism of graded vector spaces: $HL_*(\mathfrak{gl}(A)) \simeq E^1_{0,*}$. It remains to see that $E^1_{0,*} \simeq THH_{*-1}(A)$, which follows from the identification of complexes:

$F_0 L_* \overset{\sim}{\longrightarrow} TC_{*-1}(A)$

$(1, \cdots, \alpha_2 - 1) \cdots (\alpha_r, \cdots, n) \otimes (a_1 \otimes \cdots \otimes a_n) \longmapsto (a_1 \otimes \cdots \otimes a_{\alpha_2 - 1}) \otimes \cdots \otimes (a_{\alpha_r} \otimes \cdots \otimes a_n)$,

where $C_{*-1}(A)$ is the suspension of the Hochschild complex of A. (Note that Lemma (4.6) ensures that it is an isomorphism of complexes.) □

REMARK. Another proof of Theorem 4.1 was also given by Jerry Lodder in [Lo] by using a multisimplicial structure on the complex L_*.

The end of this paper is devoted to an interpretation of $HL^*(\mathfrak{gl}(A))$ in terms of cogroups in the category of connected graded dual Leibniz algebras. To this end, we have to assume that A is of finite dimension. But we can again avoid this finiteness condition by working in the dual Leibniz coalgebra framework (cf. (1.10.2), (2.5.2)). Let $HH^*(A)$ denote the dual graded vector space of $HH_*(A)$. In terms of Hochschild cohomology, $HH^*(A) = H^*(A, A^*)$ [L4].

COROLLARY 4.10. *Let k be a field of characteristic zero and let A be an associative k-algebra with unit, which is of finite dimension over k. Then there is an isomorphism of abelian cogroups in the category of connected graded dual Leibniz algebras as follows:*
$$HL^*(\mathfrak{gl}(A)) \simeq THH^{*-1}(A).$$

PROOF. For any two integers p and q, the direct sum of matrices defined as follows:

$$\mathfrak{gl}_p(A) \quad \times \quad \mathfrak{gl}_q(A) \quad \longrightarrow \quad \mathfrak{gl}_{p+q}(A)$$

$$\begin{pmatrix} \times & \times & \cdots \\ \times & \times & \cdots \\ \vdots & \vdots & \vdots \end{pmatrix}, \begin{pmatrix} + & + & \cdots \\ + & + & \cdots \\ \vdots & \vdots & \vdots \end{pmatrix} \longmapsto \begin{pmatrix} \times & 0 & \times & 0 & \cdots \\ 0 & + & 0 & + & \cdots \\ \times & 0 & \times & 0 & \cdots \\ 0 & + & 0 & + & \cdots \\ \vdots & \vdots & \vdots & \vdots & \vdots \end{pmatrix}$$

is a map of Lie algebras, and thus induces a map of graded dual Leibniz algebras:
$$C^*(\mathfrak{gl}_{p+q}(A)) \longrightarrow C^*(\mathfrak{gl}_p(A)) * C^*(\mathfrak{gl}_q(A)).$$
By taking the invariants and by taking the limit, this map defines a structure of abelian cogroup in the category of connected graded dual Leibniz algebras on $L^* = \sum_{n\geq 1} k[S_n] \otimes A^{*\otimes^n}$, the dual complex of L_*. Therefore $HL^*(\mathfrak{gl}(A))$ is an abelian cogroup in the category of connected graded dual Leibniz algebras and according to (3.13), there is a graded sub-vector space $S \subset HL^*(\mathfrak{gl}(A))$ such that $HL^*(\mathfrak{gl}(A)) \underset{CGDL-ab}{\simeq} TS$.

As $HL^*(\mathfrak{gl}(A))$ is the dual graded vector space of $HL_*(\mathfrak{gl}(A))$, it follows from (4.1) that $HL^*(\mathfrak{gl}(A)) \simeq THH^{*-1}(A)$. Therefore, one has an isomorphism of graded vector spaces $TS \simeq THH^{*-1}(A)$ and an easy induction permits us to conclude that $S \simeq HH^{*-1}(A)$. □

References

[B]   BERSTEIN, I., *On cogroups in the category of graded algebras*, Trans.Amer. Math. Soc. **115** (1965), 257–269.

[C]   CUVIER, C., *Homologie des algébres de Leibniz*, Ann. Sc. Ecol. Norm. 4ième série t.27 (1994), 1–45.

[E-H]   ECKMANN, B. and HILTON, P.-J., *Group-like structures in general categories I.*, Math. Ann. **145** (1962), 227–255.

[G-K]   GINZBURG, V. and KAPRANOV, M., *Koszul duality for operads*, Duke J. Math. **76** (1994), 203–272.

[G]   GNEDBAYE, A.V., *Cohomologie de l'algèbre de Leibniz $\mathcal{L}(S^2)$*, preprint IRMA Strasbourg (1994).

[K]   KOSZUL, J.-L., *Homologie et cohomologie des algèbres de Lie*, Bull. Soc. Math. France **78** (1950), 65–127.

[L1]   LODAY, J.-L., *Künneth-style formula for the homology of Leibniz algebras*, Math. Zeit-schrift **221** (1996), 41–47.

[L2]   LODAY, J.-L., *Cup-product for Leibniz cohomology and dual Leibniz algebras*, Math. Scand. **77** (1995), 189–196.

[L3]   LODAY, J.L., *Une version non commutative des algèbres de Lie: les algèbres de Leibniz*, Ens. Math. **39** (1993), 269–293.

[L4]   LODAY, J.L., *Cyclic Homology*, Springer Verlag, Grun. math. Wiss. 301, 1992.

[L-P]   LODAY, J.-L. and PIRASHVILI, T., *Universal enveloping algebras of Leibniz algebras and (co) homology*, Math. Ann. **296** (1993), 139–158.

[L-Q]   LODAY, J.-L. and QUILLEN, D., *Cyclic homology and the Lie algebra homology of matrices*, Comment. Math. Helvetici **59** (1984), 565–591.

[Lo]   LODDER, J.-M., *Leibniz homology and the James model*, Math. Nachrichten **175** (1995), 209–229.

[O]   OUDOM, J.-M., *La diagonale en homologie des algèbres de Leibniz*, C. R. Acad. Sci. Paris **t. 320, Série I** (1995), 1165–1170.

[M-M]   MILNOR, J. and MOORE, J.C., *On the structure of Hopf algebras*, Ann. Math. **81** (1965), 211–264.

[T]   TSYGAN, B.L., *The homology of the matrix Lie algebras over rings and the Hochschild homology*, Russ. Math. Survey **38** (1983), 198–199.

UNIVERSITÉ DE MONTPELLIER II, LABORATOIRE A.G.A.T.A, DÉPARTEMENT DES SCIENCES MATHÉMATIQUES, CASE 051, PLACE EUGÈNE BATAILLON 34095 MONTPELLIER CEDEX 5
E-mail address: oudom@math.univ-montp2.fr

Contemporary Mathematics
Volume **202**, 1997

Cohomology of monoids in monoidal categories

Hans-Joachim Baues, Mamuka Jibladze, and Andy Tonks

Introduction

It has been known for some time that the cohomology theories of many classical algebraic objects — monoids, groups, associative algebras and Lie algebras for
instance — have a common framework in terms of cohomology of internal monoids
in a symmetric monoidal category; see for example [**26**]. But there are also important examples of algebraic structures which occur as monoids in non-symmetric
monoidal categories, such as operads, monads, theories, categories, and square rings
as described below. In this article we show that these structures are still suscept-
ible to cohomological investigation, by developing the theory in the absence of the
symmetry condition. Later we shall assume that the monoidal structure is left distributive over coproducts and the category is an abelian category; this is the case
for operads, our original motivating example.

1. Monoids and Modules

We define monoids in monoidal categories and introduce the "module" objects
which will be used later as coefficients in the cohomology of such monoids. We
also give some of our motivating examples of monoidal categories and the monoids
therein.

Let us start by recalling that a monoidal category is a tuple $\mathbb{V} = (\mathbf{V}, \odot, I, a, l, r)$
where $\mathbf{V}$ is a category, $\odot : \mathbf{V} \times \mathbf{V} \to \mathbf{V}$ is a functor, I is an object of $\mathbf{V}$, and

$$
\begin{aligned}
a &= (a_{X,Y,Z} : (X \odot Y) \odot Z \to X \odot (Y \odot Z))_{X,Y,Z \in \mathbf{V}}, \\
l &= (l_X : I \odot X \to X)_{X \in \mathbf{V}}, \\
r &= (r_X : X \odot I \to X)_{X \in \mathbf{V}}
\end{aligned}
$$

are natural isomorphisms, required to satisfy certain conditions which we omit
here (see e.g. [**20**]). In many examples our monoidal categories will be strictly
associative and have strict units, in the sense that all $a_{X,Y,Z}$ and l_X, r_X are identity
morphisms. The monoidal category $\mathbb{V}$ is *abelian* if the underlying category $\mathbf{V}$ is an

---

1991 *Mathematics Subject Classification.* 18D10 18D35 18G50.
Key words and phrases. cohomology, internal monoids, operads.

abelian category. Suppose **V** has binary coproducts, denoted $X \sqcup Y$; then the monoidal structure is *left distributive* if the canonical natural transformation

$$(X_1 \otimes Y) \sqcup (X_2 \otimes Y) \to (X_1 \sqcup X_2) \otimes Y$$

is an isomorphism. Right distributivity is defined similarly.

A strict monoidal functor between monoidal categories is a functor between the underlying categories preserving all the existing structure in the obvious way.

Given such a $\mathbb{V}$, a monoid in $\mathbb{V}$, or a $\mathbb{V}$-monoid, is a triple $\mathcal{G} = (G, \mu, \eta)$ where $G \in \mathbf{V}$, $\mu : G \otimes G \to G$, $\eta : I \to G$ must satisfy the identities

$$\begin{aligned}
\mu(\mu \otimes G) &= \mu(G \otimes \mu)a_{G,G,G} && \text{(associativity),} \\
\mu(\eta \otimes G) &= l_G && \text{(left unit),} \\
\mu(G \otimes \eta) &= r_G && \text{(right unit).}
\end{aligned}$$

Basic examples of monoidal categories are the following:

EXAMPLE 1.1. Let **C** be any category with finite products. Then these products may be used to give it a monoidal structure $(\mathbf{C}, \times, 1, a, l, r)$, where $\times$ is the binary product, 1 is the terminal object (which exists as the empty product), and a, l, r are uniquely determined by the universal property of the products. A monoid in this monoidal category is what is usually called an internal monoid in a category with products.

Also in this "cartesian" situation one may define what it means for a monoid $\mathcal{G} = (G, \mu : G \times G \to G, \eta : 1 \to G)$ to be an internal group object: there must exist an endomorphism $\iota : G \to G$ satisfying

$$\mu(G \times \iota)d = \eta p = \mu(\iota \times G)d$$

where $d : X \to X \times X$ and $p : X \to 1$ are the canonical morphisms (which are only available in the cartesian case).

In particular, taking **C** to be the category **Ens** of sets and functions, one obtains just monoids and groups in the ordinary sense; or, taking the categories of spaces, simplicial sets, etc., one obtains topological or simplicial monoids and groups.

EXAMPLE 1.2. The category R-**mod** of modules over a commutative ring R may be given a monoidal structure $(R\text{-}\mathbf{mod}, \otimes_R, R, a, l, r)$ using the tensor product over R. Here a, l, r are the obvious isomorphisms. Monoids in this example are the associative R-algebras with unit.

These are in fact examples of *symmetric* monoidal categories, i.e. they admit additional structure consisting of natural isomorphisms

$$c = (c_{X,Y} : X \otimes Y \to Y \otimes X)_{X,Y}$$

satisfying further coherence conditions (see [20] for these). In the symmetric situation one may also talk about commutative monoids: (G, μ, η) is commutative if

$$\mu c_{G,G} = \mu$$

holds. In particular, in the cartesian situation of the example 1.1 one has the notion of an internal commutative, or abelian, group. We write $\mathbf{Ab}(\mathbf{C})$ for the category of abelian group objects in the cartesian monoidal category **C**.

There is also an important relaxation of the symmetric structure called braiding (the same $c_{X,Y}$, but satisfying less stringent coherence conditions; see e.g. [17] for numerous examples of monoidal categories of this kind).

We are going to define cohomology of $\mathbb{V}$-monoids; hence we must first determine what are the coefficients for such a cohomology theory. For this we recall (see e.g. [29]) that a general notion of coefficients for the cohomology of an object X in a category $\mathbf{C}$ is given by internal abelian group objects in the slice category $\mathbf{C}/X$. Here $\mathbf{C}/X$ is the category whose objects are morphisms $Y \to X$ in $\mathbf{C}$ and whose morphisms are commutative triangles of the obvious kind. In order to speak about internal abelian groups in the slice categories one has to assume that the $\mathbf{C}/X$ have finite products, or equivalently that $\mathbf{C}$ has pullbacks.

Given a monoidal category $\mathbb{V}$, there is an obvious notion of a morphism between $\mathbb{V}$-monoids, so we have the category $\mathbf{Mon}(\mathbb{V})$ of monoids and their morphisms, equipped with a forgetful functor $U : \mathbf{Mon}(\mathbb{V}) \to \mathbf{V}$. And if we assume existence of pullbacks in $\mathbf{V}$, the same will be true for $\mathbf{Mon}(\mathbb{V})$. Indeed, one has

LEMMA 1.3. For any monoidal category $\mathbb{V} = (\mathbf{V}, ...)$, the forgetful functor

$$U : \mathbf{Mon}(\mathbb{V}) \to \mathbf{V}$$

reflects any inverse limits that exist in $\mathbf{V}$.

PROOF. Consider any diagram $((\mathcal{G}_i)_{i \in I}, (f_\iota : \mathcal{G}_i \to \mathcal{G}_{i'})_{\iota : i \to i'})$ in $\mathbf{Mon}(\mathbb{V})$, where $\mathcal{G}_i = (G_i, \mu_i, \eta_i)$ are $\mathbb{V}$-monoids. Suppose we are given a limiting cone $(f_i : G \to G_i)_{i \in I}$ over this diagram, considered as a diagram in $\mathbf{V}$. One easily sees that $(G \odot G \xrightarrow{f_i \odot f_i} G_i \odot G_i \xrightarrow{\mu_i} G_i)_{i \in I}$ and $(I \xrightarrow{\eta_i} G_i)_{i \in I}$ are cones in $\mathbf{V}$, hence they determine maps $\mu : G \odot G \to G$ and $\eta : I \to G$, respectively. And one then checks that this gives a structure of a limiting cone in $\mathbf{Mon}(\mathbb{V})$. ∎

Note that for any monoid $\mathcal{G} = (G, \mu, \eta)$ in $\mathbb{V}$, there is a natural monoidal structure on $\mathbf{V}/G$, which we will denote by $\mathbb{V}/\mathcal{G} = (\mathbf{V}/G, \odot_\mu, I_\eta, a, l, r)$. Here the functor $\odot_\mu$ is determined by $(X \xrightarrow{x} G) \odot_\mu (Y \xrightarrow{y} G) = (X \odot Y \xrightarrow{x \odot y} G \odot G \xrightarrow{\mu} G)$; I_η is just $I \xrightarrow{\eta} M$; and a, l and r are those of $\mathbb{V}$ (in fact there is a one-to-one correspondence between monoid structures on an object G and those monoidal structures on $\mathbf{V}/G$ which turn the forgetful functor $U : \mathbf{V}/G \to \mathbf{V}$ into a strict monoidal functor). With respect to this monoidal structure one has the equivalence of categories $\mathbf{Mon}(\mathbb{V}/\mathcal{G}) \simeq \mathbf{Mon}(\mathbb{V})/\mathcal{G}$.

So we shall assume henceforward that our category $\mathbf{V}$ has pullbacks, and, for a $\mathbb{V}$-monoid $\mathcal{G} = (G, \mu, \eta)$ we choose the category $\mathbf{Ab}(\mathbf{Mon}(\mathbb{V})/\mathcal{G})$ of internal abelian groups in $\mathbf{Mon}(\mathbb{V})/\mathcal{G}$ and their homomorphisms to be the category of coefficients for the cohomology of $\mathcal{G}$. Fortunately, this category has a much simpler description, up to equivalence. This description involves the notion of action of a monoid on an object:

DEFINITION 1.4. A left action of a $\mathbb{V}$-monoid $\mathcal{G} = (G, \mu, \eta)$ on an object A of $\mathbf{V}$ is a morphism $u : G \odot A \to A$ satisfying

$$u(\mu \odot A) = u(G \odot u)a_{G,G,A},$$
$$u(\eta \odot A) = l_A.$$

We will also say that A is a left $\mathcal{G}$-object. Similarly, a right action of a monoid $\mathcal{G}' = (G', \mu', \eta')$ on A is a morphism $u' : A \odot G' \to A$ satisfying analogous identities. And given two such actions we say that they are compatible, or that A is an $\mathcal{G}$-$\mathcal{G}'$-biobject, if

$$u'(u \odot G') = u(G \odot u')a_{G,A,G'}.$$

For example, given any monoid $\mathcal{G} = (G, \mu, \eta)$, there is an evident $\mathcal{G}$-$\mathcal{G}$-biobject structure on G itself.

It is obvious how to define a morphism of left $\mathcal{G}$-, right $\mathcal{G}'$-, or $\mathcal{G}$-$\mathcal{G}'$-biobjects; the corresponding categories will be denoted by $^{\mathcal{G}}\mathbf{V}$, $\mathbf{V}^{\mathcal{G}'}$, and $^{\mathcal{G}}\mathbf{V}^{\mathcal{G}'}$, respectively. All these categories come with forgetful functors to $\mathbf{V}$ (which will be denoted by the same letter U); and just as in the lemma above, these forgetful functors reflect all the limits that happen to exist in $\mathbf{V}$. Hence we also can talk about internal abelian groups in $^{\mathcal{G}}\mathbf{V}^{\mathcal{G}}$. And we have

PROPOSITION 1.5. For any monoid $\mathcal{G}$ in $\mathbf{V}$, there is an equivalence of categories
$$\mathbf{Ab}(\mathbf{Mon}(\mathbb{V})/\mathcal{G}) \simeq \mathbf{Ab}(^{\mathcal{G}}\mathbf{V}^{\mathcal{G}}/G).$$

PROOF. To simplify exposition, we will prove the proposition in the particular case when the monoid in question is the terminal object 1 of $\mathbf{V}$, with its unique monoid structure. That is we will prove that there is an equivalence
$$\mathbf{Ab}(\mathbf{Mon}(\mathbb{V})) \simeq \mathbf{Ab}(^{1}\mathbf{V}^{1}).$$

By the above remarks on slice categories, this will suffice: for any monoid $\mathcal{G}$, the underlying object G (more precisely, its identity map) is clearly terminal in $\mathbf{V}/G$.

Now an object of the category $\mathbf{Ab}(\mathbf{Mon}(\mathbb{V}))$ looks like $(A, \mu : A \odot A \to A, \eta : I \to A, + : A \times A \to A, 0 : 1 \to A, - : A \to A)$. First of all note that 0 must be a morphism of monoids, in particular $\eta = (I \to 1 \xrightarrow{0} A)$, so that η is in fact determined by 0. As for μ, one has the commutative diagram

$$\begin{array}{ccc}
(A \times A) \odot (A \times A) & \xrightarrow{\mu_\times} & A \times A \\
{\scriptstyle +\odot +}\downarrow & & \downarrow{\scriptstyle +} \\
A \odot A & \xrightarrow{\mu} & A
\end{array}$$

where $\mu_\times$ is the monoid structure on $A \times A$ which, by a particular case of lemma 1.3, equals
$$(A \times A) \odot (A \times A) \xrightarrow{(p_1 \odot p_1, p_2 \odot p_2)} (A \odot A) \times (A \odot A) \xrightarrow{\mu \times \mu} A \times A.$$

Composing with $A \odot A \cong (A \times 1) \odot (1 \times A) \xrightarrow{(A \times 0) \odot (0 \times A)} (A \times A) \odot (A \times A)$ reveals that μ is equal to the composite
$$A \odot A \xrightarrow{(A \odot p, p \odot A)} (A \odot 1) \times (1 \odot A) \xrightarrow{v \times u} A \times A \xrightarrow{+} A,$$

where p is the unique morphism from A to 1, and $u : 1 \odot A \xrightarrow{0 \odot A} A \odot A \xrightarrow{\mu} A$, $v : A \odot 1 \xrightarrow{A \odot 0} A \odot A \xrightarrow{\mu} A$ are easily seen to define a 1-1-biobject structure on A, compatible with the abelian group structure. Hence μ is determined by these structures.

Conversely, given an object $(A, u : 1 \odot A \to A, v : A \odot 1 \to A, + : A \times A \to A, 0 : 1 \to A, - : A \to A)$ of $\mathbf{Ab}(^{1}\mathbf{V}^{1})$, one equips it with a $\mathbb{V}$-monoid structure via $A \odot A \xrightarrow{(A \odot p, p \odot A)} (A \odot 1) \times (1 \odot A) \xrightarrow{v \times u} A \times A \xrightarrow{+} A$ and $I \to 1 \xrightarrow{0} A$ and checks that this is compatible with the abelian group structure. ∎

Hence we are left with $\mathbf{Ab}(^{\mathcal{G}}\mathbf{V}^{\mathcal{G}}/G)$ as our category of coefficients for the cohomology of the $\mathbb{V}$-monoid $\mathcal{G}$. In the next section we will simplify the category of coefficients even more by imposing the conditions that $\mathbf{V}$ be abelian with left distributive monoidal structure.

We finish this section with the examples of monoids in non-symmetric monoidal categories which mainly motivated the results in this paper.

EXAMPLE 1.6 (Bimodules). For an associative ring R, the category R-R-**Mod** of R-R-bimodules has a non-symmetric monoidal structure given by $\otimes_R$. A monoid G in this monoidal category may be identified with an R-ring, that is, a ring equipped with a ring homomorphism from R. The coefficients for the cohomology of an R-ring G turn out to be G-G-bimodules, as we will see later.

EXAMPLE 1.7 (Monads). For any category $\mathbf{C}$, the category $\mathbf{End}(\mathbf{C})$ of endofunctors on $\mathbf{C}$ carries a monoidal structure $(\mathbf{End}(\mathbf{C}), \circ, \mathrm{Id}_{\mathbf{C}}, \mathrm{id}, \mathrm{id}, \mathrm{id})$ induced by composition of endofunctors. This is an example of a *strict* monoidal category — the associativity and unit natural transformations are all identities. Note also that as soon as $\mathbf{C}$ has coproducts, the monoidal structure on $\mathbf{End}(\mathbf{C})$ is left distributive automatically, but is almost never right distributive, nor symmetric. Monoids in $\mathbf{End}(\mathbf{C})$ are monads on $\mathbf{C}$.

There are also variations on this example: one may take various full subcategories of $\mathbf{End}(\mathbf{C})$ which are closed under the monoidal structure, e.g. the category of *finitary* endofunctors (that is, those preserving filtered colimits), or the category of cocontinuous endofunctors (preserving all colimits), or the category of endofunctors having a right adjoint. Monoids in these categories are various kinds of monads on $\mathbf{C}$.

EXAMPLE 1.8 (Theories). Monoids in the category of finitary endofunctors are finitary monads. In the case of finitary endofunctors on **Ens** the category of finitary monads is equivalent to the category of finitary algebraic theories in the sense of Lawvere [21]. In this particular case, coefficients turn out to be the general coefficients for cohomology of algebraic theories briefly mentioned in [15].

EXAMPLE 1.9 (Operads). In example 1.7, let $\mathbf{C}$ be the category of vector spaces over a characteristic zero field k. Consider the full subcategory of $\mathbf{End}(\mathbf{C})$ consisting of endofunctors which are *analytic*; recall from [16] that these are functors F admitting a decomposition into a Taylor series

$$F(V) = \bigoplus_{n \geqslant 0} F_n \otimes_{\mathfrak{S}_n} V^{\otimes n}$$

where $(F_n)_{n \geqslant 0}$ is some sequence of linear representations of symmetric groups $\mathfrak{S}_n$. Since analytic endofunctors are closed under composition, one obtains an abelian (in fact k-linear) left distributive monoidal category. The category of monads in this example is equivalent to the category of k-linear operads, as considered in e.g. [18].

Now let $\mathbf{C}$ be the category of modules over a commutative ring (or more generally any distributive symmetric monoidal category with colimits) and write $\mathfrak{S}$ for the symmetric groupoid $\coprod_{n \geqslant 0} \mathfrak{S}_n$. Then there is an analogous monoidal structure on the functor category $\mathbf{Cat}(\mathfrak{S}, \mathbf{C})$, and monoids in this category are operads in $\mathbf{C}$. We will identify coefficients in the next section.

EXAMPLE 1.10 (Square rings). Let $\mathbf{C}$ be the category of groups $\mathbf{Gr}$ or that of abelian groups $\mathbf{Ab}$, and consider the full subcategory of the endomorphism category

$$\mathbf{Degree}_n(\mathbf{C}) \quad \subset \quad \mathbf{End}(\mathbf{C})$$

whose objects are the finitary endofunctors which preserve cokernels and which have degree n. In particular functors F of degree one, or *linear* functors, are those which carry coproducts to products, i.e. the canonical natural transformation

$$(r_{1*}, r_{2*}) : F(X \sqcup Y) \to F(X) \times F(Y)$$

is an isomorphism. Functors F of degree two, or *quadratic* functors, are those for which the cross effect $F(X|Y) = \ker(r_{1*}, r_{2*})$ is linear as a bifunctor in X and Y. It is shown in [4] that there are canonical equivalences of monoidal categories

$$\mathbf{Ab} \cong \mathbf{Degree}_1(\mathbf{Ab}) \cong \mathbf{Degree}_1(\mathbf{Gr})$$

Moreover $\mathbf{Degree}_2(\mathbf{Ab})$ and $\mathbf{Degree}_2(\mathbf{Gr})$ are equivalent to categories of certain simple algebraic objects termed quadratic $\mathbb{Z}$-modules [1] and square groups [4] respectively. The category $\mathbf{Degree}_n(\mathbf{Ab})$ is equivalent to the category of modules over a certain commutative ring defined by Pirashvili [27] and calculated by Dreckmann [7].

Now unlike linear endofunctors, the quadratic ones are *not* closed under composition. However in the cases considered, the inclusion of the full subcategory of quadratic endofunctors into $\mathbf{End}(\mathbf{C})$ has a left adjoint $(\ )^{\mathrm{quad}}$. So one may define a monoidal structure on $\mathbf{Degree}_2(\mathbf{C})$ by $F \odot G = (F \circ G)^{\mathrm{quad}}$. Monoids in $\mathbf{Degree}_2(\mathbf{Gr})$ correspond under the equivalence with square groups to the *square rings* of [3]. Similarly one can define *rings of degree n* in the category $\mathbf{Degree}_n(\mathbf{Gr})$. Rings of degree 1 are just the classical rings.

EXAMPLE 1.11 (Categories). Given an object I in a category with finite limits $\mathbf{S}$, there is a monoidal structure on the slice category $\mathbf{S}/(I \times I)$: the unit object is the diagonal map $d : I \to I \times I$ and for $f : X \to I \times I$, $g : Y \to I \times I$ the object $f \odot g : Z \to I \times I$ is determined by the diagram

$$
\begin{array}{ccccc}
Z & \longrightarrow & I \times I \times I & \xrightarrow{\ I \times p \times I\ } & I \times I \\
\downarrow & & \downarrow{\scriptstyle I \times d \times I} & & \\
X \times Y & \xrightarrow{\ f \times g\ } & I \times I \times I \times I & &
\end{array}
$$

in which p is the projection to the terminal object and the square is a pullback. This is sometimes termed the "category of matrices", since for $\mathbf{S} = \mathbf{Ens}$ it is equivalent to the category of families $(X_{ij})_{i,j \in I}$ of sets, with the operation

$$(X_{ij}) \odot (Y_{ij}) = (\coprod_k X_{ik} \times Y_{kj})_{i,j \in I}.$$

Now monoids in this monoidal category may be identified with those internal categories in $\mathbf{S}$ having I as the object of objects; and morphisms of monoids are those internal functors which are the identity on objects. For any two such categories $\mathcal{C}$ and $\mathcal{D}$, the $\mathcal{C}$-$\mathcal{D}$-biobjects may be identified with *internal profunctors* from $\mathcal{D}$ to $\mathcal{C}$. When $\mathbf{S} = \mathbf{Ens}$, these are just bifunctors $\mathcal{C} \times \mathcal{D}^{\mathrm{op}} \to \mathbf{S}$. In particular, the canonical $\mathcal{C}$-$\mathcal{C}$-biobject structure on $\mathcal{C}$ itself corresponds to its hom bifunctor. Coefficients for the cohomology of an internal category $\mathcal{C}$ are *natural systems* on $\mathcal{C}$, that is, abelian group objects in the category of internal profunctors. For $\mathbf{S} = \mathbf{Ens}$ these are exactly the natural systems in the sense of [5].

Note that even this example may be fitted into the general setting of monads, as in example 1.7. Each object $X \xrightarrow{f} I \times I$ of $\mathbf{S}/(I \times I)$ determines an endofunctor of the category $\mathbf{S}/I$ as follows:

$$\mathbf{S}/I \xrightarrow{(p_1 f)^*} \mathbf{S}/X \xrightarrow{(p_2 f)_*} \mathbf{S}/I,$$

where $p_1, p_2 : I \times I \to I$ are the projections, $(p_1 f)^*$ is pullback along $p_1 f$, and $(p_2 f)_*$ is composition with $p_2 f$. For $\mathbf{S} = \mathbf{Ens}$, $\mathbf{S}/I$ may be identified with the category of I-indexed families of sets, and then the endofunctor corresponding to the "matrix" (X_{ij}) is given by

$$(V_i)_{i \in I} \mapsto (\coprod_j X_{ij} \times V_j)_{i \in I}.$$

Endofunctors of this kind are obviously closed under composition, and the monoidal structure given by the composition coincides with the "matrix multiplication" above.

EXAMPLE 1.12 (Spectra). By recent work of Elmendorf-Kriz-Mandell-May the category of spectra can be given a monoidal structure. Moreover the monoids in this category correspond to A_∞-ring spectra; compare 6.2 in [**11**].

2. Monoids and modules in the abelian left distributive case

Throughout this section $\mathbb{A} = (\mathbf{A}, \odot, I)$ will be an abelian left distributive monoidal category. For this case the coefficient objects for a monoid $\mathcal{G} = (G, \mu, \eta)$ in $\mathbb{A}$, given by abelian groups in $^{\mathcal{G}}\mathbb{A}^{\mathcal{G}}/G$ according to proposition 1.5, can be further simplified. In fact if the monoidal structure is also right distributive the coefficients are just bimodules:

PROPOSITION 2.1. Let $\mathbb{A}$ be an abelian monoidal category which is both left and right distributive, and suppose $\mathcal{G}$ is a monoid in $\mathbb{A}$. Then there is an equivalence of categories

$$\mathrm{Ab}(^{\mathcal{G}}\mathbb{A}^{\mathcal{G}}/G) \simeq {}^{\mathcal{G}}\mathbb{A}^{\mathcal{G}}.$$

This can be readily seen by the arguments below for the left distributive case.

The results in this section can be applied to the following examples.

EXAMPLES 2.2. The following are abelian left distributive monoidal categories. Let R be a commutative ring.

1. Clearly the monoid operation $\otimes_R$ on R-**mod** of example 1.2 is both left and right distributive, and applying proposition 2.1 shows that the coefficients for cohomology of R-algebras G are the G-bimodules. This is the classical case in for example [**22**].

2. Let $\mathfrak{S}$ be the symmetric groupoid (the disjoint union of the symmetric groups) and let $\mathbf{A} = \mathbf{Cat}(\mathfrak{S}, R\text{-}\mathbf{mod})$ be the category of functors from $\mathfrak{S}$ to R-modules. Then there is a monoidal structure $\odot$ on $\mathbf{A}$ such that $\mathbf{Mon}(\mathbf{A})$ is the category of *operads* in $\mathbf{A}$. See example 1.9.

3. Let $\mathbf{A}$ be the category of endofunctors of R-**mod** which preserve filtered colimits and cokernels. Then composition yields a monoidal structure and $\mathbf{Mon}(\mathbf{A})$ is the category of *monads* on R-**Mod**.

4. The category $\mathbf{Degree}_n(\mathbf{Ab})$ of example 1.10.

We may consider $(1) \subseteq (2) \subseteq (3)$ as a sequence of inclusions of monoidal categories.

DEFINITION 2.3. Let (G, μ, η) be a monoid in an abelian monoidal category $(\mathbf{A}, \odot, I)$, with $\odot$ left distributive over $\oplus$. Then a *coefficient G-module* is an object M and morphisms

$$G \odot (G \oplus M) \xrightarrow{\lambda} M, \qquad M \odot G \xrightarrow{\rho} M$$

in $\mathbf{A}$ with the following properties

1. λ is *linear* in M:

$$
\begin{array}{ccc}
G \odot (G \oplus M \oplus M) & \xrightarrow{(1 \odot p_1, 1 \odot p_2)} & G \odot (G \oplus M) \oplus G \odot (G \oplus M) \\
{\scriptstyle 1 \odot (1 \oplus +)} \downarrow & & \downarrow {\scriptstyle \lambda + \lambda} \\
G \odot (G \oplus M) & \xrightarrow{\lambda} & M
\end{array}
$$

2. λ is a *cross-action*:

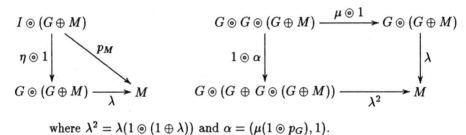

where $\lambda^2 = \lambda(1 \odot (1 \oplus \lambda))$ and $\alpha = (\mu(1 \odot p_G), 1)$.

3. ρ is a *right action*:

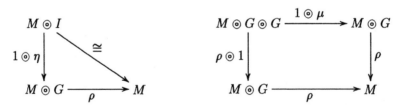

4. λ and ρ are compatible:

$$
\begin{array}{ccc}
G \odot (G \oplus M) \odot G & \xrightarrow{\lambda \odot 1} & M \odot G \\
{\scriptstyle 1 \odot (\mu \oplus 1)} \downarrow & & \searrow {\scriptstyle \rho} \\
& & \qquad M \\
G \odot (G \oplus M \odot G) & \xrightarrow[{1 \odot (1 \oplus \rho)}]{} G \odot (G \oplus M) \nearrow {\scriptstyle \lambda}
\end{array}
$$

Morphisms between coefficient G-modules are morphisms in $\mathbf{A}$ which respect all the structure. We write $\mathbf{Coef}_G$ for the category of coefficient G-modules M over a fixed monoid G in $\mathbf{A}$.

PROPOSITION 2.4. Let $\mathcal{G} = (G, \mu, \eta)$ be a monoid in an abelian left distributive monoidal category **A** as above. Then there is an equivalence of categories

$$\mathbf{Ab}(^{\mathcal{G}}\mathbf{A}^{\mathcal{G}}/G) \;\simeq\; \mathbf{Coef}_G.$$

PROOF. Let $(A, u, v, +, 0, -)$ be an object of $\mathbf{Ab}(^{\mathcal{G}}\mathbf{A}^{\mathcal{G}}/G)$. Then the map $p : A \to G$ is split by $0 : G \to A$ and so we can write $A = G \oplus M$ with $p = p_G$ and $0 = i_G$. The addition $+ : A \times_G A \to A$ becomes now $1 \oplus (1, 1) : G \oplus M \oplus M \to G \oplus M$ and the actions u, v are given by

$$G \odot (G \oplus M) \xrightarrow{(1 \odot p_G, 1)} G \odot G \oplus G \odot (G \oplus M) \xrightarrow{\mu \oplus \lambda} G \oplus M,$$

$$(G \oplus M) \odot G \xrightarrow{\;\cong\;} G \odot G \oplus M \odot G \xrightarrow{\mu \oplus \rho} G \oplus M$$

for some $\lambda : G \odot (G \oplus M) \to M$ and $\rho : M \odot G \to M$, where the biobject axioms on u and v are just the (cross-)action and compatibility laws for λ and ρ. Furthermore the compatibility of $+$ with u is equivalent to the linearity of λ. ∎

Let $\mathbf{Coef}_G$ be the category of coefficient G-modules, for (G, η, μ) a monoid in **A**. The *forgetful functor*

$$U : \mathbf{Coef}_G \longrightarrow \mathbf{A}$$

is the functor which takes a coefficient G-module (M, λ, ρ) to M regarded simply as an object of **A**. We will show that U has a left adjoint F, giving explicitly the *free coefficient G-module* $(F(V), \lambda, \rho)$ on an object V of **A**. The adjunction gives an isomorphism of abelian groups

$$\mathrm{Hom}_{\mathbf{A}}(V, M) \;\cong\; \mathrm{Hom}_{\mathbf{Coef}_G}(F(V), M)$$

which is natural in $V \in \mathbf{A}$ and $M \in \mathbf{Coef}_G$.

We give first an alternative definition of coefficient G-modules using the language of additive functors.

DEFINITION 2.5. (cf example 1.10) Let $F : \mathbf{A} \to \mathbf{A}$ be an endofunctor on an abelian category **A**. We define for objects A, B of **A** the *cross-effect* $F(A|B)$ by the kernel

$$F(A|B) \;=\; \ker(\pi : F(A \oplus B) \to F(A) \oplus F(B))$$

where $\pi = (Fp_A, Fp_B)$ is given by the projections from $A \oplus B$ to A and to B respectively. Clearly $F(A|B)$ is functorial in A and B. We say that F is an *additive* functor if $F(A|B)$ is zero for all A, B. We define natural maps P by

$$P : F(A|A) \lhook\joinrel\longrightarrow F(A \oplus A) \xrightarrow{F(+)} F(A)$$

where $+$ is the addition map $(1, 1) : A \oplus A \to A$ for A an object of **A**. The *additivisation* of F is the additive functor F^{add} defined by the cokernel

$$F^{\mathrm{add}}(A) \;=\; \mathrm{coker}(P : F(A|A) \to F(A)).$$

The *quotient map* $q : F \to F^{\mathrm{add}}$ has the universal property that any natural transformation $F \to G$ where G is additive has a unique factorisation through q.

In our situation the left distributivity of the tensor product $\odot$ in $\mathbf{A}$ says that each functor $-\odot B : A \mapsto A \odot B$ is additive. However the functors $A \odot - : B \mapsto A \odot B$ are not in general additive; for $\mathbf{A} = \mathbf{Cat}(\mathfrak{S}, R\text{-mod})$ for example the functor $A \odot -$ is additive if and only if the object A is concentrated in degree 1.

Consider the functor $L_0 : \mathbf{A} \to \mathbf{A}$ with

$$L_0(X) = G \odot (G \oplus X)$$

and the additive functor $L = L_0^{\mathrm{add}} : \mathbf{A} \longrightarrow \mathbf{A}$ defined by the additivisation of L_0. We note that for a coefficient G-module (M, λ, ρ), the linearity property (2.3)(1) says precisely that $\lambda : G \odot (G \oplus M) = L_0(M) \to M$ factors through the quotient map $q : L_0 \to L$. Furthermore the cross-action properties (2.3)(2) may be written as $\lambda(\eta \odot 1) = p_M : G \oplus M \to M$ and

(2.6)

$$
\begin{array}{ccc}
G \odot L_0(M) & \xrightarrow{\mu \odot 1} & L_0(M) \\
{\scriptstyle 1 \odot \alpha} \downarrow & & \downarrow {\scriptstyle \lambda} \\
L_0(L_0(M)) & \xrightarrow{\lambda^2} & M
\end{array}
$$

where $\lambda^2 = \lambda L_0(\lambda) = \lambda(1 \odot (1 \oplus \lambda))$ and $\alpha = (\mu(1 \odot p_G), 1)$.

LEMMA 2.7. In the presence of the linearity condition on λ, the commutativity of (2.6) is equivalent to that of

(2.8)

$$
\begin{array}{ccc}
G \odot L_0(M) & \xrightarrow{\mu \odot 1} & L_0(M) \\
{\scriptstyle 1 \odot \beta} \uparrow & & \downarrow {\scriptstyle \lambda} \\
L_0(L_0(M)) & \xrightarrow{\lambda^2} & M
\end{array}
$$

$$
\begin{array}{llll}
\text{where} & \beta & = & p_2 + (1 \odot i_G)\beta' & : & G \oplus L_0(M) & \longrightarrow & L_0(M) \\
\text{and} & \beta' & = & (\eta \odot 1)p_1(1 - \alpha p_2) & : & G \oplus L_0(M) & \longrightarrow & G \odot G.
\end{array}
$$

PROOF. Since $p_2 \alpha = 1$ the maps $(1 - \alpha p_2)\alpha$ and $\beta'\alpha$ are zero. Thus $\beta\alpha$ is the identity on $L_0(M)$ and the commutativity of (2.8) implies that of (2.6). In the opposite direction, we will show that $(1 \oplus \lambda)\alpha\beta = 1 \oplus \lambda$, so that $\lambda L_0(\lambda)(1 \odot \alpha)(1 \odot \beta) = \lambda L_0(\lambda)$ and (2.6) will imply (2.8). We have

$$
\begin{aligned}
p_1 \alpha\beta & = p_1 \alpha p_2 + p_1 \alpha(1 \odot i_G)\beta' \\
& = p_1 \alpha p_2 + \mu(1 \odot p_G)(1 \odot i_G)(\eta \odot 1)p_1(1 - \alpha p_2) \\
& = p_1 \alpha p_2 + p_1(1 - \alpha p_2) \\
& = p_1.
\end{aligned}
$$

Also $\lambda(1 \odot i_G)$ is zero by linearity and so $\lambda p_2 \alpha\beta = \lambda\beta = \lambda p_2$. Thus $(1 \oplus \lambda)\alpha\beta = 1 \oplus \lambda$ as required. ∎

We say $\mathbb{A}$ is *right compatible with cokernels* if for each $A \in \mathbf{A}$ the additive functor $A \odot - : \mathbf{A} \to \mathbf{A}$ given by $B \mapsto A \odot B$ preserves cokernels. If $\mathbb{A}$ has this property one has natural transformations $\eta_{(1)}$, $\mu_{(1)}$ and $\mu_{(2)}$ given by the following commutative diagrams, in which q_X is the quotient map from $L_0(X) = G \odot (G \oplus X)$

to the additivisation $L(X)$, q^2 is the composite $q_{L(X)}L_0(q_X)$ and $\mu' = 1 \odot (\mu \oplus 1)$ from $L_0(X) \odot G = G \odot (G \odot G \oplus X \odot G)$ to $L_0(X \odot G)$.

$$
\begin{array}{ccc}
G \oplus X \xrightarrow{\;px\;} X & L_0(L_0(X)) \xrightarrow{\;q^2\;} L(L(X)) & L_0(X) \odot G \xrightarrow{qx \odot 1} L(X) \odot G \\
\downarrow{\scriptstyle \eta \odot 1} \quad \downarrow{\scriptstyle \eta_{(1)}} & \downarrow{\scriptstyle \mu \odot \beta} \quad \downarrow{\scriptstyle \mu_{(1)}} & \downarrow{\scriptstyle \mu'} \quad\qquad \downarrow{\scriptstyle \mu_{(2)}} \\
L_0(X) \xrightarrow{\;qx\;} L(X) & L_0(X) \xrightarrow{\;qx\;} L(X) & L_0(X \odot G) \xrightarrow{qx \odot G} L(X \odot G)
\end{array}
$$

LEMMA 2.9. The natural transformations $\eta_{(1)}$, $\mu_{(1)}$ and $\mu_{(2)}$ are well-defined.

PROOF. Since X is clearly the additivisation of $G \oplus X$ in X, $\eta_{(1)}$ is well defined. Similarly $\mu_{(2)}$ is well defined since $- \odot G$ is additive. By the assumption that $\mathbb{A}$ is right compatible with cokernels it follows that $L(L(X))$ is the additivisation of $L_0(L_0(X))$ in X with q^2 the corresponding quotient map. Thus $\mu_{(1)}$ is also well defined. ∎

Using these natural transformations between additive functors we have

PROPOSITION 2.10. A coefficient G-module is equivalently specified by an object M and morphisms

$$
L(M) \xrightarrow{\;\overline{\lambda}\;} M, \qquad\qquad M \odot G \xrightarrow{\;\rho\;} M
$$

such that $\overline{\lambda}\eta_{(1)} = 1_M$, ρ is a right action as in (2.3)(3), and the following diagrams commute:

$$
\begin{array}{ccc}
L(L(M)) \xrightarrow{\;L(\overline{\lambda})\;} L(M) & \qquad & L(M) \odot G \xrightarrow{\;\overline{\lambda} \odot 1\;} M \odot G \\
\downarrow{\scriptstyle \mu_{(1)}} \qquad\qquad \downarrow{\scriptstyle \overline{\lambda}} & & \downarrow{\scriptstyle \mu_{(2)}} \qquad\qquad\quad \searrow{\scriptstyle \rho} \\
L(M) \xrightarrow{\;\overline{\lambda}\;} M & & L(M \odot G) \xrightarrow{\;L(\rho)\;} L(M) \xrightarrow{\;\overline{\lambda}\;} M
\end{array}
$$

Note that these are just the diagrams in (2.8) and (2.3)(4) made additive.

PROOF. Given $\overline{\lambda}$ we obtain λ by the composite

$$
\lambda : G \odot (G \oplus M) \xrightarrow{\;q\;} L(M) \xrightarrow{\;\overline{\lambda}\;} M.
$$

Then (M, λ, ρ) is a coefficient G-module in the sense of (2.3), as follows from the previous lemmas. Conversely any coefficient G-module M is obtained in this way since the linearity property in (2.3)(1) is equivalent to the existence of $\overline{\lambda}$ with $\lambda = \overline{\lambda}q$. ∎

We can now give an explicit construction for free coefficient modules. If the monoidal structure is both right and left distributive, the coefficient modules are just bimodules, and it is well known that the free G-bimodule is given by $F(V) = G \odot V \odot G$ with left and right actions given by the multiplication in G. With the assumption that $\mathbb{A}$ is right compatible with cokernels we have a similar explicit presentation of F in our more general situation.

Proposition 2.11. Let $G = (G, \eta, \mu)$ be a monoid in $\mathbb{A}$. Then the free coefficient G-module on an object V of $\mathbf{A}$ is given by

$$F(V) \;=\; L(V \odot G)$$

with the structure maps $\overline{\lambda}$ and ρ given by

$$\overline{\lambda} : L(L(V \odot G)) \xrightarrow{\;\;\mu_{(1)}\;\;} L(V \odot G),$$

$$\rho : L(V \odot G) \odot G \xrightarrow{\;\;\mu_{(2)}\;\;} L(V \odot G \odot G) \xrightarrow{\;\;L(1 \odot \mu)\;\;} L(V \odot G).$$

Proof. For an object V of $\mathbf{A}$ and a coefficient G-module $(M, \overline{\lambda}, \rho)$ we have natural maps $V \to UF(V)$ in $\mathbf{A}$ and $F(UM) \to M$ in $\mathbf{Coef}_G$ given by

$$V = V \odot I \xrightarrow{\;\;1 \odot \eta\;\;} V \odot G \xrightarrow{\;\;\eta_{(1)}\;\;} L(V \odot G),$$

$$L(M \odot G) \xrightarrow{\;\;L(\rho)\;\;} L(M) \xrightarrow{\;\;\overline{\lambda}\;\;} M$$

respectively, and these satisfy the triangle identities required to define an adjunction. ∎

We end by interpreting the results of this section for operads, the example promised in (2.2.2). First recall the definition of an operad from e.g. [**18, 24**] or definition 1 of [**25**].

Let $\mathfrak{S}$ be the symmetric groupoid; that is, $\mathfrak{S}$ is given by the disjoint union of the symmetric groups $\mathfrak{S}_n$, with $\mathfrak{S}_0 = \{*\}$. Let $\mathbf{C} = R\text{-}\mathbf{Mod}$ be the category of R-modules (or R-module chain complexes) for R a commutative ring, with monoidal structure $\otimes = \otimes_R$ and $I = R$. Consider the category $\mathbf{A} = \mathbf{Cat}(\mathfrak{S}, \mathbf{C})$ of $\mathfrak{S}$-objects in $\mathbf{C}$, given by functors A from the symmetric groupoid to $\mathbf{C}$, or equivalently by families $\{A_n\}_{n \geqslant 0}$ together with actions of $\mathfrak{S}_n$. The category $\mathbf{Cat}(\mathfrak{S}, \mathbf{C})$ is clearly abelian, with the sum $A \oplus B$ of $\mathfrak{S}$-objects given by the sum in $\mathbf{C}$

$$(A \oplus B)_n \;=\; A_n \oplus B_n$$

The tensor product of $\mathfrak{S}$-objects is defined as follows. Let $\mathcal{P}_n^k$ be the set of partitions of $\{1, \ldots, n\}$ into k disjoint subsets $(J_i)_{i=1}^k$, and write j_i for $|J_i|$. Then for an $\mathfrak{S}$-object B let

$$B_n^k \;=\; \bigoplus_{(J_i) \in \mathcal{P}_n^k} B_{j_1} \otimes \ldots \otimes B_{j_k}.$$

Clearly $\mathfrak{S}_k$ acts on B_n^k. In fact $\mathfrak{S}_n$ also acts on B_n^k via the $\mathfrak{S}_{j_i}$ actions. Thus the monoidal structure on $\mathbf{Cat}(\mathfrak{S}, \mathbf{C})$ can be defined by

$$(A \odot B)_n \;=\; \bigoplus_{k=0}^{\infty} A_k \otimes_{\mathfrak{S}_k} B_n^k.$$

If $A_0 = B_0 = 0$ this is a finite sum $\bigoplus_{k=1}^n$. The functor $\iota : \mathbf{C} \to \mathbf{Cat}(\mathfrak{S}, \mathbf{C})$ with $\iota(C)_1 = C$ and $\iota(C)_n = 0$ for $n \neq 1$ preserves the tensor product, and $I = \iota(R)$ defines a neutral object for $\odot$ in $\mathbf{Cat}(\mathfrak{S}, \mathbf{C})$. The monoidal structure on $\mathbf{Cat}(\mathfrak{S}, \mathbf{C})$

is not symmetric, but it is left distributive. In fact $- \odot B$ preserves all colimits and has a right adjoint $[B, -]$ given by

$$[B, C]_k \quad = \quad \bigoplus_{n=0}^{\infty} \mathbf{C}(B_n^k, C_n)_{\mathfrak{S}_n}$$

where $\mathbf{C}(-, -)_{\mathfrak{S}_n}$ is the object of $\mathfrak{S}_n$-equivariant maps in $\mathbf{C}$.

DEFINITION 2.12. An *operad* in $\mathbf{C}$ is a monoid in $\mathbf{Cat}(\mathfrak{S}, \mathbf{C})$, that is, an $\mathfrak{S}$-object A together with morphisms $\eta : I \to A$, $\mu : A \odot A \to A$ satisfying the unit and associativity laws.

Thus an operad is specified by the objects $\{A_n\}_{n \geqslant 0}$ and $\mathfrak{S}_n$-actions, together with operations

$$A_k \otimes A_{j_1} \otimes A_{j_2} \otimes \ldots \otimes A_{j_k} \xrightarrow{\quad \mu \quad} A_n$$

where $n = j_1 + \ldots + j_k$, satisfying the obvious unit and associative laws, together with certain equivariance relations as in [25].

DEFINITION 2.13. A *linear module* over an operad G is a coefficient G-module in $\mathbf{Cat}(\mathfrak{S}, \mathbf{C})$, that is, an $\mathfrak{S}$-object M together with a right action $\rho : M \odot G \to M$ and a left cross-action $\lambda : G \odot (G \oplus M) \to M$ with the properties (1)–(4) of definition 2.3.

The functor $L_0(M) = G \odot (G \oplus M)$ may be expanded by the distributivity of the tensor product $\otimes$ in $\mathbf{C}$, and we see that the additivisation $L(M)$ consists of those summands which contain precisely one factor from M. Thus a linear G-module is a family of objects $\{M_n\}_{n \geqslant 0}$ with $\mathfrak{S}_n$-actions, and operations

$$M_k \otimes G_{j_1} \otimes G_{j_2} \otimes \ldots \otimes G_{j_k} \xrightarrow{\quad \rho \quad} M_n,$$

$$G_k \otimes G_{j_1} \otimes \ldots \otimes G_{j_{i-1}} \otimes M_{j_i} \otimes G_{j_{i+1}} \otimes \ldots \otimes G_{j_k} \xrightarrow{\quad \lambda_i \quad} M_n$$

for $1 \leqslant i \leqslant k$ and $n = j_1 + \ldots + j_k$, satisfying the obvious action and compatibility laws together with equivariance relations as those for the operad structure. Compare also [23].

3. Cohomology

Let $\mathcal{G} = (G, \mu, \eta)$ be a monoid in a monoidal category $\mathbb{V} = (\mathbf{V}, \odot, I)$. We will avoid mentioning the associativity isomorphisms where possible.

We write $G^{\odot n}$ for the n-fold iterated tensor product $G \odot G \odot \cdots \odot G$, and let $\mu^n : G^{\odot n} \to G$ be given by the iterated multiplication map, with $\mu^0 = \eta$ and μ^1 the identity. We also write μ_i and η_i for the maps given by applying the multiplication and the unit between the ith and $(i+1)$st tensor factors:

$$\mu_i : G^{\odot n} \cong G^{\odot (i-1)} \odot G \odot G \odot G^{\odot (n-i-1)} \xrightarrow{\quad 1 \odot \mu \odot 1 \quad} G^{\odot (n-1)} \quad (0 < i < n)$$

$$\eta_i : G^{\odot n} \cong G^{\odot i} \odot I \odot G^{\odot (n-i)} \xrightarrow{\quad 1 \odot \eta \odot 1 \quad} G^{\odot (n+1)} \quad (0 \leqslant i \leqslant n)$$

DEFINITION 3.1. We denote by $B_\bullet(\mathcal{G})$ the two-sided bar construction [24] in the monoidal category $\mathbb{V}/\mathcal{G}$. This is the simplicial object in $\mathbb{V}/\mathcal{G}$ with

$$B_n(\mathcal{G}) = (G^{\odot(n+2)} \xrightarrow{\mu^{n+2}} G)$$

and face and degeneracy maps given by

$$d_i : G^{\odot(n+2)} \xrightarrow{\mu_{i+1}} G^{\odot(n+1)},$$

$$s_i : G^{\odot(n+2)} \xrightarrow{\eta_{i+1}} G^{\odot(n+3)}$$

for $0 \leqslant i \leqslant n$. As usual, this in fact defines a simplicial object in ${}^{\mathcal{G}}\mathbf{V}^{\mathcal{G}}/G$. There are extra degeneracy operators $s_{-1} = \eta_0 = \eta \odot G^{\odot(n+2)}$ and $s_{n+1} = \eta_{n+2} = G^{\odot(n+2)} \odot \eta$ which provide contractions of $B_\bullet(\mathcal{G})$ in $\mathbf{V}^{\mathcal{G}}/G$ and ${}^{\mathcal{G}}\mathbf{V}/G$ respectively, but *not* in ${}^{\mathcal{G}}\mathbf{V}^{\mathcal{G}}/G$.

Given an internal abelian group A in ${}^{\mathcal{G}}\mathbf{V}^{\mathcal{G}}/G$, we define

DEFINITION 3.2. The cohomology $H^*(\mathcal{G}; A)$ of a monoid $\mathcal{G} \in \mathbf{Mon}(\mathbb{V})$ with coefficients in an internal abelian group $A \in \mathbf{Ab}({}^{\mathcal{G}}\mathbf{V}^{\mathcal{G}}/G) \simeq \mathbf{Ab}(\mathbf{Mon}(\mathbb{V})/\mathcal{G})$ is the cohomology of the cochain complex associated to the cosimplicial abelian group $\mathrm{Hom}_{\mathcal{G}\mathbf{V}^{\mathcal{G}}/G}(B_\bullet(\mathcal{G}), A)$.

Now the forgetful functor $U : {}^{\mathcal{G}}\mathbf{V}^{\mathcal{G}}/G \to \mathbf{V}/G$ has a left adjoint F, where in particular

$$F(G^{\odot n} \xrightarrow{\mu^n} G) = (G^{\odot(n+2)} \xrightarrow{\mu^{n+2}} G).$$

Hence there are natural bijections

$$\mathrm{Hom}_{\mathcal{G}\mathbf{V}^{\mathcal{G}}/G}(G^{\odot(n+2)} \to G, A) \cong \mathrm{Hom}_{\mathbf{V}/G}(G^{\odot n} \to G, A)$$

and translating the cosimplicial structure of $B_\bullet(\mathcal{G})$ along these one gets

PROPOSITION 3.3. $H^*(\mathcal{G}; A)$ is isomorphic to the cohomology of the complex $C^*(\mathcal{G}; A)$ with $C^n(\mathcal{G}; A) = \mathrm{Hom}_{\mathbf{V}/G}(G^{\odot n} \xrightarrow{\mu^n} G, A)$ and differentials

$$d = \sum_{i=0}^{n} (-1)^i d^i : C^{n-1}(\mathcal{G}; A) \to C^n(\mathcal{G}; A)$$

where

$$d^0(G^{\odot(n-1)} \xrightarrow{f} A) = (G^{\odot n} \xrightarrow{1 \odot f} G \odot A \xrightarrow{u} A),$$
$$d^i(G^{\odot(n-1)} \xrightarrow{f} A) = (G^{\odot n} \xrightarrow{\mu_i} G^{\odot(n-1)} \xrightarrow{f} A) \text{ for } 0 < i < n,$$
$$d^n(G^{\odot(n-1)} \xrightarrow{f} A) = (G^{\odot n} \xrightarrow{f \odot 1} A \odot G \xrightarrow{v} A).$$

Since the forgetful functor $U : {}^{\mathcal{G}}\mathbf{V}^{\mathcal{G}}/G \to \mathbf{V}/G$ is monadic, there is also a standard way to define cohomology in this setting, the so called cotriple cohomology (see [6]). We will show that this leads to the same result:

PROPOSITION 3.4. The cohomology groups $H^*(\mathcal{G}; A)$ defined above are isomorphic to the cotriple cohomology groups w. r. t. the cotriple on ${}^{\mathcal{G}}\mathbf{V}^{\mathcal{G}}/G$ induced by the monadic adjunction $(F \dashv U) : {}^{\mathcal{G}}\mathbf{V}^{\mathcal{G}}/G \to \mathbf{V}/G$.

PROOF. This proposition is proved in appendix A.   ■

We now identify the simplification of the cochain complex in proposition 3.3 in the special case of monoids in an abelian and left distributive monoidal category **A**. In this case we know by proposition 2.4 that the coefficients $A \in \mathbf{Ab}(^{\mathcal{G}}\mathbf{A}^{\mathcal{G}}/G)$ can be replaced by coefficient G-modules $(M, \lambda, \rho) \in \mathbf{Coef}_G$.

PROPOSITION 3.5. Let M be a coefficient G-module. Then there is a cosimplicial abelian group

$$C^n(G, M) \quad = \quad \mathrm{Hom}_{\mathbf{A}}(\underbrace{G \odot \ldots \odot G}_{n \text{ factors}}, M).$$

The coface and codegeneracy maps are defined on $c \in C^n(G, M)$ by

$$d^0(c): G^{\odot(n+1)} \cong G \odot G^{\odot n} \xrightarrow{\; 1 \odot (\mu^n, c) \;} G \odot (G \oplus M) \xrightarrow{\; \lambda \;} M$$

$$d^{n+1}(c): G^{\odot(n+1)} \cong G^{\odot n} \odot G \xrightarrow{\; c \odot 1 \;} M \odot G \xrightarrow{\; \rho \;} M$$

$$d^i(c): G^{\odot(n+1)} \cong G^{\odot(i-1)} \odot G^{\odot 2} \odot G^{\odot(n-i)} \xrightarrow{\; 1 \odot \mu \odot 1 \;} G^{\odot n} \xrightarrow{\; c \;} M$$

$$s^i(c): G^{\odot(n-1)} \cong G^{\odot i} \odot I \odot G^{\odot(n-i-1)} \xrightarrow{\; 1 \odot \eta \odot 1 \;} G^{\odot n} \xrightarrow{\; c \;} M$$

where $G^{\odot n}$ is the nth tensor power, $\mu^0 = \eta$, $\mu^1 = 1$ and $\mu^n: G^{\odot n} \to G$ for $n \geqslant 2$ is given by the multiplication on G.

PROOF. We must check those cosimplicial identities which involve d^0; the others are exactly as in the classical definition of Hochschild cohomology. We have

a) $d^1 d^0 = d^0 d^0 \quad \Leftrightarrow \quad \lambda(1 \odot (\mu^n, c))(\mu \odot 1) = \lambda\left[1 \odot [\mu^{n+1}, \lambda(1 \odot (\mu^n, c))]\right]$
b) $d^{n+2} d^0 = d^0 d^{n+1} \quad \Leftrightarrow \quad \rho(\lambda(1 \odot (\mu^n, c)) \odot 1) = \lambda(1 \odot (\mu^{n+1}, \rho(c \odot 1)))$
c) $d^{i+1} d^0 = d^0 d^i \quad \Leftrightarrow \quad \lambda(1 \odot (\mu^n, c))\mu_{i+1} = \lambda(1 \odot (\mu^{n+1}, c\mu_i))$
d) $s^0 d^0 = 1 \quad \Leftrightarrow \quad \lambda(1 \odot (\mu^n, c))(\eta \odot 1) = c$
e) $s^{i+1} d^0 = d^0 s^i \quad \Leftrightarrow \quad \lambda(1 \odot (\mu^n, c))\eta_{i+1} = \lambda(1 \odot (\mu^{n-1}, c\eta_i))$

for all $c: G^{\odot n} \to M$, where we write $\mu_i: G^{\odot(k+1)} \to G^{\odot k}$ and $\eta_i: G^{\odot(k-1)} \to G^{\odot k}$ for the multiplication and unit of G applied at the ith factor. By the cross-action property we know

$$\lambda(\mu \odot (\mu^n, c)) = \lambda\left[1 \odot [(\mu(1 \odot p_G), \lambda)(1 \odot (\mu^n, c))]\right]$$
$$= \lambda\left[1 \odot [\mu(1 \odot \mu^n), \lambda(1 \odot (\mu^n, c))]\right]$$

and hence (a) follows. Also the left distributivity and the compatibility of λ and ρ give

$$\rho(\lambda \odot 1)(1 \odot (\mu^n, c) \odot 1) = \lambda(1 \odot (\mu \oplus \rho))(1 \odot (\mu^n \odot 1, c \odot 1))$$

and hence (b). By the unit law for λ we have $\lambda(\eta \odot (\mu^n, c)) = p_M(\mu^n, c) = c$ which gives (d), and (c) and (e) are clear from naturality and the monoid laws.

Finally we note that $\mathrm{Hom}(G^{\odot n}, M)$ has an abelian group structure by addition in M, and that d^0 is a group homomorphism by the linearity of λ. ∎

DEFINITION 3.6. Let M be a coefficient G-module as above. Then the *cohomology of G with coefficients in M*, $H^n(G, M)$, is given by the cohomology of the cochain complex (C^*, δ) with $C^n = C^n(G, M)$ the abelian group of homomorphisms $c : G^{\odot n} \to M$ under pointwise addition, and the boundary maps given by

$$\delta^n(c) = \sum_{i=0}^{n+1}(-1)^i d^i(c) = \lambda(1 \odot (\mu^n, c)) + \left(\sum_{i=1}^{n}(-1)^i c\mu_i\right) + (-1)^{n+1}\rho(c \odot 1)$$

Below we show that this cohomology is a special case of the cohomology in (3.2). From the usual relations between the cosimplicial maps d^i in the proposition we know that $\delta^{n+1}\delta^n$ is zero. As usual the same cohomology is obtained from the *normalised* cochain complex $C_N^*(G, M)$ defined by quotienting by the subcomplex of C^* of elements arising as codegeneracies.

DEFINITION 3.7. Assume that $\mathbb{A}$ is right compatible with cokernels so that we have a free coefficient G-module functor F as in (2.11). We define a simplicial coefficient G-module $B(G)$ termed the *bar resolution* of the monoid G. In the case of R-**mod**, example 1.2, this will be the un-normalised bar resolution described in MacLane [22, X.2]. The objects $B_n(G)$ are given by the free coefficient G-modules on $G^{\odot n}$

$$B_n(G) = F(G^{\odot n}).$$

The degeneracy maps $s_i : B_n(G) \to B_{n+1}(G)$ and face maps $d_i : B_{n+1}(G) \to B_n(G)$ are given by

$$\begin{aligned} s_i &= F(\eta_i) \quad \text{for } 0 \leqslant i \leqslant n, \\ d_i &= F(\mu_i) \quad \text{for } 1 \leqslant i \leqslant n \end{aligned}$$

where η_i and μ_i are defined on $G^{\odot k}$ by applying $\eta : I \to G$ and $\mu : G \odot G \to G$ at the ith factor. The face maps $d_0, d_{n+1} : F(G^{\odot(n+1)}) \to F(G^{\odot n})$ are the morphisms of coefficient G-modules corresponding under the adjunction to the following maps $d_0', d_{n+1}' : G^{\odot(n+1)} \to F(G^{\odot n})$ in $\mathbf{A}$:

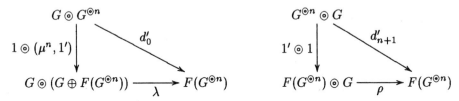

where $1' : G^{\odot n} \to F(G^{\odot n})$ in $\mathbf{A}$ corresponds to the identity on $F(G^{\odot n})$ in $\mathbf{Coef}_G$.

PROPOSITION 3.8. Let M be a coefficient G-module. Then there is a natural isomorphism

$$\psi : C^*(G, M) \cong \operatorname{Hom}_{\mathbf{Coef}_G}(B(G), M)$$

and hence the cohomology of G is determined by maps from the bar resolution

$$H^*(G, M) \cong H^*\operatorname{Hom}_{\mathbf{Coef}_G}(B(G), M).$$

PROOF. The free/forget adjunction gives natural isomorphisms

$$\psi^n : \operatorname{Hom}_{\mathbf{A}}(G^{\odot n}, M) \cong \operatorname{Hom}_{\mathbf{Coef}_G}(F(G^{\odot n}), M)$$

and we must check these respect the (co)simplicial structures. We have

$$\psi^n(s^i) = \psi^n \operatorname{Hom}_{\mathbf{A}}(\eta_i, M) = \operatorname{Hom}_{\mathbf{Coef}_G}(F(\eta_i), M) = \operatorname{Hom}_{\mathbf{Coef}_G}(s_i, M)$$

and similarly $\psi^n(d^i) = \mathrm{Hom}_{\mathbf{Coef}_G}(d_i, M)$ for $i \neq 0, n+1$. For $i = 0$ consider morphisms $c' : G^{\odot n} \to M$ in $\mathbf{A}$ and $c = \psi^n(c') : F(G^{\odot n}) \to M$ in $\mathbf{Coef}_G$, and let d'_0 and $1'$ be as in the definition of d_0 above. Then naturality of the adjunction implies $c' = c1'$ and $d_0^* c = \psi^n(cd'_0)$, and

$$d^0(c1') = \lambda(1 \odot (\mu^n, c1')) = \lambda(1 \odot (1 \oplus c))(1 \odot (\mu^n, 1')) = c\lambda(1 \odot (\mu^n, 1')) = cd'_0.$$

Thus $\psi^n(d^0 c') = d_0^* c$. One shows $\psi^n(d^{n+1} c') = d_{n+1}^* c$ in the same way. ∎

Finally we show that the definition of cohomology in (3.6) is a special case of that in (3.2).

PROPOSITION 3.9. Let $\mathcal{G} = (G, \mu, \eta)$ be a monoid in an abelian left distributive monoidal category $\mathbf{A}$ and let M be a coefficient G-module. Then there is a natural isomorphism between the cochain complexes of propositions 3.3 and 3.5:

$$\theta : C^*(\mathcal{G}; A_M) \cong C^*(G, M).$$

Here $A_M \in \mathbf{Ab}(^{\mathcal{G}}\mathbf{A}^{\mathcal{G}}/G)$ is the internal abelian group given by M under the equivalence of proposition 2.4.

PROOF. Recall first that $A_M = (p_G : G \oplus M \to G)$, and that the structure maps satisfy

$$p_M u = \lambda : G \odot (G \oplus M) \to M, \qquad p_M v = \rho(p_M \odot 1) : (G \oplus M) \odot G \to M.$$

Now a morphism $c : G^{\odot n} \to M$ in $\mathbf{A}$ determines a morphism $(\mu^n, c) : G^{\odot n} \to G \oplus M$ in $\mathbf{A}/G$, and conversely a morphism $f : G^{\odot n} \to A$ in the slice category gives a morphism $p_M f : G^{\odot n} \to M$ in $\mathbf{A}$.

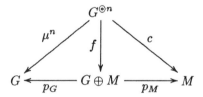

Clearly this gives isomorphisms of abelian groups

$$\theta^n : C^n(\mathcal{G}; A_M) = \mathrm{Hom}_{\mathbf{A}/G}(\mu^n, p_G) \cong \mathrm{Hom}_{\mathbf{A}}(G^{\odot n}, M) = C^n(G, M)$$

and we must check the cosimplicial structures coincide. For cochains f, c with $\theta^n f = c$ we have

$$\begin{aligned}
\theta^{n+1}(d^0 f) &= p_M u(1 \odot f) &= \lambda(1 \odot (\mu^n, c)) &= d^0 c \\
\theta^{n+1}(d^{n+1} f) &= p_M v(f \odot 1) &= \rho(p_M f \odot 1) &= d^{n+1} c
\end{aligned}$$

and the results for the other cofaces and the codegenacies are straightforward. ∎

REMARK 3.10. Particular examples of the cohomology defined by (3.2) or (3.6) above coincide with various cohomologies in the literature.

1. For $\mathbf{V} = R\text{-}\mathbf{mod}$ in example 1.2 the cohomology $H^*(G, A)$ is the same as the classical cohomology of an R-algebra G; see [22, X.3]. We saw in proposition 2.1 that the coefficients A are G-bimodules.
2. Consider the monoidal category $\mathbf{V} = R\text{-}R\text{-}\mathbf{mod}$ of bimodules over an arbitrary ring R, as in example 1.6. The cohomology $H^*(G, A)$ we obtain is the R-relative Hochschild cohomology from [13]. Indeed, direct comparison

shows that in this case our complex coincides with the one used by Gersten-haber and Schack in [13] to define the R-relative Hochschild cohomology groups.

3. For $\mathbf{V} = \mathbf{Cat}(\mathfrak{S}, R\text{-}\mathbf{Mod})$ in example 2.2.2 the cohomology $H^*(G, A)$ is the cohomology of an operad with coefficients as described in proposition 2.13. These have also appeared in [12, 23].

4. For $\mathbf{V} = \mathbf{Ens}/I \times I$ in example 1.11 the cohomology $H^*(G, A)$ coincides with the cohomology of a category G with coefficients in a natural system A, see [5].

5. For $\mathbf{V}$ the category of finitary endofunctors of $\mathbf{Ens}$ in example 1.8 the cohomology $H^*(G, A)$ is the cohomology of a finitary theory G considered briefly in [15].

4. Derivations, extensions and torsors

We now turn to the interpretation of elements in cohomology groups. We first consider abelian and left distributive monoidal categories $\mathbb{A}$ and the low degree cohomology of monoids in $\mathbb{A}$, which we interpret in terms of derivations and extensions. In the second part of this section we deal with the case of a general monoidal category $\mathbb{V}$ and the cohomology of monoids in $\mathbb{V}$ which in low degrees can be interpreted using torsors.

Recall that for the cohomology of a monoid $\mathcal{G} = (G, \mu, \eta)$ in $\mathbb{A}$ we use the coefficient G-modules (M, λ, ρ) of definition 2.3.

DEFINITION 4.1. A *derivation* (or *crossed homomorphism*) from a monoid G to a coefficient G-module M is a morphism $\Delta : G \to M$ in $\mathbf{A}$ which satisfies $\Delta\mu = \lambda(1 \odot (1, \Delta)) + \rho(\Delta \odot 1)$.

$$
\begin{array}{ccc}
G \odot G & \xrightarrow{(1 \odot (1, \Delta), \Delta \odot 1)} & G \odot (G \oplus M) \oplus M \odot G \\
{\scriptstyle \mu} \downarrow & & \downarrow {\scriptstyle \lambda + \rho} \\
G & \xrightarrow{\quad \Delta \quad} & M
\end{array}
$$

The abelian group of derivations from G to M is written $\mathrm{Der}(G, M)$.

In particular a morphism $\phi : I \to M$ in $\mathbf{A}$ defines an *inner derivation* $\mathrm{Inn}(\phi) : G \to M$ by $\mathrm{Inn}(\phi) = \lambda\phi_0 - \rho\phi_1$ where

$$\phi_0 = 1 \odot (\eta, \phi) : G \odot I \to G \odot (G \oplus M) \quad \text{and} \quad \phi_1 = \phi \odot 1 : I \odot G \to M \odot G.$$

We thus have a homomorphism

$$\mathrm{Hom}(I, M) \longrightarrow \mathrm{Der}(G, M)$$

whose image is the subgroup $\mathrm{Inn}(G, M)$ of inner derivations. The kernel consists of those ϕ with

$$\lambda(1 \odot (\eta, \phi)) = \rho(\phi \odot 1).$$

This may be thought of as the subgroup M^G of *G-invariant* morphisms $I \to M$.

PROPOSITION 4.2. There are isomorphisms

$$H^0(G, M) \cong M^G \quad \text{and} \quad H^1(G, M) \cong \mathrm{Der}(G, M)/\mathrm{Inn}(G, M)$$

and an exact sequence of abelian groups

$$0 \longrightarrow H^0(G, M) \longrightarrow \mathrm{Hom}(I, M) \longrightarrow \mathrm{Der}(G, M) \longrightarrow H^1(G, M) \longrightarrow 0.$$

PROOF. The derivation property is $\delta^1 \Delta = 0$, so derivations are just 1-cocycles. Also the inner derivation map $\phi \mapsto \mathrm{Inn}(\phi)$ is just the coboundary map δ^0. ∎

We now describe the theory of extensions of monoids (G, η, μ) in $\mathbb{A}$. Our exposition will be parallel to and will extend the classical description for the case $\mathbf{A} = R\text{-}\mathbf{Mod}$ of example 1.2, where the tensor $\otimes_R$ preserves colimits on both sides; see for example MacLane [22].

DEFINITION 4.3. An *extension* of a monoid $\mathcal{G}$ in $\mathbb{A}$ is a short exact sequence

$$0 \longrightarrow M \overset{i}{\longrightarrow} A \overset{p}{\longrightarrow} G \longrightarrow 0$$

in the abelian category $\mathbf{A}$ together with a monoid structure on A such that p is a morphism of monoids. The extension is $\mathbf{A}$-*split* if there is an $s : G \to A$ in $\mathbf{A}$ which is right inverse to p, $ps = 1_G$. The extension is termed *singular* if the following conditions hold.

1. The map $\mu_A(i \odot 1) : M \odot A \to A$ is zero on the kernel of $1 \odot p : M \odot A \to M \odot G$.
2. The maps $\mu_A(1 \odot +)$, $\mu_A(1 \odot p_1) + \mu_A(1 \odot p_2)$: $A \odot (A \oplus_G A) \longrightarrow A$ are equal.

$$
\begin{array}{ccc}
A \odot (A \oplus_G A) & \overset{1 \odot +}{\longrightarrow} & A \odot A \\
{\scriptstyle (1 \odot p_1, 1 \odot p_2)} \downarrow & & \downarrow {\scriptstyle \mu_A} \\
A \odot A \oplus_G A \odot A & \overset{\mu_A + \mu_A}{\longrightarrow} & A
\end{array}
$$

Extensions A, A' are *equivalent* if there is a morphism $\varepsilon : A \to A'$ of monoids with $\varepsilon i = i'$ and $p' \varepsilon = p$.

$$
\begin{array}{ccccc}
M & \overset{i}{\longrightarrow} & A & \overset{p}{\longrightarrow} & G \\
\| & & {\scriptstyle \varepsilon} \downarrow {\scriptstyle \cong} & & \| \\
M & \overset{i'}{\longrightarrow} & A' & \overset{p'}{\longrightarrow} & G
\end{array}
$$

Fixing a monoid G and a coefficient G-module M, we write $\mathrm{Ext}(G, M)$ for the set of equivalence classes of $\mathbf{A}$-split singular extensions.

Suppose $M \overset{i}{\longrightarrow} A \overset{p}{\longrightarrow} G$ is an $\mathbf{A}$-split singular extension with section s as above. Let $d = s\eta_G - \eta_A : I \to A$. Then by replacing s by $s - \mu_A(d \odot 1)$ if necessary we can assume that s respects the units of G and A. Also the map $s + i : G \oplus M \to A$ is a map of short exact sequences and hence an isomorphism in $\mathbf{A}$ by the 5-lemma.

$$
\begin{array}{ccccc}
M & \overset{i}{\longrightarrow} & A & \underset{s}{\overset{p}{\rightleftarrows}} & G \\
\| & & {\scriptstyle s+i} \uparrow {\scriptstyle \cong} & & \| \\
M & \overset{i_M}{\longrightarrow} & G \oplus M & \underset{i_G}{\overset{p_G}{\rightleftarrows}} & G
\end{array}
$$

Using the isomorphism $s + i$ we obtain a coefficient G-module structure (λ, ρ) on M as follows. The maps $\mu_A(s \odot (s+i) - s \odot (s+0)) : G \odot (G \oplus M) \to A \odot A \to A$ and $\mu_A(i \odot s) : M \odot G \to A \odot A \to A$ factor through $\ker(p)$ and define

$$G \odot (G \oplus M) \xrightarrow{\quad\lambda\quad} M, \qquad\qquad M \odot G \xrightarrow{\quad\rho\quad} M$$

respectively. The singularity conditions show that λ and ρ are independent of the choice of splitting s and that λ is linear in the sense of (2.3.1). The action and compatibility laws follow by the associativity of μ_A.

Conversely, suppose M is a coefficient G-module and $M \xrightarrow{\ i\ } A \xrightarrow{\ p\ } G$ is an extension of G. Then it is a singular extension if and only if the monoid structure on A extends the coefficient G-module structure on M:

$$
\begin{array}{ccc}
M \odot A \xrightarrow{1 \odot p} M \odot G \xrightarrow{\ \rho\ } M & \qquad & A \odot (A \oplus M) \xrightarrow{p \odot (p \oplus 1)} G \odot (G \oplus M) \xrightarrow{\ \lambda\ } M \\
\downarrow{\scriptstyle i \odot 1} \qquad\qquad\qquad \downarrow{\scriptstyle i} & & \downarrow{\scriptstyle 1 \odot (1 \oplus i)} \qquad\qquad\qquad\qquad \downarrow{\scriptstyle i} \\
A \odot A \xrightarrow[\ \ \mu_A\ \ ]{} A & & A \odot (A \oplus A) \xrightarrow[\ \kappa\ ]{} A \odot A \xrightarrow[\ \mu_A\ ]{} A
\end{array}
$$

where $\kappa = 1 \odot (1 + 1) - 1 \odot (1 + 0) : A \odot (A \oplus A) \to A \odot A$.

The simplest example of an **A**-split singular extension is the *trivial extension* or *semi-direct sum* given by $A = G \oplus M$ with unit $i_G \eta_G$ and multiplication

$$(\mu(p_G \odot p_G), \lambda(p_G \odot 1) + \rho(p_M \odot p_G)) : (G \oplus M) \odot (G \oplus M) \to G \oplus M.$$

Any **A**-split singular extension for which p is split by a morphism of monoids is equivalent to the semi-direct sum. More generally each splitting s of a singular extension defines a *factor set* $c_s : G \odot G \to M$, or 2-cochain of $C^*(G, M)$, by

$$\mu_A(s \odot s) \ =\ s\mu_G + ic_s$$

which is *normalised* if $s\eta_G = \eta_A$ and is zero if s is a monoid homomorphism. The factor set c_t given by a different choice of splitting t differs from c_s by a coboundary: one can define $\Delta : G \to M$ by $t = s + i\Delta$, and then

$$
\begin{aligned}
ic_t - ic_s &= \mu_A((s + i\Delta) \odot (s + i\Delta)) - \mu_A(s \odot s) - (s + i\Delta)\mu_G + s\mu_G \\
&= i\lambda(1 \odot (1, \Delta)) + i\rho(\Delta \odot 1) - i\Delta\mu_G \ =\ i\delta\Delta.
\end{aligned}
$$

This process also respects equivalent extensions: given an equivalence $\varepsilon : A \to A'$ and a splitting s for A, we take εs as a splitting for A' and the factor sets c_s and $c_{\varepsilon s}$ are equal.

THEOREM 4.4. Let G be a monoid and M a coefficient G-module. Then assigning factor sets to **A**-split singular extensions induces a bijection between the equivalence classes of such extensions and the cohomology classes of cocycles $G \odot G \to M$

$$\Phi : \mathrm{Ext}(G, M) \ \cong\ H^2(G, M)$$

under which the trivial extension corresponds to the zero cohomology class.

PROOF. We construct an inverse Ψ to Φ. Given a 2-cocycle $c : G \odot G \to M$, there is an extension given by $A = G \oplus M$ with unit $i_G \eta_G$ and multiplication μ_c as follows:

$$\mu_c = (\mu_G, \lambda(p_G \odot 1) + \rho(p_M \odot p_G) + c_G) : (G \oplus M) \odot (G \oplus M) \longrightarrow G \oplus M$$

where $\mu_G = \mu(p_G \odot p_G) = p_G\mu_c$ and $c_G = c(p_G \odot p_G)$. Clearly p_G is a monoid homomorphism, and the monoid structure on $G \oplus M$ extends the coefficient G-module structure on M. If cocycles c and d differ by a coboundary $\delta\Delta$ for some $\Delta : G \to M$, then the map $\varepsilon : G \oplus M \to G \oplus M$ given by $(p_G, p_M + \Delta p_G)$ shows that the extensions $\Psi(c)$ and $\Psi(d)$ are equivalent.

For the associativity of μ_c we note first

$$p_G\mu_c(\mu_c \odot 1) = \mu(\mu_G \odot p_G) = \mu(p_G \odot \mu_G) = p_G\mu_c(1 \odot \mu_c)$$

by the associativity of μ. Now $p_M\mu_c(\mu_c \odot 1) = \lambda(\mu_G \odot 1) + \rho(p_M\mu_c \odot p_G) + c(\mu_G, p_G)$ which is

$$\lambda(\mu_G \odot 1) + \rho(\lambda(p_G \odot 1) \odot p_G) + \rho(\rho(p_M \odot p_G) \odot p_G) + \rho(c_G \odot p_G) + c(\mu_G \odot p_G)$$

and $p_M\mu_c(1 \odot \mu_c) = \lambda(p_G \odot \mu_c) + \rho(p_M \odot \mu_G) + c(p_G \odot \mu_G)$ which by linearity of λ in M is

$$\lambda(p_G \odot (\mu_G, \lambda(p_G \odot 1))) + \lambda(p_G \odot (\mu_G, \rho(p_M \odot p_G))) + \lambda(p_G \odot (\mu_G, c_G))$$
$$+ \rho(p_M \odot \mu_G) + c(p_G \odot \mu_G)$$

Now evaluate these on the inclusions $i_G \odot i_G \odot 1$, $i_G \odot 1 \odot i_G$, $i_M \odot i_G \odot i_G$ and $i_G \odot i_G \odot i_G$. Then since λ is linear in G we see that $p_M\mu_c(\mu_c \odot 1) = p_M\mu_c(1 \odot \mu_c)$ if and only if the following relations hold:

$$
\begin{array}{rccl}
a) & \lambda(\mu \odot 1) & = & \lambda(1 \odot (\mu(1 \odot p_G), \lambda)) \\
b) & \rho(\lambda \odot 1) & = & \lambda(1 \odot (\mu \oplus \rho)) \\
c) & \rho(\rho \odot 1) & = & \rho(1 \odot \mu) \\
d) & \rho(c \odot 1) + c(\mu \odot 1) & = & \lambda(1 \odot (\mu, c)) + c(1 \odot \mu)
\end{array}
$$

But $(a), (b), (c)$ are respectively just the cross-action, compatibility and right action laws for λ and ρ, and (d) is the cocycle condition $\delta c = 0$. ∎

We now give similar interpretations of low degree cohomology of monoids in the case of a general monoidal category $\mathbb{V}$. Note that there is already a general interpretation of cotriple cohomology by Duskin [8, 9] as in the following remark, which applies to our cohomology by proposition 3.4. Let $\mathcal{G}$ be a monoid in $\mathbb{V}$ and A an internal abelian group in $\mathcal{G}\mathbb{V}^{\mathcal{G}}/G$. Let $A_{\mathbf{Mon}}$ be the corresponding abelian group in $\mathbf{Mon}(\mathbf{V})/\mathcal{G}$ according to proposition 1.5.

REMARK 4.5. Let $K(A, n)$ be the Eilenberg-MacLane object of A in degree n. Then a $K(A, n)$-torsor relative to the forgetful functor $U : \mathcal{G}\mathbb{V}^{\mathcal{G}}/G \to \mathbb{V}$ is a simplicial object $X_\bullet$ in $\mathcal{G}\mathbb{V}^{\mathcal{G}}/G$, together with a simplicial map $\chi : X_\bullet \to K(A, n)$, such that

1. $X_\bullet$ is isomorphic to the coskeleton of the nth truncation of $X_\bullet$,
2. χ satisfies the Kan fibration condition *exactly* in dimension $\geqslant n$,
3. $U(X_\bullet)$ has a contracting homotopy in $\mathbf{V}/G$.

Duskin proves in [9, section 5.2] that there is a natural bijection between the set of equivalence classes of $K(A, n)$-torsors and the nth cotriple cohomology of G with coefficients in A.

Simplification is possible since it turns out that in degrees $n = 1, 2$ elements of $H^n(\mathcal{G}; A)$ can also be interpreted using $K(A_{\mathbf{Mon}}, n-1)$-torsors. For higher degrees we make the following observations. Suppose we have a left adjoint to the forgetful functor $U : \mathbf{Mon}(\mathbf{V}/G) \to \mathbf{V}/G$, giving a free monoid functor. We construct explicitly the free monoid functor in appendix B, if the monoidal category satisfies some reasonable conditions. Thus we can assume the cotriple cohomology groups $H^*(\mathcal{G}; A_{\mathbf{Mon}})$ are defined. Suppose further that for $\mathcal{G}$ a free monoid our cohomology groups $H^n(\mathcal{G}; A)$ are trivial for $n > 1$. Then an analysis of the proof of Theorem C of [15] shows that one has isomorphisms

$$H^n(\mathcal{G}; A) \cong H^{n-1}(\mathcal{G}; A_{\mathbf{Mon}}), \ n > 1,$$

and under the assumptions above interpretation of $H^n(\mathcal{G}; A)$ by $K(A_{\mathbf{Mon}}, n-1)$-torsors is valid in all degrees.

Let us begin with degree 0; we give an explicit interpretation generalising that for the abelian case above. For any $A \xrightarrow{p} G$ in $\mathcal{G}\mathbf{V}^{\mathcal{G}}/G$, let $A^{\mathcal{G}}$ denote the set of $\mathcal{G}$-invariant elements of A, that is, $A^{\mathcal{G}}$ is the subset of those morphisms $a \in \mathrm{Hom}_{\mathbf{V}}(I, A)$ satisfying $pa = \eta : I \to G$ and

$$(G \xrightarrow{l_G^{-1}} I \odot G \xrightarrow{a \odot G} A \odot G \xrightarrow{v} A) = (G \xrightarrow{r_G^{-1}} G \odot I \xrightarrow{G \odot a} G \odot A \xrightarrow{u} A).$$

Then inspection of the complex in proposition 3.3 gives

PROPOSITION 4.6. There is a natural bijection $H^0(\mathcal{G}; A) \cong A^{\mathcal{G}}$.

Clearly the $\mathcal{G}$-invariant elements correspond to morphisms from 1_G to $A \xrightarrow{p} G$ in $\mathcal{G}\mathbf{V}^{\mathcal{G}}/G$; these are just the $K(A, 0)$-torsors of Duskin.

Turning to degree 1 we make the following definition.

DEFINITION 4.7. For $A \xrightarrow{p} G$ in $\mathbf{Ab}(\mathcal{G}\mathbf{V}^{\mathcal{G}}/G)$, a *derivation* is a morphism $\Delta : G \to A$ in $\mathbf{V}$ satisfying $p\Delta = 1_G$ and

We write $\mathrm{Der}(\mathcal{G}; A)$ for the set of such derivations, and define a map $\mathrm{Inn}(\ )$ from $\mathrm{Hom}_{\mathbf{V}/G}(I \xrightarrow{\eta} G, A \xrightarrow{p} G)$ to $\mathrm{Der}(\mathcal{G}; A)$ which sends $a : I \to A$ to the composite

$$(G \xrightarrow{(r_G^{-1}, l_G^{-1})} G \odot I \times I \odot G \xrightarrow{G \odot a \times a \odot G} G \odot A \times A \odot G \xrightarrow{u \times v} A \times A \xrightarrow{-} A).$$

PROPOSITION 4.8. There is an exact sequence of abelian groups

$$0 \to H^0(\mathcal{G}; A) \to \mathrm{Hom}_{\mathbf{V}/G}(I \xrightarrow{\eta} G, A \xrightarrow{p} G) \xrightarrow{\mathrm{Inn}(\ )} \mathrm{Der}(\mathcal{G}; A) \to H^1(\mathcal{G}; A) \to 0.$$

PROOF. Straightforward, on noting that $\mathrm{Inn}() : \mathrm{Hom}_{\mathbf{V}/G}(I \xrightarrow{\eta} G, A \xrightarrow{p} G) \to \mathrm{Der}(\mathcal{G}; A)$ may be identified with $d : C^0(\mathcal{G}; A) \to \ker(C^1(\mathcal{G}; A) \xrightarrow{d} C^2(\mathcal{G}; A))$. ∎

Clearly (4.7) and (4.8) reduce to (4.1) and (4.2) in the abelian situation above, where $A = G \oplus M$.

One readily sees that

$$\mathrm{Der}(\mathcal{G}; A) = \mathrm{Hom}_{\mathbf{Mon}(\mathbb{V})/\mathcal{G}}(1_{\mathcal{G}}, A_{\mathbf{Mon}})$$

whose elements are the $K(A_{\mathbf{Mon}}, 0)$-torsors relative to $U : \mathbf{Mon}(\mathbb{V})/\mathcal{G} \to \mathbf{V}/G$.

For degree two we make the following definition.

DEFINITION 4.9. Let $U : \mathbf{C} \to \mathbf{D}$ be a product-preserving functor between categories with finite products, and let A be an internal group object in $\mathbf{C}$. An *A-torsor* relative to U is an object T of $\mathbf{C}$ together with

• morphisms

$$T \times A \xrightarrow{\ +\ } T, \qquad\qquad T \times T \xrightarrow{\ -\ } A$$

in $\mathbf{C}$, such that $+$ is a right action and the morphisms $(p_1, +) : T \times A \to T \times T$, $(p_1, -) : T \times T \to T \times A$ are mutually inverse isomorphisms, and
• a morphism $s : 1 \to U(T)$ where 1 is the terminal object in $\mathbf{D}$.

As in Duskin [9, section 3] the A-torsors relative to U can be identified with the $K(A,1)$-torsors relative to U.

For A-torsors with $A = A_{\mathbf{Mon}}$ as above we now show

PROPOSITION 4.10. There is a one-to-one correspondence between $H^2(\mathcal{G}; A)$ and the set of isomorphism classes of $A_{\mathbf{Mon}}$-torsors relative to the forgetful functor $U : \mathbf{Mon}(\mathbb{V})/\mathcal{G} \to \mathbf{V}/G$.

More explicitly, an $A_{\mathbf{Mon}}$-torsor relative to the forgetful functor U in 4.10 is a $\mathbb{V}$-monoid T, equipped with monoid homomorphisms

$$p : T \to G, \qquad + : T \times_G A \to T, \qquad - : T \times_G T \to A$$

with properties as above, and a section $s : G \to T$, $ps = 1_G$, in $\mathbf{V}$. A morphism of torsors is a monoid homomorphism respecting p, $+$ and $-$.

PROOF. Given an $A_{\mathbf{Mon}}$-torsor T with s as above, assign to it the map

$$f_T = (G \odot G \xrightarrow{(f_1, f_2)} T \times_G T \xrightarrow{-} A),$$

where $f_1 = (G \odot G \xrightarrow{\mu_G} G \xrightarrow{s} T)$ and $f_2 = (G \odot G \xrightarrow{s \odot s} T \odot T \xrightarrow{\mu_T} T)$. One checks easily that f_T is a cocycle, that a different choice of s would give a cohomologous cocycle, and any morphism $T_1 \to T_2$ of torsors produces a 1-cochain whose coboundary is equal to $f_{T_1} - f_{T_2}$.

Conversely, for a 2-cocycle $f : G \odot G \to A$, define a new $\odot$-monoid multiplication on A by

$$\mu_f = (A \odot A \xrightarrow{(1_{A \odot A}, p \odot p)} A \odot A \times_G G \odot G \xrightarrow{\mu \times f} A \times_G A \xrightarrow{+} A).$$

One then checks that this together with $+ : A \times_G A \to A$, $- : A \times_G A \to A$ defines a $A_{\mathbf{Mon}}$-torsor T_f, and cohomologous cocycles yield isomorphic torsors.

Finally, it is straightforward to check that any torsor T is isomorphic to T_{f_T} and any cocycle f is cohomologous to f_{T_f}. ∎

EXAMPLES 4.11. In the example of categories, 1.11, one easily sees that the $A_{\mathbf{Mon}}$-torsors correspond exactly to linear extensions of categories from [5] so that (4.10) corresponds to the result of [5] that the elements of the second cohomology of a category $\mathbf{C}$ classify linear extensions of $\mathbf{C}$. In the example 1.8 one recovers extensions of theories from [15].

Note that in these examples there are also interpretations of the third cohomology, see [2, 14, 28], for example in terms of linear track extensions of categories. These suggest that at least in the presence of a free monoid functor there is an interpretation of $H^3(\mathcal{G}; A)$ by $K(A_{\mathbf{Mon}}, 2)$-torsors. In fact we might expect there to be an explicit correspondence between $K(A, n)$-torsors and $K(A_{\mathbf{Mon}}, n-1)$-torsors, without appealing to cocycles.

Appendix A. Proof of proposition 3.4

Let $\mathcal{G} = (G, \mu, \eta)$ be a monoid in a monoidal category $\mathbf{V}$ and A an internal abelian group in $^{\mathcal{G}}\mathbf{V}^{\mathcal{G}}/G$. We show that the cohomology

$$H^*(\mathcal{G}; A) = \mathrm{Hom}_{\mathcal{G}\mathbf{V}^{\mathcal{G}}/G}(B_\bullet(\mathcal{G}), A)$$

of definition 3.2 agrees with the *cotriple cohomology* of the object $(1_G : G \to G)$ in $^{\mathcal{G}}\mathbf{V}^{\mathcal{G}}/G$, with respect to the free-forget adjunction

$$\mathbf{V}/G \underset{U}{\overset{F}{\underset{\perp}{\rightleftarrows}}} {}^{\mathcal{G}}\mathbf{V}^{\mathcal{G}}/G.$$

Recall that the simplicial object $C_\bullet = C_\bullet(\mathcal{G})$ for the cotriple cohomology has $C_n = (FU)^n(1_G)$ in dimension n; as the free functor is given by

$$F(X \overset{f}{\to} G) = (G \odot X \odot G \xrightarrow{1 \odot f \odot 1} G \odot G \odot G \xrightarrow{\mu^3} G),$$

we have $C_n = (G^{\odot(2n+3)} \xrightarrow{\mu^{2n+3}} G)$. In fact by direct comparison we see that $C_\bullet(\mathcal{G})$ is edgewise subdivision of $B_\bullet(\mathcal{G})$ in the sense of [30]:

DEFINITION A.1. Let K be a simplicial object in a category $\mathbf{C}$. Then the *edgewise subdivision* $\mathrm{Sub}(K)$ of K is the simplicial object with $\mathrm{Sub}(K)_n = K_{2n+1}$ and face and degeneracy maps

$$d_i : \mathrm{Sub}(K)_n \longrightarrow \mathrm{Sub}(K)_{n-1}, \quad s_i : \mathrm{Sub}(K)_n \longrightarrow \mathrm{Sub}(K)_{n+1}$$

given by $d_i d_{2n+1-i} : K_{2n+1} \to K_{2n-1}$ and $s_{2n+2-i} s_i : K_{2n+1} \to K_{2n+3}$ respectively.

Alternatively, let Δ be the category of finite (non-empty) ordered sets and functions, so that a simplicial object is a functor $K : \Delta^{\mathrm{op}} \to \mathbf{C}$. Then $\mathrm{Sub}(K)$ is the composite $K \circ T$ where the functor $T : \Delta \to \Delta$ is defined by

$$T\{0 < 1 < \cdots < n\} = \{0 < 1 < \cdots < n < n' < (n-1)' < \cdots < 0'\}.$$

The edgewise subdivision of a cosimplicial object $L : \Delta \to \mathbf{C}$ is defined similarly.

It is not clear that a (co)simplicial object in a general category should be equivalent to its edgewise subdivision. But for present purposes it is enough to compare the cosimplicial abelian groups obtained by applying the functor $\mathrm{Hom}(-, A)$ to $B_\bullet(\mathcal{G})$ and $C_\bullet(\mathcal{G})$, and proposition 3.4 will follow from the following lemma.

LEMMA A.2 (Subdivision Lemma). Let X be a cosimplicial abelian group in a category $\mathbf{C}$. Then the map $f : X \to \mathrm{Sub}(X)$ defined by

$$f_n = d^{2n+1} d^{2n} \ldots d^{n+1} : X^n \longrightarrow X^{2n+1}$$

induces an isomorphism in cohomology

$$H^*(f) : H^*(X) \xrightarrow{\cong} H^*(\mathrm{Sub}(X)).$$

PROOF. We construct a map $g : \mathrm{Sub}(X) \to X$ and show that the induced cochain maps

$$(X, \partial) \underset{g}{\overset{f}{\rightleftarrows}} (\mathrm{Sub}(X), \delta)$$

are homotopy inverse. Let $g_0 = s^0 : X^1 \to X^0$, and define inductively

$$g_n = g'_{n-1} s^{2n} + (-1)^n s^0 g'_{n-1}$$

Here we adopt notation from [10]: if x is a cosimplicial operator given by a sum of terms $\sum \pm d^{i_1} \ldots d^{i_p} s^{j_1} \ldots s^{j_q}$ then the corresponding *derived operator* x' is given by $\sum \pm d^{i_1+1} \ldots d^{i_p+1} s^{j_1+1} \ldots s^{j_q+1}$. We note that the terms of g_n are precisely those codegeneracy maps $X^{2n+1} \to X^n$ which do not factor through any codegeneracy $s^i s^{2n-i} : X^{2n+1} \to X^{2n-1}$ of $\mathrm{Sub}(X)$.

We show inductively that g is a cochain map, that is, $g_n \delta_n = \partial_n g_{n-1}$. From the definition of g and the relation $\delta_n = d^{2n+1} d^0 - \delta'_{n-1}$ we obtain

$$
\begin{aligned}
g_n \delta_n &= g'_{n-1} s^{2n} \delta_n + (-1)^n s^0 g'_{n-1} d^{2n+1} d^0 - (-1)^n s^0 g'_{n-1} \delta'_{n-1} \\
&= g'_{n-1} s^{2n} \delta_n + (-1)^n g_{n-1} d^{2n} - (-1)^n s^0 \partial'_{n-1} g'_{n-2},
\end{aligned}
$$

by the inductive hypothesis and since $x' d^0 = d^0 x$ for any operator x. Similarly the relation $\partial_n = d^0 - \partial'_{n-1}$ leads to

$$\partial_n g_{n-1} = -g'_{n-1} \delta'_{n-1} s^{2n-2} + g'_{n-1} d^0 + (-1)^n \partial'_{n-1} s^0 g'_{n-2}.$$

The relations for δ and ∂ easily give $s^{2n} \delta_n + \delta'_{n-1} s^{2n-2} - d^0 = d^1(1 - d^{2n-1} s^{2n-2})$ and $s^0 \partial'_{n-1} + \partial'_{n-1} s^0 = 1 - (-1)^n d^n s^0$, hence $g_n \delta_n - \partial_n g_{n-1}$ is given by

$$g'_{n-1} d^1(1 - d^{2n-1} s^{2n-2}) + (-1)^n g_{n-1} d^{2n} + (d^n s^0 - (-1)^n) g'_{n-2}.$$

On substituting $g_{n-1} = g'_{n-2} s^{2n-2} + (-1)^{n-1} s^0 g'_{n-2}$ one sees that this is zero, and g is a cochain map as required.

Next consider the composite cochain map gf. For each n, $g_n f_n : X^n \to X^n$ is a sum of terms

$$g_n f_n = \sum \pm s^{k_0} s^{k_1} \ldots s^{k_n} d^{2n+1} d^{2n} \ldots d^{n+1}$$

where $0 \leqslant k_i < k_{i+1} \leqslant 2n$. There is precisely one non-codegenerate term, with $k_i = n + i$ for $0 \leqslant i \leqslant n$, and this term has positive sign and is the identity. But for cohomology we can assume that we are working in the normalised complex, where codegeneracies are quotiented out, and so we have $H^*(g) H^*(f) = 1$.

Finally, we construct inductively a homotopy h between fg and the identity. Let $h_1 = s^0 s^1 : X^3 \to X^1$ and define inductively

$$h_{n+1} = -h'_n + f'_{n-1} s^0 g'_n : X^{2n+3} \longrightarrow X^{2n+1}$$

for $n \geqslant 2$. The definition of h_n and the relation $\delta_n = -\delta'_{n-1} + d^{2n+1} d^0$ imply

$$
\begin{aligned}
h_{n+1} \delta_{n+1} &= h'_n \delta'_n - f'_{n-1} s^0 g'_n \delta'_n - h'_n d^{2n+3} d^0 + f'_{n-1} s^0 g'_n d^{2n+3} d^0, \\
\delta_n h_n &= \delta'_{n-1} h'_{n-1} - \delta'_{n-1} f'_{n-2} s^0 g'_{n-1} + d^{2n+1} d^0 h_n.
\end{aligned}
$$

Now f and g are cochain maps, and $s^0 \delta'_n + \delta'_{n-1} s^0 = 1$, so

$$f'_{n-1} s^0 g'_n \delta'_n + \delta'_{n-1} f'_{n-2} s^0 g'_{n-1} = f'_{n-1} g'_{n-1}.$$

Also $x' d^0 = d^0 x$ for any operator x, and noting the relations $d^{2n} h_n = h_n d^{2n+2}$, $g_n d^{2n+1} = d^{n+1} g_n$ and $f'_{n-1} d^{n+1} = f_n$ we obtain

$$h'_n d^{2n+3} d^0 = d^{2n+1} d^0 h_n \quad \text{and} \quad f'_{n-1} s^0 g'_n d^{2n+3} d^0 = f_n g_n.$$

Combining the above results gives

$$h_{n+1}\delta_{n+1} + \delta_n h_n - f_n g_n \;=\; h'_n\delta'_n + \delta'_{n-1}h'_{n-1} - f'_{n-1}g'_{n-1}.$$

For $n = 0$ we have $h_1\delta_1 - f_0 g_0 = s^0 s^1(d^3 d^0 - d^2 d^1) - d^1 s^0 = 1$, and since $1' = 1$ it follows that $h_{n+1}\delta_{n+1} + \delta_n h_n = f_n g_n - 1$ for all $n \geqslant 1$. Thus h is a cochain homotopy between fg and the identity as required. ∎

Appendix B. Free monoids

Let $(\mathbf{C}, \odot, I)$ be a monoidal category in which the monoid operation $\odot$ is left distributive over coproducts $\sqcup$ and preserves filtered colimits. In this case we are going to define an explicit free monoid functor which is the left adjoint of the forgetful functor

$$\mathbf{Mon}(\mathbf{C}) \xrightarrow{\quad U \quad} \mathbf{C}.$$

If $\mathbf{C} = R\text{-}\mathbf{Mod}$ then the free monoid on $V \in \mathbf{C}$ is the classical tensor algebra $T(V)$. The assumptions on $\mathbf{C}$ also hold for the monoidal category $\mathbf{C} = \mathbf{Cat}(\mathfrak{S}, R\text{-}\mathbf{Mod})$ in which monoids are operads. In this case the free monoid is the free operad on an $\mathfrak{S}$-object in $R\text{-}\mathbf{Mod}$ which is used for the definition of the bar construction of operads in [19].

Let V be an object of $\mathbf{C}$ and define a sequence of objects V_n by $V_0 = I$ and inductively $V_{n+1} = I \sqcup V \odot V_n$. The first few terms are:

$$V_0 = I, \quad V_1 = I \sqcup V, \quad V_2 = I \sqcup V \odot (I \sqcup V), \quad V_3 = I \sqcup V \odot (I \sqcup V \odot (I \sqcup V)), \dots$$

There are maps $i_n : V_{n-1} \to V_n$ given inductively by $i_{n+1} = 1 \sqcup 1 \odot i_n$, with $i_1 : I \to I \sqcup V$ the natural inclusion of the summand. We define V_∞ by the colimit

$$V_\infty \;=\; \mathrm{colim}\,(V_0 \to V_1 \to V_2 \to V_3 \to \cdots).$$

We will write i for any of the maps $V_n \to V_m$ for $n < m \leqslant \infty$.

There are also maps $\mu_{n,m} : V_n \odot V_m \longrightarrow V_{n+m}$ as follows. Let $\mu_{0,m} = 1_{V_m}$. If $n \geqslant 1$ then $V_n \odot V_m = (I \sqcup V \odot V_{n-1}) \odot V_m = V_m \sqcup V \odot V_{n-1} \odot V_m$ and we define $\mu_{n,m}$ inductively by

$$V_n \odot V_m = V_m \sqcup V \odot V_{n-1} \odot V_m \xrightarrow{\quad \mu_{n,m} = \left(i,\, j_{n+m}(1 \odot \mu_{n-1,m})\right) \quad} V_{n+m}.$$

Here $j_k : V \odot V_{k-1} \to V_k = I \sqcup V \odot V_{k-1}$ is the inclusion of the direct summand.

PROPOSITION B.1. Suppose the tensor product $\odot$ in $\mathbf{C}$ is left distributive over coproducts and preserves filtered colimits, and let V be an object of $\mathbf{C}$. Then the free monoid on V is $T(V) = (V_\infty, \eta, \mu)$, with unit η given by the map $i : I = V_0 \to T(V)$ and multiplication $\mu : T(V) \odot T(V) \to T(V)$ induced by the maps $i\mu_{n,m} : V_n \odot V_m \to T(V)$.

We also write $T_{\leqslant n}(V)$ for V_n. For $\mathbf{C} = R\text{-}\mathbf{Mod}$ the category of R-modules the tensor product is distributive on both sides and we have $T_{\leqslant n}(V) = \bigoplus_{k \leqslant n} V^{\otimes k}$. In this situation the maps i_n are the natural inclusions of summands, and the multiplication structure is given by the isomorphisms $V^{\otimes n} \otimes V^{\otimes m} \cong V^{\otimes(n+m)}$.

PROOF. To show that the multiplication is well defined on the colimit we need the relations $\mu_{n+1,m-1}(i_{n+1} \odot 1) = i_k\mu_{n,m-1} = \mu_{n,m}(1 \odot i_m)$ where $k = n + m$. For $n = 0$ this becomes $(i_m, j_m)(i_1 \odot 1) = i_m = I \odot i_m$. For $n \geq 1$ we have

$$\begin{aligned}
\mu_{n+1,m-1}(i_{n+1} \odot 1) &= (i, j_k(1 \odot \mu_{n,m-1}))(1 \sqcup 1 \odot i_n \odot 1) \\
&= (i, j_k(1 \odot \mu_{n,m-1}(i_n \odot 1))) \\
\mu_{n,m}(1 \odot i_m) &= (i, j_k(1 \odot \mu_{n-1,m}))(i_m \sqcup 1 \odot i_m) \\
&= (i, j_k(1 \odot \mu_{n-1,m}(1 \odot i_m)))
\end{aligned}$$

which are both equal to $(i, j_k(1 \odot i_{k-1})(1 \odot \mu_{n-1,m-1}))$ by the inductive hypothesis. Since $j_k(1 \odot i_{k-1}) = i_k j_{k-1} : V \odot V_{k-2} \to V_k$ this is just $(i, i_k j_{k-1}(1 \odot \mu_{n-1,m-1}))$ which equals $i_k\mu_{n,m-1}$ as required. For the identity laws $\mu(\eta \odot 1) = 1 = \mu(1 \odot \eta)$ we note that $\mu_{0,m} = 1 = \mu_{n,0}$, where $\mu_{n,0} = 1$ follows inductively from the fact that (i, j_n) is the identity on $V_n \odot V_0 = I \sqcup V \odot V_{n-1}$. For the associative law we note that $j_{n+m}(1 \odot \mu_{n-1,m}) = \mu_{n,m}(j_n \odot 1) : V \odot V_{n-1} \odot V_m \to V_{n+m}$, and $i\mu_{q,r} = \mu_{p+q,r}(i \odot 1)$ as above, so that we have inductively

$$\begin{aligned}
\mu_{p,q+r}(1 \odot \mu_{q,r}) &= (i, j_{p+q+r}(1 \odot \mu_{p-1,q+r}))(\mu_{q,r} \sqcup 1 \odot \mu_{q,r}) \\
&= (i\mu_{q,r},\ j_{p+q+r}(1 \odot \mu_{p-1,q+r}(1 \odot \mu_{q,r}))) \\
&= (i\mu_{q,r},\ j_{p+q+r}(1 \odot \mu_{p+q-1,r})(1 \odot \mu_{p-1,q} \odot 1)) \\
&= \mu_{p+q,r}((i \odot 1), (j_{p+q} \odot 1)(1 \odot \mu_{p-1,q} \odot 1)) \\
&= \mu_{p+q,r}(\mu_{p,q} \odot 1).
\end{aligned}$$

This construction is functorial. If $f : V \to W$ is a morphism in $\mathbf{C}$ then $T(f)$ is defined by maps $f_n : V_n \to W_n$ where $f_0 = 1_I$ and $f_n = 1 \sqcup f \odot f_{n-1}$. The map $T(f)$ is well defined since $i_n f_{n-1} = f_n i_n$ is clear inductively. Using this and $j_k(f \odot f_{k-1}) = f_k j_k$ we have

$$f_{n+m}\mu_{n,m} = f_{n+m}(i, j_{n+m}(1 \odot \mu_{n-1,m})) = (if_m, j_{n+m}(f \odot f_{n+m-1}\mu_{n-1,m}))$$

which if $f_{n+m-1}\mu_{n-1,m} = \mu_{n-1,m}(f_{n-1} \odot f_m)$ becomes $(i, j_{n+m}(1 \odot \mu_{n-1,m}))(f_m \sqcup f \odot f_{n-1} \odot f_m)$ which is just $\mu_{n,m}(f_n \odot f_m)$. By induction $T(f)$ is thus a monoid homomorphism.

There is a natural monoid homomorphism $\phi_A : T(A) \longrightarrow A$ for (A, η_A, μ_A) a monoid in $\mathbf{C}$ defined as follows. Let $\phi_0 = \eta_A$ and $\phi_n = (\eta_A, \mu_A(1 \odot \phi_{n-1}))$ for $n \geq 1$. Then $\phi_1 i_1 = \eta_A = \phi_0$, and $\phi_{n+1}i_{n+1} = (\eta_A, \mu_A(1 \odot \phi_n i_n)) = \phi_n$ if $\phi_n i_n = \phi_{n-1}$, so the ϕ_n give a well-defined ϕ_A on $T(A) = A_\infty$. Clearly $\phi_A \eta = \eta_A$. By the unit and associativity laws for A and by the relations $\phi_{n+m}i = \phi_m$ and $\phi_k j_k = \mu_A(1 \odot \phi_{k-1})$ we have

$$\begin{aligned}
\mu_A(\phi_n \odot \phi_m) &= \mu_A((\eta_A, \mu_A(1 \odot \phi_{n-1})) \odot \phi_m) \\
&= (\phi_m, \mu_A(1 \odot \mu_A(\phi_{n-1} \odot \phi_m))), \\
\phi_{n+m}\mu_{n,m} &= (\phi_{n+m}i, \phi_{n+m}j_{n+m}(1 \odot \mu_{n-1,m})) \\
&= (\phi_m, \mu_A(1 \odot \phi_{n+m-1})(1 \odot \mu_{n-1,m})).
\end{aligned}$$

Thus $\mu_A(\phi_A \odot \phi_A) = \phi_A\mu$ follows inductively and ϕ_A is a monoid homomorphism. For any object V of $\mathbf{C}$ we also have a natural map $\psi_V = ij_1 : V \to T(V)$. The freeness of $T(V)$ will now follow from showing that the composites

$$A \xrightarrow{\psi_A} T(A) \xrightarrow{\phi_A} A, \qquad\qquad T(V) \xrightarrow{T(\psi_V)} T(T(V)) \xrightarrow{\phi_{T(V)}} T(V)$$

are the identity. The first of these is clear: $\phi_A \psi_A = \phi_A i j_1 = \phi_1 j_1 = \mu_A(1 \odot \eta_A) = 1$.
Consider the maps $(\psi_V)_n : T_{\leqslant n}(V) \to T_{\leqslant n}(T(V))$ and $\phi_n : T_{\leqslant n}(T(V)) \to T(V)$
which define $T(\psi_V)$ and $\phi_{T(V)}$. Then $\phi_0(\psi_V)_0 = \eta = i : I \to T(V)$, and assuming
inductively that $\phi_{n-1}(\psi_V)_{n-1} = i : V_{n-1} \to T(V)$ we have

$$
\begin{aligned}
\phi_n(\psi_V)_n &= (\eta, \mu(1 \odot \phi_{n-1}))(1 \sqcup (ij_1) \odot (\psi_V)_{n-1}) \\
&= (\eta, i\mu_{1,n-1}(j_1 \odot 1)) = i(i, j_n) = i
\end{aligned}
$$

since $\mu_{1,n-1}(j_1 \odot 1) = j_n : V \odot V_{n-1} \to V_n$. Thus $\phi_{T(V)} T(\psi_V) = 1$. ∎

References

[1] H.-J. Baues, *Quadratic functors and metastable homotopy*, J. Pure & Appl. Algebra **91** (1994) 49–107.

[2] H.-J. Baues and W. Dreckmann, *The cohomology of homotopy categories and the general linear group*, K-Theory, **3** (1989), 307–338.

[3] H.-J. Baues, M. Hartl and T. Pirashvili, *Quadratic categories and square rings*, preprint Max-Planck-Inst. für Math, Bonn, 1995.

[4] H.-J. Baues and T. Pirashvili, *Quadratic endofunctors of the category of groups*, preprint Max-Planck-Inst. für Math, Bonn, 1995.

[5] H.-J. Baues and G. Wirsching, *The cohomology of small categories*, J. Pure Appl. & Algebra **38** (1985), 187–211.

[6] M. Barr and J. Beck, *Homology and standard constructions*, in: "Sem. on Triples and Categorical Homology Theory", Lecture Notes in Math. **80**, Springer, Berlin, 1969, 245–335.

[7] W. Dreckmann, private communication.

[8] J. Duskin, "Simplicial methods and the interpretation of 'triple' cohomology". Mem. AMS **3** (1975), issue 2, no. 163, v+135 pp.

[9] J. Duskin, *Higher-dimensional torsors and the cohomology of topoi: the abelian theory*, in: "Applications of sheaves", Springer Lect. Notes in Math., **753**, Springer, Berlin, 1979, 255–279.

[10] S. Eilenberg and S. Mac Lane, *On the groups $H(\Pi, n)$, I, II*, Ann. Math., **58** 55–106, **60** 49–139, (1953).

[11] A. D. Elmendorf, I. Kriz, M. Mandell and J. P. May, *Modern foundations for stable homotopy*, in: "Handbook of Algebraic Topology", edited I. M. James, North-Holland, Amsterdam, 1995, 213–254.

[12] T. Fox and M. Markl, *Distributive laws, bialgebras and cohomology*, preprint, 1995.

[13] M. Gerstenhaber and S. D. Schack, *Relative Hochschild cohomology, rigid algebras, and the Bockstein*, J. Pure & Appl. Algebra **43** (1986), 53–74.

[14] M. Jibladze, *Bicategory interpretation of the third cohomology group*, Proc. A. Razmadze Math. Inst. **97** (1992), 3–9 (Russian).

[15] M. Jibladze and T. Pirashvili, *Cohomology of algebraic theories*, Journal of Algebra **137** (1991) 253–296.

[16] A. Joyal, *Foncteurs analytiques et espèces de structures*, in: "Combinatoire enumerative", Lecture Notes in Math. **1234**, Springer, Berlin, 1986, 126–159.

[17] A. Joyal and R. Street, *Braided Monoidal Categories*, Adv. Math. **102** (1993), no. 1, 20–78.

[18] E. Getzler and J. D. S. Jones, *Operads, homotopy algebra, and iterated integrals for double loop spaces*, preprint, 1994.

[19] V. Ginzburg and M. M. Kapranov, *Koszul duality for operads*, Duke Mathematical Journal **76** (1994), 203–272.

[20] G. M. Kelly, *Basic concepts of enriched category theory*, London Mathematical Society Lecture Note Series **64**, Cambridge University Press, 1982.

[21] F. W. Lawvere, *Functorial semantics of algebraic theories*, Proc. Nat. Acad. Sci. USA **50** (1963), 869–872.

[22] S. MacLane, *Homology*, Springer, Berlin, 1963.

[23] M. Markl, *Models for operads*, Comm. in Algebra **24** (1996), 1471–1500.

[24] J. P. May, *The geometry of iterated loop spaces*, Lecture Notes in Math. **271**, Springer, Berlin, 1972.

[25] J. P. May, *Definitions: Operads, algebras and modules*. This volume.

[26] B. Pachuashvili, *Cohomologies and extensions in monoidal categories*, J. Pure & Appl. Algebra **72** (1991), 109–147.

[27] T. Pirashvili, *Polynomial functors*, Proc. Math. Inst. Tbilisi **91** (1988), 55–66.

[28] T. Pirashvili, H^3 *and models for the homotopy theory*, Proc. A. Razmadze Math. Inst. **94** (1991), 73–85 (Russian).

[29] D. Quillen, *On the (co-) homology of commutative rings*, in: "Applications of Categorical Algebra", Proc. Sympos. Pure Math., Vol. XVII, New York, 1968, 65–87.

[30] G. Segal, *Configuration spaces and iterated loop spaces*, Invent. Math. **21** (1973), 213–221.

MAX PLANCK INSTITUTE FOR MATHEMATICS, GOTTFRIED-CLAREN-STRASSE 26, 53225 BONN. GERMANY.

Contemporary Mathematics
Volume **202**, 1997

Distributive Laws, Bialgebras, and Cohomology

Thomas F. Fox and Martin Markl

Abstract

The advent of quantum group theory has led to a proliferation of new algebra types, each calling for its own deformation theory and attendant cohomology theory. We here explore two methods of unifying such constructions, using triples and operads.

1 Introduction

There are many ways of describing a class of algebraic structures. The classical approach is to define groups, rings et al. as sets having structure maps satisfying certain axioms. An individual object is given by a presentation in terms of generators and relations. Homological algebra is then a way of gleaning from a particular presentation information about the structure of the given object. However, in this setting there is no natural homology theory attached to each class of algebras. The constructions of homology for groups, rings, associative algebras, commutative algebras, and Lie algebras were each developed separately in an ad hoc manner. As new classes of algebras and bialgebras have arisen in the study of quantum groups, the appropriate homology theories had to be created, but to do so in the classical manner seems to be crude when more unified approaches are available.

Triples and cotriples give a unified way of presenting classes of algebras defined on sets or modules. They lead to a canonical homology theory for each class of algebras, and this may be used to study deformations of algebraic structures. Distributive laws are generalizations of the distributive law of multiplication over addition that we all learned in kindergarten. They were introduced in the context of triples to codify the rules governing abstract algebras having more than one operation. They may be extended to include mixed distributive laws involving triples and cotriples, and this opens the door to the study of the cohomology and deformations of bialgebras, Lie-bialgebras, Poisson Hopf-algebras, etc.

The problem with the categorical approach using triples is that for all their abstract elegance, the complexes used to define homology groups are too unwieldy for computations and do not look like the complexes used to define the corresponding classical cohomology groups. However, they point the way towards an understanding of distributive laws and cohomology in the theory of operads. With a few restrictions, algebras of general type can be presented in this setting, and a natural cohomology theory may be constructed. This has the great advantage of yielding resolutions of classical appearance. These notes are meant as a self-contained introduction to these subjects, with an emphasis on the algebras and bialgebras arising in quantum group theory.

Mathematics Subject Classification: Primary 18C15, Secondary 18G99

This work was partially supported by grants from the FCAR, NSERC, and AV ČR

We start by briefly reviewing the concepts needed for triple cohomology. We then discuss distributive laws between triples and between a triple and a cotriple and show how to compute the cohomology of the resulting algebras and bialgebras using double complexes. In section 6 we give examples, including Poisson algebras and Lie bialgebras, which help to elucidate all our abstract machinery. We then rework everything from the ground up using operads, which yields computationally useful cohomology theories. The interplay between the abstraction of categorical triples and the classicism of operads is the heart of the paper.

We would like to express our thanks to Jim Stasheff for careful reading of the manuscript and many useful remarks.

2   Algebras defined by triples

Throughout this paper, $\mathbf{k}$ will denote a commutative ring or field and $\mathcal{M}$ will denote the category of graded modules over $\mathbf{k}$. We will be interested in algebras X over $\mathbf{k}$. The usual approach to abstract algebra is to define a class or category of algebras as modules X equipped with a set of structure maps $\{\alpha_i\}$ each of the form $\alpha_i : \otimes_{\mathbf{k}}^{n_i} X \longrightarrow X$ (we will often drop the $\mathbf{k}$) and satisfying some set of equations. Triples, or "monads", give an alternate way of describing a class of algebras. The basic observation is this: An algebra is always a quotient of a free algebra in a canonical way, so to build a class of algebras one may start with the construction of free algebras and then define general algebras as quotients of frees (with certain restrictions). A triple is just a functor that acts like a free-algebra functor. It turns out that all classical categories of algebras may be defined by triples on the category $\mathcal{M}$. The free algebra construction may then be iterated to yield a canonical resolution to use when defining homology. This not only unifies extant homology theories but builds a homology theory for each new type of algebras.

Recall that a *triple* $\mathbf{T} = (T, \mu, \eta)$ on $\mathcal{M}$ is a functor $T : \mathcal{M} \longrightarrow \mathcal{M}$ equipped with natural transformations $\eta : 1 \longrightarrow T$ and $\mu : T^2 \longrightarrow T$ satisfying $\mu \cdot \mu T = \mu \cdot T\mu$ and $\mu \cdot T\eta = \mu \cdot \eta T = 1$. A $\mathbf{T}$-*algebra* (A, α) is a $\mathbf{k}$-module A equipped with a multiplication map $\alpha : TA \longrightarrow A$ satisfying

$$(1) \qquad\qquad \alpha \cdot \mu = \alpha \cdot T\alpha \qquad \alpha \cdot \eta = 1$$

A $\mathbf{T}$-algebra map $(A, \alpha) \longrightarrow (B, \beta)$ is a $\mathbf{k}$-linear map $\psi : A \longrightarrow B$ satisfying $\beta \cdot T\psi = \psi \cdot \alpha$, so a map of algebras is just a linear map that preserves the operations, as one would expect. We will use $\mathbf{T}$-*alg* to denote the category of all $\mathbf{T}$-algebras.

If we want to construct associative algebras, TA is just the tensor algebra generated by the module A. Note that we view TA as just a module – the module underlying the free algebra generated by A. Then $\alpha : TA \longrightarrow A$ defines a product on A, while equations (1) ensure that α is unitary, associative, and defined by its action on $A \otimes A$, so $\mathbf{T}$-algebras are just associative algebras in the normal sense. If we let T be the symmetric algebra functor, we will get commutative algebras, while the free Lie algebra functor yields the category of Lie algebras, etc. Hence any construction carried out in this general setting applies equally well to all these classical special cases, as well as to newly created classes of algebras (see e.g. [1]). We will also need the duals of triples. A *cotriple* $\mathbf{G} = (G, \delta, \varepsilon)$ is an endofunctor G with natural transformations $\delta : G \longrightarrow G^2$ and $\varepsilon : G \longrightarrow 1$ satisfying $G\delta \cdot \delta = \delta G \cdot \delta$ and $G\varepsilon \cdot \delta = \varepsilon G \cdot \delta = 1$.

Suppose we are given a category of algebras $\mathcal{A}$ with a pair of adjoint functors

$F : \mathcal{M} \to \mathcal{A}$ and $U : \mathcal{A} \to \mathcal{M}$, the free algebra and underlying module respectively. These give rise to a triple $\mathbf{T}$ on $\mathcal{M}$ with $T = UF$, and a cotriple $\mathbf{G}$ on $\mathcal{A}$ with $G = FU$. Notice that the natural transformation $\varepsilon : GA \to A$ is the multiplication on A. The question of whether or not $\mathcal{A}$ is equivalent to $\mathbf{T}$-*alg* is the question of "tripleability" [7], with which we will not concern ourselves since all naturally occurring categories of algebras are tripleable.

Triples and cotriples lend themselves to the construction of homology and co-homology groups. The usual definition of the triple cohomology groups $H^n(A, B)$ is as follows: The cotriple $\mathbf{G}$ on $\mathcal{A}$ yields the following cotriple-generated simplicial resolution of an algebra A:

(2) $A \leftarrow GA \Leftarrow G^2 A \Lleftarrow G^3 A \Lleftarrow G^4 A \dots$

where the i-th face $G^{n+1}A \to G^n A$ is $G^{n-i}\varepsilon G^i$, for $0 \le i \le n$. In the general theory, one applies a contravariant abelian group-valued functor E to this complex, yielding a complex of abelian groups

(3) $EA \to EGA \rightrightarrows EG^2 A \rightrightarrows EG^3 A \rightrightarrows \dots$

The homology of the associated cochain complex defines the cotriple cohomology groups (denoted $H^n(A, E)_G$ in [6]).

We are interested in the cohomology of A with coefficients in an A-module B, so the appropriate candidate for E is the functor $\mathrm{Der}(-, B)$ ([6] 1.3,1.4). Thus the cohomology groups $H^n(A, B)$ are defined by the "homogeneous" cochain complex of abelian groups given below:

(4) $0 \to \mathrm{Der}(GA, B) \to \mathrm{Der}(G^2 A, B) \to \mathrm{Der}(G^3 A, B) \dots$

$$\partial^{n-1} f = \sum_{i=0}^{n} (-1)^i f \cdot G^i \varepsilon G^{n-i}$$

However, since F is the left adjoint of U, $G = FU$ and $T = UF$, we can convert this to a complex of modules using

$\mathrm{Der}(G^{n+1}A, B) = \mathrm{Der}(FUG^n A, B) \cong \mathrm{Hom}(UG^n A, UB) = \mathrm{Hom}(UT^n A, UB)$

We will drop the superfluous applications of U, and denote $\mathrm{Hom}(-, -)$ by $(-, -)$. If the switch from Der to Hom seems strange, remember that any linear map lifts to a unique derivation from the free algebra. In fact T acts as a linear map $(T^n A, B) \to (T^{n+1}A, TB)$ taking a cochain $f : T^m A \to B$ to its unique lifting as a derivation $T^{m+1}A \to TB$. This action is extended to α by defining $T\alpha : T^2 A \to TA$ to be the unique lifting of α as an algebra map. We now find that the complex (4) above is isomorphic to Beck's "non-homogeneous" complex [8]:

(5) $0 \to (A, B) \to (TA, B) \to (T^2 A, B) \to \dots$

It is easier to give the boundaries for this complex if we first define the following compositions:

2.1 **Definition:** Let $\alpha : TA \to A$ and $\beta : TB \to B$ be two $\mathbf{T}$-algebras over a triple $\mathbf{T} = (T, \mu, \eta)$. For $f \in (T^m A, B)$ define $f \circ \alpha$, $f \circ \mu T^k$ $(0 \le k < m)$ and $\beta \circ f$ in $(T^{m+1}A, B)$ by $f \circ \alpha = f \cdot T^m \alpha$, $f \circ \mu T^k = f \cdot T^{m-k-1}\mu T^k$, and $\beta \circ f = \beta \cdot Tf$.

2.2 **Definition:** If (A, α) and (B, β) are $\mathbf{T}$-algebras, and B is an A-module, the cotriple cohomology groups $H^m(A, B)$ are the homology groups of the cochain complex (5) with the boundaries $\partial^m : (T^m A, B) \to (T^{m+1} A, B)$ given by

$$(6) \qquad \partial^m f = (-1)^m f \circ \alpha + \sum_{k=1}^{m} (-1)^{m+k} f \circ \mu T^{k-1} - \beta \circ f$$

In particular $\partial^0 f = f \cdot \alpha - \beta \cdot Tf$, so the 0-cocycles are precisely the derivations from A to B, while $H^1(A, A)$ classifies infinitesimal deformations of the **T**-algebra A.

2.3   Remarks: We have presumed that B is an A-module, and the reader may take this at face value, but the categorical approach is necessary for the generalization to follow. We could ask that B be an abelian group object in the category **T**-*alg*, but there are none in the classical categories of algebras. To get from **k**-algebras to A-modules in the usual sense, we replace $\mathcal{A}$ by the slice category $\mathcal{A}/A$ of algebras over A. Then an A-module is an abelian group object $B \to A$ in this category [7]. In particular, we have an algebra map $\varphi : A \to B$ (the identity of the group structure) splitting $B \to A$. Hence $B \cong A \oplus M$, where M is the kernel of $B \to A$. Now $\mathrm{Der}(A, B)$ may be thought of as the **k**-derivations from A to B along φ with image in the A-module M.

This is only the classical piece of a more general construction of cohomology groups which deals with higher order derivations between algebras, [13]. There the group (A, B) is the cofree coalgebra generated by $\mathrm{Hom}_{\mathbf{k}}(A, B)$. The definition of $T : (A, B) \to (TA, TB)$ is quite delicate in this setting, but there is a composition of these "enriched" cochains defined as follows: If $f \in (T^m A, B)$ and $g \in (T^n B, C)$, then define $g \circ f$ in $(T^{m+n} A, C)$ by

$$g \circ f = g \cdot T^n f$$

It is quite easy to see that composition satisfies the Leibniz formula $\partial(g \circ f) = \partial g \circ f + (-1)^n g \circ \partial f$ and thus lifts to the level of homology. Hence there is a "composition" (circle product) making $H^*(A, A)$ a graded ring, the "cohomology ring" of the algebra A. There is also an internal grading, since we are dealing with graded modules to begin with, but this plays no role in the triple theoretic cohomology. It will play a substantial role in the operadic theory below.

3   Triples and distributive laws

$$x(y + z) = xy + xz$$

From an abstract point of view, this distributive law says that the operation of multiplication by an element of a ring is a homomorphism of the abelian group underlying the ring. If T_1 is the commutative monoid triple on *Sets* and T_2 is the abelian group triple, then the equation above generates a natural transformation $T_1 T_2 \to T_2 T_1$ taking a product of sums to a sum of products. This allows the functor T_2 to act on the category of monoids as we will see below, so one can then think of a ring as a monoid endowed with an abelian group structure, that is, a ring is an abelian group in the category of monoids. Of course one can also think of a ring as a set with a single structure map from the free ring generated by the set, i.e. you can build the ring structure in steps or all at once, and it is the interplay between these approaches that is arbitrated by the distributive law.

There are many other examples of the same sort of relationship between two algebraic structures. For example, in a Poisson algebra the commutative multiplication and Lie bracket are related by the distributive law

$$[x, yz] = [x, y]z + y[x, z]$$

If $\mathbf{T}_1$ is the Lie algebra triple on a category of modules and $\mathbf{T}_2$ is the commutative algebra triple (the symmetric algebra), then the distributive law $T_1 T_2 \rightarrow T_2 T_1$ takes a bracket of products to a product of brackets. This allows one to think of a Poisson algebra as a Lie algebra equipped with a commutative multiplication, that is, a commutative ring in the category of Lie algebras. The "distributive laws" of [9] generalize these ideas to algebra types defined by more than one operation.

Given triples $\mathbf{T}_1 = (T_1, \mu_1, \eta_1)$ and $\mathbf{T}_2 = (T_2, \mu_2, \eta_2)$ on $\mathcal{M}$, we want to consider $\mathbf{T}_2$ objects in $\mathbf{T}_1$-alg . This only makes sense if we may extend $\mathbf{T}_2$ to act as a triple on the category $\mathbf{T}_1$-alg, so given $\alpha : T_1 A \rightarrow A$ in $\mathbf{T}_1$-alg we would like $T_2 A$ to also be a $\mathbf{T}_1$-algebra, hence we need a map $T_1 T_2 A \rightarrow T_2 A$. This may be achieved by supposing that there is a natural transformation $\lambda : T_1 T_2 A \rightarrow T_2 T_1 A$ and letting $T_2'(A, \alpha) = (T_2 A, T_2 \alpha \cdot \lambda)$. This T_2' will be the functor part of the lifting of $\mathbf{T}_2$ to $\mathbf{T}_1$-alg, so $(T_2 A, T_2 \alpha \cdot \lambda)$ must be a $\mathbf{T}_1$-algebra, and equations (1) put two conditions on λ. First we need $T_2 \alpha \cdot \lambda \cdot \mu_1 T_2 = T_2 \alpha \cdot \lambda \cdot T_1 T_2 \alpha \cdot T_1 \lambda$. The latter equals $T_2 \alpha \cdot T_2 T_1 \alpha \cdot \lambda T_1 \cdot T_1 \lambda$ (by naturality) $= T_2 \alpha \cdot T_2 \mu_1 \cdot \lambda T_1 \cdot T_1 \lambda$ (α is a $\mathbf{T}_1$-algebra, meaning that $\alpha \cdot \mu_1 = \alpha \cdot T_1 \alpha$, and T_2 is natural), so we would like $\lambda \cdot \mu_1 T_2 = T_2 \mu_1 \cdot \lambda T_1 \cdot T_1 \lambda$. Second, we need $T_2' \alpha \cdot \eta_1 T_2 = 1$, which can be ensured by assuming $\lambda \cdot \eta_1 T_2 = T_2 \eta_1$. These conditions on λ make T_2' a functor $\mathbf{T}_1$-$alg \rightarrow \mathbf{T}_1$-$alg$ by defining $T_2' f = T_2 f$ on maps. That this carries algebra maps to algebra maps follows from the naturality of λ.

Now we will get pretty tired of writing expressions like $\lambda T_1 \cdot T_1 \lambda$, so we use Λ to denote the natural transformation $\Lambda : T_1^m T_2^n \rightarrow T_2^n T_1^m$ built from repeated applications of λ. This is well defined, since all ways of moving the T_1s to the back are equal by the naturality of λ. With this notation, the two conditions on λ are

(7) $$\Lambda \cdot \mu_1 T_2 = T_2 \mu_1 \cdot \Lambda \qquad \Lambda \cdot \eta_1 T_2 = T_2 \eta_1$$
$$(T_2 A \text{ is a } \mathbf{T}_1\text{-algebra})$$

To make the functor T_2' into a triple $\mathbf{T}_2' = (T_2', \mu_2', \eta_2')$, we let $\mu_2' = \mu_2$ and $\eta_2' = \eta_2$. However, since these must be $\mathbf{T}_1$-algebra maps, we need two more conditions on λ

(8) $$\Lambda \cdot T_1 \mu_2 = \mu_2 T_1 \cdot \Lambda \qquad \Lambda \cdot T_1 \eta_2 = \eta_2 T_1$$
$$(\mu_2 \text{ and } \eta_2 \text{ are } \mathbf{T}_1\text{-algebra maps})$$

3.1 Definition: A *distributive law* of $\mathbf{T}_1$ over $\mathbf{T}_2$ is a natural transformation $\lambda : T_1 T_2 \rightarrow T_2 T_1$ satisfying (7) and (8).

We may now construct the category of $\mathbf{T}_2'$-algebras in $\mathbf{T}_1$-alg. This is denoted $\mathbf{T}_2\mathbf{T}_1$-alg. Its elements (A, α_1, α_2) are $\mathbf{T}_1$-algebras $\alpha_1 : T_1 A \rightarrow A$ equipped with a second structure map $\alpha_2 : T_2 A \rightarrow A$ making it a $\mathbf{T}_2'$-algebra, so α_2 must satisfy $\alpha_2 \mu_2 = \alpha_2 \cdot T_2 \alpha_2$ and $\alpha_2 \cdot \eta_2 = 1$. Hence (A, α_2) is just a $\mathbf{T}_2$-algebra that has been lifted to $\mathbf{T}_1$-alg. Since α_2 must be a $\mathbf{T}_1$-algebra map from $T_2'(A, \alpha_1)$ to (A, α_1), we also have

(9) $$\alpha_2 \cdot T_2 \alpha_1 \cdot \lambda = \alpha_1 \cdot T_1 \alpha_2$$

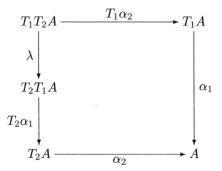

Note that the triple $\mathbf{T}_2$ may be lifted to $\mathbf{T}_1$-*alg only* if such a distributive law exists, [7]. The distributive law λ also allows one to construct the composite triple $\mathbf{T}_{21} = (T_{21}, \mu_{21}, \eta_{21})$ on $\mathcal{M}$ [4]. It is defined by

$$T_{21}A = T_2 T_1 A \qquad \mu_{21} = \mu_2 T_1 \cdot T_2^2 \mu_1 \cdot \Lambda \qquad \eta_{21} = \eta_2 T_1 \cdot \eta_1$$

That this does define a triple follows from the conditions we have already put on λ. One now may look at the category $\mathbf{T}_{21}$-*alg* of $\mathbf{T}_{21}$-algebras in $\mathcal{M}$. A $\mathbf{T}_{21}$-algebra (A, ξ) has a single structure map $\xi : T_2 T_1 A \to A$, but as we can see from the last diagram, any element of $\mathbf{T}_2 \mathbf{T}_1$-*alg* already has such a map. The following theorem is exactly what one would expect — it doesn't matter if you define algebras step by step or all at once.

3.2   Theorem: $\mathbf{T}_{21}$-alg *is isomorphic to* $\mathbf{T}_2 \mathbf{T}_1$-alg.

Proof: ([9]) $\mathbf{T}_2 \mathbf{T}_1$-*alg* $\to \mathbf{T}_{21}$-*alg* is defined by $(A, \alpha_1, \alpha_2) \mapsto (A, \alpha_2 \cdot T_2 \alpha_1)$, while its inverse $\mathbf{T}_{21}$-*alg* $\to \mathbf{T}_2 \mathbf{T}_1$-*alg* is defined by $(A, \xi) \mapsto (A, \xi \cdot \eta_2 T_1, \xi \cdot T_2 \eta_1)$ □

4   Effect on triple cohomology

The interplay between composite structures has its effect on cohomology and deformations, where one might want to consider the two operations separately. Suppose (A, α_1, α_2) is a $\mathbf{T}_2 \mathbf{T}_1$-algebra, and (B, β_1, β_2) is an A-module (or an abelian-group object in the category $\mathbf{T}_2 \mathbf{T}_1$-*alg* – see Remarks 2.3); both $\mathbf{T}_1$ and $\mathbf{T}_2'$ yield non-homogeneous complexes

$$0 \to (A, B) \to (T_1 A, B) \to (T_1^2 A, B) \to \ldots$$
$$0 \to (A, B) \to (T_2 A, B) \to (T_2^2 A, B) \to \ldots$$

Furthermore, each $T_2^n A$ is a $\mathbf{T}_1$-algebra, and so there is a non-homogeneous complex

$$0 \to (T_2^n A, B) \to (T_1 T_2^n A, B) \to (T_1^2 T_2^n A, B) \to \ldots$$

This looks like the beginning of a double complex (below): However, $T_1 A$ is not, in general, a $\mathbf{T}_2'$-algebra (the free monoid generated by a ring is not naturally a group), so the vertical arrows in the complex below are not really the usual boundaries for a non-homogeneous complex.

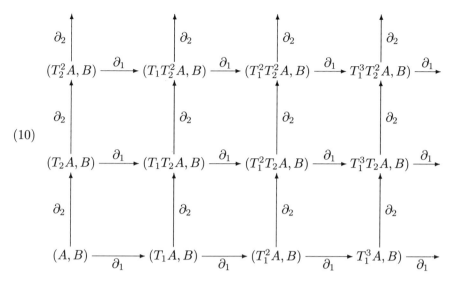

(10)

Once again it is much easier to give the boundaries for this complex if we first define a circle product involving the cochains, μ_2, α_2, and β_2.

4.1 Definition: Let $f \in (T_1^j T_2^i A, B)$. Then define

$$f \circ \alpha_2 = f \cdot T_1^j \cdot T_2^{i-1} \alpha_2$$
$$\beta_2 \circ f = \beta_2 \cdot T_2 f \cdot \Lambda^{ij}$$
$$f \circ \mu_2 = f \cdot T_1^j T_2^{i-1} \mu_2$$

As before, T takes a cochain to its unique lifting as a derivation. We have denoted by Λ^{ij} the variant of Λ which keeps the first i instances of T_2 fixed and shifts j instances of T_1 to the front of the remaining T_2 :

$$\Lambda^{ij} : T_1^j T_2^{i+1} \to T_2 T_1^j T_2^i$$

The vertical complexes now have the same boundary formula as the usual non-homogeneous complex, though they have been twisted, a fact that is hidden in the definition of the circle product. One may easily check that these are indeed cochain complexes, the key being that this new circle product is associative (as are all the circle products discussed in this paper). It is then a formality to check that this circle product satisfies the Leibniz formula. Note that there is a circle product on the entire double complex, which (of course) also satisfies the Leibniz formula.

4.2 Definition: Let $f \in (T_1^j T_2^i A, B)$ and $g \in (T_1^n T_2^m B, C)$. Then define

$$g \circ f = (-1)^{ij} g \cdot T_1^n T_2^m f \cdot \Lambda^{ij} \in (T_1^{n+j} T_2^{m+i} A, C)$$

The homology of the total complex of this double complex above defines the cohomology of the $\mathbf{T}_2 \mathbf{T}_1$-algebra A with coefficients in B. By Beck's theorem above, we could also consider A and B as $\mathbf{T}_{21}$-algebras and look at the usual triple cohomology using the cotriple $\mathbf{T}_{21}$. The chain complex defining these cohomology groups looks like

(11) $$0 \to (A, B) \to (T_{21} A, B) \to (T_{21}^2 A, B) \to (T_{21}^3 A, B) \dots$$

This looks a far cry from the double complex (10), but the main result of [9] is that (11) is homotopy equivalent to the diagonal complex of (10), i.e.

$$0 \to (A, B) \to (T_2 T_1 A, B) \to (T_2^2 T_1^2 A, B) \to (T_2^3 T_1^3 A, B) \ldots$$

This combined with the Eilenberg-Zilber theorem shows that the cohomology of these three complexes are the same. Note that all of these chain complexes are those associated to obvious simplicial complexes, this being necessary for the application of the Eilenberg-Zilber theorem [18].

5   Bialgebras using triples

So far we have looked at objects with *algebraic* structures, i.e. their operations are maps of the form $\otimes^n A \to A$. Bialgebras also have a *coalgebraic* operation — a map $A \to A \otimes A$ — so we must deal with these types of structures before considering bialgebras. In general an abstract coalgebra has operations of the form $C \to \otimes^n C$. Coalgebras are defined by cotriples on $\mathcal{M}$ just as algebras are defined by triples.

For example, let $\mathcal{C}$ denote the category of coassociative coalgebras in $\mathcal{M}$. There is a pair of adjoint functors $K : \mathcal{M} \to \mathcal{C}$, $U : \mathcal{C} \to \mathcal{M}$, where U is the obvious forgetful functor, and K is the cofree-coalgebra functor [14]. The composite $S = UK$ induces a cotriple $\mathbf{S} = (S, \delta, \varepsilon)$ on $\mathcal{M}$, and the category $\mathcal{C}$ is equivalent to the category of $\mathbf{S}$-coalgebras. An $\mathbf{S}$-coalgebra (C, c) is a module C equipped with a map $c : C \to SC$ satisfying the usual identities: $\varepsilon \cdot c = 1$ and $\delta \cdot c = Sc \cdot c$. These ensure that an $\mathbf{S}$-coalgebra is determined by a coassociative "comultiplication" or "diagonal" $C \to C \otimes C$. Of course, using the cocommutative variant of $\mathbf{S}$ gives cocommutative coalgebras, using the Lie version of $\mathbf{S}$ gives Lie coalgebras, etc. Note that these are all subtriples of the triple which defines non-associative non-cocommutative coalgebras, just as free algebras of special types are generally quotients of the free non-associative non-commutative algebra.

The cohomology of coalgebras is defined in a manner dual to that used for algebras [32]. The simplicial cocomplex used is generated by repeated applications of S, that is if $b : B \to SB$ and $c : C \to SC$ are coalgebras the groups $H^n(B, C)$ are defined by a complex $(B, S^*C) = \operatorname{Hom}_k(B, S^*C)$ (S^* denotes the iterated functor S) whose boundary maps depend on the comultiplications on B and C. As in the algebraic case, the cochains form abelian groups. This is achieved by taking B to be a C-comodule, i.e. an abelian cogroup in the category $C/\mathcal{C}$. Such an object has a map $B \to C$ and the function $S : (B, C) \to (SB, SC)$ takes a linear map to its unique lifting as a coderivation along that map, though $Sc : SC \to S^2C$ is the unique lifting of c to a map of coalgebras. The group $H^0(B, C)$ is the group of coderivations from B to C, while $H^1(C, C)$ classifies infinitesimal deformations of the $\mathbf{S}$-coalgebra C (see [12]). Once again there is a circle product defined on the level of enriched cochains. It satisfies the usual Leibniz formula and is defined by

$$g \circ f = S^m g \cdot f$$

where $f \in (B, S^m C)$ and $g \in (C, S^n E)$. The boundaries for the coalgebra cochain complex above are then given by

$$d^n f = c \circ f + \sum_{i=1}^{n} (-1)^i \delta S^{i-1} \circ f + (-1)^{n+1} f \circ b$$

Now suppose that we are also given a triple $\mathbf{T} = (T, \mu, \eta)$ on $\mathcal{M}$, hence the category $\mathbf{T}$-*alg* of $\mathbf{T}$-algebras in $\mathcal{M}$. For example, T could be the tensor algebra triple or the Lie algebra triple, yielding the category of associative or Lie algebras respectively. To ensure harmony between $\mathbf{S}$ and $\mathbf{T}$, that is, to make sure that the triple $\mathbf{T}$ lifts to $\mathbf{S}$-*coalg* and $\mathbf{S}$ lifts to $\mathbf{T}$-*alg*, we insist that there be a "mixed" distributive law between them [33].

5.1 Definition: If $\mathbf{T}$ is a triple and $\mathbf{S}$ is a cotriple, a *mixed distributive law* $\lambda : TS \rightarrow ST$ *between* $\mathbf{T}$ *and* $\mathbf{S}$ is a natural transformation satisfying the following four conditions:

$$\text{(i)} \quad \Lambda \cdot \mu S = S\mu \cdot \Lambda \quad \text{(ii)} \quad \delta T \cdot \Lambda = \Lambda \cdot T\delta$$

$$\text{(iii)} \quad \lambda \cdot \eta S = S\eta \quad \text{(iv)} \quad \varepsilon T \cdot \lambda = T\varepsilon$$

We have again used Λ to denote the variant of λ which commutes appropriate compositions of T and S. These equations have the following consequences: Given a $\mathbf{T}$-algebra A, equations *(i)* and *(iii)* ensure that SA is again a $\mathbf{T}$-algebra, while *(ii)* and *(iv)* ensure that δ and ε are $\mathbf{T}$-algebra maps respectively. Dually, given an $\mathbf{S}$-coalgebra C, equations *(ii)* and *(iv)* ensure that TC is again an $\mathbf{S}$-coalgebra, while *(i)* and *(iii)* ensure that μ and η are $\mathbf{S}$-coalgebra maps.

Given $\mathbf{T}$ and $\mathbf{S}$ and the distributive law λ as above, a $\mathbf{TS}$-bialgebra (B, β, b) is a module B equipped with two structure maps $\beta : TB \rightarrow B$ and $b : B \rightarrow SB$, making it a $\mathbf{T}$-algebra and an $\mathbf{S}$-coalgebra. Further, the structure maps β and b are required to be compatible in the sense that they must satisfy $S\beta \cdot \lambda \cdot Tb = b \cdot \beta$, that is, the following diagram must commute:

(12)

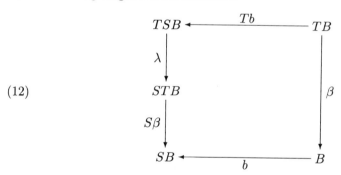

$\mathbf{TS}$-bialgebras may be thought of as $\mathbf{S}$-coalgebras in $\mathbf{T}$-*alg*, or as $\mathbf{T}$-algebras in $\mathbf{S}$-*coalg*. A map between two $\mathbf{TS}$-bialgebras is just a linear map that is both a $\mathbf{T}$-algebra map and an $\mathbf{S}$-coalgebra map. The resulting category is denoted $\mathbf{TS}$-*bialg* or $\mathbf{ST}$-*bialg*. Note that $S\beta \cdot \lambda : TSB \rightarrow SB$ makes SB a $\mathbf{T}$-algebra, and $\lambda \cdot Tb : TB \rightarrow STB$ makes TB an $\mathbf{S}$-coalgebra. Then (12) says β is a coalgebra map and b is an algebra map. The reader should compare this with (9) above.

If (A, α, a) is also a bialgebra, the bialgebra cohomology groups of A with coefficients in B are defined via a double complex (T^*A, S^*B) (below) whose boundaries depend on the structure maps of A and B, as well as λ, [33]. The cochains in this case are biderivations between the two bialgebras, while the boundaries of the double complex are just the usual boundaries for algebra and coalgebra cohomology if we put the $\mathbf{S}$-coalgebra structure on $T^m A$ and the $\mathbf{T}$-algebra structure on $S^n B$ as above. To define the cohomology groups for a particular category of bialgebras, one need only define the cotriple $\mathbf{S}$, the triple $\mathbf{T}$, and the distributive law λ.

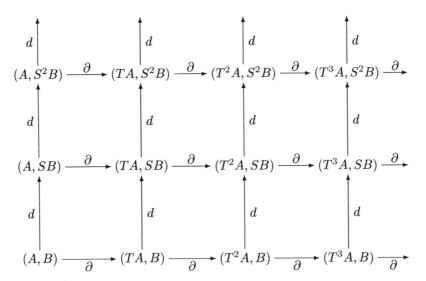

As in the algebraic and coalgebraic situations, there is a circle product on the double complex and it carries over to the total complex. Let $C^k(A, B)$ and $\mathbf{d}$ respectively denote the k-cochains and boundary of the total complex. We have from [12]

$$C^k(A, B) = \oplus_{m+n=k}(T^m A, S^n B)$$
$$\mathbf{d}^{m,n} = (\partial + (-1)^{n+1}d) : (T^m A, S^n B) \longrightarrow (T^{m+1} A, S^n B) \oplus (T^m A, S^{n+1} B)$$

There is again a circle product on the total complex defined by

$$\circ : (T^m B, S^n C) \otimes (T^i A, S^j B) \longrightarrow (T^{i+m} A, S^{j+n} C)$$
$$g \circ f = (-1)^{in} S^j g \cdot \Lambda \cdot T^m f$$

This allows us to write the equations for a **TS** bialgebra (B, β, b) as follows:

$$\beta \circ \mu = \beta \circ \beta \qquad \delta \circ b = b \circ b \qquad \beta \circ b = b \circ \beta$$

For the reader's convenience we note the following useful equations, which follow by naturality and the equations defining a bialgebra. Here $f \in (T^m B, S^n C)$, $g \in (T^i A, S^j B)$:

$$f \circ \mu S \circ g = f \circ \mu \circ g \qquad f \circ \delta T \circ g = f \circ \delta \circ g$$
$$f \circ \mu T^{i+m} = \mu T^i S^n \circ f \qquad \delta S^{i+n} \circ f = f \circ \delta S^i T^m$$

6   Examples

6.1   Example: The first useful example of a distributive law was, of course, the distributive law of multiplication over addition in a ring. In this setting rings are viewed as sets which have two algebraic structures, multiplication and addition, given by the semigroup triple $\mathbf{T}_1$ on *Sets* and the abelian group triple $\mathbf{T}_2$, with their natural distributive law $T_1 T_2 \rightarrow T_2 T_1$. That the cohomology of rings viewed as $\mathbf{T}_{21}$-algebras over sets is Shukla cohomology is the main result of [2].

It is instructive to formulate diagram (9) for this simple example using the more usual presentation of multiplication and addition. First write

$$
\begin{aligned}
T_1 X &= X + X \times X + X \times X \times X + \cdots \\
T_2 X &= 1 + X + X * X + X * X * X + \cdots
\end{aligned}
$$

where we have used $*$ to denote the symmetrized product. One usually thinks of the multiplication $T_1 X \rightarrow X$ as being defined by its quadratic piece $\cdot : X \times X \rightarrow X$ (we hope that the reader understands that the dot '·' denotes the operation). Similarly the addition $T_2 X \rightarrow X$ is defined by $+ : X * X \rightarrow X$. Then (9) becomes

(13)

$$
\begin{array}{ccc}
(X * X) \times (X * X) & \xrightarrow{\ +\times+\ } & X \times X \\
{\scriptstyle \lambda}\big\downarrow & & \big\downarrow {\scriptstyle \cdot} \\
(X \times X) * (X \times X) & & \\
{\scriptstyle \cdot\,*\,\cdot}\big\downarrow & & \\
X * X & \xrightarrow{\quad + \quad} & X
\end{array}
$$

where λ is given by

(14) $$ (w * x, y * z) \mapsto (w, y) * (w, z) + (x, y) * (x, z) $$

Of course $X \times X$ and $X * X$ are only pieces of T_1 and T_2 respectively, but the diagram above gives the idea of how to define λ in general. Note that λ defines a monoid structure on $X * X$ with respect to which $+$ becomes a map of monoids. Note also that (14) is a bit of a cheat, since we have used the addition in $(X \times X) * (X \times X)$. Clearly the distributive law should be viewed as a map

(15) $$ \lambda : (X * X) \times (X * X) \longrightarrow (X \times X) * (X \times X) * (X \times X) * (X \times X) $$

This may seem like a trivial point, but it shows that one should not expect the distributive law to be defined between quadratic terms only.

6.2   Example: As mentioned above a Poisson algebra is a module equipped with an associative commutative product and a Lie bracket connected by the equation

(16) $$ [x, yz] = [x, y]z + y[x, z]. $$

If $\mathbf{T}_1$ is the Lie algebra triple on a category of modules and $\mathbf{T}_2$ is the commutative algebra triple, then (16) above yields a distributive law

(17) $$ \lambda : T_1 T_2 \rightarrow T_2 T_1 $$

taking a bracket of products to a product of brackets. Regard $T_1 A$ as a quotient of the exterior algebra $A + A \wedge A + A \wedge A \wedge A \cdots$ and write

$$ T_2 A = k + A + A * A + A * A * A + \cdots $$

Then (16) shows how to define a map $A \wedge (A * A) \longrightarrow (A \wedge A) * A + A * (A \wedge A)$ which is a piece of λ. In this case the diagram corresponding to (13) becomes

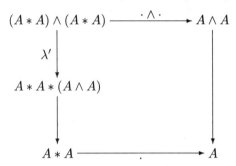

The map λ' above is the piece of λ coming from the equation $[wx, yz] = wy[x, z] + wz[x, y] + xy[w, z] + xz[w, y]$ which we must check is well-defined by (16). As in the previous example we have cheated a bit by using the algebra of $A * A$ to define the left edge of the diagram above. The lower left corner should be $A * A * A$.

As mentioned in section 3, another way of looking at this problem is as follows: Given a Lie algebra $\alpha : T_1 A \longrightarrow A$, the map $T_2 \alpha \lambda$ should define a Lie algebra structure on $T_2 A$. (More traditionally speaking, given a Lie algebra $[-, -] : A \wedge A \longrightarrow A$, the map $(A * A) \wedge (A * A) \longrightarrow (A * A)$ should define a Lie algebra structure on $(A * A)$.) We must check that (16) gives a well-defined map, and that it satisfies the requisite identities, including the Jacobi equation. A simple induction shows we need only check products of the forms $[wx, yz]$ and $[w, xyz]$. The commutativity of the product then gives well-definedness, and the fact that α already defines a Lie algebra takes care of the rest. It has been shown by the second author that this example is typical of all "quadratic" algebras (see section 9 below or [24]).

6.3 Example: Before turning to bialgebras, we mention an odd-looking example that seems to be an extreme case. Consider a module A having two associative multiplications $\cdot : A \otimes A \longrightarrow A$ and $\langle -, - \rangle : A \otimes A \longrightarrow A$ connected by the equations

(18) $x \cdot \langle y, z \rangle = \langle x \cdot y, z \rangle$ $\langle x, y \rangle \cdot z = \langle x, y \cdot z \rangle$

These algebras were introduced by the second author in [24] and called there *non-symmetric Poisson algebras*. Letting TA denote the tensor algebra generated by A, the multiplications $\cdot$ and $\langle -, - \rangle$ yield maps α_1 and $\alpha_2 : TA \longrightarrow A$, and we have a distributive law $\lambda : TTA \longrightarrow TTA$ defined by (18). To understand this let both AA and $A \cdot A$ denote $A \otimes A$. Then

$$TTA = A + AA + A \cdot A + AAA + A \cdot A \cdot A + A \cdot (AA) + (AA) \cdot A + \cdots$$

The distributive law is determined by $A \cdot (AA) + (AA) \cdot A \mapsto (A \cdot A)A + A(A \cdot A)$ using (18). This is an example of a distributive law between "two" triples whose inverse is again a distributive law.

6.4 Example: As an example of the bialgebra situation, we consider classical associative/coassociative bialgebras. If **T** is the tensor algebra triple and **S** is the coassociative coalgebra cotriple, a bialgebra B has a multiplication $\beta : TB \longrightarrow B$ and a comultiplication $b : B \longrightarrow SB$ defined by $\cdot : B \otimes B \longrightarrow B$ and $\Delta : B \longrightarrow B \otimes B$. The multiplication and diagonal are classically connected by the equation

$$\Delta(x \cdot y) = (\Delta x) \cdot (\Delta y)$$

The mixed distributive law $\lambda B : TSB \longrightarrow STB$ is then generated by the usual middle interchange $(B \otimes B) \otimes (B \otimes B) \longrightarrow (B \otimes B) \otimes (B \otimes B)$, which is exactly what is needed to put an algebra or coalgebra structure on $B \otimes B$. To write this completely in terms of the triple and cotriple would necessitate a discussion of the cofree coalgebra construction, which we do not want to go into here (see [14] for details). Suffice to say that SB may be realized as $B \times (B \star B) \times (B \star B \star B) \times \cdots$ where this is a submodule of $B \times (B \otimes B) \times (B \otimes B \otimes B) \times \cdots$. The map $b : B \longrightarrow SB$ sends x to $(x, \Delta x, (\Delta \otimes 1)\Delta x, \ldots)$, which is well-defined since we are looking at coassociative coalgebras, and the distributive law λ then looks like

$$(19) \quad (B \times B \star B \cdots) + (B \times B \star B \cdots) \otimes (B \times B \star B \cdots) \longrightarrow$$
$$(B + B \otimes B \cdots) \times ((B + B \otimes B \cdots) \star (B + B \otimes B \cdots)) \cdots$$

The middle interchange defines λ on the quadratic pieces of TSB, i.e. it can be thought of as a map $(B \star B) \otimes (B \star B) \longrightarrow (B \otimes B) \star (B \otimes B)$ and (12) becomes

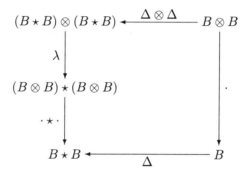

6.5 Example: We may also look at Lie bialgebras, i.e. Lie algebras B which also have a Lie-diagonal $\Delta : B \longrightarrow B \otimes B$ [27]. In this case, the triple **T** is the Lie algebra triple, while the cotriple **S** defines Lie coalgebras. The usual way of presenting the connection between the diagonal and bracket product is $\Delta[x, y] = [\Delta x, y] + [x, \Delta y]$ or

$$(20) \ \Delta[x, y] = \sum [x_{(1)}, y] \otimes x_{(2)} + [x, y_{(1)}] \otimes y_{(2)} + x_{(1)} \otimes [x_{(2)}, y] + y_{(1)} \otimes [x, y_{(2)}]$$

where $\Delta x = \sum x_{(1)} \otimes x_{(2)}$ and $\Delta y = \sum y_{(1)} \otimes y_{(2)}$. This serves as a guide to the definition of $\lambda : TSB \longrightarrow STB$, which may be written as a map

$$(21) \quad (B \times B \star B \cdots) + (B \times B \star B \cdots) \wedge (B \times B \star B \cdots) \longrightarrow$$
$$(B + B \wedge B \cdots) \times ((B + B \wedge B \cdots) \star (B + B \wedge B \cdots)) \cdots$$

Equation (20) shows the second coordinate (the diagonal) of λ acting on the quadratic piece $(B \star B) \wedge (B \star B)$ of TSB:

$$(B \star B) \wedge (B \star B) \longrightarrow (B \wedge B) \star B + B \star (B \wedge B)$$

In this case (12) becomes

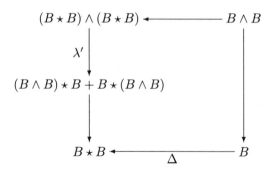

6.6 Example: We end these examples by considering trialgebras, i.e. modules equipped with three operations. We will look at the situation involving two triples $\mathbf{T}_1$ and $\mathbf{T}_2$ and a cotriple $\mathbf{S}$. An $\mathbf{ST}_2\mathbf{T}_1$-*trialgebra* A will have three structure maps $\alpha_1 : T_1 A \rightarrow A$, $\alpha_2 : T_2 A \rightarrow A$ and $a : A \rightarrow SA$. The question is what relationship must there be between the three functors for this to determine a reasonable class of algebras. First of all we must assume there is a distributive law $\lambda : T_1 T_2 \rightarrow T_2 T_1$, so that we may look at the category of $\mathbf{T}_2\mathbf{T}_1$-*alg* of $\mathbf{T}_2\mathbf{T}_1$-algebras. Then we look for $\mathbf{S}$-coalgebras in $\mathbf{T}_2\mathbf{T}_1$-*alg* so we must be able to lift the cotriple $\mathbf{S}$.

The easiest situation occurs when there are two more mixed distributive laws $\lambda_1 : T_1 S \rightarrow ST_1$ and $\lambda_2 : T_2 S \rightarrow ST_2$. We may then sensibly look at the category $(\mathbf{T}_2\mathbf{S})\mathbf{T}_1$-*trialg* of $\mathbf{T}_2\mathbf{S}$-bialgebras in $\mathbf{T}_1$-*alg*, the category $\mathbf{T}_2(\mathbf{ST}_1)$-*trialg* of $\mathbf{T}_2$ algebras in $\mathbf{ST}_1$-*bialg*, the category $\mathbf{S}(\mathbf{T}_2\mathbf{T}_1)$-*trialg* of $\mathbf{S}$ coalgebras in $\mathbf{T}_2\mathbf{T}_1$-*alg* , or the category $\mathbf{ST}_{21}$-*trialg* of $\mathbf{S}$ coalgebras in $\mathbf{T}_{21}$-*alg* . These four categories are, of course, identical, and the last makes sense because we have a distributive law

$$(22) \qquad T_2\lambda_1 \cdot \lambda_2 T_1 : T_{21}S \rightarrow ST_{21}$$

The cohomology of such trialgebras is straightforward, being given by a tricomplex which the reader can easily construct. Unfortunately, the only natural examples of trialgebras we know do not fit this neat scheme. In particular, a distributive law $T_{21}S \rightarrow ST_{21}$ need not decompose as in (22).

Recall that a Poisson bialgebra B is a Poisson algebra equipped with a coassociative diagonal $\Delta : B \rightarrow B \otimes B$ satisfying

$$(23) \qquad \Delta(x \cdot y) \quad = \quad (\Delta x) \cdot (\Delta y)$$
$$(24) \qquad \Delta[x, y] \quad = \quad [x_{(1)}, y_{(1)}] \otimes x_{(2)} y_{(2)} + x_{(1)} y_{(1)} \otimes [x_{(2)}, y_{(2)}]$$

One usually thinks of B as a classical commutative/coassociative bialgebra having an added Lie multiplication, but this is a bit deceptive (see [31]). Let $\mathbf{T}_1$, $\mathbf{T}_2$, and λ be as in (17) above, and let $\mathbf{S}$ be the cofree coalgebra cotriple. Then equation (23) determines a distributive law $\lambda_2 : T_2 S \rightarrow ST_2$ defined by the usual middle interchange (c.f. (19)), but there is no distributive law of the form $\lambda_1 : T_1 S \rightarrow ST_1$ (this explains why there is no category of bialgebras having a Lie multiplication and coassociative comultiplication). Rather, equation (24) determines a distributive law $T_{21}S \rightarrow ST_{21}$, so Poisson bialgebras should be thought of as coassociative coalgebras in the category of Poisson algebras. Their cohomology groups are determined by a double complex of the form (T_{21}^*, S^*), each row (T_{21}^*, S^n) is homotopic to a double complex $(T_1^* T_2^*, S^n)$, and there are well-defined bicomplexes $(T_1^m T_2^*, S^*)$, but the whole mess does not determine a tricomplex since the double complexes $(T_1^* T_2^m, S^*)$ remain undefined. The implication for deformation theory (quantization) is that one may not look for deformations of a Poisson bialgebra that leave *one* of the multiplications fixed.

Finally we must point out that quasi-whatever bialgebras do not fit into the scenario of triples and distributive laws, since free and cofree objects do not exist unless one fixes the inner automorphisms involved in the definitions of such objects.

7   Algebras defined by operads

We now turn our attention to an alternative description of algebras and their cohomology. As in the first part of the paper, we work over the category $\mathcal{M}$ of graded **k**-modules, but here the graded structure of $\mathcal{M}$ becomes more manifest because we must immediately address the question of signs. In the triple case the differential of an n-cochain was always a summation of $(n+1)$ pieces (see formula (4)) and the sign convention was the obvious one. This is no longer true for the operadic cohomology which we are going to introduce now, where the signs are a much more delicate matter (for example the Chevalley-Eilenberg cohomology , which is the operadic cohomology of Lie algebras). A solution of this problem is the systematic use of the Koszul sign convention (recalled below) which give us a proper sign convention almost for free. *From now on, we will always assume that the ground field **k** is of characteristic zero.*

For two graded vector spaces, $V, W \in \mathcal{M}$, let $\mathcal{M}^p(V, W)$ denote the set of linear homogeneous maps $f : V \longrightarrow W$ of degree p, and denote by $\mathcal{M}(V, W)$ the graded vector space $\bigoplus_p \mathcal{M}^p(V, W)$. For $V \in \mathcal{M}$, let $\uparrow V$ (resp. $\downarrow V$) be the suspension (resp. the desuspension) of V, i.e. the graded vector space defined by $(\uparrow V)_p = V_{p-1}$ (resp. $(\downarrow V)_p = V_{p+1}$). By $\#V$ we denote the graded dual of V, i.e. the graded vector space $\bigoplus_p (\#V)_p$ with $(\#V)_p = \mathcal{M}^p(V, \mathbf{k}) = \mathcal{M}(V_{-p}, \mathbf{k})$, the space of linear maps from V_{-p} to **k**. For a graded vector space V we have the natural maps $\uparrow: V \longrightarrow \uparrow V$ and $\downarrow: V \longrightarrow \downarrow V$. We will consider $\mathcal{M}$ as a *symmetric monoidal category* with the symmetry isomorphism given by $x \otimes y \mapsto (-1)^{|x| \cdot |y|} y \otimes x$. The symmetry isomorphism is based on the *Koszul sign convention*, meaning that whenever we commute two "things" of degrees p and q, respectively, we multiply the sign by $(-1)^{pq}$. We will systematically use this convention throughout the paper. Let us recall some basic definitions of [26].

7.1   Definition: By an *operad* we mean an operad in the symmetric monoidal category $\mathcal{M}$, i.e. a sequence $\mathcal{P} = \{\mathcal{P}(n); n \geq 1\}$ of graded vector spaces such that:

(i) Each $\mathcal{P}(n)$ is equipped with a **k**-linear (left) action of the symmetric group Σ_n on n elements, $n \geq 1$.

(ii) For any $m_1, \ldots, m_l \geq 1$ we have degree zero linear maps (called the *composition maps*)

$$\gamma = \gamma_{m_1, \ldots, m_l} : \mathcal{P}(l) \otimes \mathcal{P}(m_1) \otimes \cdots \otimes \mathcal{P}(m_l) \longrightarrow \mathcal{P}(m_1 + \cdots + m_l).$$

These data have to satisfy the usual axioms including the existence of a unit $1 \in \mathcal{P}(1)$, for which we refer again to [26]. We sometimes write $\mu(\nu_1, \cdots, \nu_l)$ or $\mu(\nu_1 \otimes \cdots \otimes \nu_l)$ instead of $\gamma(\mu \otimes \nu_1 \otimes \cdots \otimes \nu_l)$.

Cooperads are defined in the dual manner. Hence a *cooperad* is a sequence $\mathcal{Q} = \{\mathcal{Q}(n); n \geq 1\}$ of graded vector spaces such that each $\mathcal{Q}(n)$ has an action of the symmetric group Σ_n, $n \geq 1$, and, for any $n \geq 1$, the map

$$\nu(n) : \mathcal{Q}(n) \longrightarrow \bigoplus \mathcal{Q}(l) \otimes \mathcal{Q}(m_1) \otimes \cdots \otimes \mathcal{Q}(m_l)$$

is given, where the summation is taken over all $l, m_1, \ldots, m_l \geq 1$ with $m_1 + \cdots + m_l = n$. These maps have to satisfy some axioms which are exactly the duals of the axioms of an operad. As we will need cooperads only marginally, we just say that, under an obvious finite-dimension assumption, the dual of an operad is a cooperad and vice versa.

By a *map* $\chi : \mathcal{P} \longrightarrow \mathcal{Q}$ of operads we mean a sequence $\chi = \{\chi(n) : \mathcal{P}(n) \longrightarrow \mathcal{Q}(n); \ n \geq 1\}$ of degree zero Σ_n-invariant linear maps, such that $\chi(1)(1) = 1$ and the sequence χ commutes in the obvious sense with the composition maps. We denote by *Oper* the category of operads and their maps in the above sense. As a matter of fact, we will tacitly assume that all operads $\mathcal{P}$ considered in our paper are such that $\mathcal{P}(n)$ is of finite type for any $n \geq 1$, and that the unitary ring $\mathcal{P}(1)$ is isomorphic to the ground field $\mathbf{k}$, with two very important exceptions – the endomorphism operad $\mathcal{E}_V$ and the dual endomorphism operad $\mathcal{G}_W$ defined below.

For a graded vector space V, define the operad $\mathcal{E}_V$ as follows: Let $\mathcal{E}_V(n) = \mathcal{M}(V^{\otimes n}, V)$ and define the composition maps by the usual composition, i.e.

$$\gamma_{m_1,\ldots,m_l}(f \otimes \varphi_1 \otimes \cdots \otimes \varphi_l) = f(\varphi_1 \otimes \cdots \otimes \varphi_l),$$

for $f \in \mathcal{E}_V(l)$ and $\varphi_i \in \mathcal{E}_V(m_i)$, $l \geq 1$ and $m_1, \ldots, m_l \geq 1$. The action of the symmetric group is given, for $f \in \mathcal{E}_V(n)$ and $\sigma \in \Sigma_n$, by

$$(25) \qquad (\sigma f)(x_1, \ldots, x_n) = \epsilon(\sigma; x_1, \ldots, x_n) \cdot f(x_{\sigma(1)}, \ldots, x_{\sigma(n)}),$$

where $\epsilon(\sigma; x_1, \ldots, x_n)$ (the *Koszul sign*) is determined by the equation

$$(26) \qquad x_1 \wedge \cdots \wedge x_n = \epsilon(\sigma; x_1, \cdots, x_n) \cdot x_{\sigma(1)} \wedge \cdots \wedge x_{\sigma(n)},$$

which has to be satisfied in the free graded commutative algebra $\wedge(x_1, \ldots, x_n)$. The operad $\mathcal{E}_V$ will be called the *operad of endomorphisms* on V.

Dually, for a (graded) vector space W, let $\mathcal{G}_W$ be the operad defined as follows: Put $\mathcal{G}(n) = \mathcal{M}(W, W^{\otimes n})$ and define the composition maps as

$$\gamma_{m_1,\ldots,m_l}(f \otimes \varphi_1 \otimes \cdots \otimes \varphi_l) = (-1)^{|f| \cdot (|\varphi_1| + \cdots + |\varphi_l|)} \cdot (\varphi_1 \otimes \cdots \otimes \varphi_l) \circ f,$$

for $f \in \mathcal{G}_W(l)$ and $\varphi_i \in \mathcal{G}_W(m_i)$, $l \geq 1$ and $m_1, \ldots, m_l \geq 1$. The action of the symmetric group is defined as in (25). We call $\mathcal{G}_W$ the *dual endomorphism operad* on the space W.

By an *algebra* over an operad $\mathcal{P}$ (or by a $\mathcal{P}$-*algebra*), we mean a graded vector space A together with a map $a : \mathcal{P} \longrightarrow \mathcal{E}_A$ of operads. We will write $A = (A, a)$. If not necessary, we make no distinction between the algebra and its underlying space. A $\mathcal{P}$-algebra is thus given by a sequence of degree zero Σ_n-invariant linear maps $a(n) : \mathcal{P}(n) \longrightarrow \mathcal{M}^*(A^{\otimes n}, A)$ satisfying certain axioms, hence the elements of $\mathcal{P}$ act as n-multilinear operations on A. A *homomorphism* of two $\mathcal{P}$-algebras $A = (A, a)$ and $B = (B, b)$ is a homogeneous degree zero map $g : A \longrightarrow B$ which commutes in the obvious sense with the $\mathcal{P}$-algebra structures a and b. We denote by $\mathcal{P}$-*alg* the category of $\mathcal{P}$-algebras and their homomorphisms.

Dually, by a $\mathcal{P}$-*coalgebra* we mean a (graded) vector space C together with an operad map $c : \mathcal{P} \longrightarrow \mathcal{G}_C$; we write $C = (C, c)$. We say that a $\mathcal{P}$-coalgebra C is *connected* if for any $x \in C$ there exists $n \geq 2$ such that $c(n)(\mu)(x) = 0$ for all $\mu \in \mathcal{P}(n)$.

Let us recall the following well-known classical facts. If A is an "ordinary" algebra (such as associative, commutative, Lie, &c.), then an A-module structure on a vector space M is the same as an algebra structure on the direct sum $A \oplus M$ such that both the projection $A \oplus M \longrightarrow A$ and the inclusion $A \longrightarrow A \oplus M$ are algebra homomorphisms and that the product of $(0 \oplus m)$ and $(0 \oplus m')$ in $A \oplus M$ is trivial for all $m, m' \in M$. This motivates the following definition.

Let $A = (A, a)$ be a $\mathcal{P}$-algebra. An A-*module* is a (graded) vector space M together with a $\mathcal{P}$-algebra structure $m : \mathcal{P} \longrightarrow \mathcal{E}_{A \oplus M}$ on $A \oplus M$ such that both the projection $A \oplus M \longrightarrow A$ and the inclusion $A \longrightarrow A \oplus M$ are algebra homomorphisms and that for all $n \geq 1$ and $\mu \in \mathcal{P}$ we have $m(\mu)(x_1, \ldots, x_n) = 0$ if $x_i \in M \subset A \oplus M$ for at least two i, $1 \leq i \leq n$. It is possible to show that an A-module is an abelian group object in the slice category $\mathcal{P}$-alg/A, but we will not need this result (see the remarks in 2.3).

8   Operadic cohomology

We define, for an algebra A over a quadratic (see below for the definition) operad $\mathcal{P}$ and for an A-module M, a cohomology theory $H_{\mathcal{P}}(A; M)$ based on a very small chain complex. This cohomology is a natural generalization of such classical constructions as Hochschild cohomology, Harrison cohomology and Chevalley-Eilenberg cohomology. The ideas of the construction are implicit in the papers [17, 16] and it is quite possible that they circulate among people as folklore, but, as Murray Gerstenhaber pointed out to us, the problem with folklore is that we are not all of the same folk, so we decided to give an explicit construction here.

The construction is motivated by the following observation, which is certainly well known to people working in rational homotopy theory, though we are not able to find a suitable reference (for related ideas see [22, 23, 29, 30]). Our observation is that the Hochschild cohomology can be computed as the cohomology of the algebra of coderivations of a certain coassociative coalgebra. Similarly, the Chevalley-Eilenberg cohomology of a Lie algebra can be computed as the cohomology of the algebra of coderivations of a certain cocommutative coalgebra and, finally, the Harrison cohomology of a commutative algebra can be computed as the cohomology of the algebra of coderivations of a certain Lie coalgebra. The plan of our construction will be based on a similar scheme.

The basic ingredient of our definition is the notion of Koszul duality for operads, introduced in [17]. This construction gives, for any quadratic operad $\mathcal{P}$, another operad, denoted by $\mathcal{P}^!$ and called the *Koszul dual* of $\mathcal{P}$. Our cohomology will be then defined as the cohomology of the algebra of coderivations of a certain (almost) cofree $\mathcal{P}^!$-coalgebra (see below for the definition).

Let us recall first some notions from [17]. By a *collection* we mean a sequence $E = \{E(n); \ n \geq 2\}$ such that each $E(n)$ is a $\mathbf{k}$-linear Σ_n-space. The obvious forgetful functor from the category of operads into the category of collections has a left adjoint $\mathcal{F}$, and we call the operad $\mathcal{F}(E)$ the *free operad* generated by the collection E. We will assume that all collections considered in this paper have the property that the graded vector space $E(n)$ is of finite type, $n \geq 2$.

Let $\mathcal{P}$ be an operad. A sequence $I = \{I(n);\ n \geq 1\}$ of Σ_n-invariant linear subspaces $I(n) \subset \mathcal{P}(n)$ is called an *ideal* if, for $\lambda \in \mathcal{P}(n)$ and $\mu_i \in \mathcal{P}(m_i)$, $1 \leq i \leq l$, the composition $\lambda(\mu_1 \otimes \cdots \otimes \mu_l)$ belongs to $I(m_1 + \cdots + m_l)$ if either $\lambda \in I(l)$ or if for at least one i, $1 \leq i \leq l$, we have $\mu_i \in I(m_i)$. For an ideal I in $\mathcal{P}$, it makes sense to speak about the quotient operad $\mathcal{P}/I$. For a given sequence $R = \{R(n);\ n \geq 2\}$ of linear invariant subspaces $R(n) \subset \mathcal{P}(n)$, denote by (R) the ideal generated by R.

Every Σ_2-invariant linear space E defines a collection $\{E(2) = E,\ E(n) = 0$ for $n \geq 3\}$ (denoted also by E), and we can consider the free operad $\mathcal{F}(E)$ on E. Choose a Σ_3-invariant linear subspace $R \subset \mathcal{F}(E^\backslash(3)$ and form the operad $\langle E; R \rangle = \mathcal{F}(E)/(R)$ We say that an operad $\mathcal{P}$ is *quadratic* if $\mathcal{P} = \langle E; R \rangle$ for some E and R as above.

Let V be a **k**-linear Σ_n-space of finite type. We equip its dual $\#V$ with the Σ_n-action given by $\sigma\varphi(v) = \text{sgn}(\sigma) \cdot \varphi(\sigma^{-1}v)$. If E is a linear Σ_2 space of finite type, then there exists, for each $n \geq 2$, a natural Σ_n-invariant identification $\#\mathcal{F}(E)(n) \cong F(\#E)(n)$, given by the pairing $\langle -|- \rangle_n : F(E)(n) \otimes F(\#E)(n) \longrightarrow$ **k**, which is characterized by the following conditions.

(i) $\langle -|- \rangle_2$ is the evaluation between $F(E)(2) = E$ and $F(\#E)(2) = \#E$.

(ii) For $\mu \in F(E)(k)$, $\nu \in F(E)(l)$, $\varphi \in F(\#E)(k)$ and $\psi \in F(\#E)(l)$ we have

$$\langle \mu(1^{\otimes(i-1)} \otimes \nu \otimes 1^{\otimes(k-i)})|\varphi(1^{\otimes(j-1)} \otimes \psi \otimes 1^{\otimes(k-j)}) \rangle_{k+l-1} =$$
$$= \begin{cases} (-1)^{(l+1)(i+1)} \cdot \langle \mu|\varphi \rangle_k \cdot \langle \nu|\psi \rangle_l, & \text{for } i = j, \\ 0, & \text{otherwise.} \end{cases}$$

(iii) $\langle \sigma\mu|\sigma\varphi \rangle_l = \text{sgn}(\sigma) \cdot \langle \mu|\varphi \rangle_l$, for $\mu \in F(E)(l)$, $\varphi \in F(\#E)(l)$ and $\sigma \in \Sigma_l$.

Let $\mathcal{P} = \langle E; R \rangle$ be a quadratic operad. Using the above identification, we can view the subspace $R \subset \mathcal{F}(E)(3)$ as a subspace (denoted by the same symbol) of $\#F(\#E)(3)$ and we can take its annihilator $R^\perp \subset F(\#E)(3)$. The operad $\mathcal{P}^! = \langle \#E; R^\perp \rangle$ was introduced in [17] as the *Koszul dual* of the operad $\mathcal{P}$.

For a graded vector space W, let $C_{\mathcal{P}}(W) = \bigoplus_{n \geq 1} C_{\mathcal{P}}^n(W)$, where

$$C_{\mathcal{P}}^n(W) = (\#\mathcal{P}(n) \otimes W^{\otimes n})^{\Sigma_n}$$

Here Σ_n acts on $W^{\otimes n}$ by permuting the factors (but taking into account the Koszul convention as in (25)) and $(\#\mathcal{P}(n) \otimes W^{\otimes n})^{\Sigma_n}$ is the space of invariants of the diagonal action of Σ_n on $\#\mathcal{P}(n) \otimes W^{\otimes n}$. We equip $C_{\mathcal{P}}(W)$ with the obvious $\mathcal{P}$-coalgebra structure. Denote by $\pi : C_{\mathcal{P}}(W) \longrightarrow W$ the canonical projection. The coalgebra $C_{\mathcal{P}}(W)$ is connected and it has the following universal property (which we state without proof, which is a standard one). For any connected $\mathcal{P}$-coalgebra C and for any homogeneous degree zero linear map $\psi : C \longrightarrow W$, there exists exactly one coalgebra homomorphism $g : C \longrightarrow C_{\mathcal{P}}(W)$ making the following diagram commutative.

$$C \xrightarrow{g} C_{\mathcal{P}}(W)$$

with ψ and π to W.

Let us point out, however, that the coalgebra $C_{\mathcal{P}}(W)$ is the cofree *connected* coalgebra on W, not an honest cofree coalgebra; see also the remarks in Example 6.4.

Dually, for any graded vector space V we can define, following [17], the *free $\mathcal{P}$-algebra* on V by $A_{\mathcal{P}}(V) = \bigoplus_{n \geq 1} A_{\mathcal{P}}^n(V)$, with

$$A_{\mathcal{P}}^n(V) = (\mathcal{P}(n) \otimes V^{\otimes n})_{\Sigma_n}$$

where $(-)_{\Sigma_n}$ denote the space of coinvariants. This algebra has the classical and absolutely obvious universal property. Moreover, any $\mathcal{P}$-algebra structure $a : \mathcal{P} \to \mathcal{E}_V$ on V defines, by dualization, the *canonical map* (which we denote again by a) $a : A_{\mathcal{P}}(V) \to V$. One would expect that there exists also a dual analog of this map for coalgebras. More precisely, let $c : \mathcal{P} \to \mathcal{G}_W$ be a $\mathcal{P}$-coalgebra structure on W. We would like to dualize this map to get a map (again denoted by the same symbol) $c : W \to C_{\mathcal{P}}(W)$. An immediate observation shows that such a dualization hits, in general, infinitely many components of $C_{\mathcal{P}}(W)$, so the target space is $\prod_{n \geq 1} C_{\mathcal{P}}^n(W)$ rather than $C_{\mathcal{P}}(W) = \bigoplus_{n \geq 1} C_{\mathcal{P}}^n(W)$. This is related with the fact that $C_{\mathcal{P}}(W)$ is not the honest cofree coalgebra; see also the comments above. But it still makes sense to take, for any $n \geq 1$, the component $c^n : W \to C_{\mathcal{P}}^n(W)$ of this dual, and by the *canonical map* in the coalgebra case we mean the *sequence* $\{c^n : W \to C_{\mathcal{P}}^n(W)\}_{n \geq 1}$ of these maps. The functor $A_{\mathcal{P}}(-)$ gives rise to a triple on the category $\mathcal{M}$ and the algebras over this triple are exactly $\mathcal{P}$-algebras, see [16]. This connection gives formal meaning to the analogy between the triple and operadic definition of algebras.

Let $C = (C, c)$ be a $\mathcal{P}$-coalgebra and $\theta : C \to C$ a homogeneous degree p linear map. We say that θ is a degree p *coderivation* of C (into itself) if

$$c(n)(\mu) \circ \theta = (-1)^{|\mu| \cdot |\theta|} \sum_{0 \leq i \leq n-1} (\mathbb{1}^{\otimes i} \otimes \theta \otimes \mathbb{1}^{\otimes (n-i-1)}) \circ c(n)(\mu),$$

for any $n \geq 2$ and $\mu \in \mathcal{P}(n)$. We denote by $\mathrm{Coder}^p(C)$ the linear space of all degree p coderivations of C.

Consider again the coalgebra $C_{\mathcal{P}}(W)$ introduced above. The grading $C_{\mathcal{P}}(W) = C_{\mathcal{P}}^n(W)$ induces on $\mathrm{Coder}^p(C_{\mathcal{P}}(W))$ a second grading, namely $\mathrm{Coder}^p(C_{\mathcal{P}}(W)) = \bigoplus_n \mathrm{Coder}^{p,n}(C_{\mathcal{P}}(W))$ with

$$\mathrm{Coder}^{p,n}(C_{\mathcal{P}}(W)) = \{\theta \in \mathrm{Coder}^p(C_{\mathcal{P}}(W)); \ (\pi \circ \theta)(C_{\mathcal{P}}^q(W)) = 0 \text{ for } q \neq n+1\}.$$

8.1   Lemma: *The map* $\omega : \mathrm{Coder}^{p,n}(C_{\mathcal{P}}(W)) \to \mathcal{M}^p(C_{\mathcal{P}}^{n+1}(W), W)$, *given by* $\omega(\theta) = \pi \circ \theta$, *is an isomorphism for all p, n.*

Proof: The lemma is a consequence of the universal property of $C_{\mathcal{P}}(W)$ mentioned above and we leave the proof to the reader. $\square$

In the rest of the paper, all operads will be assumed to be quadratic.

8.2   Theorem: *Let V be a graded vector space and let $W = \downarrow V$. Then there is a natural one-to-one correspondence between $\mathcal{P}$-algebra structures $a : \mathcal{P} \to \mathcal{E}_V$ on V and coderivations $d \in \mathrm{Coder}^{1,1}(C_{\mathcal{P}^{!}}(W))$ of the $\mathcal{P}^{!}$-coalgebra $C_{\mathcal{P}^{!}}(W)$ with $d^2 = 0$.*

Proof: To fix the notation, let $\mathcal{P} = \langle E; R \rangle$. First we show that there is a one-to-one natural map

$$\Omega : Oper(\mathcal{F}(E), \mathcal{E}_V) \longrightarrow \mathrm{Coder}^{1,1}(C_{\mathcal{P}^{!}}(W)).$$

To this end, observe that, since $\mathcal{F}(E)$ is free, $Oper(\mathcal{F}(E), \mathcal{E}_V)$ is naturally isomorphic to $\mathcal{M}_{\Sigma_2}(E, \mathcal{M}^*(V^{\otimes 2}, V))$, while, by Lemma 8.1, we have

$$\mathrm{Coder}^{1,1}(C_{\mathcal{P}^{!}}(W)) \cong \mathcal{M}^1(C_{\mathcal{P}^{!}}^2(W), W) \cong \mathcal{M}^1((\#\mathcal{P}^{!}(2) \otimes W^{\otimes 2})^{\Sigma_2}, W)$$
$$\cong \mathcal{M}^1((E \otimes W^{\otimes 2})^{\Sigma_2}, W)$$

Let now $\Xi : \mathcal{M}^0_{\Sigma_2}(E, \mathcal{M}(V^{\otimes 2}, V)) \to \mathcal{M}^1((E \otimes W^{\otimes 2})^{\Sigma_2}, W)$ be the isomorphism defined by

$$\Xi(f)(e \otimes w_1 \otimes w_2) = (-1)^{|w_1|} \downarrow f(e)(\uparrow w_1 \otimes \uparrow w_2).$$

We define Ω by the commutativity of the diagram

$$
\begin{array}{ccc}
Oper(\mathcal{F}(E), \mathcal{E}_V) & \xrightarrow{\ \Omega\ } & \mathrm{Coder}^{1,1}(C_{\mathcal{P}!}(W)) \\
\cong \Big\downarrow & & \Big\downarrow \cong \\
\mathcal{M}^0_{\Sigma_2}(E, \mathcal{M}(V^{\otimes 2}, V)) & \xrightarrow[\ \Xi\ ]{} & \mathcal{M}^1((E \otimes W^{\otimes 2})^{\Sigma_2}, W)
\end{array}
$$

It is easy to verify that $a \in Oper(\mathcal{F}(E), \mathcal{E}_V)$ defines an algebra structure on V, i.e. factors through $\mathcal{F}(E)/(R)$, if and only if the coderivation $d = \Omega(a)$ satisfies $d^2 = 0$, which finishes the proof. □

Suppose now that $A = (A, a)$ is a $\mathcal{P}$-algebra and $M = (M, m)$ an A-module. Let $U = \downarrow A$, $X = \downarrow M$ and $W = U \oplus X$. Notice that the decomposition $W^{\otimes n} = \bigoplus_{i+j=n}(W^{\otimes n})^{i,j}$, where $(W^{\otimes n})^{i,j}$ is the subspace formed by i-tuple products of elements of U and j-tuple products of elements of X, induces the decomposition $C^n_{\mathcal{P}!}(W) = \bigoplus_{i+j=n} C^{i,j}_{\mathcal{P}!}(W)$ for any $n \geq 1$. Let $C^{p,n}_{\mathcal{P}}(A; M)$ be the subspace of $\mathrm{Coder}^{p,n}(C_{\mathcal{P}!}(W))$ consisting of those coderivations θ for which $(\pi \circ \theta)(C^{i,j}_{\mathcal{P}!}(W)) = 0$ whenever $j \geq 1$, and $(\pi \circ \theta)(C^{n+1}_{\mathcal{P}!}(W)) \subset X \subset W$.

Suppose now that $d \in \mathrm{Coder}^{1,1}(C_{\mathcal{P}!}(W))$ corresponds to the algebra structure m on $A \oplus M$ as in Proposition 8.2. Let us denote, for $\theta \in \mathrm{Coder}^{p,n}(C_{\mathcal{P}!}(W))$, by $\nabla(\theta) = d \circ \theta - (-1)^p \theta \circ d$ the (graded) commutator. It is easy to verify that ∇ maps $C^{p,n}_{\mathcal{P}}(A; M)$ to $C^{p+1,n+1}_{\mathcal{P}}(A; M)$ and that $\nabla^2 = 0$. We can then formulate the following definition.

8.3   Definition: Let A be an algebra over a quadratic operad $\mathcal{P}$ and let M be an A-module. Define the cohomology $H^{p,n}_{\mathcal{P}}(A; M)$ of A with coefficients in M as

$$H^{p,n}_{\mathcal{P}}(A; M) := H^{p,n}(C^{*,*}_{\mathcal{P}}(A; M), \nabla).$$

We call p and n the *inner* and the *simplicial* degrees, respectively.

8.4   Remark: The map ω of Lemma 8.1 induces an isomorphism

$$(27) \quad C^{p,n}_{\mathcal{P}}(A; M) \cong \mathcal{M}^p(C^{n+1}_{\mathcal{P}!}(U), X) =$$
$$= \mathcal{M}^p(C^{n+1}_{\mathcal{P}!}(\downarrow A), \downarrow M) \cong \mathcal{M}^p_{\Sigma_{n+1}}(\#\mathcal{P}^!(n+1) \otimes (\downarrow A)^{\otimes(n+1)}, \downarrow M).$$

We could have *defined* $C^{p,n}_{\mathcal{P}}(A; M)$ by the equation above, without having referred to coderivations of some coalgebra. The problem is that we do not know an easy way to describe the differential in terms of $\mathcal{M}^p_{\Sigma_{n+1}}(\#\mathcal{P}(n+1) \otimes (\downarrow A)^{\otimes(n+1)}, \downarrow M)$, the only way we know is the one based on the somewhat explicit description of the Koszul dual operad $\mathcal{P}^!$ using the language of trees as in [16, 17].

8.5   Examples: If A is an associative algebra, i.e. an algebra over the associative operad Ass, we know ([17]) that the Koszul dual $Ass^!$ is again Ass and the description above gives us the usual definition of the Hochschild cohomology. For a Lie algebra, i.e. for an algebra over the Lie operad Lie, we have $Lie^! = Comm$, the commutative operad, and the description above gives the Chevalley-Eilenberg cohomology. Finally, for a commutative algebra, i.e. for an algebra over the commutative operad *Comm*, we have $Comm^! = Lie$ and the description above gives us the Harrison cohomology.

It can be shown, using the computation of [17], that for a so-called *Koszul operad* (see again [17] for the definition) we have $H_{\mathcal{P}}^{p,\geq 1}(A; M) = 0$ whenever the algebra A is free. This is, by [5], enough to infer the following proposition.

8.6 Proposition: *For an algebra A over a Koszul operad $\mathcal{P}$ and for an A-module M, the cohomology $H_{\mathcal{P}}^{p,n}(A; M)$ coincides with the Barr-Beck triple cohomology of A with coefficients in M.*

This proposition relates the operadic cohomology with the "triple cohomology" introduced in Definition 2.2.

9 Distributive laws and operads

In this paragraph we discuss distributive laws from the operadic point of view. The definitions and results quoted here were taken from the paper [24] of the second author. We formulate them for quadratic operads only, since the range of our applications is limited by our definition of the cohomology. Similar definitions can be made for more general types of operads as well.

Suppose that a Σ_2-space E has an invariant decomposition $E = E_1 \oplus E_2$. This decomposition induces the decomposition

$$\mathcal{F}(E)(3) = \mathcal{F}(E)(3)_{11} \oplus \mathcal{F}(E)(3)_{12} \oplus \mathcal{F}(E)(3)_{21} \oplus \mathcal{F}(E)(3)_{22},$$

where $\mathcal{F}(E)(3)_{ij}$ is the Σ_3-invariant subspace of $\mathcal{F}(E)(3)$ generated by the compositions of the form $\mu(1, \nu)$ (or $\mu(\nu, 1)$) with $\mu \in E_i$ and $\nu \in E_j$, for $i, j = 1$ or 2. Notice that $\mathcal{F}(E)(3)_{ii}$ can be identified with the image of the map $F(E_i)(3) \to \mathcal{F}(E)(3)$ induced by the inclusion $E_i \subset E$, $i = 1, 2$. Let us consider a Σ_3-invariant map $\mathcal{D} : \mathcal{F}(E)(3)_{12} \to \mathcal{F}(E)(3)_{21}$. Every such a map defines an invariant subspace $R_{\mathcal{D}} \subset \mathcal{F}(E)(3)$ generated by elements of the form $x - \mathcal{D}(x)$, for $x \in \mathcal{F}(E)(3)_{12}$.

Let $\mathcal{P} = \langle E; R \rangle$ be a quadratic operad for which there exists a Σ_2-invariant decomposition $E = E_1 \oplus E_2$, an invariant linear map $\mathcal{D} : \mathcal{F}(E)(3)_{12} \to \mathcal{F}(E)(3)_{21}$ and Σ_3-invariant subsets $R_i \subset \mathcal{F}(E)(3)_{ii}$, $i = 1, 2$, such that $R = R_1 \oplus R_{\mathcal{D}} \oplus R_2$. In this case we write

$$\mathcal{P} = \langle E_1, E_2; R_1, \mathcal{D}, R_2 \rangle.$$

We can clearly form the suboperads $\mathcal{P}_i = \langle E_i; R_i \rangle \subset \mathcal{P}$, for $i = 1, 2$. Denote, for any $n \geq 1$ and $l \leq n$, by $\mathcal{P}(n)_l$ the invariant subspace of $\mathcal{P}(n)$ generated by the elements of the form $\mu(\nu_1, \ldots, \nu_l)$, for $\mu \in \mathcal{P}_2(l)$ and $\nu_s \in \mathcal{P}_1(m_s)$, for a sequence $m_1, \ldots, m_l \geq 1$, $m_1 + \cdots + m_l = n$, and $1 \leq s \leq l$. The inclusions $\mathcal{P}_i \subset \mathcal{P}$, $i = 1, 2$ induce, for any $n \geq 2$, a map

(28) $$\xi(n) : \bigoplus_{1 \leq l \leq n} \mathcal{P}(n)_l \to \mathcal{P}(n).$$

9.1 Definition: We say that $\mathcal{D}$ is a distributive law (or sometimes also that $\mathcal{P} = \langle E_1, E_2; R_1, \mathcal{D}, R_2 \rangle$ is an operad with a distributive law) if the map $\xi(n)$ is an isomorphism for each $n \geq 2$.

We refer to [24] for how the distributive law in the sense of the above definition induces the transformation λ from the triple definition (Definition 3.1) of the distributive law. Observe that $\xi(n)$ is always an isomorphism for $n = 2, 3$. The second author proved in [24] the following coherence theorem.

9.2   Theorem: *The map $\xi(n)$ is an isomorphism for any $n \geq 2$ if and only if it is an isomorphism for $n = 4$.*

The following lemma can be verified directly.

9.3   Lemma: *Let $\mathcal{P} = \langle E_1, E_2; R_1, \mathcal{D}, R_2 \rangle$ be a quadratic operad with a distributive law $\mathcal{D}$. Then its Koszul dual operad $\mathcal{P}^!$ is again a (quadratic) operad with a distributive law, namely*

$$\mathcal{P}^! = \langle \#E_2, \#E_1; R_2^\perp, \#\mathcal{D}, R_1^\perp \rangle,$$

where $\#\mathcal{D} : F(\#E)(3)_{12} \to F(\#E)(3)_{21}$ is the dual of $\mathcal{D} : \mathcal{F}(E)(3)_{12} \to \mathcal{F}(E)(3)_{21}$ under the natural identification $F(\#E)(3)_{ij} \cong \#\mathcal{F}(E)(3)_{ji}$.

9.4   Example: Below we give, following some ideas of [16], an innocuous generalization of Poisson algebras already discussed in Example 6.2 and describe the Koszul dual of the corresponding operad (Proposition 9.5). This will also be an example of a nontrivially graded operad. By an (m, n)-*algebra* we mean a (graded) vector space P together with two bilinear maps, $- \cup - : P \otimes P \to P$ of degree m, and $[-, -] : P \otimes P \to P$ of degree n (m and n are natural numbers), such that, for any homogeneous $a, b, c \in P$,

(i) $a \cup b = (-1)^{|a| \cdot |b| + m} \cdot b \cup a$,

(ii) $[a, b] = -(-1)^{|a| \cdot |b| + n} \cdot [b, a]$,

(iii) $- \cup -$ is associative in the sense that

$$a \cup (b \cup c) = (-1)^{m \cdot (|a| + 1)} \cdot (a \cup b) \cup c,$$

(iv) $[-, -]$ satisfies the following form of the Jacobi identity:

$$(-1)^{|a| \cdot (|c| + n)} \cdot [a, [b, c]] + (-1)^{|b| \cdot (|a| + n)} \cdot [b, [c, a]] + (-1)^{|c| \cdot (|b| + n)} \cdot [c, [a, b]] = 0,$$

(v) the operations $- \cup -$ and $[-, -]$ are compatible in the sense that

$$(-1)^{m \cdot |a|} [a, b \cup c] = [a, b] \cup c + (-1)^{(|b| \cdot |c| + m)} [a, c] \cup b.$$

Obviously $(0, 0)$-algebras are exactly (graded) Poisson algebras, $(0, -1)$-algebras are Gerstenhaber algebras introduced under this name in [15], while $(0, n - 1)$-algebras are n-algebras of [16]. For the relation between the cohomology of configuration spaces and n-algebras we refer to [11]. We may think of an (m, n)-structure on P as a Lie algebra structure on $\uparrow^n P$ together with an associative commutative algebra structure on $\uparrow^m P$ such that both structures are related via the compatibility axiom (v).

Let us give an operadic description of (m, n)-algebras. To simplify the notation, let $\chi = (-1)^n$ and $\lambda = (-1)^m$. Let E be the space spanned by two elements, μ (for

$-\cup-$) of degree m and ℓ (for $[-,-]$) of degree n, with the action of Σ_2 defined by $s\mu = \lambda\mu$ and $s\ell = -\chi\ell$, s being the generator of Σ_2. We can easily see that

$$\mathcal{F}(E)(3) = \mathrm{Span}(X_{ij}, Y_{ij}, Z_{ij}; \ i,j = 1 \text{ or } 2)$$

with

(29) $\qquad X_{ij} = e_i(e_j \otimes 1), \ Y_{ij} = e_i(1 \otimes e_j) \text{ and } Z_{ij} = e_i(e_j \otimes 1)(\mathbb{1} \otimes s),$

where $e_i = \ell$ for $i = 1$ and $e_i = \mu$ for $i = 2$. The action of the group Σ_3 on $\mathcal{F}(E)(3)$ is described by the following table (which is, unfortunately, broken into two pieces, due to the limited width of the page), whose meaning is clear, we hope.

	X_{11}	Y_{11}	Z_{11}	X_{12}	Y_{12}	Z_{12}
(123)	X_{11}	Y_{11}	Z_{11}	X_{12}	Y_{12}	Z_{12}
(312)	$-\chi Z_{11}$	$-\chi X_{11}$	Y_{11}	λZ_{12}	$-\chi X_{12}$	$-\chi\lambda Y_{12}$
(231)	$-\chi Y_{11}$	Z_{11}	$-\chi X_{11}$	$-\chi Y_{12}$	$-\chi\lambda Z_{12}$	λX_{12}
(213)	$-\chi X_{11}$	$-\chi Z_{11}$	$-\chi Y_{11}$	λX_{12}	$-\chi Z_{12}$	$-\chi Y_{12}$
(321)	Y_{11}	X_{11}	$-\chi Z_{11}$	$-\chi\lambda Y_{12}$	$-\chi\lambda X_{12}$	λZ_{12}
(132)	Z_{11}	$-\chi Y_{11}$	X_{11}	Z_{12}	λY_{12}	X_{12}

	X_{21}	Y_{21}	Z_{21}	X_{22}	Y_{22}	Z_{22}
(123)	X_{21}	Y_{21}	Z_{21}	X_{22}	Y_{22}	Z_{22}
(312)	$-\chi Z_{21}$	λX_{21}	$-\lambda\chi Y_{21}$	λZ_{22}	λX_{22}	Y_{22}
(231)	λY_{21}	$-\lambda\chi Z_{21}$	$-\chi X_{21}$	λY_{22}	Z_{22}	λX_{22}
(213)	$-\chi X_{21}$	λZ_{21}	λY_{21}	λX_{22}	λZ_{22}	λY_{22}
(321)	$-\lambda\chi Y_{21}$	$-\lambda\chi X_{21}$	$-\chi Z_{21}$	Y_{22}	X_{22}	λZ_{22}
(132)	Z_{21}	$-\chi Y_{21}$	X_{21}	Z_{22}	λY_{22}	X_{22}

The space of relations R is the Σ_3-invariant subspace of $\mathcal{F}(E)(3)$ generated by the elements

$$X_{11} - \chi Y_{11} - \chi Z_{11} \qquad \text{(for the Jacobi identity)}$$
$$X_{22} - \lambda Y_{22} \qquad \text{(for the associativity)}$$
$$Y_{12} - \lambda Z_{21} - X_{21} \qquad \text{(for the compatibility)}.$$

From this we get easily that the operad $\mathcal{P}(m,n)$ for the category of (m,n)-algebras can be described as

$$\mathcal{P}(m,n) = \langle E_1, E_2; R_1, \mathcal{D}, R_2 \rangle$$

with

$$E_1 = \mathrm{Span}(\ell), \ E_2 = \mathrm{Span}(\mu), \ R_1 = \mathrm{Span}(X_{11} - \chi Y_{11} - \chi Z_{11}),$$
$$R_2 = \mathrm{Span}(X_{22} - \lambda Y_{22}, Y_{22} - Z_{22})$$

and the distributive law $\mathcal{D} : \mathcal{F}(E)(3)_{12} \to \mathcal{F}(E)(3)_{21}$ defined by

$$\mathcal{D}(X_{12}) = Y_{21} + Z_{21}, \ \mathcal{D}(Y_{12}) = X_{21} + \lambda Z_{21} \text{ and } \mathcal{D}(Z_{12}) = X_{21} - \chi Y_{21}.$$

We leave it to the reader to verify that the condition of Theorem 9.2 is satisfied.

Let us describe the Koszul dual of the operad $\mathcal{P}(m,n)$. For $\bar{E} = \#E$, $\bar{E} = \bar{E}_1 \oplus \bar{E}_2$ with $\bar{E}_1 = \mathrm{Span}(\bar{\ell})$, $\deg(\bar{\ell}) = -n$, and $\bar{E}_2 = \mathrm{Span}(\bar{\mu})$, $\deg(\bar{\mu}) = -m$, with the action of Σ_2 given by $s\bar{\mu} = \bar{\mu}$ and $s\bar{\ell} = -\bar{\ell}$. The evaluation between E and $\bar{E} = \#E$ is given by

(30) $\qquad \langle \mu | \bar{\ell} \rangle = \langle \ell | \bar{\mu} \rangle = 1 \text{ and } \langle \mu | \bar{\mu} \rangle = \langle \ell | \bar{\ell} \rangle = 0.$

As above, $\mathcal{F}(\bar{E})(3)$ has the basis

(31) $\bar{X}^{ij} = \bar{e}^i(\bar{e}^j \otimes 1)$, $\bar{Y}^{ij} = \bar{e}^i(1 \otimes \bar{e}^j)$ and $\bar{Z}^{ij} = \bar{e}^i(\bar{e}^j \otimes 1)(\mathbb{1} \otimes s)$,

where $\bar{e}^i = \bar{\ell}$ for $i = 1$ and $\bar{e}^i = \bar{\mu}$ for $i = 2$. The pairing between $\mathcal{F}(E)(3)$ and $\mathcal{F}(\bar{E})(3)$ is given by

(32) $\langle X_{ij} | \bar{X}^{kl} \rangle = \delta_i^k \delta_j^l$, $\langle Y_{ij} | \bar{Y}^{kl} \rangle = -\delta_i^k \delta_j^l$ and $\langle Z_{ij} | \bar{Z}^{kl} \rangle = -\delta_i^k \delta_j^l$

while the pairing is trivial on other combinations of the basis elements and δ_*^* denotes the Kronecker delta. From this we get immediately that

$$R_2^\perp = \mathrm{Span}(\bar{X}^{11} - \lambda \bar{Y}^{11} - \lambda \bar{Z}^{11}), \quad R_1^\perp = \mathrm{Span}(\bar{X}^{22} - \chi \bar{Y}^{22}, \bar{Y}^{22} - \bar{Z}^{22})$$

and that $\#\mathcal{D}$ is given by

$$\#\mathcal{D}(\bar{X}^{12}) = \bar{Y}^{12} + \bar{Z}^{21}, \quad \#\mathcal{D}(\bar{Y}^{12}) = \bar{X}^{21} + \chi \bar{Z}^{21} \text{ and } \#\mathcal{D}(\bar{Z}^{12}) = \bar{X}^{21} - \lambda \bar{Y}^{21}.$$

We have proved the following proposition.

9.5 Proposition: *For any natural numbers m and n we have*

$$\mathcal{P}(m,n)^! = \mathcal{P}(-n,-m).$$

In particular, the category of $(m,-m)$-algebras is Koszul self-dual, $\mathcal{P}(m,-m)^! = \mathcal{P}(m,-m)$, for any m.

9.6 Example: Recall from Example 6.3 that a nonsymmetric Poisson algebra consists of a vector space P and two associative multiplications, $\cdot\,, \langle -, - \rangle : P \otimes P \to P$ such that:

$$\langle a \cdot b, c \rangle = a \cdot \langle b, c \rangle \text{ and } \langle a, b \cdot c \rangle = \langle a, b \rangle \cdot c,$$

for any $a, b, c \in P$. The operadic description appears as follows: Let E_1 (resp. E_2) be the free Σ_2-space generated by a symbol ℓ (resp. μ), ℓ corresponding to $\langle -, - \rangle$ and μ corresponding to $\cdot\,$. If $E = E_1 \oplus E_2$. Then $\mathcal{F}(E)(3)$ is generated, as a Σ_3-module, by $X_{ij} = e_i(e_j \otimes 1)$ and $Y_{ij} = e_i(1 \otimes e_j)$, where $e_i = \ell$ for $i = 1$ and $e_i = \mu$ for $i = 2$. Define $\mathcal{D} : \mathcal{F}(E)(3)_{12} \to \mathcal{F}(E)(3)_{21}$ by

$$\mathcal{D}(X_{12}) = Y_{21} \text{ and } \mathcal{D}(Y_{12}) = X_{21}.$$

It was verified in [24] that this map defines a distributivity law and that nonsymmetric Poisson algebras are algebras over the operad $\mathcal{P} = \langle E_1, E_2; R_1, \mathcal{D}, R_2 \rangle$, where R_1 (resp. R_2) is the associativity axiom for ℓ (resp. μ). It is also almost immediate to see that the category of these algebras is Koszul self-dual.

10   Effect on cohomology (operad case)

Let $\mathcal{P} = \langle E_1, E_2; R_1, \mathcal{D}, R_2 \rangle$ be an operad with a distributive law. Recall that this means, by definition, that the map

$$\xi(n) : \bigoplus_{1 \le l \le n} \mathcal{P}(n)_l \longrightarrow \mathcal{P}(n)$$

of (28) is an isomorphism for each $n \geq 2$. We use this isomorphism to *identify* $\mathcal{P}(n)$ with $\bigoplus_{1 \leq l \leq n} \mathcal{P}(n)_l$. This identification induces, for any (graded) vector space W, the decomposition

$$(33) \qquad\qquad C_{\mathcal{P}}^n(W) = \bigoplus C_{\mathcal{P}}^{\alpha,\beta}(W)$$

where the summation runs over all $\alpha + \beta = n$, $1 \leq \alpha \leq n$, and

$$(34) \qquad\qquad C_{\mathcal{P}}^{\alpha,\beta}(W) = (\#\mathcal{P}(\alpha+\beta)_\alpha \otimes W^{\otimes(\alpha+\beta)})^{\Sigma_{(\alpha+\beta)}}.$$

On the other hand, take the $\mathcal{P}_1$-coalgebra $C_{\mathcal{P}_1}(W)$, forget the coalgebra structure, and form the $\mathcal{P}_2$-coalgebra $C_{\mathcal{P}_2}(C_{\mathcal{P}_1}(W))$. This coalgebra again decomposes as

$$C_{\mathcal{P}_2}(C_{\mathcal{P}_1}(W)) = \bigoplus C_{\mathcal{P}_2}(C_{\mathcal{P}_1}(W))^{\alpha,\beta},$$

where the summation runs again over all $\alpha + \beta = n$, $1 \leq \alpha \leq n$, with

$$C_{\mathcal{P}_2}(C_{\mathcal{P}_1}(W))^{\alpha,\beta} =$$
$$\bigoplus \{\#\mathcal{P}_2(\alpha) \otimes (\#\mathcal{P}_1(m_1) \otimes W^{\otimes m_1})^{\Sigma_{m_1}} \otimes \cdots \otimes (\#\mathcal{P}_\alpha(m_\alpha) \otimes W^{\otimes m_\alpha})^{\Sigma_{m_\alpha}}\}^{\Sigma_\alpha},$$

where the summation runs over all $m_1, \ldots, m_\alpha \geq 1$ with $m_1 + \cdots + m_\alpha = \alpha + \beta$. We can show that the map

$$\#\mathcal{P}(\alpha+\beta)_\alpha \otimes W^{\otimes(\alpha+\beta)} \longrightarrow \#(\mathcal{P}_2(\alpha) \otimes \mathcal{P}_1(m_1) \otimes \cdots \otimes \mathcal{P}_1(m_\alpha)) \otimes W^{\otimes(\alpha+\beta)}$$

induced by the composition map $\mathcal{P}_2(\alpha) \otimes \mathcal{P}_1(m_1) \otimes \cdots \otimes \mathcal{P}_1(m_\alpha) \longrightarrow \mathcal{P}(\alpha+\beta)$ gives an identification $C_{\mathcal{P}_2}(C_{\mathcal{P}_1}(W))^{\alpha,\beta} \cong C_{\mathcal{P}}^{\alpha,\beta}(W)$, which in turn induces an isomorphism

$$(35) \qquad\qquad C_{\mathcal{P}}(W) \cong C_{\mathcal{P}_2}(C_{\mathcal{P}_1}(W))$$

of $\mathcal{P}_2$-coalgebras. The decomposition (33) then induces the decomposition

$$\operatorname{Coder}^{p,n}(C_{\mathcal{P}}(W)) = \bigoplus_{i+j=n} \operatorname{Coder}^{p,i,j}(C_{\mathcal{P}}(W))$$

with

$$\operatorname{Coder}^{p,i,j}(C_{\mathcal{P}}(W)) =$$
$$\{\theta \in \operatorname{Coder}^p(C_{\mathcal{P}}(W)); \ \pi \circ \theta(C_{\mathcal{P}}^{\alpha,\beta}(W)) = 0 \text{ for } (\alpha,\beta) \neq (i+1,j)\}.$$

Notice that the two most extreme pieces of the decomposition above have a particularly easy description:

$$(36) \qquad \begin{aligned} \operatorname{Coder}^{p,n,0}(C_{\mathcal{P}}(W)) &= \operatorname{Coder}^{p,n}(C_{\mathcal{P}_2}(W)) \text{ and} \\ \operatorname{Coder}^{p,0,n}(C_{\mathcal{P}}(W)) &= \operatorname{Coder}^{p,n}(C_{\mathcal{P}_1}(W)). \end{aligned}$$

All the decompositions above apply, by Lemma 9.3, also to $\mathcal{P}^!$ in place of $\mathcal{P}$, with $\mathcal{P}_1^! = (\mathcal{P}_2)^!$ and $\mathcal{P}_2^! = (\mathcal{P}_1)^!$. In particular, (36) gives

$$(37) \quad \begin{aligned} \operatorname{Coder}^{1,1}(C_{\mathcal{P}^!}(W)) &= \operatorname{Coder}^{1,1,0}(C_{\mathcal{P}^!}(W)) \oplus \operatorname{Coder}^{1,0,1}(C_{\mathcal{P}^!}(W)) \\ &= \operatorname{Coder}^{1,1}(C_{\mathcal{P}_1^!}(W)) \oplus \operatorname{Coder}^{1,1}(C_{\mathcal{P}_2^!}(W)). \end{aligned}$$

Let $a : \mathcal{P} \longrightarrow \mathcal{E}_V$ be a $\mathcal{P}$-algebra structure on a (graded) vector space V. The composition of a and the inclusion $\mathcal{P}_i \hookrightarrow \mathcal{P}$ induces on V a $\mathcal{P}_i$-algebra structure a_i, for $i = 1, 2$. Let $d \in \text{Coder}^{1,1}(C_{\mathcal{P}^!}(W))$ (resp. $d_i \in \text{Coder}^{1,1}(C_{\mathcal{P}_i^!}(W))$) correspond to a (resp. to a_i) as in Proposition 8.2. We immediately have the following lemma.

10.1   Lemma: *Under the identification of (37), $d = d_1 + d_2$.*

Let A be a $\mathcal{P}$-algebra over an operad with a distributive law as above, and let M be an A-module. Then

$$C_{\mathcal{P}}^{p,n}(A; M) = \bigoplus_{i+j=n} C_{\mathcal{P}}^{p,i,j}(A; M)$$

with $C_{\mathcal{P}}^{p,i,j}(A; M) = C_{\mathcal{P}}^{p,i+j}(A; M) \cap \text{Coder}^{p,i,j}(C_{\mathcal{P}^!}(W))$. Notice that we have, much as in (36),

$$C_{\mathcal{P}}^{p,n,0}(A; M) = C_{\mathcal{P}_1}^{p,n}(A; M) \text{ and } C_{\mathcal{P}}^{p,0,n}(A; M) = C_{\mathcal{P}_2}^{p,n}(A; M).$$

We may easily verify that

$$d_1(C_{\mathcal{P}}^{p,i,j}(A; M)) \subset C_{\mathcal{P}}^{p,i+1,j}(A; M) \text{ and } d_2(C_{\mathcal{P}}^{p,i,j+1}(A; M)) \subset C_{\mathcal{P}}^{p,i,j}(A; M),$$

which immediately gives the following statement.

10.2   Theorem: *The cohomology $H_{\mathcal{P}}^{*,*}(A; M)$ of a $\mathcal{P}$-algebra A with coefficients in an A-module M, where $\mathcal{P} = \mathcal{P}(E_1, E_2; R_1, \mathcal{D}, R_2)$ is a quadratic operad with a distributivity law, can be computed as the cohomology of the total complex of the bicomplex below*

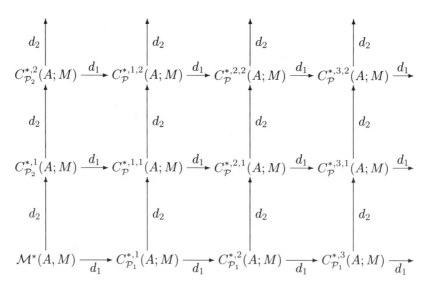

Notice that the bottom row (resp. the extreme left column) of the above bicomplex computes the cohomology $H_{\mathcal{P}_1}^{*,*}(A; M)$ (resp. $H_{\mathcal{P}_2}^{*,*}(A; M)$).

The bicomplex in the above theorem is the operadic analog of the bicomplex in (10). Having in mind Theorem 8.6, it is useful to quote the following result of the second author [24].

10.3   Theorem: *If the operads $\mathcal{P}_1$, $\mathcal{P}_2$ are Koszul, then the operad $\mathcal{P}$ is Koszul as well.*

10.4 **Example:** In this example we construct the operadic cohomology of non-symmetric Poisson algebras from examples 6.3 and 9.6. Let $P = (P, \cdot, \langle -, - \rangle)$ be such an algebra. A P-module is a graded vector space M equipped with two actions, $\nu, \lambda : P \otimes M \oplus M \otimes P \to M$ such that, for all $a, b \in P$ and $x \in M$,

(i) $\nu(a, \nu(b, x)) = \nu(a \cdot b, x)$, $\nu(a, \nu(x, b)) = \nu(\nu(a, x), b)$
and $\nu(\nu(x, a), b) = \nu(x, a \cdot b)$,

(ii) $\lambda(a, \lambda(b, x)) = \lambda(\langle a, b \rangle, x)$, $\lambda(a, \lambda(x, b)) = \lambda(\lambda(a, x), b)$ and $\lambda(\lambda(x, a), b) = \lambda(x, \langle a, b \rangle)$,

(iii) $\lambda(a, \nu(b, x)) = \nu(\langle a, b \rangle, x)$, $\nu(a, \lambda(b, x)) = \lambda(a \cdot b, x)$, $\lambda(a, \nu(x, b)) = \nu(\lambda(a, x), b)$,
$\nu(a, \lambda(x, b)) = \lambda(\nu(a, x), b)$, $\lambda(x, a \cdot b) = \nu(\lambda(x, a), b))$ and $\nu(x, \langle a, b \rangle) = \lambda(\nu(x, a), b))$.

As we already observed in Example 9.6, the category of such algebras is Koszul self-dual, so $\mathcal{P}^! = \mathcal{P}$ for the corresponding operad. The distributive law gives the isomorphism $C_{\mathcal{P}^!}(W) \cong C_{\mathcal{P}_1^!}(C_{\mathcal{P}_2^!}(W))$ of (35) and, because $\mathcal{P}_i = \mathcal{P}_i^! = Ass$, for $i = 1, 2$, $C_{\mathcal{P}_1^!}(C_{\mathcal{P}_2^!}(W)) \cong T(T(W))$, the "double" tensor coalgebra. Notice the perfect symmetry $C_{\mathcal{P}_1^!}(C_{\mathcal{P}_2^!}(W)) \cong C_{\mathcal{P}_2^!}(C_{\mathcal{P}_1^!}(W))$ (see also the comments in Example 6.3).

The observation above makes it clear that a typical element of the space $C_{\mathcal{P}^!}^n(W)$ is of the form $w_1 \circ_1 w_2 \circ_2 \cdots \circ_{n-1} w_n$, where $\circ_i \in \{*, \star\}$ with $*$ corresponding, via the Koszul duality, to $\langle -, - \rangle$ and $\star$ corresponding to $\cdot$. The space $C_{\mathcal{P}^!}^{\alpha,\beta}(W)$ of (34) then consists to those elements of $C_{\mathcal{P}^!}^{\alpha+\beta}(W)$ for which the cardinality of the set $\{i; \circ_i = *\}$ is $\alpha - 1$.

To make life easier, we suppose P to be trivially graded, so we may neglect the internal degrees and identify $C_{\mathcal{P}}^{n+1}(P, M)$ with $\mathcal{M}(C_{\mathcal{P}^!}^n(P), M)$ using (27). For $f \in C_{\mathcal{P}}^{n-1}(P, M)$ the differential d is then given by the formula

$$df(a_0 \circ_0 \cdots \circ_{n-1} a_n) = \sum_{j=0}^{n+1} (-1)^j d^j f(a_0 \circ_0 \cdots \circ_{n-1} a_n)$$

where

$$d^0 f(a_0 \circ_0 \cdots \circ_{n-1} a_n) = \begin{cases} \lambda(a_0, f(a_1 \circ_1 \cdots \circ_{n-1} a_n)), & \text{if } \circ_0 = * \\ \nu(a_0, f(a_1 \circ_1 \cdots \circ_{n-1} a_n)), & \text{if } \circ_0 = \star \end{cases}$$

$$d^j f(a_0 \circ_0 \cdots \circ_{n-1} a_n) = \begin{cases} f(a_0 \circ_0 \cdots \langle a_{j-1}, a_j \rangle \cdots \circ_{n-1} a_n), & \text{if } \circ_j = * \\ f(a_0 \circ_0 \cdots a_{j-1} \cdot a_j \cdots \circ_{n-1} a_n), & \text{if } \circ_j = \star \end{cases}$$

$$d^{n+1} f(a_0 \circ_0 \cdots \circ_{n-1} a_n) = \begin{cases} \lambda(f(a_0 \circ_0 \cdots \circ_{n-2} a_{n-1}), a_n), & \text{if } \circ_{n-1} = * \\ \nu(f(a_0 \circ_0 \cdots \circ_{n-2} a_{n-1}), a_n), & \text{if } \circ_{n-1} = \star \end{cases}$$

with $1 \leq j \leq n$. The differential d thus defined obviously decomposes as $d = d_1 + d_2$ where d_1 (resp. d_2) corresponds to the first (resp. second) cases of the formulas above. This decomposition reflects the bicomplex description of the cohomology as predicted by Theorem 10.2.

In the first draft of the paper we did some computations related to the cohomology of Poisson algebras (Example 9.4), but we found them rather technical and not very stimulating, hence we decided not to include them in this revision of the paper. The computations are available on request from the second author.

11 Bialgebras (operadic approach)

In this section we are going to develop an operadic definition of various types of bialgebras and prove a technical statement about induced structures (Theorem 11.11).
The conceptual trouble here is related to the fact that bialgebras are not algebras
over some operad (the presence of co-algebraic operations makes this impossible),
but rather algebras over PROPs [19] or, more precisely, over their **k**-linear versions
called *theories* in [21]. A theory is a system $\{\mathbf{A}(m,n)\}_{m,n\geq 1}$ of **k**-linear spaces together with some operations of composition and an action of the symmetric group.
We may view a $\mathbf{A}(m,n)$ as a space of operations with m "inputs" and n "outputs."
Roughly speaking, an operad is then a theory generated by $\{\mathbf{A}(m,1)\}_{m\geq 1}$ while a
cooperad is a theory generated by $\{\mathbf{A}(1,m)\}_{n\geq 1}$.

Suppose now that $\mathcal{P}$ and $\mathcal{Q}$ are two operads. Temporarily denote by $\mathbf{F} =$
$\{\mathbf{F}(m,n)\}_{m,n\geq 1}$ the free theory generated by $\mathcal{P}$ and by the cooperad $\#\mathcal{Q}$. A *mixed
distributive law* is a relation M in $\mathbf{F}$ of a special type. A bialgebra is then an algebra
over the quotient-theory $\mathbf{F}/(M)$. We now give precise meanings for these notions
without referring to theories.

Let $N, s_1, \ldots, s_n, t_1, \ldots, t_m$ be natural numbers, $s_1 + \cdots s_n = t_1 + \cdots + t_m = N$.
Then $\Sigma_{s_1} \times \cdots \times \Sigma_{s_n}$ and $\Sigma_{t_1} \times \cdots \times \Sigma_{t_m}$ are subgroups of Σ_N, hence we may consider
$\mathbf{k}[\Sigma_N]$ as a $\Sigma_{s_1} \times \cdots \times \Sigma_{s_n}$- $\Sigma_{t_1} \times \cdots \times \Sigma_{t_m}$-bimodule. Also $\mathcal{P}(t_1) \otimes \cdots \otimes \mathcal{P}(t_m)$
(resp. $\mathcal{Q}(s_1) \otimes \cdots \otimes \mathcal{Q}(s_n)$) is a natural left $\Sigma_{t_1} \times \cdots \times \Sigma_{t_m}$ (resp. right $\Sigma_{s_1} \times \cdots \times \Sigma_{s_n}$)
module and it makes sense to form the product

$$\mathcal{Q}(s_1) \otimes \cdots \otimes \mathcal{Q}(s_n) \otimes_{\Sigma_{s_1} \times \cdots \times \Sigma_{s_n}} \mathbf{k}[\Sigma_N] \otimes_{\Sigma_{t_1} \times \cdots \times \Sigma_{t_m}} \mathcal{P}(t_1) \otimes \cdots \otimes \mathcal{P}(t_m).$$

We simplify the notation by writing

$$- \overline{\otimes}\, \mathbf{k}[\Sigma_N]\, \overline{\otimes}\, - \quad \text{instead of} \quad - \otimes_{\Sigma_{s_1} \times \cdots \times \Sigma_{s_n}} \mathbf{k}[\Sigma_N] \otimes_{\Sigma_{t_1} \times \cdots \times \Sigma_{t_m}} -.$$

By a *mixed distributive law* (between $\mathcal{P}$ and $\mathcal{Q}$) we mean a sequence

$$M = \{M(m,n)\}_{m,n\geq 1}$$

of maps

$$(38) \qquad M = M(m,n) : \mathcal{P}(m) \otimes \mathcal{Q}(n) \longrightarrow$$
$$\bigoplus \{\mathcal{Q}(t_1) \otimes \cdots \otimes \mathcal{Q}(t_m)\, \overline{\otimes}\, \mathbf{k}[\Sigma_N]\, \overline{\otimes}\, \mathcal{P}(s_1) \otimes \cdots \otimes \mathcal{P}(s_n)\},$$

where the summation is taken over all $N \geq 1$ and $s_1 + \cdots + s_n = t_1 + \cdots + t_m = N$.
The spaces above are in fact subspaces of $\mathbf{F}(m,n)$, so we may understand the latter
as things having m "inputs" and n "outputs." To make this even more mnemonic,
we write the target space on the right-hand side as

$$(39) \qquad \bigoplus \begin{pmatrix} \mathcal{Q}(t_1) \\ \vdots \\ \mathcal{Q}(t_m) \end{pmatrix} \overline{\otimes}\, \mathbf{k}[\Sigma_N]\, \overline{\otimes} \begin{pmatrix} \mathcal{P}(s_1) \\ \vdots \\ \mathcal{P}(s_n) \end{pmatrix}.$$

The map M, considered as a map of theories, must satisfy some conditions which
mean that it is compatible with the structure maps of the theory. We formulate
these conditions in operadic terms.

First, we have an obvious action of the symmetric group Σ_m "on inputs" of
$\mathcal{P}(m) \otimes \mathcal{Q}(n)$ given by $(\sigma, p \otimes q) \mapsto \sigma p \otimes q$, where the action of Σ_m on $\mathcal{P}(m)$ is

the one given by the operad structure. Similarly we have an action of Σ_N on the "inputs" of the space in (39) given as follows: For $\sigma \in \Sigma_m$, let $\sigma^{-1}_{t_1,\ldots,t_n} \in \Sigma_N$ be the permutation sending $(1,\ldots,N)$ to

$$(t_1 + \cdots + t_{\sigma(1)-1} + 1, \ldots, t_1 + \cdots + t_{\sigma(1)}, \ldots, t_1 + \cdots + t_{\sigma(m)-1} + 1, \ldots, t_1 + \cdots + t_{\sigma(m)}).$$

We may think of $\sigma^{-1}_{t_1,\ldots,t_n}$ as of the permutation permuting the blocks

$$(1,\ldots,t_1)(t_1+1,\ldots,t_1+t_2)\ldots(t_1+\cdots+t_{m-1}+1,\ldots,N)$$

via σ^{-1}. The action of σ is then given by

$$\left(\sigma, \begin{pmatrix} q_1 \\ \vdots \\ q_m \end{pmatrix} \overline{\otimes} \, \mu \, \overline{\otimes} \begin{pmatrix} p_1 \\ \vdots \\ p_n \end{pmatrix}\right) \longmapsto$$

$$\epsilon(\sigma; q_1, \ldots, q_m) \cdot \begin{pmatrix} q_{\sigma(1)} \\ \vdots \\ q_{\sigma(m)} \end{pmatrix} \overline{\otimes} \, \sigma^{-1}_{t_1,\ldots,t_n} \mu \, \overline{\otimes} \begin{pmatrix} p_1 \\ \vdots \\ p_n \end{pmatrix}.$$

The action of Σ_n "on outputs" is defined in a similar way. The first condition we require is that the map M is invariant with respect to both actions.

The second condition is the compatibility with composition maps. We formulate it again first "for inputs". Suppose that $p \in \mathcal{P}(m)$ is of the form $p = \mu(\nu_1,\ldots,\nu_k)$, with $\mu \in \mathcal{P}(k)$, $\nu_i \in \mathcal{P}(l_i)$ and $\sum l_i = m$. Let us write, for $q \in \mathcal{Q}(n)$

$$M(\mu \otimes q) = \sum_i \begin{pmatrix} q^i_1 \\ \vdots \\ q^i_k \end{pmatrix} \overline{\otimes} \, \sigma^i \, \overline{\otimes} \begin{pmatrix} p^i_1 \\ \vdots \\ p^i_n \end{pmatrix}.$$

The second condition then says that

$$(40) \qquad M(p \otimes q) = \sum_i \xi^i \cdot \begin{pmatrix} M(\nu_1, q^i_1) \\ \vdots \\ M(\nu_k, q^i_k) \end{pmatrix} \otimes \sigma^i \otimes \begin{pmatrix} p^i_1 \\ \vdots \\ p^i_n \end{pmatrix},$$

with $\xi^i = (-1)^{(|\nu_2|\cdot|q^i_1| + |\nu_3|\cdot(|q^i_1| + |q^i_2|) + \cdots + |\nu_k|\cdot(|q^i_1| + \cdots + |q^i_{k-1}|))}$, and similarly for "outputs." A little caution is needed to interpret formula (40) properly, because the right-hand side is not of the form required in (38). If $1 \le j \le k$ and $q^i_j = \mathcal{Q}(s^i_j)$ then each $M(\nu_j, q^i_j)$ is a sum of elements of the form

$$\begin{pmatrix} q^{i,j}_1 \\ \vdots \\ q^{i,j}_{l_j} \end{pmatrix} \overline{\otimes} \, \sigma^{i,j} \, \overline{\otimes} \begin{pmatrix} p^{i,j}_1 \\ \vdots \\ p^{i,j}_{s^i_j} \end{pmatrix}.$$

To interpret formula (40) we must compose the "outputs"

$$\begin{pmatrix} p^{i,j}_1 \\ \vdots \\ p^{i,j}_{s^i_j} \end{pmatrix}$$

of $M(\nu_j, q_j^i)$ with the elements of the "output" column

$$\begin{pmatrix} p_1^i \\ \vdots \\ p_n^i \end{pmatrix}$$

of $M(\mu \otimes q)$ via the permutation σ^i using the structure maps of the operad $\mathcal{Q}$. We suggest that the reader look at formula (43) for an explicit and easy example of this kind of manipulation. Let us sum up the above remarks in a compact definition.

11.1 Definition: We say that a system $\{M = M(m, n)\}_{m,n \geq 1}$ of maps as in (38) is a mixed distributivity law if:

  (i) it is compatible with the symmetric group actions both on the "input" and "output" sides

  (ii) it is compatible with the compositions both on the "input" and the "output" sides.

Suppose now that our operads $\mathcal{P}$ and $\mathcal{Q}$ are quadratic, $\mathcal{P} = \langle E; R \rangle$, $\mathcal{Q} = \langle F; S \rangle$.

11.2 Proposition: *A mixed distributive law for quadratic operads is uniquely determined by its component $M(2, 2)$ (which we denote $\mathcal{R}$),*

$$(41) \qquad \mathcal{R} : E \otimes F \longmapsto \bigoplus \begin{pmatrix} \mathcal{Q}(t_1) \\ \mathcal{Q}(t_2) \end{pmatrix} \overline{\otimes}\, \mathbf{k}[\Sigma_N]\, \overline{\otimes} \begin{pmatrix} \mathcal{P}(s_1) \\ \mathcal{P}(s_2) \end{pmatrix}.$$

On the other hand, a map $\mathcal{R}$ as above determines a mixed distributive law if and only if it is compatible with the Σ_2-actions and if the obvious extensions of $\mathcal{R}$ to $R \otimes F$ and $E \otimes S$ are zero.

Proof: The compatibility condition (40) enables us to express the mixed distributivity law M on $\mathcal{P}(m) \otimes \mathcal{Q}(n)$ via its values on the space of generators, i.e. via $\mathcal{R} = M(2, 2)$.

   On the other hand, given $\mathcal{R}$, we may extend it inductively to a mixed distributive law between the free operads $\mathcal{F}(E)$ and $\mathcal{F}(F)$. It is not hard to see that this extension induces a mixed distributive law between $\mathcal{P}$ and $\mathcal{Q}$ if and only if it sends $R \otimes F$ and $E \otimes S$ to zero.   □

11.3 Definition: We call a map $\mathcal{R}$ having the properties stated in the previous proposition a *replacement rule*.

   Let $\mathcal{P}$ and $\mathcal{Q}$ be quadratic operads and $\mathcal{R}$ a replacement rule between $\mathcal{P}$ and $\mathcal{Q}$. Suppose we have a graded vector space B, a $\mathcal{P}$-algebra structure $a : \mathcal{P} \to \mathcal{E}_B$ and a $\mathcal{Q}$-coalgebra structure $c : \mathcal{Q} \to \mathcal{G}_B$ on B. These two structures give rise to a natural map $J_l : E \otimes F \to \mathcal{M}(B^{\otimes 2}, B^{\otimes 2})$ given by $e \otimes f \mapsto c(f) \circ a(e)$. Similarly, they define another natural map

$$J_r : \begin{pmatrix} \mathcal{Q}(t_1) \\ \mathcal{Q}(t_2) \end{pmatrix} \overline{\otimes}\, \mathbf{k}[\Sigma_N]\, \overline{\otimes} \begin{pmatrix} \mathcal{P}(s_1) \\ \mathcal{P}(s_2) \end{pmatrix} \longrightarrow \mathcal{M}^*(B^{\otimes 2}, B^{\otimes 2})$$

given by

$$\begin{pmatrix} q_1 \\ q_2 \end{pmatrix} \overline{\otimes}\, \sigma\, \overline{\otimes} \begin{pmatrix} p_1 \\ p_2 \end{pmatrix} \longmapsto (c(q_1) \otimes c(q_2)) \circ \sigma \circ (a(p_1) \otimes a(p_2)),$$

where we interpret $\sigma \in \mathbf{k}[\Sigma_N]$ as a map from $B^{\otimes n}$ onto itself given by

$$(b_1, \ldots, b_n) \mapsto \epsilon(\sigma; b_1, \cdots, b_n) \cdot (b_{\sigma(1)}, \ldots, b_{\sigma(n)}).$$

11.4 Definition: Let $\mathcal{P}$ and $\mathcal{Q}$ be quadratic operads and let $\mathcal{R}$ be a replacement rule between $\mathcal{P}$ and $\mathcal{Q}$. We say that $B = (B, a, c)$ is a $\mathcal{P}$-$\mathcal{Q}$-bialgebra if

(i) $a : \mathcal{P} \to \mathcal{E}_B$ is a $\mathcal{P}$-algebra structure on B.

(ii) $c : \mathcal{Q} \to \mathcal{G}_B$ is a $\mathcal{Q}$-coalgebra structure on B.

(iii) $J_l(e \otimes f) = J_r(\mathcal{R}(e \otimes f))$ in $\mathcal{M}(B^{\otimes 2}, B^{\otimes 2})$.

11.5 Example: In this example we discuss associative/coassociative bialgebras of Example 6.4 from the operadic point of view. Let $Ass = \langle E; R \rangle$ be the associative algebra operad, i.e. E is the free Σ_2-space generated by one symbol μ and R is the ideal generated by the associativity relation $\mu(\mu, 1) - \mu(1, \mu) = 0$. Define

$$\mathcal{R} : E \otimes E \longmapsto \left(\begin{array}{c} E \\ E \end{array} \right) \overline{\otimes} \mathbf{k}[\Sigma_4] \overline{\otimes} \left(\begin{array}{c} E \\ E \end{array} \right)$$

by

(42)
$$\mathcal{R}(\mu \otimes \mu) = \left(\begin{array}{c} \mu \\ \mu \end{array} \right) \overline{\otimes} T_{1324} \overline{\otimes} \left(\begin{array}{c} \mu \\ \mu \end{array} \right).$$

Here T_{1324} denotes the map $B^{\otimes 4} \to B^{\otimes 4}$ induced by the permutation $(1234) \mapsto (1324)$. Let us verify that it satisfies the conditions of Proposition 11.2. The Σ_2-Σ_2-equivariance of $\mathcal{R}$ is clear. For the extension M of R we have

(43)
$$M(\mu(\mu, 1) \otimes \mu) = \left(\begin{array}{c} M(\mu \otimes \mu) \\ \mu \end{array} \right) \overline{\otimes} T_{1324} \overline{\otimes} \left(\begin{array}{c} \mu \\ \mu \end{array} \right)$$

$$= \left(\begin{array}{c} \mu \\ \mu \\ \mu \end{array} \right) \overline{\otimes} T_{135246} \overline{\otimes} \left(\begin{array}{c} \mu(\mu, 1) \\ \mu(\mu, 1) \end{array} \right),$$

because $M(\mu \otimes \mu) = \mathcal{R}(\mu \otimes \mu)$ and we expanded $\mathcal{R}(\mu \otimes \mu)$ as in (42). Similarly,

$$M(\mu(1, \mu) \otimes \mu) = \left(\begin{array}{c} \mu \\ \mu \\ \mu \end{array} \right) \overline{\otimes} T_{135246} \overline{\otimes} \left(\begin{array}{c} \mu(1, \mu) \\ \mu(1, \mu) \end{array} \right).$$

This computation shows that M takes $R \otimes E$ to zero, and similarly for $E \otimes R$. The corresponding Ass-Ass-bialgebras are bialgebras in the usual sense. We leave to the reader to verify that the same replacement rule defines $Comm$-Ass-bialgebras, Ass-$Comm$-bialgebras and even $Comm$-$Comm$-bialgebras (the latter are rather boring, due to the Structure Theorem).

11.6 Example: Here we discuss Lie bialgebras of Example 6.5 as bialgebras given by a replacement rule. Let $Lie = \langle E; R \rangle$ be the Lie algebra operad; let ℓ be an antisymmetric element generating E. Then R is generated by the Jacobi identity. Let us define the replacement rule

$$\mathcal{R} : E \otimes E \longmapsto$$

$$\left(\begin{array}{c} E \\ \mathbf{k} \end{array} \right) \overline{\otimes} \mathbf{k}[\Sigma_3] \overline{\otimes} \left(\begin{array}{c} \mathbf{k} \\ E \end{array} \right) \oplus \left(\begin{array}{c} E \\ \mathbf{k} \end{array} \right) \overline{\otimes} \mathbf{k}[\Sigma_3] \overline{\otimes} \left(\begin{array}{c} E \\ \mathbf{k} \end{array} \right) \oplus \left(\begin{array}{c} \mathbf{k} \\ E \end{array} \right) \overline{\otimes} \mathbf{k}[\Sigma_3] \overline{\otimes} \left(\begin{array}{c} E \\ \mathbf{k} \end{array} \right) \oplus \left(\begin{array}{c} \mathbf{k} \\ E \end{array} \right) \overline{\otimes} \mathbf{k}[\Sigma_3] \overline{\otimes} \left(\begin{array}{c} \mathbf{k} \\ E \end{array} \right)$$

between *Lie* and *Lie* by

$$\mathcal{R}(\ell,\ell) = \binom{\ell}{1}\overline{\otimes}1\overline{\otimes}\binom{1}{\ell} \oplus \binom{\ell}{1}\overline{\otimes}T_{132}\overline{\otimes}\binom{\ell}{1} \oplus \binom{1}{\ell}\overline{\otimes}1\overline{\otimes}\binom{\ell}{1} \oplus \binom{1}{\ell}\overline{\otimes}T_{213}\overline{\otimes}\binom{1}{\ell}.$$

We leave it to the reader to verify that this really does define a mixed distributive law (the computation is rather long, involving many terms).

11.7 Example: In this example we describe a nonsymmetric analog of Lie bialgebras. Let $Ass = \langle E; R\rangle$ be the associative algebra operad as in Example 11.5 and define the replacement rule

$$\mathcal{R}: E \otimes E \longmapsto \binom{E}{\mathbf{k}}\overline{\otimes}\mathbf{k}[\Sigma_3]\overline{\otimes}\binom{\mathbf{k}}{E} \oplus \binom{\mathbf{k}}{E}\overline{\otimes}\mathbf{k}[\Sigma_3]\overline{\otimes}\binom{E}{\mathbf{k}}$$

between *Ass* and *Ass* by

$$\mathcal{R}(\mu,\mu) = \binom{\mu}{1}\overline{\otimes}1\overline{\otimes}\binom{1}{\mu} \oplus \binom{1}{\mu}\overline{\otimes}1\overline{\otimes}\binom{\mu}{1}.$$

We call these *Ass-Ass*-bialgebras *mock bialgebras*. So, a mock bialgebra is a vector space B with an associative multiplication $\cdot$ and a coassociative comultiplication Δ such that

$$\Delta(x \cdot y) = \sum x_{(1)} \otimes x_{(2)} \cdot y + x \cdot y_{(1)} \otimes y_{(2)}.$$

Mock bilagebras will illustrate the necessity of the homogeneity assumption in Theorem 11.11.

11.8 Example: In this example we discuss Poisson bialgebras already introduced in Example 6.6. Let $Poiss = \mathcal{P}(0,0) = \langle E; R\rangle$ be the operad for Poisson algebras, let ℓ (resp. μ) denote the antisymmetric (resp. symmetric) generator of E, see Example 9.4. Let $Ass = \langle F; S\rangle$ be the operad for associative algebras, let ν be the generator of F. Define the replacement rule

$$\mathcal{R}: E \otimes F \longmapsto \binom{F}{F}\overline{\otimes}\mathbf{k}[\Sigma_4]\overline{\otimes}\binom{E}{E}$$

between *Poiss* and *Ass* by

$$\mathcal{R}(\ell\otimes\nu) = \binom{\nu}{\nu}\overline{\otimes}T_{1324}\overline{\otimes}\binom{\ell}{\mu} \oplus \binom{\nu}{\nu}\overline{\otimes}T_{1324}\overline{\otimes}\binom{\mu}{\ell}$$

while $\mathcal{R}(\mu\otimes\nu)$ is defined by the same formula as in the bialgebra case, see Example 11.5.

11.9 Definition: We say that a replacement rule $\mathcal{R}$ is *homogeneous* if, on the right-hand side of (41), $t_1 = t_2$ and $s_1 = s_2$.

Observe that $\mathcal{R}$ is homogeneous if and only if on the right-hand side of (38) we always have $t_1 = t_2 = \cdots = t_m$ and $s_1 = s_2 = \cdots = s_n$. The replacement rules in examples 11.5 and 11.8 are homogeneous, the replacement rules in examples 11.6 and 11.7 are not.

11.10 Theorem: *Let B be a $\mathcal{P}$-$\mathcal{Q}$-bialgebra. Then*

(i) *The Q-coalgebra $C_Q(B)$ has a natural P-algebra structure. Moreover, in the homogeneous case this P-algebra structure is such that each $C_Q^n(B)$ forms a subalgebra and the canonical map $c^n : B \mapsto C_Q^n(B)$ is a P-algebra homomorphism for any $n \geq 1$.*

(ii) *The P-algebra $A_P(B)$ has a natural Q-coalgebra structure such that the canonical map $a : A_P(B) \longrightarrow B$ is a Q-coalgebra homomorphism.*

Proof: Let us prove (i). To define a P-algebra structure on $C_Q(B)$ means to define, for each $m, n \geq 1$, a map

$$(44) \quad P(m) \otimes (\bigoplus \# Q(t_1) \otimes B^{\otimes t_1})^{\Sigma_{t_1}} \otimes \cdots \otimes (\bigoplus \# Q(t_m) \otimes B^{\otimes t_m})^{\Sigma_{t_m}}) \longrightarrow$$
$$\longrightarrow (\bigoplus \# Q(n) \otimes B^{\otimes n})^{\Sigma_n}.$$

The mixed distributive law gives, after a proper dualization, the map

$$P(m) \otimes (\bigoplus \# Q(t_1) \otimes B^{\otimes t_1})^{\Sigma_{t_1}} \otimes \cdots \otimes (\bigoplus \# Q(t_m) \otimes B^{\otimes t_m})^{\Sigma_{t_m}} \longrightarrow$$
$$\longrightarrow \bigoplus \# Q(n) \otimes (\bigoplus P(s_1) \otimes B^{\otimes s_1})_{\Sigma_{s_1}} \otimes \cdots \otimes (\bigoplus Q(s_n) \otimes B^{\otimes s_n})_{\Sigma_{s_n}},$$

while the P-algebra structure on B gives, for any $1 \leq i \leq n$, a map $(P(s_i) \otimes B^{\otimes s_i})_{\Sigma_{s_i}} \longrightarrow B$. The composition of these two maps give the requisite map of (44). We leave the verification of the desired properties to the reader. The proof of (ii) is the verbatim dual of the arguments above. □

The theorem we have just proven relates the operadic notion of a replacement rule with the triple definition of a mixed distributive law; see Definition 5.1.

11.11   Theorem: *Let B be a P-Q-bialgebra.*

(i) *If the replacement rule is homogeneous, then both $C_P(B)$ and $A_{Q^!}(B)$ have natural (B, a)-module structures.*

(ii) *Both $A_P(B)$ and $A_{Q^!}(B)$ have natural (B, c)-comodule structures.*

Proof: Let us prove (i). Observe first that the P-algebra structure on $C_P^n(B)$ of Theorem 11.10 induces, via the canonical map $c^n : B \longrightarrow C_P^n(B)$, a (B, a)-module structure on $C_P^n(B)$ for any $n \geq 1$. This gives rise to an obvious (B, a)-module structure on the direct sum $C_P(B) = \bigoplus_{n \geq 1} C_P^n(B)$. The homogeneity assumption was necessary to build up the action from the partial actions on the homogeneous parts, otherwise there is no way to guarantee the convergence of such an action, see Example 11.14.

The existence of a (B, a)-module structure on $A_{Q^!}(B)$ is a much more delicate matter. To fix the notation, let $Q = \langle F; S \rangle$. The first observation is that a replacement rule $\mathcal{R}$ between P and Q induces a mixed distributive law between P and $\mathcal{F}(F) = \langle F; 0 \rangle$, the free operad on F, which is homogeneous if $\mathcal{R}$ is. We thus have a (B, a)-module structure on $C_{\mathcal{F}(F)}(B)$. The second important observation is that there is a canonical isomorphism of graded spaces $C_{\mathcal{F}(F)}^*(B) \cong A_{\mathcal{F}(\# F)}^*(B)$, therefore we have a natural (B, a)-module structure also on $A_{\mathcal{F}(\# F)}(B)$. The third observation is that the algebra $A_{Q^!}(B)$ is a quotient of $A_{\mathcal{F}(\# F)}(B)$. So it remains to prove that the (B, a)-module structure on $A_{\mathcal{F}(\# F)}(B)$ induces a (B, a)-module structure on the quotient $A_{Q^!}(B)$. We leave this rather technical verification to the reader. The proof of the second half of the theorem is an exact dual. The homogeneity assumption is not needed here, because the action always 'converges', see Example 11.14.   □

In the following examples we focus our attention on the module structure on $A_{Q^!}(B)$ which, we think, has not yet been observed except in the bialgebra case where it is rather obvious, as we will see in the next example.

11.12   Example: Let us begin with the classical bialgebras (see examples 6.4 and 11.5). Here $\mathcal{P} = \mathcal{Q} = Ass$, therefore $\mathcal{Q}^!$ is again Ass and the free $\mathcal{Q}^!$-algebra $A_{Q^!}(B)$ on a vector space B is the usual tensor algebra, $A_{Q^!}(B) = T(B) = \bigoplus_{k\geq 1} \bigotimes^k B$. Let $B = (B, \mu, \Delta)$ be a bialgebra. We introduce the *iterated diagonal* $\Delta^{[n]} : B \to B^{\otimes n}$, $n \geq 1$, by $\Delta^{[1]} = \mathbb{1}$ and $\Delta^{[n+1]} = (\Delta \otimes \mathbb{1}^{\otimes n-1}) \circ \Delta^{[n]}$. Using the Sweedler notation $\Delta^{[n]}(b) = \sum b_{(1)} \otimes \cdots \otimes b_{(n)}$ we define a left action $\bullet : B \otimes T(B) \to T(B)$ by

$$b \bullet (x_1 \otimes \cdots \otimes x_n) = \sum b_{(1)} \cdot x_1 \otimes \cdots \otimes b_{(n)} \cdot x_n,$$

where $\cdot$ denotes the multiplication μ. The right action is defined in a similar way and these together give a (B, μ)-module structure on $T(B)$.

11.13   Example: A bit less trivial is the case of Ass-$Comm$-bialgebras. We have $\mathcal{Q}^! = Lie$ and $A_{Q^!}(B)$ is the free Lie algebra $L(B)$ on B. Let $B = (B, \mu, \Delta)$ be an Ass-$Comm$-bialgebra (i.e. a cocommutative bialgebra). Let us define a left action $\bullet$ of (B, a) on $L(B)$ inductively by saying that $b \bullet u = b \cdot u$ for $b, u \in B$ while

$$b \bullet [u, v] = \sum [b_{(1)} \bullet u, b_{(2)} \bullet v]$$

for $u, v \in L(B)$. The right action is defined by a similar rule. The well-definedness of the action means that it preserves the antisymmetry (this is almost obvious) and the ideal generated by the Jacobi identity. Let us inspect how $b \in B$ acts on it! We have, for $u, v, w \in B$,

$$b \bullet ([u, [v, w]] + [v, [w, u]] + [w, [u, v]]) =$$
$$= [b_{(1)} \cdot u, [b_{(2)} \cdot v, b_{(3)} \cdot w]] + [b_{(1)} \cdot v, [b_{(2)} \cdot w, b_{(3)} \cdot u]] + [b_{(1)} \cdot w, [b_{(2)} \cdot u, b_{(3)} \cdot v]]$$
$$= [b_{(1)} \cdot u, [b_{(2)} \cdot v, b_{(3)} \cdot w]] + [b_{(2)} \cdot v, [b_{(3)} \cdot w, b_{(1)} \cdot u]] + [b_{(3)} \cdot w, [b_{(1)} \cdot u, b_{(2)} \cdot v]] = 0.$$

Here the cocommutativity of Δ which implies that

$$\sum b_{(1)} \otimes b_{(2)} \otimes b_{(3)} = \sum b_{(\sigma(1))} \otimes b_{(\sigma(2))} \otimes b_{(\sigma(3))}$$

for any $\sigma \in \Sigma_3$ is absolutely crucial.

11.14   Example: Let (B, μ, Δ) be the mock bialgebra introduced in Example 11.7. As in Example 11.5, $\mathcal{P} = \mathcal{Q} = \mathcal{Q}^!$ and $A_{Q^!}(B) = T(B)$, the tensor algebra. Because of the nonhomogeneity of the replacement rule, this example is really weird, and we describe first the $\mathcal{P} =$ associative algebra structure on $C_{\mathcal{Q}}(B) = T(B)$. It is given by the multiplication $\star : T(B) \otimes T(B) \to T(B)$ defined by

$$(x_1 \otimes \cdots \otimes x_n) \star (y_1 \otimes \cdots \otimes y_m) = x_1 \otimes \cdots \otimes x_n \cdot y_1 \otimes \cdots \otimes y_m,$$

for $(x_1 \otimes \cdots \otimes x_n) \in T^n(B)$ and $(y_1 \otimes \cdots \otimes y_m) \in T^m(B)$. If we try to blindly repeat the construction of a (B, a)-module structure on $A_{Q^!}(B) = T(B)$ as it is given in the proof of Theorem 11.11, we arrive at the formula

$$b \bullet (x_1 \otimes \cdots \otimes x_n) = b \cdot x_1 \otimes \cdots \otimes x_n +$$

$$+ \sum b_{(1)} \otimes b_{(2)} \cdot x_1 \otimes \cdots \otimes x_n + \sum b_{(1)} \otimes b_{(2)} \otimes b_{(3)} \cdot x_1 \otimes \cdots \otimes x_n + \cdots,$$

which is not an element of $C_{\mathcal{P}}(B) = \bigoplus_{n \geq 1} C_{\mathcal{P}}^n(B)$ but rather of $\prod_{n \geq 1} C_{\mathcal{P}}^n(B)$. We see that the homogeneity is a necessary assumption in Theorem 11.11 (i).

On the other hand, there are no problems with the existence of the structures of Theorem 11.10 (ii). The coalgebra structure, say $\delta : T(B) \longrightarrow T(B) \otimes T(B)$, is given by

$$\delta(x_1 \otimes \cdots \otimes x_n) = \sum_{1 \leq i \leq n} x_1 \otimes (x_i)_{(1)} \otimes (x_i)_{(2)} \otimes \cdots \otimes x_n$$

where we interpret $x_1 \otimes (x_i)_{(1)} \otimes (x_i)_{(2)} \otimes \cdots \otimes x_n$ as an element of $T^i(B) \otimes T^{n-i+1}(B)$. The left (B, c)-comodule structure on $T(B)$, say $\nu_l : T(B) \longrightarrow B \otimes T(B)$, is given by

$$\nu_l(x_1 \otimes \cdots \otimes x_n) = \sum_{1 \leq i \leq n} (x_1 \cdot \cdots \cdot x_i)_{(1)} \otimes (x_1 \cdot \cdots \cdot x_i)_{(2)} \otimes x_{i+1} \cdots \otimes x_n,$$

where we interpret $(x_1 \cdot \cdots \cdot x_i)_{(1)} \otimes (x_1 \cdot \cdots \cdot x_i)_{(2)} \otimes x_{i+1} \cdots \otimes x_n$ as an element of $B \otimes T^{n-i+1}(B)$. The right action is defined in a similar way.

12   Operadic cohomology of bialgebras

In this section we define a cohomology of a $\mathcal{P}$-$\mathcal{Q}$-bialgebra $B = (B, a, c)$, where $\mathcal{P}$ and $\mathcal{Q}$ are quadratic operads with a homogeneous replacement rule $\mathcal{R}$. The homogeneity assumption is not absolutely necessary, but in the general case we do not have a nice definition based on a bicomplex. There are also some problems with the grading because the resulting cohomology is not a direct sum of the homogeneous components but rather the direct product, this phenomenon being already observed on the cohomology of A(m)-algebras [20].

We need first the notion of the operadic cohomology of a coalgebra $C = (C, c)$ with coefficients in a C-comodule N. The definition is exactly the dual of the definition of the operadic cohomology of an algebra with coefficients in a module (Definition 8.3). We thus merely sketch the definition.

Let $A = (A, a)$ be a $\mathcal{P}$-algebra and $\theta : A \longrightarrow A$ a homogeneous degree p linear map. We say that θ is a *degree p derivation* if, for any $n \geq 2$ and $\mu \in \mathcal{P}(n)$,

$$\theta \circ a(n)(\mu) = (-1)^{|\mu| \cdot |\theta|} \cdot \sum_{0 \leq i \leq n-1} a(n)(\mu) \circ (1^{\otimes i} \otimes \theta \otimes 1^{\otimes (n-i-1)}).$$

Let us denote by $\mathrm{Der}^p(A)$ the space of all degree p derivations of the algebra A. Recall that, for a graded vector space W, $A_{\mathcal{P}}(W)$ denotes the free $\mathcal{P}$-algebra on W. As in the case of coderivations, $\mathrm{Der}^p(A_{\mathcal{P}}(W))$ has a second grading, namely

$$\mathrm{Der}^{p,n}(A_{\mathcal{P}}(W)) = \{\theta \in \mathrm{Der}^p(A_{\mathcal{P}}(W)); \, \theta(W) \subset A_{\mathcal{P}}^{n+1}(W)\}.$$

We have the following analog of Theorem 8.2.

12.1   Theorem: *Let V be a graded vector space and let $W = \uparrow V$. Then there exists a one-to-one correspondence between $\mathcal{P}$-coalgebra structures $c : \mathcal{P} \longrightarrow \mathcal{G}_V$ on*

V and degree one derivations $\delta \in \mathrm{Der}^{1,1}(A_{\mathcal{P}^!}(W))$ of the free $\mathcal{P}^!$-coalgebra $A_{\mathcal{P}^!}(W)$ with $\delta^2 = 0$.

Let $C = (C, c)$ be a $\mathcal{P}$-coalgebra. By a C-comodule we mean a graded vector space N together with a $\mathcal{P}$-coalgebra structure $\nu : \mathcal{P} \to \mathcal{G}_{C \oplus N}$ on $C \oplus N$ such that both the projection $C \oplus N \to C$ and the inclusion $C \oplus C \oplus N$ are $\mathcal{P}$-algebra homomorphisms and $\nu(n)(p) : C \oplus N \to (C \oplus N)^{\otimes n}$, for each $n \geq 1$ and $p \in \mathcal{P}(n)$, maps N to the subspace of $(C \oplus N)^{\otimes n}$ spanned by monomials which contain exactly one element of N.

Finally, let $C = (C, c)$ be a $\mathcal{P}$-coalgebra and N a C-comodule. Let $V = \uparrow C$, $Y = \uparrow N$ and $Z = V \oplus X$. The decomposition $Z^{\otimes n} = \bigoplus_{i+j=n}(Z^{\otimes n})^{i,j}$, where $(Z^{\otimes n})^{i,j}$ is the subspace spanned by monomials containing i elements of V and j elements of Y, induces the decomposition $A_{\mathcal{P}^!}^n(Z) = \bigoplus_{i+j=n} A_{\mathcal{P}^!}^{i,j}(Z)$ for any $n \geq 1$. Let $D_{\mathcal{P}}^{p,n}(N, C)$ be the subspace of $\mathrm{Der}^{p,n}(A_{\mathcal{P}^!}(Z))$ consisting of those derivations θ for which $\theta(Y) \subset A_{\mathcal{P}^!}^{i,0}(Z)$ and $\theta(V) = 0$. The graded commutator with the derivation $\delta \in \mathrm{Der}^{1,1}(C_{\mathcal{P}^!}(W))$ which corresponds to the coalgebra structure on $C \oplus N$ via the correspondence of Theorem 12.1 induces a differential $\Delta : D_{\mathcal{P}}^{p,n}(N, C) \to D_{\mathcal{P}}^{p+1,n+1}(N, C)$. We have the following analog of Definition 8.3.

12.2 Definition: Let C be a coalgebra over a quadratic operad $\mathcal{P}$ and let N be an C-comodule. The cohomology $H_{\mathcal{P}}^{p,n}(N; C)$ of C with coefficients in N is defined as

$$H_{\mathcal{P}}^{p,n}(N; C) := H^{p,n}(D_{\mathcal{P}}^{*,*}(N, C), \Delta).$$

We call p and n the *inner* and *simplicial* degrees, respectively. As in (27) we observe that

$$(45) \qquad D_{\mathcal{P}}^{p,n}(N, C) \cong \mathcal{M}^p(\uparrow N, A_{\mathcal{P}^!}^{n+1}(\uparrow C)).$$

Now we proceed to the definition of the operadic bialgebra cohomology of a $\mathcal{P}$-$\mathcal{Q}$-coalgebra $B = (B, a, c)$ with a homogeneous distributive law $\mathcal{R}$. Put

$$C_{\mathcal{P}, \mathcal{Q}}^{p,m,n}(B, B) = \mathcal{M}^{p-m-n}(C_{\mathcal{P}^!}^{m+1}(B), A_{\mathcal{Q}^!}^{n+1}(B)).$$

By Theorem 11.10 (i) there exists a (B, a)-module structure on $A_{\mathcal{Q}^!}^{n+1}(B)$ (recall that $\mathcal{R}$ is supposed to be homogeneous) and (27) gives the identification

$$(46) \quad C_{\mathcal{P}, \mathcal{Q}}^{p,m,n}(B, B) \cong \mathcal{M}^{p-n}(C_{\mathcal{P}^!}^{m+1}(\downarrow B), \downarrow A_{\mathcal{Q}^!}^{n+1}(B)) = C_{\mathcal{P}}^{p-n,m}(B, A_{\mathcal{Q}^!}^{n+1}(B)).$$

Similarly, by Theorem 11.10 (ii) there exists a (B, c)-comodule structure on the space $C_{\mathcal{P}^!}^{m+1}(B)$ and we have the identification

$$(47) \quad C_{\mathcal{P}, \mathcal{Q}}^{p,m,n}(B, B) \cong \mathcal{M}^{p-m}(\uparrow C_{\mathcal{P}^!}^{m+1}(B), A_{\mathcal{Q}^!}^{n+1}(\uparrow B)) = D_{\mathcal{Q}}^{p-m,n}(C_{\mathcal{P}^!}^{m+1}(B), B).$$

We may thus consider the following bicomplex.

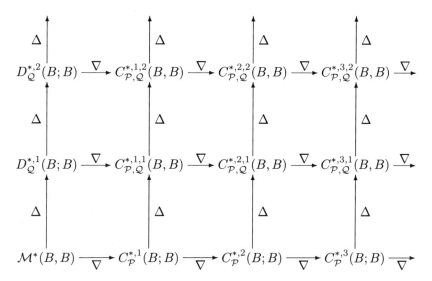

The n-th row of the bicomplex is the complex defining the operadic cohomology of (B, a) with coefficients in $A_{\mathcal{Q}!}^{n+1}(B)$ (46) and the m-th column is the complex from the definition of the operadic cohomology of the coalgebra (B, c) with coefficients in $C_{\mathcal{P}!}^{m+1}(B)$ (47). It is not hard to verify that the differentials in the above bicomplex commute, so the following definition makes sense.

12.3 Definition: The cohomology $H_{\mathcal{P},\mathcal{Q}}^{*,*}(B; B)$ of a $\mathcal{P}$-$\mathcal{Q}$-bialgebra $B = (B, a, c)$ is the cohomology of the bicomplex above,

$$H_{\mathcal{P},\mathcal{Q}}^{p,*}(B; B) = H^{p,*}(\bigoplus_{m+n=*} C_{\mathcal{P},\mathcal{Q}}^{p,m,n}(B, B), \nabla + \Delta).$$

12.4 Example: In this example we discuss the cohomology of bialgebras introduced in examples 6.4 and 11.5. We again consider the nongraded case only, so we may neglect the inner grading. The cohomology $H_{Ass,Ass}^{*}(B, B)$ of associative coassociative bialgebras coincides with the bialgebra cohomology introduced by Gerstenhaber and Schack in [15]. At the (m, n)-position of the bicomplex is then the space $\mathcal{M}(\bigotimes^{m+1}(B), \bigotimes^{n+1}(B))$.

The cohomology of Ass-$Comm$-bialgebras, i.e. associative coassociative bialgebras with the cocommutative comultiplication looks very similar except that the space at the (m, n)-position of the bicomplex is $\mathcal{M}(\bigotimes^{m+1}(B), L^{n+1}(B))$, where $L^{n+1}(B)$ is the subspace of the free Lie algebra $L(B)$ on B consisting of elements of length n.

In the dual case of the cohomology of $Comm$-Ass-bialgebras the element at the (m, n)-position is $\mathcal{M}(C_{Lie}^{m+1}(B), \bigotimes^{n+1}(B)) \cong \mathcal{M}_{\mathrm{Harr}}(\bigotimes^{m+1}(B), \bigotimes^{n+1}(B))$ where $\mathcal{M}_{\mathrm{Harr}}(\bigotimes^{m+1}(B), \bigotimes^{n+1}(B))$ is the subspace of $\mathcal{M}^{*}(\bigotimes^{m+1}(B), \bigotimes^{n+1}(B))$ of linear maps which are zero on decomposables of the shuffle product, see [28]. The differential in the last description is given by the restriction of the differential from the Gerstenhaber-Schack complex. A similar bicomplex was used in [25] to describe restricted deformations of triangular quasi-Hopf algebras.

References

[1] D. Balavoine. *Déformations et rigidité geometrique des algèbres de Leibniz.* Université Montpellier II, preprint (1994).

[2] M. Barr. *Shukla cohomology and triples.* J. Algebra **5** (1967) 222–231.

[3] M. Barr. *Harrison homology, Hochschild homology and triples.* J. Algebra **8** (1968) 314–323.

[4] M. Barr. *Composite cotriples and derived functors.* In Seminar on Triples and Categorical Homology Theory, Lecture Notes in Math. **80**, Springer-Verlag, Berlin, Heidelberg, New York (1969) 336–356.

[5] M. Barr. *Cartan-Eilenberg cohomology and triples.* J. Pure Appl. Algebra (to appear).

[6] M. Barr and J. Beck. *Homology and standard constructions.* In Seminar on Triples and Categorical Homology Theory, Lecture Notes in Math. **80**, Springer-Verlag, Berlin, Heidelberg, New York (1969) 245–335.

[7] M. Barr and C. Wells. Toposes, Triples and Theories. Springer-Verlag, Berlin, Heidelberg, New York, 1985.

[8] J. Beck. *Triples, algebras and cohomology.* Dissertation, Columbia University, New York (1967).

[9] J. Beck. *Distributive laws.* In Seminar on Triples and Categorical Homology Theory, Lecture Notes in Math. **80**, Springer-Verlag, Berlin, Heidelberg, New York (1969) 119–140.

[10] H. Cartan and S. Eilenberg. Homological Algebra. Princeton University Press, Princeton, New Jersey, 1956.

[11] F. R. Cohen. *Artin's braid groups, classical homotopy theory and sundry other curiosities.* Contemp. Math. **78** (1988) 167–206.

[12] T. Fox. *Algebraic deformations and bicohomology.* Canad. Bull. Math. **32** (1989) 182–189.

[13] T. Fox. *An introduction to algebraic deformation theory.* J. Pure Applied Algebra **84** (1993) 17–41.

[14] T. Fox. *The construction of cofree coalgebras.* J. Pure Applied Algebra **84** (1993) 191–198.

[15] M. Gerstenhaber and S. D. Schack. *Algebras, bialgebras, quantum groups, and algebraic deformations.* Contemp. Math. **134** (1992), 51–92.

[16] E. Getzler and J.D.S. Jones. *n-algebras.* (in preparation).

[17] V. Ginzburg and M. Kapranov. *Koszul duality for operads.* Duke Math. J. **76** (1994) 203–272.

[18] S. MacLane. Homology. Springer-Verlag, Berlin, Heidelberg, New York, 1963.

[19] S. MacLane. *Categorical algebra.* Bull. Amer. Math. Soc. **71** (1965) 40–106.

[20] M. Markl. *A cohomology theory for A(m)-algebras and applications.* J. Pure Appl. Algebra **83** (1992) 141–175.

[21] M. Markl. *Cotangent cohomology of a theory and deformations.* J. Pure Appl. Algebra (to appear).

[22] M. Markl. *Free and commutative models of a space.* Publ. IRMA, Lille, Numéro Spécial: Géometrie et Topologie 1988-89, 117–127

[23] M. Markl. *On the formality of products and wedges.* Rend. Circ. Mat. Palermo, Ser. II, **16** (1987) 199–206.

[24] M. Markl. *Distributive laws and the Koszulness.* Ann. Institut Fourier (to appear). Available as preprint `hepth/9409192`.

[25] M. Markl. *A note on quasi-Hopf algebras.* (Unpublished note, 1990).

[26] P.May. The Geometry of Iterated Loop Spaces, Lecture Notes in Math. **271**, Springer-Verlag, Berlin, Heidelberg, New York (1972).

[27] W. Michaelis. *Lie coalgebras.* Advances in Math. **38** (1980) 1–54.

[28] R. Ree. *Lie elements and an algebra associated with shuffles.* Annals of Math. **68** (1958) 210–220

[29] J. D. Stasheff. *The intrinsic bracket on the deformation complex of an associative algebra.* J. Pure Appl. Algebra **89** (1993) 231–235

[30] D. Tanré. *Cohomologie de Harrison et type d'homotopie rationnelle.* Lecture Notes in Math. **1183**, Springer-Verlag, Berlin, Heidelberg, New York (1983), 361–370.

[31] T. Tjin. *An introduction to quantized Lie groups and algebras.* Internat. J. Modern Phys. **A7** (1992) 6175–6213.

[32] D. VanOsdol. *Coalgebras, sheaves, and cohomology.* Proc. Amer. Math. Soc. **33** (1972) 257–263.

[33] D. VanOsdol. *Bicohomology theory.* Trans. Amer. Math. Soc. **183** (1973) 449–476.

T. Fox:
McGill University, Dept. of Mathematics and Statistics
805 Sherbrooke St. W., Montréal, Québec, Canada H2V 4J4
email: `fox@math.mcgill.ca`

M. Markl:
Mathematical Institute of the Academy, Žitná 25
115 67 Prague 1, The Czech Republic
email: `markl@mbox.cesnet.cz`

Contemporary Mathematics
Volume **202**, 1997

Deformations of Algebras over a Quadratic Operad

David Balavoine

0. Introduction

In this paper we study formal deformations (in the sense of M. Gerstenhaber [G2]) of algebras over a quadratic operad, by using the theory of square zero elements in graded Lie algebras developed by P. Lecomte and M. de Wilde [L–W]. The general framework we construct then gives us the cohomology theory controlling formal deformations of such algebras, in the sense of [G2]. In particular, this framework can be applied to classical structures such as associative, commutative, Lie and Poisson algebras. Moreover, it permits us to consider the case of a new structure of algebras (Leibniz algebras), we studied first in a different manner in ([BA1],[BA2]) using a method due to T. Fox [FO].

Let us recall on an example what is the square zero element theory. Let k be a field of characteristic zero and V a vector space over k. For $p \in \mathbb{N}$, set $M^p(V) = Hom_k(V^{\otimes p+1}, V)$, and $M(V) = \bigoplus_{p \geq 0} M^p(V)$. For $f \in M^p(V)$, $g \in M^q(V)$, let $I_f(g) \in M^{p+q}(V)$ be the map defined by:

$$I_f(g)(x_0 \otimes \cdots \otimes x_{p+q}) = \sum_{k=0}^{p} (-1)^{qk} f(x_0 \otimes \cdots \otimes x_{k-1} \otimes g(x_k \otimes \cdots \otimes x_{k+q}) \otimes \cdots \otimes x_{p+q}).$$

For $f, g \in M^p(V) \times M^q(V)$ put $[f, g] = I_f(g) + (-1)^{pq+1} I_g(f)$; then $(M(V), [,])$ becomes a graded Lie algebra (this structure was originally found by M. Gerstenhaber in [G1]). An element $\varphi \in M^1(V)$ defines a product on V, and it is immediately seen that φ is an associative product if, and only if, $[\varphi, \varphi] = 0$. We then say that φ is a square zero element. Let $A = (V, \varphi)$ be the corresponding associative algebra, and let $\delta_\varphi : M^p(V) \to M^{p+1}(V)$ be the operator defined by $\delta_\varphi(f) = [f, \varphi]$. The condition for φ to be of square zero implies that δ_φ is a differential; furthermore the cohomology of $(M(V), \delta_\varphi)$, denoted by $H_\varphi^*(M(V))$, is nothing but the Hochschild cohomology of A, $HH^{*+1}(A, A)$.

This presentation enables us to study formal deformations of associative algebras. Namely, let $A = (V, \varphi)$ be an associative algebra. A formal deformation of φ is a formal power series, $\varphi_t = \sum_{i=0}^{+\infty} \varphi_i t^i$, with coefficients in $M^1(V)$, such that

1991 *Mathematics Subject Classification.* Primary 17B55; Secondary 16E40, 16S80, 17B70.
Key words and phrases. Operads, deformation of algebras, cohomology of algebras.

$\varphi_0 = \varphi$ and $[\varphi_t, \varphi_t] = \sum_{n=0}^{+\infty} (\sum_{i+j=n} [\varphi_i, \varphi_j]) \, t^n = 0$. Comparing the n-th terms, we have the following conditions of existence of a deformation of φ,

$$(c_n) : \sum_{i+j=n} [\varphi_i, \varphi_j] = 0.$$

If we rewrite the (c_n)'s conditions by using the δ_φ operator, then (c_n) has to be replaced by,

$$(c'_n) : 2\delta_\varphi(\varphi_n) = - \sum_{\substack{i+j=n \\ i,j \neq 0}} [\varphi_i, \varphi_j].$$

For $n = 1$, $(c'_1) \iff \delta_\varphi(\varphi_1) = 0$, i.e. φ_1 is a 1-cocycle for $H^*_\varphi(M(V))$ (hence a 2-cocycle for $HH^*(A, A)$). Using the graded Lie algebra structure on $M(V)$, we see that:

$$\delta_\varphi (\sum_{\substack{i+j=n \\ i,j \neq 0}} [\varphi_i, \varphi_j]) = 0,$$

because we have the equality,

$$\delta_\varphi (\sum_{\substack{i+j=n \\ i,j \neq 0}} [\varphi_i, \varphi_j]) = 2 \sum_{\substack{i+j+k=n \\ i,j,k \neq 0}} [\varphi_i, [\varphi_j, \varphi_k]]$$

$$= \frac{2}{3} \sum_{\substack{i+j+k=n \\ i,j,k \neq 0}} \{ [\varphi_i, [\varphi_j, \varphi_k]] + [\varphi_j, [\varphi_k, \varphi_i]] + [\varphi_k, [\varphi_i, \varphi_j]] \} = 0$$

(Jacobi's identity). So if $H^2_\varphi(M(V)) = HH^3(A, A) = 0$, then any Hochschild 2-cocycle φ_1 can be extended to a formal deformation of φ, which is a basic result of [G2]. However, notice that we can't deduce from the graded Lie algebra structure on $M(V)$ the fact that if $HH^2(A, A) = 0$, then any deformation of φ is equivalent to the trivial deformation. Nevertheless, with the operadic approach we'll be able to show a theorem for algebraic rigidity.

What we have done for associative algebras can be done, in fact, for any algebra over any quadratic operad. Namely, in this paper, for a quadratic operad $\mathcal{P}$ and a vector space V we construct a graded Lie algebra $(L_\mathcal{P}(V) = \bigoplus_{n \geq 0} L^n_\mathcal{P}(V), [\,,\,])$, such that $\varphi \in L^1_\mathcal{P}(V)$, satisfying $[\varphi, \varphi] = 0$, determines a $\mathcal{P}$-algebra structure on V. An essential point is the fact that the graded Lie algebra structure on $L_\mathcal{P}(V)$ is strongly related with the Koszul dual of $\mathcal{P}$, in the sense of Ginzburg-Kapranov [G–K], denoted by $\mathcal{P}^!$. In the classical examples (associative, Lie, Com), this relation had not been observed, because the associated operads (**Ass, Lie, Com**) either were self-dual (**Ass**) or have a simple dual (**Lie**$^!$ = **Com**). As an example of an operad whose dual is not trivial, we propose the example of the operad associated to Leibniz algebras denoted by **Leib** in the sequel.

The existence of a graded Lie algebra structure on $L_\mathcal{P}(V)$ then enables us to show that, if $A = (V, \varphi)$ is a $\mathcal{P}$-algebra, then the cohomology theory $H^*_\mathcal{P}(A, A)$ defined in [G–K] is the cohomology theory controlling formal deformations of A as a $\mathcal{P}$-algebra, in the sense of [G2].

This paper is divided in four parts: in the first part we recall basic definitions and results on operads and we fix the notations; then, in the second part, we construct the graded lie algebra $L_\mathcal{P}(V)$; in the third part we apply the results of the second part to classical examples and to the new case of Leibniz algebras; finally in the fourth part we study deformations of a $\mathcal{P}$-algebra.

Acknowledgments: I would like to thank M. Markl for both help and discuss about the second part of this paper. I would also thank the Mathematical Institute of Prague for hospitality.

1. Notations, basic definitions and results

1.1. Notations. All objects are assumed to be defined over a fixed ground field k of characteristic zero. The tensor product over k will be noted $\otimes$, or when it will be necessary $\otimes_k$. For any natural $n \in \mathbb{N}^*$, Σ_n will denote the group of permutations of a finite set having n elements, and 1_n will be the notation for the identity permutation; for $\sigma \in \Sigma_n$, $\epsilon(\sigma) \in \{-1, 1\}$ will be the sign of σ, and $\mathbf{sgn}_n$ will denote the sign representation of Σ_n. We recall that a $\sigma(p, q)$-shuffle is a permutation $\sigma \in \Sigma_{p+q}$ such that: $\sigma(1) < \cdots < \sigma(p)$, and $\sigma(p+1) < \cdots < \sigma(p+q)$. The number of $\sigma(p, q)$-shuffle, for p and q fixed, is equal to C_{p+q}^p (binomial coefficient). In particular, a $\sigma(p, 0)$-shuffle is nothing but the identity permutation (it is the same for a $\sigma(0, q)$-shuffle).

1.2. Operads. Here we review background material, referring the reader to Ginzburg-Kapranov's paper [G–K], and to the introduction of this Proceedings.

We recall that a k-linear operad is a sequence $\mathcal{P} = \{\mathcal{P}(n)\}_{n \in \mathbb{N}^*}$ of right $k[\Sigma_n]$-modules equipped with compositions maps,

$$\gamma_{l; m_1, \ldots, m_l} : \mathcal{P}(l) \otimes \mathcal{P}(m_1) \otimes \cdots \otimes \mathcal{P}(m_l) \to \mathcal{P}(m_1 + \cdots + m_l),$$

satisfying axioms of equivariance, associativity and unit (for a unit element $1 \in \mathcal{P}(1)$). A morphism of operads $a : \mathcal{P} \to \mathcal{Q}$ is a sequence $a = \{a(n)\}_{n \in \mathbb{N}^*}$ of morphisms of $k[\Sigma_n]$-modules, preserving the composition maps. In the sequel, we'll note

$$\gamma_{l; m_1, \ldots, m_l}(\mu_l \otimes \nu_{m_1} \otimes \cdots \otimes \nu_{m_l}) = \mu_l(\nu_{m_1}, \ldots, \nu_{m_l}),$$

and we'll consider operads $\mathcal{P}$ such that $\mathcal{P}(1) = k$, the unit corresponding to the unit of k.

Let $\mathcal{P}$ be a k-linear operad, a $\mathcal{P}$-algebra or an algebra over $\mathcal{P}$ is a k-vector space V such that there exists a morphism of operads $a : \mathcal{P} \to End(V)$, where $End(V)$ is the endomorphism operad: $End(V)(n) = Hom_k(V^{\otimes n}, V)$; the right action of Σ_n is given, for $\sigma \in \Sigma_n$, $f \in End(V)(n)$ and $v_1, \ldots, v_n \in V$, by:

$$f\sigma(v_1 \otimes \cdots \otimes v_n) = f(v_{\sigma^{-1}(1)} \otimes \cdots \otimes v_{\sigma^{-1}(n)});$$

the composition maps are given by,

$$\gamma_{l; m_1, \ldots, m_l}(f_l \otimes g_{m_1} \otimes \cdots \otimes g_{m_l}) = f_l \circ (g_{m_1} \otimes \cdots \otimes g_{m_l}).$$

In the sequel, $a(\mu)(v_1 \otimes \cdots \otimes v_n)$ will be denoted by $\mu(v_1, \ldots, v_n)$, and we always suppose that any operad $\mathcal{P}$ is such that, for any $n \in \mathbb{N}^*$, $\mathcal{P}(n)$ is **finite dimensional.**

Suppose now that V is a graded k-vector space, then we can form $End^0(V)$, the set of homogeneous degree zero k-linear maps from $V^{\otimes n}$ to V. For $\sigma \in \Sigma_n$ and

$v_1, \ldots, v_n \in V$, let $\epsilon(\sigma, v_1, \ldots, v_n) \in \{-1, 1\}$ be the Koszul's sign, determined by the equation,

$$v_1 \wedge \cdots \wedge v_n = \epsilon(\sigma, v_1, \ldots, v_n)\, v_{\sigma(1)} \wedge \cdots \wedge v_{\sigma(n)}$$

in the free graded commutative algebra $\Lambda(v_1, \ldots, v_n)$. Taking the same composition maps as for $End(V)$ and putting the right action of Σ_n given, for $\sigma \in \Sigma_n$, $f \in End^0(V)(n)$ and $v_1, \ldots, v_n \in V$, by

$$(f \cdot \sigma)(v_1 \otimes \cdots \otimes v_n) = \epsilon(\sigma, v_1, \ldots, v_n) f(v_{\sigma^{-1}(1)} \otimes \cdots \otimes v_{\sigma^{-1}(n)}),$$

we get an operad structure on $End^0(V)$, and V is a graded $\mathcal{P}$-algebra if there exists a morphism of operads, $\bar{a} : \mathcal{P} \to End^0(V)$. Again, we'll note $\bar{a}(\mu)(v_1 \otimes \cdots \otimes v_n) = \mu(v_1, \ldots, v_n)$.

1.3. Usual operads. In all the paper, **Ass**, **Com**, **Lie** and **Leib** will denote respectively the operads associated to associative, commutative, Lie and Leibniz algebras. For the reader, we recall that a Leibniz algebra is a k-vector space V equipped with a k-bilinear product (called Leibniz bracket), $[,] : V \times V \to V$, satisfying the following Leibniz identity: for $x, y, z \in V$,

$$[x, [y, z]] = [[x, y], z] - [[x, z], y].$$

If the bracket is skew-symmetric, then V is a Lie algebra (for further details see J.-L. Loday [LO1]).

1.4. Elementary operations.

DEFINITION 1.4.1. Let $\mathcal{P}$ be a k-linear operad, for any natural $n, m \in \mathbb{N}^*$ and for any $1 \leq i \leq m$, define $\circ_i$ to be the map

$$\circ_i : \mathcal{P}(m) \otimes \mathcal{P}(n) \to \mathcal{P}(m + n - 1),$$

such that, for $\mu, \nu \in \mathcal{P}(m) \times \mathcal{P}(n)$, $\mu \circ_i \nu = \mu(1, \ldots, \nu, \ldots, 1)$ (ν is placed at the i-th place).

We call $\circ_i$ the i-th elementary operation. The $\circ_i$'s satisfy the following associativity conditions [MAR2]: for $l, m, n \in \mathbb{N}^*$, $\lambda \in \mathcal{P}(l)$, $\mu \in \mathcal{P}(m)$ and $\nu \in \mathcal{P}(n)$,

$$(1) \qquad (\lambda \circ_i \mu) \circ_j \nu = \begin{cases} (\lambda \circ_j \nu) \circ_{i+n-1} \mu & 1 \leq j \leq i-1 \\ \lambda \circ_i (\mu \circ_{j-i+1} \nu) & i \leq j \leq m+i-1 \\ (\lambda \circ_{j-m+1} \nu) \circ_i \mu & i+m \leq j \end{cases}$$

Furthermore, we have the equivariant conditions: for $\mu, \nu \in \mathcal{P}(m) \times \mathcal{P}(n)$, $\sigma, \tau \in \Sigma_m \times \Sigma_n$ and $i \in \{1, \ldots, m\}$,

$$(1') \qquad\qquad (\mu \circ_i \nu) \cdot (\sigma \circ_i \tau) = \mu\sigma \circ_{\sigma^{-1}(i)} \nu\tau;$$

here $\sigma \circ_i \tau$ stands for the permutation of Σ_{m+n-1} defined by

$$\sigma \circ_i \tau = \hat{\sigma} \circ (1 \times \cdots \times \tau \times \cdots \times 1)$$

(τ is at the i-th place), where $\hat{\sigma}$ permutes the blocks in which $1 \times \cdots \times \tau \times \cdots \times 1$ acts in the same way as σ permutes $\{1, \ldots, n\}$.

REMARK 1.4.2. Thanks to the elementary operations, one can rebuild the composition maps by the following formula: for $l \in \mathbb{N}^*$, $m_1, \ldots, m_l \in \mathbb{N}^*$, $\mu_l \in \mathcal{P}(l)$ and $\nu_{m_1}, \ldots, \nu_{m_l} \in \mathcal{P}(m_1) \times \cdots \times \mathcal{P}(m_l)$,

$$\mu_l(\nu_{m_1}, \ldots, \nu_{m_l}) = (\cdots (\mu_l \circ_l \nu_{m_l}) \circ_{l-1} \nu_{m_{l-1}}) \circ_{l-2} \cdots) \circ_1 \nu_{m_1}.$$

1.5. Quadratic operads. We recall without proof the following proposition:

PROPOSITION 1.5.1. *Let E be a $k[\Sigma_2]$-module, then there exists an operad $\mathcal{F}(E)$ (the free operad generated by E) such that the following property holds: for any operad $\mathcal{Q}$ and for any morphism of $k[\Sigma_2]$-modules $a : E \to \mathcal{Q}(2)$, there is a unique morphism of operads, $\hat{a} : \mathcal{F}(E) \to \mathcal{Q}$, such that $\hat{a}(2) = a$.*

We recall the following definition [G–K]:

DEFINITION 1.5.2. *Let E be a $k[\Sigma_2]$-module and R a $k[\Sigma_3]$-submodule of $\mathcal{F}(E)(3)$. Denote by (R) the ideal generated by R, we call $\mathcal{F}(E)/(R)$ the quadratic operad generated by E with relations R.*

1.5.3. Notation. In the sequel, we'll note $\mathcal{F}(E)/(R) = \mathcal{P}(k, E, R)$.

1.5.4. Examples. The operads **Ass**, **Com**, **Lie** and **Leib** are quadratic.

We have the useful following proposition:

PROPOSITION 1.5.5. *Let $\mathcal{P} = \mathcal{P}(k, E, R)$ be a quadratic operad and V a k-vector space. A $\mathcal{P}$-algebra structure on V is entirely determined by a morphism of $k[\Sigma_2]$-modules $a : E \to End(V)(2)$, such that for any $r \in R$, $\hat{a}(3)(r) = 0$, where $\hat{a}$ is the unique morphism of operads associated to a.*

PROOF. If V is a $\mathcal{P}$-algebra, let $f : \mathcal{P} \to End(V)$ be the underlying morphism of operads. Set $a = f(2)$ and let $\pi : \mathcal{F}(E) \to \mathcal{F}(E)/(R)$ be the canonical projection. Clearly $\hat{a} = f \circ \pi$. In particular, for any $r \in R$, $\hat{a}(3)(r) = 0$, because $\pi(r) = 0$. Conversely, let a be a map satisfying the conditions of the proposition, then $\hat{a} : \mathcal{F}(E) \to End(V)$ can be factored through R, so that to define a map $f : \mathcal{P} \to End(V)$ which determines a $\mathcal{P}$-algebra structure on V. Let's prove this fact: let $r \in (R)$, then r is a linear combination, with coefficients in $k[\Sigma_n]$, of elements of type $r_i \circ_i x_i$ and $y_j \circ_j t_j$, where r_i and t_j are in R and x_i, y_j in $\mathcal{F}(E)(n-2)$. Since $\hat{a}(n)(r_i \circ_i x_i) = \hat{a}(3)(r_i) \circ_i \hat{a}(n-2)(x_i) = 0$ and $\hat{a}(n)(y_j \circ_j t_j) = \hat{a}(n-2)(y_j) \circ_j \hat{a}(3)(t_j) = 0$, the factorization can be done. $\square$

1.6. Free $\mathcal{P}$-algebra.

1.6.1. The none graded case. Let $\mathcal{P}$ be an operad and V a k-vector space. We recall that the free $\mathcal{P}$-algebra generated by V is the graded vector space, $\mathcal{F}_{\mathcal{P}}(V) = \bigoplus_{n \geq 1} \mathcal{P}(n) \otimes_{\Sigma_n} V^{\otimes n}$, where $V^{\otimes n}$ is equipped with the natural following left action of Σ_n:

$$\sigma(v_1 \otimes \cdots \otimes v_n) = v_{\sigma^{-1}(1)} \otimes \cdots \otimes v_{\sigma^{-1}(n)}.$$

The $\mathcal{P}$-algebra structure is given for $\mu \in \mathcal{P}(n)$, $\mu_1, \ldots, \mu_n \in \mathcal{P}(m_1) \times \cdots \times \mathcal{P}(m_n)$ and $x_1, \ldots, x_n \in V^{\otimes m_1} \times \cdots \times V^{\otimes m_n}$, by:

$$\mu(\mu_1 \otimes x_1, \ldots, \mu_n \otimes x_n) = \mu(\mu_1, \ldots, \mu_n) \otimes x_1 \otimes \cdots \otimes x_n.$$

1.6.2. The graded case. Set $\hat{V}^{\otimes n} = V^{\otimes n} \otimes \mathbf{sgn}_n$, so that $\hat{V}^{\otimes n}$ is $V^{\otimes n}$ as a vector space, but with the left action of Σ_n given by,

$$\sigma(v_1 \otimes \cdots \otimes v_n) = \epsilon(\sigma) v_{\sigma^{-1}(1)} \otimes \cdots \otimes v_{\sigma^{-1}(n)}.$$

Then $\mathcal{F}_{\mathcal{P}}^{gr}(V) = \bigoplus_{n \geq 1} \mathcal{P}(n) \otimes_{\Sigma_n} \hat{V}^{\otimes n}$ is a graded $\mathcal{P}$-algebra, called the free graded $\mathcal{P}$-algebra generated by V. In the sequel, we'll note $\mathcal{F}_{\mathcal{P}}^{gr\,n}(V) = \mathcal{P}(n) \otimes_{\Sigma_n} \hat{V}^{\otimes n}$.

1.6.3. Examples. If $\mathcal{P} = \mathbf{Ass}$, then $\mathcal{F}_{\mathcal{P}}^{gr}(V) = \bigoplus_{n \geq 1} V^{\otimes n}$ is the free associative algebra (without unit) generated by V. If $\mathcal{P} = \mathbf{Com}$, then $\mathcal{F}_{\mathcal{P}}^{gr}(V) = \bigoplus_{n \geq 1} \Lambda^n(V)$ is the free graded commutative algebra (without unit) generated by V.

1.7. Koszul dual of a quadratic operad.

1.7.1. Linear dual of a $k[\Sigma_n]$-module. Following [G–K], for a right $k[\Sigma_n]$-module F, $F^\vee$ will denote $Hom_k(F, k)$ with the right action of Σ_n given, for $\sigma \in \Sigma_n$, $\phi \in F^\vee$, $x \in F$, by

$$(\phi \cdot \sigma)(x) = \epsilon(\sigma)\,\phi(x \cdot \sigma^{-1});$$

in other words as a $k[\Sigma_n]$-module, we have:

$$F^\vee = Hom_k(F, k) \otimes \mathbf{sgn}_n.$$

Let E be a $k[\Sigma_2]$-module, then $\mathcal{F}(E^\vee)(3) \simeq (\mathcal{F}(E))^\vee(3)$ via the evaluation map $<,>: \mathcal{F}(E^\vee)(3) \otimes \mathcal{F}(E)(3) \to k$ characterized [MAR1], for $\varphi, \psi \in E^\vee$, $\mu, \nu \in E$, by

(2)
$$\begin{cases} < \varphi(1, \psi), \mu(1, \nu) > = - < \varphi, \mu > < \psi, \nu > \\ < \varphi(\psi, 1), \mu(\nu, 1) > = + < \varphi, \mu > < \psi, \nu > \\ < \varphi(\psi, 1), \mu(1, \nu) > = 0 \\ < \varphi(1, \psi), \mu(\nu, 1) > = 0 \end{cases}$$

Furthermore, for $\varphi, \mu \in \mathcal{F}(E^\vee)(3) \times \mathcal{F}(E)(3)$, $\sigma \in \Sigma_3$ we have:

(3)
$$< \varphi \cdot \sigma, \mu \cdot \sigma > = \epsilon(\sigma) < \varphi, \mu > .$$

Let $R \subset \mathcal{F}(E)(3)$ be a $k[\Sigma_3]$-submodule, and $R^\perp \subset \mathcal{F}(E^\vee)(3)$ be the annihilator of R in $(\mathcal{F}(E))^\vee(3) \simeq \mathcal{F}(E^\vee)(3)$.

DEFINITION 1.7.2. The Koszul dual of the quadratic operad $\mathcal{P} = \mathcal{P}(k, E, R)$ is the quadratic operad $\mathcal{P}^! = \mathcal{P}(k, E^\vee, R^\perp)$.

REMARK 1.7.3. We have $\mathcal{P}^!(2) = E^\vee$ and $(\mathcal{P}^!)^\vee(3) \simeq R$.

1.7.4. Example. $\mathbf{Ass}^! = \mathbf{Ass}$; $\mathbf{Com}^! = \mathbf{Lie}$; $\mathbf{Lie}^! = \mathbf{Com}$ [G–K]; the Koszul dual of $\mathbf{Leib}$ has been calculated by J.-L. Loday [LO2]. A $\mathbf{Leib}^!$ algebra, or a dual Leibniz algebra, is a vector space V equipped with a product $\times$ such that, for $v_1, v_2, v_3 \in V$:

$$(v_1 \times v_2) \times v_3 = v_1 \times (v_2 \times v_3) + v_1 \times (v_3 \times v_2).$$

2. The graded Lie algebra $L_{\mathcal{P}}(V)$

2.1. Notations. In this section, we denote by V a finite dimensional k-vector space and by $\mathcal{P} = \mathcal{P}(k, E, R)$ a quadratic operad, where E is a finite dimensional k-vector space such that there is an involution map $I : E \to E$, making E a $k[\Sigma_2]$-module. The linear dual of V (i.e. $Hom_k(V, k)$) will be denoted by $\#V$. For $n \in \mathbb{N}^*$, $V^{\otimes n}$ will be equipped with the left action of Σ_n as in 1.6.1, while $(\#\hat{V})^{\otimes n}$ will stand for $(V^{\otimes n})^{\vee}$ as a $k[\Sigma_n]$-module. In the notations of 1.6.2, we have $(\#\hat{V})^{\otimes n} = \hat{W}^{\otimes n}$, where $W = \#V$.

2.2. The set $L_{\mathcal{P}}^n(V)$. In all examples where we can build a graded Lie algebra $L(V)$ describing a given structure on V, the n-degree part $L^n(V)$ corresponds to the space of cochains of degree $n + 1$ of the complex calculating the cohomology of V equipped with this structure, with coefficients in V. But if A is a $\mathcal{P}$-algebra, one knows that the homology of A is given by the complex, $(Hom_k(C_*^{\mathcal{P}}(A), k), \delta)$, where $C_n^{\mathcal{P}}(A) = (\mathcal{P}^!)^{\vee}(n) \otimes_{\Sigma_n} A^{\otimes n}$ ([G–K],[LO3]). So it is natural to choose the set $Hom_k(C_{n+1}^{\mathcal{P}}(V), V)$ for $L_{\mathcal{P}}^n(V)$.

There is another reason for making this choice: according to proposition 1.5.5, a $\mathcal{P}$-algebra structure on V is given by a $k[\Sigma_2]$-linear map, $a : \mathcal{P}(2) \to Hom_k(V^{\otimes 2}, V)$, such that $\hat{a}(3)(r) = 0$ for any $r \in R$. We can view a as a map, $\bar{a} : \mathcal{P}(2) \otimes_{\Sigma_2} V^{\otimes 2} \to V$, via $\bar{a}(\mu \otimes v_1 \otimes v_2) = a(\mu)(v_1 \otimes v_2)$. Thus a $\mathcal{P}$-algebra structure on V is defined by an element $a \in L_{\mathcal{P}}^1(V)$ satisfying a certain condition we want to express as the nullity of $[a, a]$, where $[,]$ is a certain Lie bracket to construct. But if such a bracket exists, then $[a, a] \in L_{\mathcal{P}}^2(V) \simeq Hom_k(R \otimes_{\Sigma_3} V^{\otimes 3}, V)$ (see remark 1.7.2); so it is natural to expect a relation between $[a, a]$ and $\hat{a}(3)|_R$.

For all these reasons, we take

$$L_{\mathcal{P}}^n(V) = Hom_k((\mathcal{P}^!)^{\vee}(n+1) \otimes_{\Sigma_{n+1}} V^{\otimes n+1}, V),$$

and we denote by $L_{\mathcal{P}}(V)$ the direct sum of $L_{\mathcal{P}}^n(V)$. If we dualize, then

$$L_{\mathcal{P}}^n(V) \simeq Hom_k(\#V, \mathcal{P}^!(n+1) \otimes_{\Sigma_{n+1}} (\#\hat{V})^{\otimes n+1}),$$

i.e.

$$L_{\mathcal{P}}^n(V) \simeq Hom_k(\#V, \mathcal{F}_{\mathcal{P}!}^{gr_{n+1}}(\#V))$$

(see 1.6.2, for the notation).

The graded Lie structure we put on $L_{\mathcal{P}}(V)$ is strongly related with the notion of graded derivation, that now we define in the operadic framework.

2.3. Graded derivations.

DEFINITION 2.3.1. Let $\mathcal{Q}$ be an operad, and $A = \bigoplus_{n \in \mathbb{N}} A^n$ a graded $\mathcal{Q}$-algebra. A degree n derivation of A is an homogeneous degree n linear map, $d : A \to A$, such that, for $m \in \mathbb{N}^*$, $\mu \in \mathcal{Q}(n)$, $a_1, \ldots, a_m \in A$, homogeneous,

$$(4) \quad d(\mu(a_1, \ldots, a_m)) = \sum_{i=1}^{m} (-1)^{n(|a_1| + \cdots + |a_{i-1}|)} \mu(a_1, \ldots, a_{i-1}, d(a_i), a_{i+1}, \ldots, a_m),$$

where $|a|$ stands for the degree of a.

We denote by $Der^n(A)$ the set of all degree n derivations of A.

LEMMA 2.3.2. *Let $\mathcal{Q}$ be a quadratic operad, and A an algebra over $\mathcal{Q}$. Then a homogeneous degree n linear map, $d : A \to A$, is a derivation if, and only if, for $\mu \in \mathcal{Q}(2)$, $a_1, a_2 \in A$, homogeneous, $d(\mu(a_1, a_2)) = \mu(d(a_1), a_2) + (-1)^{n|a_1|}\mu(a_1, d(a_2))$.*

PROOF. The if part of the lemma is clear. Conversely, suppose that d satisfies the condition of the lemma, let's prove by induction (on the integer m) that (4) holds for any $m \geq 1$. For $m = 1$ it is clear, and for $m = 2$ it is the condition of the lemma. Let's prove the induction's step $m \Longrightarrow m + 1$. Let μ be an element of $\mathcal{Q}(m + 1)$; since $\mathcal{Q}$ is quadratic, μ is a linear combination, with coefficients in $k[\Sigma_{m+1}]$, of elements of type $\mu_i \circ_i \nu_i$, where $i \in \{1, \ldots, p\}$, $\mu_i \in \mathcal{Q}(p)$ and $\nu_i \in \mathcal{Q}(q)$ with $p + q = m + 2$. Since obviously formula (4) is equivariant under the action of the symmetric group (it is because we have the Koszul's sign convention), it suffices to show that (4) holds for elements of that type. We have:

$$(5) \quad d(\mu_i \circ_i \nu_i)(a_1 \otimes \cdots \otimes a_{m+1}) = d(\mu_i(a_1, \ldots, \nu_i(a_i, \ldots, a_{i+q-1}), \ldots, a_{m+1})).$$

Applying the hypothesis that (4) is true for any $p \leq m$, the right part of (5) becomes:

$$(6) \quad \begin{cases} \displaystyle\sum_{k<i} (-1)^{n(|a_1|+\cdots+|a_{k-1}|)}\mu_i(a_1, \ldots d(a_k), \ldots, \nu_i(a_i, \ldots, a_{i+q-1}), \ldots, a_{m+1}) + \\[2mm] (-1)^{n(|a_1|+\cdots+|a_{i-1}|)}\mu_i(a_1, \ldots, d(\nu_i(a_i, \ldots, a_{i+q-1})), \ldots, a_{m+1}) + \\[2mm] \displaystyle\sum_{k \geq i+q} (-1)^{n(|a_1|+\cdots+|a_{k-1}|)}\mu_i(a_1, \ldots, \nu_i(a_i, \ldots, a_{i+q-1}), \ldots, d(a_k), \ldots, a_{m+1}) \end{cases}$$

In (6), we replace $d(\nu_i(a_i, \ldots, a_{i+q-1}))$ by

$$\sum_{i \leq j \leq i+q-1} (-1)^{|a_i|+\cdots+|a_{j-1}|}\nu_i(a_i, \ldots, d(a_j), \cdots, a_{i+q-1}),$$

we bring the terms together; the induction then follows.   □

Now let's take $\mathcal{Q} = \mathcal{P}^!$ and $A = \mathcal{F}^{gr}_{\mathcal{P}^!}(\#V)$. We have the following proposition due to M. Markl [MAR2]:

PROPOSITION 2.3.3. *With the above notations, we have the isomorphism of vector spaces: $Der^n(\mathcal{F}^{gr}_{\mathcal{P}^!}(\#V)) \simeq L^n_{\mathcal{P}}(V)$.*

PROOF. For the purpose of this article, we need an explicit and constructive proof of this proposition, in order to apply it on particular cases. So, we give our own proof and for that we construct two maps, Ω and ω,

$$\Omega : L^n_{\mathcal{P}}(V) \to Der^n(\mathcal{F}^{gr}_{\mathcal{P}^!}(\#V))$$
$$\omega : Der^n(\mathcal{F}^{gr}_{\mathcal{P}^!}(\#V)) \to L^n_{\mathcal{P}}(V).$$

such that $\Omega \circ \omega$ and $\omega \circ \Omega$ are the respective identity mappings.

The map ω is simply the restriction: let d be a degree $n + 1$ derivation of $\mathcal{F}^{gr}_{\mathcal{P}^!}(\#V)$, then d maps $\mathcal{F}^{gr_1}_{\mathcal{P}^!}(\#V) = \#V$ onto $\mathcal{F}^{gr_{n+1}}_{\mathcal{P}^!}(\#V)$; we set $\omega(d) = {}^t d|_{\#V}$ where ^{t} stands for the transpose.

For $f \in L^n_{\mathcal{P}}(V)$, $\Omega(f)$ will be the unique degree n derivation of $\mathcal{F}^{gr}_{\mathcal{P}!}(\#V)$ such that $\Omega(f)|_{\#V} = {}^t f$. Namely, for any m we define

$$\bar{\Omega}(f) : \mathcal{P}^!(m) \otimes_k (V^{\otimes m})^\vee \to \mathcal{F}^{gr\, m+n}_{\mathcal{P}!}(\#V),$$

in the following way:
- if $m = 1$, then $\bar{\Omega}(f) = {}^t f$
- if $m > 1$, we define $\bar{\Omega}(f)$ on homogeneous elements, $\mu \otimes v_1^* \otimes \cdots \otimes v_m^*$, by

$$\bar{\Omega}(f)(\mu \otimes v_1^* \otimes \cdots \otimes v_m^*) = \sum_{k=1}^m (-1)^{n(k-1)} \mu(v_1^* \otimes \cdots \otimes v_{k-1}^* \otimes {}^t f(v_k^*) \otimes v_{k+1}^* \otimes \cdots \otimes v_m^*),$$

where if

$${}^t f(v_k^*) = \sum_l \nu_l^k \otimes w_{l,1}^k \otimes \cdots \otimes w_{l,n+1}^k,$$

then

$$\mu(v_1^* \otimes \cdots \otimes v_{k-1}^* \otimes {}^t f(v_k^*) \otimes v_{k+1}^* \otimes \cdots \otimes v_m^*) =$$

$$\sum_l (\mu \circ_k \nu_l^k) \otimes_{\Sigma_{m+n}} (v_1^* \otimes \cdots \otimes v_{k-1}^* \otimes w_{l,1}^k \otimes \cdots \otimes w_{l,n+1}^k \otimes v_{k+1}^* \otimes \cdots \otimes v_m^*).$$

LEMMA. *Let σ be a permutation of Σ_m, then $\bar{\Omega}(f)(\mu \otimes \sigma(v_1^* \otimes \cdots \otimes v_m^*)) = \bar{\Omega}(f)(\mu\sigma \otimes v_1^* \otimes \cdots \otimes v_m^*)$. Thus, for any m, $\bar{\Omega}(f)$ defines a map from $\mathcal{F}^{gr\, m}_{\mathcal{P}!}(\#V)$ to $\mathcal{F}^{gr\, m+n}_{\mathcal{P}!}(\#V)$ we denote by $\Omega(f)$.*

PROOF OF THE LEMMA. Set $\tau = \sigma^{-1}$; we have:

$$\bar{\Omega}(f)(\mu \otimes \sigma(v_1^* \otimes \cdots \otimes v_m^*)) =$$

$$\epsilon(\sigma) \sum_{k=1}^m (-1)^{n(k-1)} \mu(v_{\tau(1)}^* \otimes \cdots \otimes v_{\tau(k-1)}^* \otimes {}^t f(v_{\tau(k)}^*) \otimes v_{\tau(k+1)}^* \otimes \cdots \otimes v_{\tau(m)}^*).$$

By definition,

$$\mu(v_{\tau(1)}^* \otimes \cdots \otimes v_{\tau(k-1)}^* \otimes {}^t f(v_{\tau(k)}^*) \otimes v_{\tau(k+1)}^* \otimes \cdots \otimes v_{\tau(m)}^*)$$

equals

$$\sum_l \mu_{k,l}^\tau \otimes_\Sigma (v_{\tau(1)}^* \otimes \cdots \otimes v_{\tau(k-1)}^* \otimes w_{l,1}^{\tau(k)} \otimes \cdots \otimes w_{l,n+1}^{\tau(k)} \otimes v_{\tau(k+1)}^* \otimes \cdots \otimes v_{\tau(m)}^*),$$

where $\mu_{k,l}^\tau$ stands for $(\mu \circ_k \nu_l^{\tau(k)})$, Σ for Σ_{m+n}.

Let's denote by σ_k the permutation $\sigma \circ_k 1_{n+1} \in \Sigma_{m+n}$ then:

$$\mu(v_{\tau(1)}^* \otimes \cdots \otimes v_{\tau(k-1)}^* \otimes {}^t f(v_{\tau(k)}^*) \otimes v_{\tau(k+1)}^* \otimes \cdots \otimes v_{\tau(m)}^*) =$$

$$\sum_l \mu_{k,l}^\tau \sigma_k \otimes_\Sigma \sigma_k^{-1}(v_{\tau(1)}^* \otimes \cdots \otimes v_{\tau(k-1)}^* \otimes w_{l,1}^{\tau(k)} \otimes \cdots \otimes w_{l,n+1}^{\tau(k)} \otimes v_{\tau(k+1)}^* \otimes \cdots \otimes v_{\tau(m)}^*)$$

$$= \epsilon(\sigma_k) \sum_l \nu_{k,l}^\tau \otimes_\Sigma (v_1^* \otimes \cdots \otimes v_{\tau(k)-1}^* \otimes w_{l,1}^{\tau(k)} \otimes \cdots \otimes w_{l,n+1}^{\tau(k)} \otimes v_{\tau(k)+1}^* \otimes \cdots \otimes v_m^*),$$

(where now $\nu_{k,l}^\tau$ stands for $(\mu\sigma \circ_{\tau(k)} \nu_l^{\tau(k)})$) since we have, according (1'),

$$(\mu \circ_k \nu_l^{\tau(k)})\sigma_k = \mu\sigma \circ_{\tau(k)} \nu_l^{\tau(k)}.$$

So, we have:

$$\mu(v_{\tau(1)}^* \otimes \cdots \otimes v_{\tau(k-1)}^* \otimes {}^t f(v_{\tau(k)}^*) \otimes v_{\tau(k+1)}^* \otimes \cdots \otimes v_{\tau(m)}^*) =$$

$$\epsilon(\sigma_k)\mu\sigma(v_1^* \otimes \cdots \otimes v_{\tau(k)-1}^* \otimes {}^t f(v_{\tau(k)}^*) \otimes v_{\tau(k)+1}^* \otimes \cdots \otimes v_m^*).$$

It remains to calculate $\epsilon(\sigma_k)$. For this, let's denote by $\{1, \ldots, k-1, a_1, \ldots, a_{n+1},$ $k+1, \ldots, m\}$ the elements on which σ_k acts. Set $s = |\tau(k) - k|$. This number represents the shifting undergone by $\{a_1, \ldots, a_{n+1}\}$ with regard to its initial position:

$$\sigma_k \left|\begin{array}{l} 1, \ldots, k-1, a_1, \ldots, a_{n-s}, \ldots, a_{n+1}, k+1, \ldots, m \\ \ldots, a_1, \ldots, a_s, \ldots, a_{n+1}, \ldots\ldots\ldots\ldots\ldots\ldots\ldots \end{array}\right.$$

If we want to bring the $\{a_1, \ldots, a_{n+1}\}$'s in coincidence, we must do $(n+1)s$ transpositions; finally we get the permutation σ' deduced from σ after having transposed k, s times (because $\sigma_k = \sigma \circ_k 1_{n+1}$). If we compare the sign, then we have: $(-1)^{(n+1)s}\epsilon(\sigma_k) = (-1)^s\epsilon(\sigma)$, i.e. $\epsilon(\sigma_k) = (-1)^{n(k-\tau(k))}\epsilon(\sigma)$.

Therefore, we have the result since $\epsilon(\sigma)\epsilon(\sigma_k)(-1)^{n(k-1)} = (-1)^{n(\tau(k)-1)}$ (it suffices to set $k' = \tau(k)$). $\square$

REMARK. For $\sigma \in \Sigma_n$ and $1 \leq i \leq n$, we have proved that

$$\epsilon(\sigma \circ_i 1_m) = (-1)^{(m-1)(i-\sigma^{-1}(i))}\epsilon(\sigma).$$

More generally, for $\sigma \in \Sigma_n, \tau \in \Sigma_m$ and $1 \leq i \leq n$, since

$$\sigma \circ_i \tau = \hat{\sigma} \circ (1 \times \cdots \times \tau \times \cdots \times 1),$$

since $\hat{\sigma} = \sigma \circ_i 1_m$ and since obviously $\epsilon(1 \times \cdots \times \tau \times \cdots \times 1) = \epsilon(\tau)$, we have the following formula:

(7) $$\epsilon(\sigma \circ_i \tau) = (-1)^{(m-1)(i-\sigma^{-1}(i))}\epsilon(\sigma)\epsilon(\tau).$$

Now, we must verify that $\Omega(f)$ is really a derivation. But according to lemma 2.3.2, it suffices to verify that for $p, q \in \mathbb{N}^*$, $\mu, \mu_1, \mu_2 \in \mathcal{P}^!(2) \times \mathcal{P}^!(p) \times \mathcal{P}^!(q)$ and $v_1^*, \ldots, v_{p+q}^* \in \#V$,

$$\Omega(f)(\mu(\mu_1 \otimes v_1^* \otimes \cdots \otimes v_p^*, \mu_2 \otimes v_{p+1}^* \otimes \cdots \otimes v_{p+q}^*)) =$$

$$\mu(\Omega(f)(\mu_1 \otimes v_1^* \otimes \cdots \otimes v_p^*), \mu_2 \otimes v_{p+1}^* \otimes \cdots \otimes v_{p+q}^* +$$

$$(-1)^{np}\mu(\mu_1 \otimes v_1^* \otimes \cdots \otimes v_p^*, \Omega(f)(\mu_2 \otimes v_{p+1}^* \otimes \cdots \otimes v_{p+q}^*)).$$

But,
$$\Omega(f)(\mu(\mu_1 \otimes v_1^* \otimes \cdots \otimes v_p^*, \mu_2 \otimes v_{p+1}^* \otimes \cdots \otimes v_{p+q}^*)) = \Omega(f)(\mu(\mu_1, \mu_2) \otimes v_1^* \otimes \cdots \otimes v_{p+q}^*) =$$

$$(*) \begin{cases} \displaystyle\sum_{k=1}^{p} (-1)^{n(k-1)} \mu(\mu_1, \mu_2)(v_1^* \otimes \cdots \otimes {}^t f(v_k^*) \otimes \cdots \otimes v_p^* \otimes v_{p+1}^* \otimes \cdots \otimes v_{p+q}^*) \\ + (-1)^{np} \displaystyle\sum_{k=p+1}^{p+q} (-1)^{n(k-p-1)} \mu(\mu_1, \mu_2)(v_1^* \otimes \cdots \otimes {}^t f(v_k^*) \otimes \cdots \otimes v_{p+q}^*) \end{cases}$$

We have:

$$\mu(\mu_1,\mu_2)(v_1^* \otimes \cdots \otimes {}^t f(v_k^*) \otimes \cdots \otimes v_p^* \otimes v_{p+1}^* \otimes \cdots \otimes v_{p+q}^*) =$$

$$\mu(\mu_1(v_1^* \otimes \cdots \otimes v_{k-1}^* \otimes {}^t f(v_k^*) \otimes v_{k+1}^* \otimes \cdots \otimes v_p^*), \mu_2 \otimes v_{p+1}^* \otimes \cdots \otimes v_{p+q}^*).$$

Actually, if we replace ${}^t f(v_k^*)$ by the above expression, we see immediately that this equality follows from the associativity condition:

$$\mu(\mu_1,\mu_2) \circ_k \nu_l = \mu(\mu_1 \circ_k \nu_l, \mu_2).$$

Samely, we have

$$\mu(\mu_1,\mu_2)(v_1^* \otimes \cdots \otimes v_p^* \otimes v_{p+1}^* \otimes \cdots \otimes {}^t f(v_k^*) \otimes \cdots \otimes v_{p+q}^*) =$$

$$\mu(\mu_1 \otimes v_1^* \otimes \cdots \otimes v_p^*, \mu_2(v_{p+1}^* \otimes \cdots \otimes {}^t f(v_k^*) \otimes \cdots \otimes v_{p+q}^*)).$$

Using these equalities in $(*)$, it follows that $\Omega(f)$ is a derivation. We leave the reader to verify that $\omega = \Omega^{-1}$. $\qquad\square$

2.4. The main results. Thanks to proposition 2.3.3,

$$L_{\mathcal{P}}(V) \simeq Der(\mathcal{F}_{\mathcal{P}!}^{gr}(\#V))$$

has a natural structure of graded Lie algebra with bracket given, for $f, g \in L_{\mathcal{P}}^n(V) \times L_{\mathcal{P}}^m(V)$, by

$$[f,g] = i_f(g) + (-1)^{nm+1} i_g(f),$$

where $i_f(g) = {}^t(\Omega(g) \circ \Omega(f)|_{\#V})$. We have the following theorem implicitly contained in [G–K]:

THEOREM 2.4.1. Let $\mathcal{P} = \mathcal{P}(k, E, R)$ be a quadratic operad where both E and V are finite dimensional. Then, there is an one to one correspondence between $\mathcal{P}$-algebra structures on V and one degree derivations $d \in Der^1(\mathcal{F}_{\mathcal{P}!}^{gr}(\#V))$, satisfying $d^2 = 0$.

COROLLARY 2.4.2. Let $\mathcal{P} = \mathcal{P}(k, E, R)$ be a quadratic operad where both E and V are finite dimensional. Then, there is an one to one correspondence between $\mathcal{P}$-algebra structures on V and elements $\varphi \in L_{\mathcal{P}}^1(V)$, satisfying $[\varphi, \varphi] = 0$.

PROOF OF THE COROLLARY. The corollary follows from theorem 2.4.1 and proposition 2.3.3, since to any derivation d one can associate an element $\varphi \in L_{\mathcal{P}}^1(V)$ such that $[\varphi, \varphi] = 2 d^2$. $\qquad\square$

PROOF OF THE THEOREM. According to proposition 1.5.5, there is an one to one correspondence between $\mathcal{P}$-algebra structures on V and maps of $k[\Sigma_2]$-modules

$$a : E \to Hom_k(V^{\otimes 2}, V),$$

such that for any $r \in R$, $\hat{a}(3)(r) = 0$ (we recall that $\hat{a}$ stands for the unique map of operads from $\mathcal{F}(E)$ to $End(V)$ such that $\hat{a}(2) = a$). Let's denote by θ_1, θ_2 and θ_3 the canonical isomorphisms of vector spaces:

$$Hom_{\Sigma_2}(E, Hom_k(V^{\otimes 2}, V)) \overset{\theta_1}{\simeq} Hom_k(E \otimes_{\Sigma_2} V^{\otimes 2}, V)$$

$$Hom_{\Sigma_3}(R, Hom_k(V^{\otimes 3}, V)) \overset{\theta_2}{\simeq} Hom_k(R \otimes_{\Sigma_3} V^{\otimes 3}, V).$$

$$Hom_{\Sigma_3}(\mathcal{F}(E)(3), Hom_k(V^{\otimes 3}, V)) \overset{\theta_3}{\simeq} Hom_k(\mathcal{F}(E)(3) \otimes_{\Sigma_3} V^{\otimes 3}, V).$$

Set $\varphi = {}^t\theta_1(a) : \#V \to E^\vee \underset{\Sigma_2}{\otimes} (\#\hat{V})^{\otimes 2}$. Put $d = \Omega(\theta_1(a))$ and $\psi = \omega(d^2)$, and denote by $b \in Hom_{\Sigma_3}(R, Hom_k(V^{\otimes 3}, V))$ the map $\theta_2^{-1}(\psi)$.

LEMMA. $b = \hat{a}(3)|_R$.

PROOF OF THE LEMMA. Let $\pi : \mathcal{F}(E^\vee)(3) \to \mathcal{P}^!(3)$ be the canonical projection. We denote by $\bar{\circ}_i : E^\vee \otimes E^\vee \to \mathcal{F}(E^\vee)(3)$, and $\circ_i : E^\vee \otimes E^\vee \to \mathcal{P}^!(3)$, the elementary operations defined in 1.4.1, so that, for $\mu, \nu \in E^\vee$, $\pi(\mu\bar{\circ}_i\nu) = \mu \circ_i \nu$. Let $(e_i)_{1 \le i \le r}$ and $(v_j)_{1 \le j \le s}$ be the respective basis of E and V. We denote also by $(v_j^*)_{1 \le j \le s}$ the dual basis of $(v_j)_{1 \le j \le s}$. To simplify, in the sequel we'll use Einstein's convention on summations.

The map a is defined by structural constants, (A^l_{ijk}) such that $a(e_i)(v_j \otimes v_k) = A^l_{ijk}v_l$. Thus $\varphi(v_l^*) = A^l_{ijk} e_i^* \otimes v_j^* \otimes v_k^*$. Since d is a derivation we have

$$d^2(v_l^*) = A^l_{ijk}(e_i^* \otimes \varphi(v_j^*) \otimes v_k^* - e_i^* \otimes v_j^* \otimes \varphi(v_k^*)),$$

i.e. if we modify the notations,

$$d^2(v_l^*) = (A^l_{pmk}A^m_{qij}e_p^* \circ_1 e_q^* - A^l_{pim}A^m_{qjk} e_p^* \circ_2 e_q^*) \otimes v_i^* \otimes v_j^* \otimes v_k^*.$$

Let $\phi : \#V \to \mathcal{F}(E^\vee)(3) \underset{\Sigma_3}{\otimes} (\#\hat{V})^{\otimes 3}$ be the map defined, for $1 \le l \le s$, by

$$\phi(v_l^*) = (A^l_{pmk}A^m_{qij} e_p^*\bar{\circ}_1e_q^* - A^l_{pim}A^m_{qjk} e_p^*\bar{\circ}_2e_q^*) \otimes v_i^* \otimes v_j^* \otimes v_k^*,$$

and set $c = \theta_3^{-1}({}^t\phi)$. We have $\pi \circ \phi = d^2$, or, in other terms, $b = c|_R$. Let's prove that $\hat{a}(3) = c$. For this, since $\mathcal{F}(E)(3)$ is generated as a $k[\Sigma_3]$-module by the $e_p\bar{\circ}_ie_q$'s (we still use the bar for denoting the operations on $\mathcal{F}(E)$), it suffices to show that the two maps agree onto these elements.

But, according to 1.7.1 (formulas (2)), $(e_p\bar{\circ}_1e_q)^* = e_p^*\bar{\circ}_1e_q^*$ and $(e_p\bar{\circ}_2e_q)^* = -e_p^*\bar{\circ}_2e_q^*$; thus we have:

$$\phi(v_l^*) = (A^l_{pmk}A^m_{qij} (e_p\bar{\circ}_1e_q)^* + A^l_{pim}A^m_{qjk} (e_p\bar{\circ}_2e_q)^*) \otimes v_i^* \otimes v_j^* \otimes v_k^*.$$

and thus,

$$c(e_p\bar{\circ}_1e_q)(v_i \otimes v_j \otimes v_k) = A^l_{pmk}A^m_{qij} v_l, \text{ and } c(e_p\bar{\circ}_2e_q)(v_i \otimes v_j \otimes v_k) = A^l_{pim}A^m_{qjk} v_l.$$

But,
$$\hat{a}(3)(e_p\bar{\circ}_1e_q)(v_i \otimes v_j \otimes v_k) =$$
$$a(e_p)(a(e_q)(v_i \otimes v_j) \otimes v_k) = a(e_p)(A^m_{qij}v_m \otimes v_k) = A^l_{pmk}A^m_{qij} v_l.$$
In the same way,
$$\hat{a}(3)(e_p\bar{\circ}_2e_q)(v_i \otimes v_j \otimes v_k) =$$
$$a(e_p)(v_i \otimes a(e_q)(v_j \otimes v_k)) = a(e_p)(A^m_{qjk}v_i \otimes v_m) = A^l_{pim}A^m_{qjk} v_l.$$
Therefore, $b = c|_R = \hat{a}(3)|_R$. ☐

END OF THE PROOF. The map a defines on V a $\mathcal{P}$-algebra structure if, and only if, $\hat{a}(3)|_R = 0$, thus if, and only if, $b = 0$, i.e. if, and only if, $d^2 = 0$ since $b = \theta_2^{-1}(\omega(d^2))$. ☐

REMARK 2.4.3. Let $\mathcal{P}$ be a quadratic operad, and (V, a) a $\mathcal{P}$-algebra. According to theorem 2.4.1, one can associate to a a differential $d : \mathcal{F}_{\mathcal{P}!}^{gr,*}(\#V) \to \mathcal{F}_{\mathcal{P}!}^{gr,*+1}(\#V)$. But this means that we have a differential,

$$^t d : (\mathcal{P}^!)^\vee(n+1) \otimes_{\Sigma_{n+1}} V^{\otimes n+1} \to (\mathcal{P}^!)^\vee(n) \otimes_{\Sigma_n} V^{\otimes n}.$$

Thus, we have a complex, and we can show that the homology of this complex coincides with the homology of V with coefficients in k, defined in [G–K], and denoted by $H_*^\mathcal{P}(V, k)$. Furthermore, according to corollary 2.4.2, we can also associate to a an element $\varphi \in L_\mathcal{P}^1(V)$, such that $[\varphi, \varphi] = 0$; thus, we still have a complex, $(L_\mathcal{P}^*(V), \delta_\varphi)$, where $\delta_\varphi(f) = [f, \varphi]$. In the sequel, we'll note $H_\mathcal{P}^*(V, V)$, or simply $H_\mathcal{P}^*(V)$, the homology groups of this complex.

2.4.4. Alternative description. There is another way of defining a graded Lie algebra structure on $L_\mathcal{P}(V)$ inspired by the work of M. Gerstenhaber and A. A. Voronov [G–V]. First, notice that we have an isomorphism of $k[\Sigma_n]$-modules,

$$\mathcal{P}^!(n) \otimes \mathbf{sgn}_n \overset{\alpha}{\simeq} Hom_k((\mathcal{P}^!)^\vee(n), k),$$

given by:

$$\alpha(\mu \otimes 1)(\mu^*) = < \mu^*, \mu >,$$

where $< , >$ is the evaluation between $\mathcal{P}^!(n)$ and $(\mathcal{P}^!)^\vee(n)$ (see also our conventions in 1.2). Let V be any vector space, then since $End(V)(n)$ is obviously k-projective (k is a field) and so $k[\Sigma_n]$-projective (because Σ_n is a finite group whose cardinal is invertible in k), if we tensored both sides by $End(V)(n)$ we have a k-isomorphism (here we put on any right $k[\Sigma_n]$-module F the natural left $k[\Sigma_n]$-module structure given, for any $x \in F$ and any $\sigma \in \Sigma_n$, by $\sigma x = x\sigma^{-1}$):

$$\mathcal{P}^!(n) \otimes \mathbf{sgn}_n \otimes_{\Sigma_n} End(V)(n) \overset{\alpha \otimes 1}{\simeq} Hom_k((\mathcal{P}^!)^\vee(n), k) \otimes_{\Sigma_n} End(V)(n).$$

Set $\overline{End}(V)(n) = \mathbf{sgn}_n \otimes End(V)(n)$, i.e. $\overline{End}(V)(n)$ coincides with $End(V)(n)$ as a vector space, but with the right action of Σ_n given, with obvious notations, by

$$(f \cdot \sigma)(v_1 \otimes \cdots \otimes v_n) = \epsilon(\sigma) f(v_{\sigma^{-1}(1)} \otimes \cdots \otimes v_{\sigma^{-1}(n)}).$$

Since $\mathcal{P}^!(n)$ is finite dimensional, we have a k-isomorphism,

$$Hom_k((\mathcal{P}^!)^\vee(n), k) \otimes_{\Sigma_n} End(V)(n) \simeq L_\mathcal{P}^{n-1}(V),$$

and so a k-isomorphism,

$$\mathcal{P}^!(n) \otimes_{\Sigma_n} \overline{End}(V)(n) \overset{\beta}{\simeq} L_\mathcal{P}^{n-1}(V),$$

where β is given with obvious notations by:

$$\beta(\mu \otimes f)(\mu^* \otimes v_1 \otimes \cdots \otimes v_n) = \frac{1}{n!} \sum_{\tau \in \Sigma_n} \epsilon(\tau) < \mu^*, \mu\tau > f(v_{\tau^{-1}(1)} \otimes \cdots \otimes v_{\tau^{-1}(n)}).$$

Let's denote by $\overline{End}(V)$ the collection of $\overline{End}(V)(n)$; then $\overline{End}(V)$ is almost an operad: the elementary operations of $\overline{End}(V)$ are the same than those of $End(V)$, but since the right action of Σ_n is altered by the sign, we don't have the equality

$(f \circ_i g)(\sigma \circ_i \tau) = f\sigma \circ_{\sigma^{-1}(i)} g\tau$, but instead for $f \in \overline{End}(V)(n)$, $g \in \overline{End}(V)(m)$, the equality

$$(8) \qquad (f \circ_i g)(\sigma \circ_i \tau) = (-1)^{(m-1)(i-\sigma^{-1}(i))} f\sigma \circ_{\sigma^{-1}(i)} g\tau$$

(simple application of (7)). Now, consider the collection $\mathcal{Q} = \mathcal{P}^! \otimes \overline{End}(V)$, i.e. for all $n \in \mathbb{N}^*$, $\mathcal{Q}(n) = \mathcal{P}^!(n) \otimes \overline{End}(V)(n)$. Again $\mathcal{Q}$ is almost an operad: the elementary operations $\circ_i$ of $\mathcal{Q}$ are given by

$$(\mu \otimes f) \circ_i (\nu \otimes g) = \mu \circ_i \nu \otimes f \circ_i g;$$

the right action of Σ_n is given by

$$(\mu \otimes f) \cdot \sigma = \mu\sigma \otimes \sigma^{-1} f.$$

What is important here is the fact that the elementary operations of $\mathcal{Q}$ still satisfies the conditions of associativity (1). With these notations, we have the isomorphism of Σ_n-modules, $L_{\mathcal{P}}^{n-1}(V) \simeq \mathcal{Q}(n)_{\Sigma_n}$, where $\mathcal{Q}(n)_{\Sigma_n}$ denotes the coinvariants. In general, $L_{\mathcal{P}}^{n-1}(V)$ is **not** an operad (and not almost an operad); this is because the elementary operations don't commute with the right action of Σ_n. Namely, for $\mu \otimes f \in \mathcal{Q}(n)$ and $\nu \otimes g \in \mathcal{Q}(m)$, we have:

$$(\mu\sigma \otimes \sigma^{-1} f) \circ_i (\nu\tau \otimes \tau^{-1} g) = \mu\sigma \circ_i \nu\tau \otimes \sigma^{-1} f \circ_i \tau^{-1} g$$
$$= (-1)^{\xi}(\mu \circ_{\sigma(i)} \nu)(\sigma \circ_{\sigma(i)} \tau) \otimes (\sigma \circ_{\sigma(i)} \tau)^{-1}(f \circ_{\sigma(i)} g)$$
$$= (-1)^{\xi}((\mu \otimes f) \circ_{\sigma(i)} (\nu \otimes g)) \cdot (\sigma \circ_{\sigma(i)} \tau)$$

where ξ stands for $(m-1)(i-\sigma(i))$. Therefore, the elementary operations are not well defined.

For any operad $\mathcal{O}$ we have a natural structure of graded Lie algebra on $L = \mathcal{O}[-1]$ (i.e. $L(n) = \mathcal{O}(n+1)$) given, for $\mu \in L(n)$, $\nu \in L(m)$, by,

$$[\mu, \nu] = \mu \circ \nu + (-1)^{nm+1} \nu \circ \mu,$$

where

$$\mu \circ \nu = \sum_{i=1}^{n+1} (-1)^{m(i-1)} \mu \circ_i \nu$$

(see [G–V]). This graded Lie bracket depends only of the associativity conditions (1); therefore, we have on $\mathcal{Q}[-1]$ a natural structure of graded Lie algebra given, for $\mu \otimes f \in \mathcal{Q}(n+1)$, $\nu \otimes g \in \mathcal{Q}(m+1)$, by

$$[\mu \otimes f, \nu \otimes g] = (\mu \otimes f) \circ (\nu \otimes g) + (-1)^{nm+1}(\nu \otimes g) \circ (\mu \otimes f),$$

where

$$(\mu \otimes f) \circ (\nu \otimes g) = \sum_{i=1}^{n+1} (-1)^{m(i-1)}(\mu \otimes f) \circ_i (\nu \otimes g)$$
$$= \sum_{i=1}^{n+1} (-1)^{m(i-1)} \mu \circ_i \nu \otimes f \circ_i g.$$

Let's denote by $\pi : \mathcal{Q}(n) \to \mathcal{Q}(n)_{\Sigma_n}$ the natural epimorphism, in $\mathcal{Q}(n)_{\Sigma_n}$. We have the following proposition:

PROPOSITION. *With the above notations, the map*

$$\bar{\circ} : L_{\mathcal{P}}^n(V) \otimes L_{\mathcal{P}}^m(V) \to L_{\mathcal{P}}^{n+m}(V)$$

defined, if we identify $L_{\mathcal{P}}^n(V)$ with $\mathcal{Q}(n+1)_{\Sigma_{n+1}}$, by

$$\pi(\mu \otimes f) \bar{\circ} \pi(\nu \otimes g) = \sum_{i=1}^{n+1} (-1)^{m(i-1)} \pi((\mu \otimes f) \circ_i (\nu \otimes g)) =$$

$$\sum_{i=1}^{n+1} (-1)^{m(i-1)} \pi(\mu \circ_i \nu \otimes f \circ_i g),$$

is well defined. Furthermore, the bracket defined by,

$$[\pi(\mu \otimes f), \pi(\nu \otimes g)] = \pi(\mu \otimes f) \bar{\circ} \pi(\nu \otimes g) + (-1)^{nm+1} \pi(\nu \otimes g) \bar{\circ} \pi(\mu \otimes f),$$

makes $L_{\mathcal{P}}(V)$ a graded Lie algebra.

PROOF OF THE PROPOSITION. The only point to establish is the validity of the definition of the composition $\bar{\circ}$; the fact that $L_{\mathcal{P}}(V)$ is a graded Lie algebra follows immediately from the equality: $[\pi(\mu \otimes f), \pi(\nu \otimes g)] = \pi([\mu \otimes f, \nu \otimes g])$. Therefore, for any $\sigma, \tau \in \Sigma_{n+1} \times \Sigma_{m+1}$, we must show the equality between

$$\sum_{i=1}^{n+1} (-1)^{m(i-1)} \pi((\mu \otimes f) \circ_i (\nu \otimes g))$$

and

$$\sum_{i=1}^{n+1} (-1)^{m(i-1)} \pi((\mu\sigma \otimes \sigma^{-1} f) \circ_i (\nu\tau \otimes \tau^{-1} g)).$$

But, according to (8) we have:

$$\pi((\mu\sigma \otimes \sigma^{-1} f) \circ_i (\nu\tau \otimes \tau^{-1} g)) = (-1)^{m(i-\sigma(i))} \pi((\mu \otimes f) \circ_{\sigma(i)} (\nu \otimes g)).$$

So,

$$\sum_{i=1}^{n+1} (-1)^{m(i-1)} \pi((\mu\sigma \otimes \sigma^{-1} f) \circ_i (\nu\tau \otimes \tau^{-1} g)) =$$

$$\sum_{i=1}^{n+1} (-1)^{m(\sigma(i)-1)} \pi((\mu \otimes f) \circ_{\sigma(i)} (\nu \otimes g)) = \sum_{i=1}^{n+1} (-1)^{m(i-1)} \pi((\mu \otimes f) \circ_i (\nu \otimes g)).$$

$\square$

So, we have two natural structures of graded Lie algebra on $L_{\mathcal{P}}(V)$: the first is given via derivations of $\mathcal{F}_{\mathcal{P}!}^{gr}(\#V)$; the second is given by the last proposition. Of course, we can show that these two structures are the same (the proof is a little bit technical, so we don't reproduce it), the correspondence been given by

$$i_{\pi(\mu \otimes f)}(\pi(\nu \otimes g)) = \pi(\mu \otimes f) \bar{\circ} \pi(\nu \otimes g).$$

The advantage of the second method is the fact that it works for any vector space V (in none finite dimension, we would have to replace derivations by coderivations), and the fact that the knowledge of the elementary operations entirely determines the bracket, making the calculus of this bracket easier (compare the two methods with the case of Leibniz algebras in 3.2.3 and 3.2.7).

3. Applications to classical structures and Leibniz algebras

In this section we show how the previous formalism, developed in the second part, can be applied to classical algebraic structures such as associative algebras [L–W], Lie algebras [N–R] and commutative algebras. This last case seems to be well-known to the specialists, but apparently there is no paper treating this case explicitly, except a short sentence in [G2] page 74.

Then we study the case of Leibniz algebras which is interesting in itself since the Koszul dual of the associated operad **Leib** is not trivial. For the calculus, we always supposed implicitly that V is finite dimensional, keeping in mind that this hypothesis is not necessary (cf. the end of part 2).

3.1. Classical structures.

3.1.1. Associative algebras. Let $\mathcal{P}$ be the operad **Ass**; we have $\mathcal{P}^! = $ **Ass**, so that $\mathcal{F}_{\mathcal{P}^!}^{gr}(\#V) = \bigoplus_{n \geq 1}(\#V)^{\otimes n}$ is the free associative algebra (without unit) generated by $\#V$. Set $\overline{T}(\#V) = \bigoplus_{n \geq 1}(\#V)^{\otimes n}$; thus we have

$$L_{\mathcal{P}}^n(V) = Hom_k(V^{\otimes n+1}, V) \simeq Der^n(\overline{T}(\#V)),$$

i.e. the set of all homogeneous degree n linear map $d : \overline{T}(\#V) \to \overline{T}(\#V)$ such that, for all $p, q \in \mathbb{N}^*$ and $x, y \in (\#V)^{\otimes p} \times (\#V)^{\otimes q}$, $d(x \otimes y) = d(x) \otimes y + (-1)^{np}x \otimes d(y)$.

A straightforward calculus shows that the graded Lie algebra structure we put on $L_{\mathcal{P}}^n(V)$ is really those given by M. Gerstenhaber in [G1], i.e. for all f, g, we have $i_f(g) = I_f(g)$ (see the introduction).

3.1.2. Lie algebras. Let's denote again by $\mathcal{P}$ the operad **Lie**; we have $\mathcal{P}^! = $ **Com**, so that

$$\mathcal{F}_{\mathcal{P}^!}^{gr}(\#V) = \bigoplus_{n \geq 1} \Lambda^n(\#V)$$

is the free graded commutative algebra (without unit) generated by $\#V$. Therefore, $L_{\mathcal{P}}^n(V)$ is the set of all alternating maps from $V^{\otimes n+1}$ to V, and thus is isomorphic, as a vector space, to $\Lambda^{n+1}(\#V) \otimes V$.

Let's prove, with the above notations, that $\Omega(\alpha \otimes X) = \alpha \wedge i_X$, where $\alpha \in \Lambda^{n+1}(\#V)$ and $X \in V$ ($i_X : \Lambda^m(\#V) \to \Lambda^{m-1}(\#V)$ is the inner product by X). Since i_X is a graded derivation (notice that it lows the degree by one), $\alpha \wedge i_X \in Der^n(\Lambda(\#V))$. Therefore, it suffices to show that $\alpha \wedge i_X$ and $\Omega(\alpha \otimes X)$ agree on $\#V$. But this is the case since

$$\alpha \wedge i_X(v^*) = v^*(X)\,\alpha = \Omega(\alpha \otimes X)(v^*).$$

Thus we have

$$\Omega(\beta \otimes Y) \circ \Omega(\alpha \otimes X)(v^*) = \beta \wedge i_Y(v^*(X)\alpha) = v^*(X)\beta \wedge i_Y(\alpha),$$

and therefore,

$$i_{(\alpha \otimes X)}(\beta \otimes Y) = \beta \wedge i_Y(\alpha) \otimes X.$$

The Lie bracket on $L_{\mathcal{P}}(V)$ is thus given, for all $\alpha \otimes X \in L_{\mathcal{P}}^a(V), \beta \otimes Y \in L_{\mathcal{P}}^b(V)$, by

$$[\alpha \otimes X, \beta \otimes Y] = \beta \wedge i_Y(\alpha) \otimes X + (-1)^{ab+1}\alpha \wedge i_X(\beta) \otimes Y.$$

This is exactly the description, given in [L–W], of the Richardson-Nijenhuis Lie algebra [N–R].

3.1.3. Commutative algebras. Let's denote by $\mathcal{P}$ the operad **Com**; we have $\mathcal{P}^! = \mathbf{Lie}$ and $\mathcal{F}^{gr}_{\mathcal{P}^!}(\#V)$ is the free graded Lie algebra generated by $\#V$. Set for $n \geq 1$, $C^n(V) = L^n_{\mathbf{Com}}(V)$, $\mathcal{L}^n(\#V) = \mathcal{F}^{gr_n}_{\mathcal{P}^!}(\#V)$, $C(V) = L_{\mathbf{Com}}(V)$, and $\mathcal{L}(\#V) = \mathcal{F}^{gr}_{\mathcal{P}^!}(\#V)$. To describe the graded Lie algebra structure on $C(V)$, we first must give a practical description of the free graded Lie algebra generated by $\#V$. Set $W = \#V$ and denote by $T(W) = \bigoplus_{n \geq 0} W^{\otimes n}$ the tensor algebra on W. It is a graded Lie algebra for the bracket given on homogeneous elements $x, y \in W^{\otimes n} \times W^{\otimes m}$ by

$$(9) \qquad\qquad [x, y] = xy + (-1)^{nm+1} yx.$$

By abstract nonsense, $\mathcal{L}(W)$ is identified with the graded sub-Lie algebra of $(T(W), [,])$ generated by W. Let's denote by Δ, the graded diagonal map, i.e. the unique map $\Delta : T(W) \to T(W) \otimes T(W)$, such that:

i) for all $x \in W$, $\Delta(x) = 1 \otimes x + x \otimes 1$,
ii) Δ is a morphism of algebras for the following associative structure on $T(W) \otimes T(W)$: for all $x, y, x', y' \in W^{\otimes n} \times W^{\otimes m} \times W^{\otimes n'} \times W^{\otimes m'}$,

$$(x \otimes y) \cdot (x' \otimes y') = (-1)^{mn'} xx' \otimes yy'.$$

Using shuffles, for $w_1, \ldots, w_m \in W$ we have explicitly

$$\Delta(w_1 \cdots w_m) = \sum_{\substack{\sigma(p,q)-\text{shuffles} \\ p+q=m}} \epsilon(\sigma) w_{\sigma(1)} \cdots w_{\sigma(p)} \otimes w_{\sigma(p+1)} \cdots w_{\sigma(p+q)}.$$

Let $\gamma : \overline{T}(W) \to \mathcal{L}(W)$ be the map defined by:

i) for all $w \in W$, $\gamma(w) = w$,
ii) for all $w_1, \ldots, w_m \in W$, $\gamma(w_1 \cdots w_m) = [w_1, [w_2, \ldots, [w_{m-1}, w_m]] \ldots]$.

The map γ can be defined by induction in the following manner:

$$\gamma(w_1 \cdots w_m) = ad_{w_1}(\gamma(w_2 \cdots w_m)).$$

We have the classical following lemma (cf. for instance Wigner [W]):

LEMMA 1. *With the above notations, the following conditions are equivalent:*
a) $x \in \mathcal{L}^m(W) \subset W^{\otimes m}$, $m \in \mathbb{N}^*$,
b) $\Delta(x) = 1 \otimes x + x \otimes 1$,
c) $\gamma(x) = mx$.

Now, let's see how lemma 1 will enables us to explicit the graded Lie algebra structure on $C(V)$. We recall that, as a vector space, $C^n(V) \simeq Der^n(\mathcal{L}(W))$. Let f be an element of $C^n(V)$; since $\mathcal{L}^{n+1}(W) \subset W^{\otimes n+1}$, we have (3.1.1)

$$^t f \in Hom_k(W, W^{\otimes n+1}) \simeq Der^n(\overline{T}(W)).$$

Let's consider $\Omega(f) \in Der^n(\overline{T}(W))$ the unique n degree derivation of $\overline{T}(W)$ associated to $^t f$ ($\Omega(f)$ **is a derivation of the associative algebra** $\overline{T}(W)$). Set $\overline{\Omega}(f) = \Omega(f)|_{\mathcal{L}(W)} : \mathcal{L}(W) \to \overline{T}(W)$.

LEMMA 2. *With the above notations we have:*
i) $\overline{\Omega}(f)|_W = {}^t f$,
ii) $\overline{\Omega}(f) \in Der^n(\mathcal{L}(W))$.

PROOF OF LEMMA 2. The assertion i) is clear. To prove ii), let's prove first that $\overline{\Omega}(f)(\mathcal{L}(W)) \subset \mathcal{L}(W)$. Since $\Omega(f)$ is a derivation of the associative algebra $\overline{T}(W)$, for $x, y \in \overline{T}(W)$ we have:

$$(10) \qquad \Omega(f)([x,y]) = [\Omega(f)(x), y] + (-1)^{n|x|}[x, \Omega(f)(y)],$$

where $|x|$ stands for the degree of x (formula (10) is a consequence of formula (9) and the fact that $\Omega(f)$ is a derivation). $\qquad\square$

Let x be an element of $\mathcal{L}^m(W)$; according to lemma 1, $x = \frac{1}{m}\gamma(x)$. By induction, using (10), we have $\Omega(f)([w_1, [w_2, \ldots, [w_{m-1}, w_m]] \ldots]) =$

$$\sum_{k=1}^{m} (-1)^{n(k-1)}[w_1, [w_2, \ldots, [w_{k-1}, [\Omega(f)(w_k), [w_{k+1}, \ldots, [w_{m-1}, w_m]] \ldots].$$

Therefore this shows that $\Omega(f)(\gamma(x)) \in \mathcal{L}^{n+m}(W)$, and thus that $\Omega(f)(x) \in \mathcal{L}^{n+m}(W)$. Furthermore, $\overline{\Omega}(f)$ is a derivation of the graded Lie algebra $\mathcal{L}(W)$ since we have (10). $\qquad\square$

The isomorphism between $C^n(V)$ and $Der^n(\mathcal{L}(W))$ is thus given by:

$$C^n(V) \xrightarrow[\simeq]{\overline{\Omega}} Der^n(\mathcal{L}(W)).$$

The diagonal map Δ is an homogeneous degree zero linear map: $\Delta|_{W^{\otimes n}} : W^{\otimes n} \to \bigoplus_{p+q=n} W^{\otimes p} \otimes W^{\otimes q}$. Let $\mu : T(V) \otimes T(V) \to T(V)$ be the map defined on homogeneous elements by: $\mu = {}^t(\Delta|_{W^{\otimes n}}) : \bigoplus_{p+q=n} V^{\otimes p} \otimes V^{\otimes q} \to V^{\otimes n}$. Namely, for all $v_1, \ldots, v_p, v_{p+1}, \ldots, v_{p+q} \in V$,

$$(11) \qquad \mu(v_1 \cdots v_p \otimes v_{p+1} \cdots v_{p+q}) = \sum_{\sigma(p,q)-\text{shuffles}} \epsilon(\sigma) v_{\sigma^{-1}(1)} \cdots v_{\sigma^{-1}(p+q)}.$$

Set $\overline{T}(V) = \bigoplus_{n>0} V^{\otimes n}$ and $(\overline{T}^2(V))^n = V^{\otimes n} \cap \mu(\overline{T}(V) \otimes \overline{T}(V))$. We have the following lemma:

LEMMA 3. *With the above notations we have:*

$$C^n(V) \simeq Hom_k(V^{\otimes n+1}/(\overline{T}^2(V))^{n+1}, V).$$

PROOF OF LEMMA 3. We have $T(W) \otimes T(W) = 1 \otimes T(W) \oplus T(W) \otimes 1 \oplus \overline{T}(W) \otimes \overline{T}(W)$; let's denote by $\pi : T(W) \otimes T(W) \to \overline{T}(W) \otimes \overline{T}(W)$ the canonical epimorphism.

Since $\mathcal{L}^{n+1}(W) \subset W^{\otimes n+1}$, for all $f \in C^n(V)$ we have ${}^t f \in Hom_k(W, W^{\otimes n+1})$. Furthermore, $\pi(\Delta({}^t f)) = 0$: actually, according to Lemma 1, $\Delta({}^t f) = 1 \otimes {}^t f + {}^t f \otimes 1$; if we transpose, we have $f \circ {}^t\Delta \circ i = 0$, where $i : \overline{T}(V) \otimes \overline{T}(V) \to T(V) \otimes T(V)$ is the inclusion; since ${}^t\Delta = \mu$, we have finally $f((\overline{T}^2(V))^{n+1}) = 0$. Therefore, f can be factored so as to define a map $\overline{f} \in Hom_k(V^{\otimes n+1}/(\overline{T}^2(V))^{n+1}, V)$. The application $f \mapsto \overline{f}$ is the expected isomorphism. $\qquad\square$

We finally deduce the following theorem:

THEOREM. *Let V be a k-vector space, $T(V)$ the tensor algebra on V and $\mu : T(V) \otimes T(V) \to T(V)$ the map defined by formula (11). Set $\overline{T}(V) = \bigoplus_{n>0} V^{\otimes n}$, $(\overline{T}^2(V))^n = V^{\otimes n} \cap \mu(\overline{T}(V) \otimes \overline{T}(V))$ and $C^n(V) = Hom_k(V^{\otimes n+1}/(\overline{T}^2(V))^{n+1}, V)$. For all $f, g \in C^n(V) \times C^m(V)$, denote by $i_f(g) \in C^{n+m}(V)$ the map defined, for $x_0, \ldots, x_{m+n} \in V$, by*

$$i_f(g)(x_0 \otimes \cdots \otimes x_{m+n} \ \ mod \ (\overline{T}^2(V))^{n+m+1}) =$$

$$\sum_{k=0}^n (-1)^{mk} f(x_0 \otimes \cdots \otimes x_{k-1} \otimes g(x_k \otimes \cdots \otimes x_{k+m}) \otimes \cdots \otimes x_{n+m}).$$

Then $C(V) = \bigoplus_{n \geq 1} C^n(V)$ equipped with the bracket defined for all $f, g \in C^n(V) \times C^m(V)$, by $[f, g] = i_f(g) + (-1)^{nm+1} i_g(f)$, is a graded Lie algebra.

If $\varphi \in C^1(V)$ satisfies $[\varphi, \varphi] = 0$, then φ makes V a commutative algebra. Furthermore, if one denotes by $\delta_\varphi : C^n(V) \to C^{n+1}(V)$ the operator defined by $\delta_\varphi(f) = [f, \varphi]$, then δ_φ is a differential, and the cohomology of $(C(V), \delta_\varphi)$ is the Harrison cohomology of the commutative algebra (V, φ) with coefficients in itself, shifted by one (i.e. $H^*(C(V), \delta_\varphi) = Harr^{*+1}(V, V)$).

PROOF. According to lemma 3, we can identify f and $\overline{f}$. By proposition 2.3.3, $C(V)$ is a graded Lie algebra for the bracket given, for $f, g \in C^n(V) \times C^m(V)$, by $[f, g] = i_f(g) + (-1)^{nm+1} i_g(f)$, where $i_f(g) = {}^t(\overline{\Omega}(g) \circ \overline{\Omega}(f)|_W)$. But $\overline{\Omega}(g) \circ \overline{\Omega}(f)|_W = \Omega(g) \circ \Omega(f)|_W$. Therefore, if we remark that $I_f(g) = {}^t(\Omega(g) \circ \Omega(f))$ (see the introduction for the definition of $I_f(g)$), then via the identification of lemma 3, we see that $I_f(g)$ must be identified with $i_f(g)$. The end of the proof follows from corollary 2.4.2 and from the definition of Harrison cohomology [HA]. □

3.2. Leibniz algebras.
In this section we apply our results to the operad $\mathcal{P} = \textbf{Leib}$.

3.2.1. Dual Leibniz algebras. According to corollary 2.4.2, we must consider $\mathcal{P}^!$; we recall that a $\mathcal{P}^!$ algebra or a dual Leibniz algebra is a vector space V equipped with a product $\times$, satisfying the following identity: for all $v_1, v_2, v_3 \in V$,

$$(v_1 \times v_2) \times v_3 = v_1 \times (v_2 \times v_3) + v_1 \times (v_3 \times v_2).$$

Furthermore, if W is a vector space, J.-L. Loday [LO2] showed that

$$\mathcal{F}_{\mathcal{P}^!}(W) = \bigoplus_{n \geq 1} W^{\otimes n},$$

with the product $\times$ given, on homogeneous elements, by:

$$(12) \quad (w_1 \otimes \cdots \otimes w_p) \times (w_{p+1} \otimes \cdots \otimes w_{p+q}) = (1 \otimes sh_{p-1,q})(w_1 \otimes \cdots \otimes w_{p+q}),$$

where $sh_{p,q}(w_1 \otimes \cdots \otimes w_{p+q}) = \sum_{\sigma(p,q)\text{-shuffles}} w_{\sigma^{-1}(1)} \otimes \cdots \otimes w_{\sigma^{-1}(p+q)}.$

Let's now consider the graded case: as a graded vector space, $\mathcal{F}_{\mathcal{P}^!}^{gr}(\#V) = \bigoplus_{n \geq 1} (\#V)^{\otimes n}$, but because of the left action of Σ_n on $(\#V)^{\otimes n}$, we must introduce the sign; thus the product $\times$ is given, on homogeneous elements, by:

(12') $(v_1^* \otimes \cdots \otimes v_p^*) \times (v_{p+1}^* \otimes \cdots \otimes v_{p+q}^*) = (1 \otimes Sh_{p-1,q})(v_1^* \otimes \cdots \otimes v_{p+q}^*),$

where now, $Sh_{p,q}(v_1^* \otimes \cdots \otimes v_{p+q}^*) = \displaystyle\sum_{\sigma(p,q)\text{-shuffles}} \epsilon(\sigma) v_{\sigma^{-1}(1)}^* \otimes \cdots \otimes v_{\sigma^{-1}(p+q)}^*.$ In fact, $\mathcal{F}_{\mathcal{P}!}^{gr}(\#V)$ is the free graded dual Leibniz algebra generated by $\#V$ (the product $\times$ satisfies the following relation: for $a, b, c \in \mathcal{F}_{\mathcal{P}!}^{gr}(\#V)$ homogeneous, $(a \times b) \times c = a \times (b \times c) + (-1)^{|b||c|} a \times (c \times b)$).

REMARK 3.2.2. We have $v_1^* \times (v_2^* \otimes \cdots \otimes v_n^*) = v_1^* \otimes \cdots \otimes v_n^*.$

Set $\overline{T}(\#V) = \bigoplus_{n \geq 1} (\#V)^{\otimes n}$ equipped with the above graded dual Leibniz algebra structure. We have

$$L_{\mathcal{P}}^n(V) = Hom_k(V^{\otimes n+1}, V) \simeq Der^n(\overline{T}(\#V)),$$

i.e. $L_{\mathcal{P}}^n(V)$ is isomorphic to the set of all homogeneous degree n maps, $d \in Hom_k^n(\overline{T}(\#V))$ satisfying, for all $a, b \in \overline{T}(\#V)$ homogeneous,

$$d(a \times b) = d(a) \times b + (-1)^{n|a|} a \times d(b).$$

In the sequel, we'll denote the sets $L_{\mathcal{P}}^n(V)$ and $\bigoplus_{n \geq 0} L_{\mathcal{P}}^n(V)$ by $L^n(V)$ and $L(V)$.

3.2.3. Calculus of $i_f(g)$. let's calculate, for

$$f, g \in Hom_k(V^{\otimes n+1}, V) \times Hom_k(V^{\otimes m+1}, V),$$

the element $i_f(g) \in Hom_k(V^{\otimes n+m+1}, V)$. For this, let $(v_i)_{1 \leq i \leq s}$ be a basis of V, and denote by $(v_i^*)_{1 \leq i \leq s}$ the dual basis. Since $i_f(g) = {}^t(\Omega(g) \circ \Omega(f)|_{\#V})$, to calculate $i_f(g)$ it suffices to calculate $\Omega(g) \circ \Omega(f)(v_l^*)$ and then to transpose.

Set $f(v_{i_1} \otimes \cdots \otimes v_{i_{n+1}}) = A_{i_1 \ldots i_{n+1}}^k v_k$ and $g(v_{j_1} \otimes \cdots \otimes v_{j_{m+1}}) = B_{j_1 \ldots j_{m+1}}^k v_k$. If we transpose then we have:

$$\Omega(f)(v_l^*) = A_{i_1 \ldots i_{n+1}}^l v_{i_1}^* \otimes \cdots \otimes v_{i_{n+1}}^*,$$

and

$$\Omega(g)(v_k^*) = B_{j_1 \ldots j_{m+1}}^k v_{j_1}^* \otimes \cdots \otimes v_{j_{m+1}}^*.$$

We have:
$$\Omega(g) \circ \Omega(f)(v_l^*) =$$
$$A_{i_1 \ldots i_{n+1}}^l \Omega(g)(v_{i_1}^* \times (v_{i_2}^* \otimes \cdots \otimes v_{i_{n+1}}^*)) =$$
$$A_{i_1 \ldots i_{n+1}}^l \{\Omega(g)(v_{i_1}^*) \times (v_{i_2}^* \otimes \cdots \otimes v_{i_{n+1}}^*) + (-1)^m v_{i_1}^* \times \Omega(g)(v_{i_2}^* \otimes \cdots \otimes v_{i_{n+1}}^*)\}.$$

Let's calculate $\Omega(g)(v_{i_1}^*) \times (v_{i_2}^* \otimes \cdots \otimes v_{i_{n+1}}^*)$. For this, we have to calculate $(v_{j_1}^* \otimes \cdots \otimes v_{j_{m+1}}^*) \times (v_{i_2}^* \otimes \cdots \otimes v_{i_{n+1}}^*)$, which is simple thanks to formula (12'). We finally find, with a change in the indexes, that

$$\Omega(g)(v_{i_1}^*) \times (v_{i_2}^* \otimes \cdots \otimes v_{i_{n+1}}^*) =$$

$$\sum_{\sigma(m,n)\text{-shuffles}} \epsilon(\sigma) A_{s j_{\sigma(m+2)} \ldots j_{\sigma(m+n+1)}}^l B_{j_{\sigma(1)} \ldots j_{\sigma(m+1)}}^s v_{j_1}^* \otimes \cdots \otimes v_{j_{m+n+1}}^*,$$

where j_{m+k} stands for i_k. The term $v_{i_1}^* \times \Omega(g)(v_{i_2}^* \otimes \cdots \otimes v_{i_{n+1}}^*)$ is equal to

$$v_{i_1}^* \times (\Omega(g)(v_{i_2}^*) \times (v_{i_3}^* \otimes \cdots \otimes v_{i_{n+1}}^*) + (-1)^m v_{i_2}^* \times \Omega(g)(v_{i_3}^* \otimes \cdots \otimes v_{i_{n+1}}^*)).$$

We repeat the process, then we transpose, and we finally find the following formula:

$$i_f(g) = \sum_{k=1}^{n+1}(-1)^{m(k-1)} \sum_{\sigma(m,n-k+1)\text{-shuffles}} \epsilon(\sigma) f \circ_k^\sigma g,$$

where for any $\sigma \in \Sigma_{m+n-k+1}$, the map $f \circ_k^\sigma g$ is defined, for all $x_1, \ldots, x_{m+n+1} \in V$, by

$$f \circ_k^\sigma g(x_1 \otimes \cdots \otimes x_{m+n+1}) =$$
$$f(x_1 \otimes \cdots \otimes x_{k-1} \otimes g(x_k \otimes x_{\sigma(k+1)} \otimes \cdots \otimes x_{\sigma(m+k)}) \otimes x_{\sigma(m+k+1)} \otimes \cdots \otimes x_{\sigma(m+n+1)}).$$

Let's see now some particular points:
i) $n = 0$ **and** $m \in \mathbb{N}^*$.
There is only one $\sigma(m,0)$-shuffle which is 1_m. Thus $i_f(g) = f \circ g$.

REMARK 3.2.4. This formula remains valid in the general case (i.e. when $\mathcal{P}$ is any quadratic operad).

ii) $n = 1$ **and** $m \in \mathbb{N}^*$.
We have

$$i_f(g) = \sum_{k=1}^{2}(-1)^{m(k-1)} \sum_{\sigma(m,2-k)\text{-shuffles}} \epsilon(\sigma) f \circ_k^\sigma g.$$

There are $m + 1$ $\sigma(m,1)$-shuffles, and they are characterized by an integer $2 \leq k \leq m+2$, such that if $k \neq 2$, then for all $2 \leq i \leq k-1$, $\sigma(i) = i$, for all $k \leq j \leq m+1$, $\sigma(j) = j+1$ and $\sigma(m+2) = k$ (sign $(-1)^{m-k}$). If $k = 2$, then for all $2 \leq i \leq m+1$, $\sigma(i) = i+1$ and $\sigma(m+2) = 2$ (sign $(-1)^m$).

Furthermore, there is only one $\sigma(m,0)$-shuffle which is 1_m. Therefore, for $x_1, \ldots, x_{m+2} \in V$, we have:
$$i_f(g)(x_1 \otimes \cdots \otimes x_{m+2}) = \sum_{k=2}^{m+2}(-1)^{m-k}f(g(x_1 \otimes \cdots \otimes \hat{x}_k \otimes \cdots \otimes x_{m+2}) \otimes x_k) +$$
$$(-1)^m f(x_1 \otimes g(x_2 \otimes \cdots \otimes x_{m+2})).$$

In particular, for $f \in L^1(V)$ and for $x_1, x_2, x_3 \in V$ we find that

$$\frac{1}{2}[f,f](x_1 \otimes x_2 \otimes x_3) = -f(f(x_1 \otimes x_3) \otimes x_2) + f(f(x_1 \otimes x_2) \otimes x_3) - f(x_1 \otimes f(x_2 \otimes x_3)).$$

Therefore, if f is a square zero element then, for $x_1, x_2, x_3 \in V$,

$$f(x_1 \otimes f(x_2 \otimes x_3)) = f(f(x_1 \otimes x_2) \otimes x_3) - f(f(x_1 \otimes x_3) \otimes x_2),$$

i.e. f defines a Leibniz bracket on V as expected.

iii) $n \in \mathbb{N}^*$ **and** $m = 0$.
We have

$$i_f(g) = \sum_{k=1}^{n+1}(-1)^{0(k-1)} \sum_{\sigma(0,n+1-k)\text{-shuffles}} \epsilon(\sigma) f \circ_k^\sigma g.$$

So, for $x_1, \ldots, x_{n+1} \in V$ it is clear that

$$i_f(g)(x_1 \otimes \cdots \otimes x_{n+1}) = \sum_{k=1}^{n+1} f(x_1 \otimes \cdots \otimes g(x_k) \otimes \cdots \otimes x_{n+1}).$$

REMARK 3.2.5. In the general case, for $m = 0$ and $n \in \mathbb{N}^*$ we have the following formula: for all $\mu \in (\mathcal{P}^!)^\vee(n+1)$, for all $x_1, \ldots, x_{n+1} \in V$,

$$i_f(g)(\mu \otimes x_1 \otimes \cdots \otimes x_{n+1}) = \sum_{k=1}^{n+1} f(\mu \otimes x_1 \otimes \cdots \otimes g(x_k) \otimes \cdots \otimes x_{n+1}).$$

iv) $n \in \mathbb{N}^*$ and $m = 1$.

We have

$$i_f(g) = \sum_{k=1}^{n+1} (-1)^{(k-1)} \sum_{\sigma(1,n+1-k)\text{-shuffles}} \epsilon(\sigma) f \circ_k^\sigma g.$$

There are $n + 2 - k$ $\sigma(1, n - k + 1)$-shuffles. They are characterized by an integer $k + 1 \leq j \leq n + 2$ such that $\sigma(k+1) = j$, for all $k + 1 < p \leq j$, $\sigma(p) = p - 1$ and for all $p > j$, $\sigma(p) = p$ (sign $(-1)^{k-j+1}$). Therefore, for $x_1, \ldots, x_{n+2} \in V$,

$$i_f(g)(x_1 \otimes \cdots \otimes x_{n+2}) =$$

$$\sum_{1 \leq i < j \leq n+2} (-1)^j f(x_1 \otimes \cdots \otimes x_{i-1} \otimes g(x_i \otimes x_j) \otimes x_{i+1} \otimes \cdots \otimes \hat{x}_j \otimes \cdots \otimes x_{n+2}).$$

We have seen that a square zero element $\varphi \in L^1(V)$ determines a Leibniz algebra structure on V. The operator $\delta_\varphi : L^n(V) \to L^{n+1}(V)$, defined by $\delta_\varphi(f) = [f, \varphi]$ is a differential. Let's explicit it: we have

$$\delta_\varphi(f) = i_f(\varphi) + (-1)^{n+1} i_\varphi(f);$$

set $\varphi(x \otimes y) = [x, y]$, if we use the above expressions, for $x_1, \ldots, x_{n+2} \in V$ we finally find that

$$\delta_\varphi(f)(x_1 \otimes \cdots \otimes x_{n+2}) =$$

$$-[x_1, f(x_2 \otimes \cdots \otimes x_{n+2})] + \sum_{i=2}^{n+2} (-1)^{i+1} [f(x_1 \otimes \cdots \otimes \hat{x}_i \otimes \cdots \otimes x_{n+2}), x_i] +$$

$$\sum_{1 \leq i < j \leq n+2} (-1)^j f(x_1 \otimes \cdots \otimes x_{i-1} \otimes [x_i, x_j] \otimes x_{i+1} \otimes \cdots \otimes \hat{x}_j \otimes \cdots \otimes x_{n+2}),$$

i.e. we find that the differential is the opposite of the differential defined by J.-L. Loday in [LO1]. So we have the following theorem:

THEOREM 3.2.6. *Let V be a k-vector space and set $L^n(V) = Hom_k(V^{\otimes n+1}, V)$. For $f, g \in L^n(V) \times L^m(V)$, denote by $i_f(g) \in L^{n+m}(V)$ the map defined by*

$$i_f(g) = \sum_{k=1}^{n+1} (-1)^{m(k-1)} \sum_{\sigma(m,n-k+1)\text{-shuffles}} \epsilon(\sigma) f \circ_k^\sigma g,$$

where for any $\sigma \in \Sigma_{m+n-k+1}$, the map $f \circ_k^\sigma g$ is defined, for all $x_1, \ldots, x_{m+n+1} \in V$, by

$$f \circ_k^\sigma g(x_1 \otimes \cdots \otimes x_{m+n+1}) =$$

$$f(x_1 \otimes \cdots \otimes x_{k-1} \otimes g(x_k \otimes x_{\sigma(k+1)} \otimes \cdots \otimes x_{\sigma(m+k)}) \otimes x_{\sigma(m+k+1)} \otimes \cdots \otimes x_{\sigma(m+n+1)}).$$

Then $L(V) = \bigoplus_{n \geq 1} L^n(V)$ equipped with the bracket defined for $f, g \in L^n(V) \times L^m(V)$, by

$$[f, g] = i_f(g) + (-1)^{nm+1} i_g(f),$$

is a graded Lie algebra.

If $\varphi \in L^1(V)$ satisfies $[\varphi, \varphi] = 0$, then φ defines on V a Leibniz algebra structure. Furthermore, if we denote by $\delta_\varphi : L^n(V) \to L^{n+1}(V)$ the map defined by $\delta_\varphi(f) = [f, \varphi]$ then δ_φ is a differential and the homology of $(L(V), \delta_\varphi)$ is nothing but the cohomology of the Leibniz algebra (V, φ) with coefficients in its adjoint representation, shifted by one, i.e. $H^*(L(V), \delta_\varphi) = HL^{*+1}(V, V)$ (this cohomology has been defined in [LO1]).

REMARK 3.2.7. We have seen in 2.4.4 that there was an alternative way for defining the graded Lie algebra structure on $L_\mathcal{P}(V)$. We would like to see, in this remark, how it works for the case of Leibniz algebras. We have $\mathbf{Leib}^!(n) \simeq k[\Sigma_n]$; the generators of $\mathbf{Leib}^!(n)$ are of the form

$$x_{\sigma(1)} \times (x_{\sigma(2)} \times (\cdots \times (x_{\sigma(n-1)} \times x_{\sigma(n)}) \cdots))$$

In the sequel we identify this generator with σ, so that the action of Σ_n is given for any $\tau \in \Sigma_n$ and any $\sigma \in \mathbf{Leib}^!(n)$, by $\sigma \cdot \tau = \tau^{-1}\sigma$ (if we take the naive action, i.e. $\sigma \cdot \tau = \sigma\tau$, we don't have the equality (1')). Therefore, for any $\sigma \in \mathbf{Leib}^!(n)$, we have the equality:

$$\sigma = 1_n \cdot \sigma^{-1}.$$

We recall that $L_\mathcal{P}^{n-1}(V) \simeq \mathcal{P}^!(n) \otimes_{\Sigma_n} \overline{End(V)}(n)$ (cf. 2.4.4); in the case where $\mathcal{P} = \mathbf{Leib}$, we have

$$\mathcal{P}^!(n) \otimes_{\Sigma_n} \overline{End(V)}(n) \overset{\Gamma}{\simeq} Hom_k(V^{\otimes n}, V),$$

where

$$\Gamma(\sigma \otimes f) = \Gamma(1_n \cdot \sigma^{-1} \otimes f) = \Gamma(1_n \otimes f \cdot \sigma) = f \cdot \sigma,$$

i.e.

$$\gamma(\sigma \otimes f)(v_1 \otimes \cdots \otimes v_n) = \epsilon(\sigma)f(v_{\sigma^{-1}(1)} \otimes \cdots \otimes v_{\sigma^{-1}(n)}).$$

In the sequel we will identify f with $1_n \otimes f$, so that $\sigma \otimes f$ corresponds to $f \cdot \sigma$. Now, let's calculate $i_f(g) = f\bar\circ g$: we have

$$f\bar\circ g = \pi(1_{n+1} \otimes f) \bar\circ \pi(1_{m+1} \otimes g) = \sum_{i=1}^{n+1} (-1)^{m(i-1)}\pi(1_{n+1} \circ_i 1_{m+1} \otimes f \circ_i g).$$

But,

$$1_{n+1} \circ_i 1_{m+1} = \alpha \times (\beta \times \gamma)$$

where α stands for

$$x_1 \times (x_2 \times (\cdots \times (x_{i-1},$$

while β stands for

$$x_i \times (x_{i+1} \times (\cdots \times (x_{m+i}) \cdots)$$

and γ for

$$x_{m+i+1} \times (x_{m+i+2} \times (\cdots \times (x_{m+n+1}) \cdots).$$

So, it remains to calculate $\beta \times \gamma$, i.e. to calculate

$$(x_i \otimes \cdots \otimes x_{m+i}) \times (x_{m+i+1} \otimes \cdots \otimes x_{m+n+1})$$

(see remark 3.2.2). According to (12), we have:

$$\beta \times \gamma = \sum_{\substack{\sigma(m, n-i+1) \\ \text{-shuffles}}} x_i \times (x_{\sigma^{-1}(i+1)} \times (\cdots \times (x_{\sigma^{-1}(m+n)} \times x_{\sigma^{-1}(m+n+1)}) \cdots).$$

Let's define, for $\sigma \in \Sigma_{m+n-i+1}$, the permutation α_i^σ by

$$\alpha_i^\sigma(j) = \quad j \qquad \text{if } j \le i$$
$$\alpha_i^\sigma(j) = \quad \sigma^{-1}(j) \quad \text{if } j > i.$$

With this notation, we have $1_{n+1} \circ_i 1_{m+1} = \sum_{\sigma(m,n-i+1) \text{-shuffles}} \alpha_i^\sigma$. Therefore, $\pi(1_{n+1} \circ_i 1_{m+1} \otimes f \circ_i g)$ must be identified with the map F_i, such that:

$$F_i(x_1 \otimes \cdots \otimes x_{m+n+1}) =$$

$$\sum_{\substack{\sigma(m,n-i+1) \\ \text{-shuffles}}} \epsilon(\alpha_i^\sigma) f \circ_i g(x_{(\alpha_i^\sigma)^{-1}(1)} \otimes \cdots \otimes x_{(\alpha_i^\sigma)^{-1}(m+n+1)}),$$

i.e. since $\epsilon(\alpha_i^\sigma) = \epsilon(\sigma)$,

$$F_i = \sum_{\sigma(m,n-i+1)\text{-shuffles}} \epsilon(\sigma) f \circ_i^\sigma g.$$

Hence, $i_f(g) = \sum_{i=1}^{n+1} (-1)^{m(i-1)} F_i$; this is exactly the same formula we found in 3.2.3, as expected.

4. Application to formal deformations
of an algebra over a quadratic operad

In all this section, we denote by $k[[t]]$ the ring of formal power series with coefficients in k. For an operad $\mathcal{P}$ we will denote by $\mathcal{P}(t)$ the extension of $\mathcal{P}$ in the category of modules over $k[[t]]$, i.e. for any $n \in \mathbb{N}^*$, $\mathcal{P}(t)(n) = \mathcal{P}(n) \otimes_k k[[t]]$. Notice that if $\mathcal{P}$ is a quadratic operad, then $\mathcal{P}(t)$ is again a quadratic operad, but in the category of modules over $k[[t]]$.

If (V, a) is $\mathcal{P}$-algebra, we'll denote by $V[[t]]$ the extension of V, i.e. $V[[t]] = V \otimes_k k[[t]]$; we see immediately that $V[[t]]$ has a natural structure of $\mathcal{P}(t)$-algebra: we formally extend the map $a : \mathcal{P} \to End(V)$ into $\bar{a} : \mathcal{P}(t) \to End(V)(t)$, since

$$Hom_{k[[t]]}(V[[t]]^{\otimes_{k[[t]]} n}, V[[t]]) \simeq End(V)(n) \otimes_k k[[t]].$$

DEFINITION 4.1. A formal deformation of a $\mathcal{P}$-algebra (V, a) consists of a morphism of operads (in the category of modules over $k[[t]]$), $F : \mathcal{P}(t) \to End(V)(t)$ such that $F|_{\mathcal{P}} = a$, which means that $F(\mu) = a(\mu)$, for any $n \in \mathbb{N}^*$ and any $\mu \in \mathcal{P}(n)$.

Another formulation is the following:

DEFINITION 4.1'. A deformation of (V, a) consists of a formal power series with coefficients in the set of morphisms of operads from $\mathcal{P}$ to $End(V)$, $F = \sum_{i=0}^{+\infty} f_i t^i$ such that $f_0 = a$ and such that F extending formally to $\mathcal{P}(t)$ makes $V[[t]]$ into a $\mathcal{P}(t)$-algebra.

4.2. The quadratic case. In this section $\mathcal{P} = \mathcal{P}(k, E, R)$ will denote a quadratic operad, (V, a) will be an algebra over $\mathcal{P}$ with $a : \mathcal{P}(2) \to End(V)(2)$ (see proposition 1.5.1).

According to proposition 1.5.5 extended to the category of $k[[t]]$-modules, F is entirely determined by

$$F(2) = \sum_{i=0}^{+\infty} f_i(2) t^i,$$

with $f_i(2) \in Hom_{\Sigma_2}(\mathcal{P}(2), End(V)(2))$ such that $\hat{F}(3)(r) = 0$ for any $r \in R$.

Let's determine $\hat{F}(3)$: for this, since $\mathcal{F}(E)(3)$ is generated as a $k[\Sigma_3]$-module by the $\mu \circ_i \nu$'s ($i \in \{1, 2\}$), it suffices to evaluate $\hat{F}(3)$ onto these elements. We have:

$$\hat{F}(3)(\mu \circ_i \nu) = F(2)(\mu) \circ_i F(2)(\nu) =$$

$$(\sum_{k=0}^{+\infty} f_k(2)(\mu)\, t^k) \circ_i (\sum_{l=0}^{+\infty} f_k(2)(\mu)\, t^l) = \sum_{n=0}^{+\infty}(\sum_{k+l=n} f_k(2)(\mu) \circ_i f_l(2)(\nu))t^n$$

(we recall that $f_k(2)(\mu) \circ_1 f_l(2)(\nu)(x \otimes y \otimes z) = f_k(2)(\mu)(f_l(2)(\nu)(x \otimes y) \otimes z)$ etc.).

Let's denote by $f_{kl}(3) : \mathcal{F}(E)(3) \to End(V)(3)$ the unique map of $k[\Sigma_3]$-modules such that

$$f_{kl}(3)(\mu \circ_i \nu) = f_k(2)(\mu) \circ_i f_l(2)(\nu),$$

then $\hat{F}(3) = \sum_{n=0}^{+\infty}(\sum_{k+l=n} f_{kl}(3))t^n$, and we see that condition $\hat{F}(3)(r) = 0$ is equivalent to the conditions (c_n):

$$(c_n) : \text{for any } r \in R, \sum_{k+l=n} f_{kl}(3)(r) = 0.$$

If $n = 0$ then (c_0) is equivalent to $\hat{a}(3)(r) = 0$ for any r, which is true. If $n = 1$ then (c_1) is equivalent to the condition

$$(f_{10}(3) + f_{01}(3))(r) = 0$$

for any $r \in R$. In fact, this last condition is a cocycle condition, since we have the following lemma:

LEMMA 4.3. *Let* $\delta_1 : Hom_k(\mathcal{P}(2) \otimes_{\Sigma_2} V^{\otimes 2}, V) \to Hom_k(R \otimes_{\Sigma_3} V^{\otimes 3}, V)$ *be the differential defined in remark 2.4.3, then* $\delta_1(f) = d(f)|_{R \otimes_{\Sigma_3} V^{\otimes 3}}$ *where*

$$d(f) : \mathcal{F}(E)(3) \otimes_{\Sigma_3} V^{\otimes 3} \to V$$

is the unique map such that:

$$d(f)(\mu(\nu, 1) \otimes x \otimes y \otimes z) =$$

$$a(\mu)(f(\nu \otimes x \otimes y) \otimes z) + f(\mu \otimes \nu(x, y) \otimes z)$$

and

$$d(f)(\mu(1, \nu) \otimes x \otimes y \otimes z) =$$
$$a(\mu)(x \otimes f(\nu \otimes y \otimes z)) + f(\mu \otimes x \otimes \nu(y, z)).$$

PROOF. The lemma simply results from the fact that for $f, g \in L^1(V)$ we have

$$i_g(f) = h|_{R \otimes_{\Sigma_3} V^{\otimes 3}}$$

where

$$h(\mu(\nu, 1) \otimes x \otimes y \otimes z) = g(\mu \otimes f(\nu \otimes x \otimes y) \otimes z)$$

and

$$h(\mu(1, \nu) \otimes x \otimes y \otimes z) = g(\mu \otimes x \otimes f(\nu \otimes y \otimes z)).$$

$\square$

Now, if we identify $Hom_k(\mathcal{P}(2) \otimes_{\Sigma_2} V^{\otimes 2}, V)$ and $Hom_k(R \otimes_{\Sigma_3} V^{\otimes 3}, V)$ with $Hom_{\Sigma_2}(\mathcal{P}(2), End(V)(2))$ and $Hom_{\Sigma_3}(R, End(V)(3))$, we see that

$$\delta_1(f)(\mu \circ_i \nu) = a(\mu) \circ_i f(\nu) + f(\mu) \circ_i a(\nu).$$

This enables us to rewrite the condition (c_1): since

$$f_{10}(3)(\mu \circ_i \nu) = f_1(2)(\mu) \circ_i f_0(2)(\nu) = f_1(2)(\mu) \circ_i a(\nu)$$

and in the same way

$$f_{01}(3)(\mu \circ_i \nu) = a(\nu) \circ_i f_1(2)(\nu),$$

we see that (c_1) means that $\delta_1(f_1(2)) = 0$, i.e. $f_1(2)$ is a 1-cocycle for the cohomology $H^*_{\mathcal{P}}(V)$ as expected.

The other conditions can be rewritten in the following way: the condition (c_n) is equivalent to the condition (c'_n):

(c'_n) : for any $r \in R$,

$$\delta_1(f_n(2))(r) = - \sum_{\substack{i+j=n \\ i,j \neq n}} f_{ij}(3)(r).$$

If we want to establish the analog of the classical theorem of [G2] (i.e. $H^3_{\mathcal{P}}(V) = 0$ implies that any 1-cocycle $f_1(2)$ can be extended to a formal deformation) we have to calculate $\delta_2(-\sum_{i+j=n} f_{ij}(3))$, and we have to prove that the result is zero (providing that (c'_k) holds for $k < n$). But this is not straightforward, because the calculus of δ_2 is more complicated than those of δ_1 (essentially because $(\mathcal{P}^!)^{\vee}(4)$ is not easy to describe). But this problem can be solved thanks to the square zero element theory. We recall that to a is associated an element $\varphi \in \mathcal{L}^1_{\mathcal{P}}(V)$ such that $[\varphi, \varphi] = 0$ (corollary 2.4.2). But to each $f_i(2)$ is also associated an element $\varphi_i \in \mathcal{L}^1_{\mathcal{P}}(V)$. Since $[\varphi_k, \varphi_l] = i_{\varphi_k}(\varphi_l) + i_{\varphi_l}(\varphi_k)$ we find that

$$\theta_2^{-1}([\varphi_k, \varphi_l])(\mu \circ_i \nu) = f_k(2)(\mu) \circ_i f_l(2)(\nu) + f_l(2)(\mu) \circ_i f_k(2)(\nu)$$

(see proof of theorem 2.4.1 for the notation θ_2) so that (c_n) is equivalent to the condition

$$(13) \qquad\qquad \delta_\varphi(\varphi_n) = -\frac{1}{2} \sum_{\substack{i+j=n \\ i,j \neq n}} [\varphi_i, \varphi_j].$$

Therefore, it is easy to calculate $\delta_2(-\sum_{i+j=n} f_{ij}(3))$; it is the same as to calculate $[\sum_{\substack{i+j=n \\ i,j \neq n}} [\varphi_i, \varphi_j], \varphi]$ (with (13) holding for any $k < n$). But this has already be done in the introduction of this paper, and the result is zero as expected. So we have proved the following theorem:

THEOREM 4.4. *Let $\mathcal{P}$ be a quadratic operad, and (V, a) an algebra over this operad. If $H^2_{\mathcal{P}}(V) = 0$ than any 1-cocycle $f(2)$ is the infinitesimal part of a deformation of a, which means that there exists a deformation of a, $F = \sum_{i=0}^{+\infty} f_i(2)\, t^i$, such that $f_1(2) = f(2)$.*

Now, let us see the problem of algebraic rigidity.

4.5. Algebraic rigidity.

DEFINITION 4.5.1. A formal deformation $F = \sum_{i=0}^{+\infty} f_i\, t^i$ is said to be trivial if there exists a morphism of $k[[t]]$-modules $a : V[[t]] \to V[[t]]$ invertible, such that, for any $n \in \mathbb{N}^*$ and any $v_1, \ldots, v_n \in V[[t]]$, we have:

$$(14) \qquad a_t \circ F(\mu)(v_1 \otimes \cdots \otimes v_n) = \mu(a_t(v_1), \ldots, a_t(v_n)).$$

Since $\mathcal{P}$ is quadratic and since $V[[t]] = V \otimes_k k[[t]]$, condition (14) can be replaced by the following one, holding for $\mu \in \mathcal{P}(2)$ and $v_1, v_2 \in V$:

$$(15) \qquad a_t \circ F(\mu)(v_1 \otimes v_2) = \mu(a_t(v_1), a_t(v_2)).$$

We can choose the map a_t of the form $\sum_{i=0}^{+\infty} a_i\, t^i$ with $a_i \in Hom_k(V, V)$; it is invertible if, and only if, a_0 is invertible. So we don't restrict ourself if we choose $a_0 = Id$. Condition (15) is equivalent to the system of equations (e_n):

$$(e_n) : \sum_{i+j=n} \mu(a_i(v_1), a_j(v_2)) = \sum_{i+j=n} a_i \circ f_j(2)(\mu)(v_1 \otimes v_2).$$

If $n = 0$, (e_0) is a tautology. If $n = 1$, (e_1) is equivalent to the condition,

$$f_1(2)(\mu)(v_1 \otimes v_2) = \mu(a_1(v_1), v_2) + \mu(v_1, a_1(v_2)) - a_1(\mu(v_1, v_2)).$$

It is in fact a cocycle condition, since we have the following lemma:

LEMMA 4.5.2. Let $\delta_0 : Hom_k(V, V) \to Hom_{\Sigma_2}(\mathcal{P}(2), End(V)(2))$ be the differential defined in remark 2.4.3, then

$$\delta_0(f)(\mu)(v_1 \otimes v_2) = f(\mu(v_1, v_2)) - \mu(f(v_1), v_2) - \mu(v_1, f(v_2)).$$

PROOF. We have $\delta_0(f) = i_f(\varphi) - i_\varphi(f)$. According to remark 3.2.5,

$$i_\varphi f(\mu)(v_1 \otimes v_2) = \mu(f(v_1), v_2) + \mu(v_1, f(v_2)),$$

and according to remark 3.2.3,

$$i_f(\varphi)(\mu)(v_1 \otimes v_2) = f(\mu(v_1, v_2)),$$

so the lemma follows. $\qquad \square$

Therefore, the lemma shows that $f_1(2)$ is a 1-coboundary for the cohomology $H_{\mathcal{P}}^*(V)$.

We have the following theorem:

THEOREM 4.5.3. Let $\mathcal{P}$ be a quadratic operad, and (V, a) an algebra over $\mathcal{P}$. If $H_{\mathcal{P}}^1(V) = 0$ then all deformations of (V, a) are trivial.

PROOF. The proof is exactly the same as the proof proposed by Gerstenhaber in [G2] for associative algebras. $\qquad \square$

4.6. Conclusion.
If now we replace the operad $\mathcal{P}$ by the operads **Ass, Com, Lie, Leib**, we refind the classical results of algebraic rigidity and extension of the 2-cocycle already established in respectively [G2], [G–S], [N–R], [BA1].

REFERENCES

[BA1]   D. Balavoine, *Déformations des algèbres de Leibniz*, C. R. Acad. Sci. Paris, t. 319, Série I, 1994, pp. 783–788.

[BA2]   ——, *Déformations et rigidité géométrique des algèbres de Leibniz*, Comm. in Alg. 24 (3), 1996, pp. 1017–1034.

[FO]   T.F. Fox, *An introduction to algebraic deformation theory*, J. Pure Appl. Algebra. 84, 1993, pp. 17–41.

[G1]   M. Gerstenhaber, *The cohomology structure of an associative ring*, Ann. of Math. 78, 1963, pp. 267–288.

[G2]   ——, *On the deformation of rings and algebras*, Ann. of Math. 79, 1964, pp. 59–103.

[G–S]   M. Gerstenhaber and S. D. Schack, *Algebraic cohomology and deformation theory*, Deformation theory of algebras and structures NATO-ASI Series C, vol. 297 (M. Gerstenhaber and M. Hazewinkel, eds.), Kluwer Academic Publishers Group, Dordrecht, 1988, pp. 11–264.

[G–V]   M. Gerstenhaber and A. A. Voronov, *Homotopy G-algebras and module space operad*, preprint Max-Planck Institut für Mathematik, Bonn MPI/94–71.

[G–K]   V. Ginzburg, M. M. Kapranov, *Koszul duality for operads*, Duke J. Math., vol. 76, n° 1, 1994, pp. 203–272.

[HA]   D. K. Harrison, *Commmutative algebras and cohomology*, Trans. AMS. 104, 1962, pp. 191–204.

[LO1]   J.-L. Loday, *Une version non commutative des algèbres de Lie : les algèbres de Leibniz*, L'Ens. Math. 39, 1993, pp. 269–293.

[LO2]   ——, *Cup-product for Leibniz cohomology and dual Leibniz algebras*, Math. Scand. (to appear).

[LO3]   ——, *La renaissance des opérades*, Séminaire Bourbaki exposé n° 792, novembre 1994, Astérisque (to appear).

[L–W]   P. Lecomte and M. De Wilde, *Formal deformations of the Poisson-Lie algebra of a symplectic manifold and star products. Existence, equivalence, derivations, Deformation theory of algebras and structures, NATO-ASI Series C, vol. 247* (1988), Kluwer Academic Publishers, 897–960.

[MAR1] M. Markl, *Distributive laws and the cohomology*, Annales de l'Institut Fourier 46 (4), 1996, pp. 307–323.

[MAR2] ——, *Models for operads*, Comm. in Alg. 24 (4), 1996, pp. 1471–1500.

[N–R]   A. Nijenhuis and R. Richardson, *Deformations of Lie algebra structures*, J. Math. Mech. 171, 1967, pp. 89–105.

[R]   C. Roger, *Algèbres de Lie graduées et quantification, Actes du Colloque Souriau (symplectic geometry and mathematical physic)*, Progress in Maths (Birkhäuser) n° 99, 1991, pp. 374–421.

[W]   D. Wigner, *An identity in the free Lie algebra*, Proc. Amer. Math. Soc. 106, 1989, pp. 639–640.

LABORATOIRE A.G.A.T.A. UNIVERSITÉ DE MONTPELLIER II. DÉPARTEMENT DES SCIENCES MATHÉMATIQUES, CASE 051. PLACE EUGÈNE BATAILLON. 34095 MONTPELLIER CEDEX 5
E-mail address: `balavoin@math.univ-montp2.fr`

Contemporary Mathematics
Volume **202**, 1997

Q-rings and the homology of the symmetric groups

Terrence P. Bisson & André Joyal

ABSTRACT.

The goal of this paper is to study the rich algebraic structure supported by the homology mod 2 of the symmetric groups. We propose to organise the algebra of homology operations around a single concept, that of Q-ring. We are guided by an analogy with the representation theory of the symmetric groups and the concept of λ-ring. We show that $H_* \Sigma_*$ is the free Q-ring on one generator. It is a Hopf algebra generated by its subgroup $\mathcal{K}$ of primitive elements. This subgroup is an algebra (for the composition of operations) that we call the Kudo-Araki algebra. It is closely related to the Dyer-Lashof algebra but is better behaved: the dual coalgebra is directly representing the substitution of Ore polynomials. Many results on the homology of E_∞-spaces can be expressed in the language of Q-rings. We formulate the Nishida relations by using a Q-ring structure on a semidirect extension $\mathcal{A}$ of Milnor's dual of the Steenrod algebra. We show that the Nishida relations lead to a commutation operator between $\mathcal{K}$ and $\mathcal{A}$.

Contents

1991 *Mathematics Subject Classification.* Primary 55S05; Secondary 55S12, 20B30, 20J06.
Key words and phrases. symmetric groups, homology, Steenrod algebra.
Support for this work was provided by the Canadian NSERC and by the Quebec FCAR

0. INTRODUCTION

This paper is partly expository and wholly algebraic. Our goal is to study the rich algebraic structure supported by the homology mod 2 of the symmetric groups. We propose to organise the algebra of homology operations around a single concept, that of Q-ring. We are guided by an analogy with the representation theory of the symmetric groups and the concept of λ-ring. Recall that the direct sum $R(\Sigma_*) = \oplus_n R(\Sigma_n)$ of the character groups of the symmetric groups is the free λ-ring on one generator. It is also a self-dual Hopf algebra isomorphic to the ring of symmetric functions. For the direct sum $H_*\Sigma_* = \oplus_n H_*\Sigma_n$ of the mod 2 homology of the symmetric groups there is a similar pattern. We show that $H_*\Sigma_*$ is the free Q-ring on one generator. It is a Hopf algebra generated by its subgroup $\mathcal{K}$ of primitive elements. This subgroup is an algebra (for the composition of operations) that we call the Kudo-Araki algebra. It is closely related to the Dyer-Lashof algebra but is better behaved: the dual coalgebra is directly representing substitution of Ore polynomials.

Many results on the homology of E_∞-spaces can be expressed in the language of Q-rings. For example, if $E_\infty(X)$ is the free E_∞-space generated by a space X then $H_*E_\infty(X)$ is the free Q-ring generated by H_*X.

We formulate the Nishida relations by using a Q-ring structure on a semidirect extension $\mathcal{A}$ of Milnor's dual of the Steenrod algebra. We introduce an abstract concept of commutation operator between an algebra and a coalgebra and show that the Nishida relations lead to a commutation operator between the algebra $\mathcal{K}$ and the coalgebra (opposite to) $\mathcal{A}$.

The present work was inspired by our work on geometric Dyer-Lashof operations in unoriented cobordism, as sketched in Bisson, Joyal [1995a,b]; this geometric underpinning will be presented in a sequel. In the present paper we do not discuss distributivity relations between different Q-ring structures, such as occur in the homology of E_∞-ring spaces. We leave this for a sequel, together with an extension of our theory to mod p homology. We view our contribution to many of the results presented here as one of conceptualisation and simplification. We hope that our approach will shed new light on the work done in the field during the last forty years.

We now present a more detailed description of the paper. Except for this introduction all discussions relating our work to homology are reserved to Addenda placed at the end of each section. Throughout, Z_2 denotes the integers mod 2.

A Q-ring is a commutative ring R with a ring homomorphism $Q_t : R \to R[[t]]$ called the total square, satisfying the following conditions:
   (i) $Q_0(a) = a^2$ for every a in R
   (ii) $Q_t \circ Q_s$ is symmetric in t and s
       where Q_t is extended to $R[[s]]$ by $Q_t(s) = s(s + t)$.

If we put $Q_t(x) = \sum_{n \geq 0} q_n(x)t^n$ then we obtain a sequence of operations $q_n : R \to R$ called the individual squares. The q_n were introduced by Kudo and Araki; they generate an algebra under composition that we call the Kudo-Araki algebra and denote by $\mathcal{K}$.

A Q-ring is *graded* when R is Z-graded and grade$(q_n(a)) = 2 \cdot$ grade$(a) + n$ for homogenous a. The mod 2 cohomology $H^*(X)$ of any space has a graded Q-ring structure obtained from the Steenrod operations $Sq^i : H^*(X) \to H^{*+i}(X)$. If $x \in H^r(X)$ we have $q_i(x) = Sq^{r-i}(x)$ for $0 \le i \le r$, and $q_i(x) = 0$ for $r < i$. For this structure the group $H^n(X)$ is homogenous of grade $-n$. The mod 2 homology H_*X of an E_∞-space X has a graded Q-ring structure classically expressed in terms of Dyer-Lashof operations $Q^i : H_*(X) \to H_{*+i}(X)$. If $x \in H_r(X)$ then $q_n(x) = Q^{r+n}(x)$. The Dyer-Lashof operations Q^i are obtained by reorganising the graded components of the q_i's so as to form homogenous operations. They generate an algebra called the Dyer-Lashof algebra that is *not* isomorphic to the Kudo-Araki algebra. Here we stress the importance and naturality of $\mathcal{K}$. The concept of Q-ring provides an alternative way of handling Dyer-Lashof operations. For example, let BO_* denote the disjoint union, for $n \ge 0$, of the spaces BO_n which classify n-dimensional real vector bundles. Then BO_* has an E_∞-structure obtained from the Whitney sum of vector bundles, and the ring H_*BO_* is isomorphic to $Z_2[\beta_0, \beta_1, \dots]$ where $\beta_n \in H_*BO_1 = H_*RP^\infty$ is the nonzero element of dimension n. The Q-structure on H_*BO_* is characterized by the identity

$$Q_t(\beta)(x(x+t)) = \beta(x)\beta(x+t)$$

where $\beta(x) = \sum \beta_i x^i$.

The classifying space BC of any symmetric monoidal category is an E_∞-space. Its homology $H_*\mathcal{C} = H_*BC$ is a Q-ring. For example, the category of finite sets and bijections has a monoidal structure given by disjoint union. It is equivalent to the disjoint union Σ_* of the symmetric groups Σ_n. Hence $H_*\Sigma_*$, the direct sum of the homology of the symmetric groups, is a Q-ring. It is $Q\langle x \rangle$, the free Q-ring on one generator.

The mod 2 homology H_*S of a spectrum S is a Q-ring. For example, the homology of the sphere spectrum $S^0 = \Omega^\infty S^\infty$ is $\{x\}^{-1}Q\langle x \rangle = Q\langle x, x^{-1} \rangle$, the free Q-ring on an invertible generator x. The homology of the representing spectrum KO of real K-theory is $Z_2[\beta_0, \beta_0^{-1}, \beta_1, \dots]$ with the Q-structure given as above. When X is a finite cell complex we have $H^n(X) = H_{-n}(X')$ where X' is the Spanier-Whitehead dual of X in the category of spectra. The Q-ring structure on the cohomology of X is isomorphic to the Q-ring structure on the homology of X' (as learned in a conversation with H. Miller).

Let $Q\langle V \rangle$ denote the free Q-ring generated by a Z_2-vector space V. For any space X the canonical map

$$Q\langle H_*X \rangle \to H_*E_\infty(X)$$

is an isomorphism, where $E_\infty(X)$ is the E_∞-space freely generated by X. If V is a coalgebra then so is $Q\langle V \rangle$; the coproduct $Q\langle V \rangle \to Q\langle V \rangle \otimes Q\langle V \rangle$ is the map of Q-rings which extends the coproduct $\delta : V \to V \otimes V$ (here we use that the tensor product $R \otimes S$ of two Q-rings is a Q-ring). For instance, the homology of any space X has a coalgebra structure obtained from the diagonal $X \to X \times X$. When $V = H_*X$ the diagonal structure on $H_*E_\infty(X) = Q\langle H_*X \rangle$ is obtained from the diagonal structure on H_*X. In particular, the diagonal structure on $H_*\Sigma_* = Q\langle x \rangle$ is given by the unique map of Q-rings $Q\langle x \rangle \to Q\langle x \rangle \otimes Q\langle x \rangle$ such that $\delta(x) = x \otimes x$.

We introduce a linear version of the concept of Q-ring called Q-*module*. It is a Z_2-vector space M equipped with an additive map $Q_t : M \to M[[t]]$ such that $Q_t \circ Q_s$ is symmetric in t and s. A Q-module is the same thing as a left module over the Kudo-Araki algebra $\mathcal{K}$. The tensor product $M \otimes N$ of two Q-modules is a Q-module with operations given by the Cartan formula $q_n(x \otimes y) = \sum q_i(x) \otimes q_{n-i}(y)$. So $\mathcal{K}$ has a coalgebra structure defined by the map of algebras $\delta : \mathcal{K} \to \mathcal{K} \otimes \mathcal{K}$ such that $\delta(q_n) = \sum q_i \otimes q_{n-i}$. Hence (like the Steenrod algebra) $\mathcal{K}$ is a cocommutative bialgebra. The dual coalgebra $\mathcal{K}' = \mathcal{W}$ splits as a direct sum

$$\mathcal{W} = \bigoplus_n \mathcal{W}(n)$$

of polynomial algebras. Here $\mathcal{W}(n)$ is the algebra $Z_2[w_0, \ldots, w_{n-1}]$ of Dickson invariants; it is generated by the coefficients of the generic Ore polynomial

$$W_n(x) = x^{2^n} + \sum_{i=0}^{n-1} w_i \, x^{2^i}.$$

There is a coalgebra structure Δ on $\mathcal{W}$ representing the operation of substitution of Ore polynomials. By definition, the map $\Delta : \mathcal{W}(m+n) \to \mathcal{W}(m) \otimes \mathcal{W}(n)$ is the ring homomorphism such that

$$\Delta(W_{m+n}) = (W_m \otimes 1) \circ (1 \otimes W_n).$$

We prove that the Kudo-Araki algebra is anti-isomorphic to the dual algebra $\mathcal{W}'$.

To construct the free Q-ring on one generator we use the *enveloping algebra* $\tilde{\Lambda} M = \tilde{\Lambda}(M, q_0)$ of an endomorphism $q_0 : M \to M$ of a Z_2-vector space. By definition, $\tilde{\Lambda}(M, q_0)$ is the quotient of the symmetric algebra $S(M)$ by the ideal generated by the relation $x^2 = q_0(x)$ for $x \in M$. When $q_0 = 0$ the enveloping algebra is the exterior algebra ΛM. If M is a Q-module then $\tilde{\Lambda} M$ is a Q-ring. The free Q-ring on one generator $Q\langle x \rangle$ is $\tilde{\Lambda}\mathcal{K}$. We have $\mathcal{K} = Z_2[q_0] \otimes \mathcal{K}^\flat$ for a subgroup $\mathcal{K}^\flat \subset \mathcal{K}$ and it follows that $Q\langle x \rangle$ is the symmetric algebra on $\mathcal{K}^\flat$. Any enveloping algebra $\tilde{\Lambda}(M, q_0)$ has a Hopf algebra structure with comultiplication given by $\sigma(x) = x \otimes 1 + 1 \otimes x$ for $x \in M$. For this structure the subgroup of primitive elements is M. In particular, $Q\langle x \rangle$ is a Hopf algebra with comultiplication the map of Q-rings $\sigma : Q\langle x \rangle \to Q\langle x \rangle \otimes Q\langle x \rangle$ such that $\sigma(x) = x \otimes 1 + 1 \otimes x$. For this structure the subgroup of primitive elements is $\mathcal{K}$.

To formulate the Nishida relations we use right comodules over the *extended Milnor coalgebra* $\mathcal{A} = Z_2[a_0^{-1}, a_0, a_1, \ldots]$. The comultiplication $\Delta : \mathcal{A} \to \mathcal{A} \otimes \mathcal{A}$ is given by

$$\Delta(a) = (a \otimes 1) \circ (1 \otimes a)$$

where $\circ$ represents composition of formal power series and where $a(x) = \sum a_i x^{2^i}$. We refer to a right $\mathcal{A}$-comodule M as a *Milnor coaction* and will denote the coaction map by $\psi : M \to M \otimes \mathcal{A}$. We also introduce the *positive bialgebra* $\mathcal{A}^+$ and show that $\mathcal{K}$ is anti-isomorphic to a subalgebra of the convolution algebra $[\mathcal{A}^+, Z_2]$. The homology of any space has an $\mathcal{A}^+$-coaction which determines the action of the Steenrod operations. There is a unique Q-structure on the algebra $\mathcal{A}$ such that

$$Q_t(a)(x(x+t)) = a(x)a(x+t).$$

If M is a Q-module then "change of parameters" gives a certain natural Q-structure Q_t' on $M \otimes \mathcal{A}$; if $x \in M$ and $r \in \mathcal{A}$ then we have $Q_t'(x \otimes r) = Q_{a(t)}(x)Q_t(r)$. We say that the *Nishida relations* hold for a Q-module M if the coaction $\psi : M \to M \otimes \mathcal{A}$ is a map of Q-modules where $M \otimes \mathcal{A}$ is equipped with Q_t'. If E is an E_∞-space then $H_* E$ is a Q-ring with a Milnor ring coaction, and the Nishida relations hold for $H_* E$. If M is an $\mathcal{A}$-comodule then there is a unique Milnor ring coaction on $Q\langle M \rangle$ extending the given coaction on M and satisfying the Nishida relations. It follows that for any space X the Milnor coaction on $H_* E_\infty(X)$ is obtained from the coaction on $H_*(X)$. In particular, this determines the Milnor coaction on $H_* \Sigma_* = Q\langle x \rangle$.

To study the Nishida relations we use an abstract theory of commutation operators between an algebra and a coalgebra, as sketched in appendix C. The *Nishida operator* is a commutation operator

$$\tilde{\rho} : \mathcal{K} \otimes \mathcal{A} \to \mathcal{A} \otimes \mathcal{K}$$

between $\mathcal{K}$ and the coalgebra $(\mathcal{A}, \Delta^o, \epsilon)$ opposite to $\mathcal{A}$. We use the operator in the form of a commutation operator ρ between a monad and a comonad. An action by $\mathcal{K}$ *ρ-commutes* with a coaction by $\mathcal{A}$ iff the Nishida relations are satisfied.

We would like to thank André Lebel for reading the manuscript and Paul Libbrecht for the typography.

Addendum to Section 0.
The homology of the symmetric groups has a rich history of which we can only sketch the chronology. It begins with Steenrod's description of cohomology operations obtained from elements in $H_*(\Sigma_*)$ (expressed most simply in Steenrod [1953] and [1957]); see also Steenrod, Epstein [1962] or May [1970], for instance. The homology and cohomology of symmetric products of spaces attracted considerable interest during the next few years; see Nakaoka [1957] and Dold [1958], for instance. The full computation of the homology of the symmetric groups was achieved in Nakaoka [1960]; he showed that the homology of the infinite symmetric group has the structure of a commutative Hopf algebra and proved that it is a polynomial algebra.

Kudo and Araki [1956] used a variant of Steenrod's methods to define operations on the homology of H_n-spaces. Applications of these operations were developed in Browder [1960], Dyer, Lashof [1962], Milgram [1965], Barratt, Kahn, Priddy [1971], May [1971], and Priddy [1972], for instance, especially in connection with the homology of iterated loop spaces.

Connections between homology operations and Dickson invariants appear in Madsen [1975] and Mui [1975]. Some good sources for background on E_∞-spaces and some indications of the history and applications of Dyer-Lashof operations are May [1977b], Adams [1978], and Madsen, Milgram [1979]. The construction of an E_∞-space from a symmetric monoidal category appears in Segal [1974].

Some good sources for background on λ-rings, the representation theory of the symmetric group, and symmetric functions are Grothendieck [1958], Atiyah [1966], Knutson [1973], Hoffman [1979], and Macdonald [1979]. For background on Hopf algebras and their comodules see Sweedler [1969].

1. Q-rings and modules.

We formally introduce the concept of Q-ring and Q-module and give a few basic algebraic examples.

DEFINITION 1. A Q-*ring* is a commutative ring R together with a ring homomorphism $Q_t : R \to R[[t]]$ called the *total square*, satisfying the following conditions for every a in R:
  i) $Q_0(a) = a^2$;
  ii) $Q_t \circ Q_s(a)$ is symmetric in s and t in $R[[s,t]]$, where Q_t is extended to $R[[s]]$
  by $Q_t(s) = s(s+t)$.

A Q-ring is always a Z_2-algebra since i) implies that the map $a \mapsto a^2$ is a ring homomorphism. If we put $Q_t(a) = \sum_{n \geq 0} q_n(a)t^n$ we obtain a sequence of operations $q_n : R \to R$ $(n \geq 0)$ called the *individual squares*. A *map of Q-rings* is a ring homomorphism preserving the operations q_n. The fact that Q_t is a ring homomorphism means that the Cartan formula holds; for all $a, b \in R$

$$q_n(ab) = \sum_{i+j=n} q_i(a)q_j(b).$$

We shall use upper and lower indexing $M_n = M^{-n}$ for Z-graded vector spaces and say that $x \in M_n$ is of dimension n and that $x \in M^n$ is of codimension n. A Q-ring is *graded* when R is graded and $\dim(q_n(a)) = 2 \cdot \dim(a) + n$. Notice that if t is given codimension 1 then the total square $Q_t(a)$ is a formal power series homogenous of dimension $2 \cdot \dim(a)$ for homogenous a.

For any Z_2-vector space M we shall denote by $M[[t]]$ the set of formal power series in t with coefficients in M. It is a module over the power series ring $Z_2[[t]]$. There is a substitution $u(f(t))$ for formal power series $u(t) \in M[[t]]$ and $f(t) \in tZ_2[[t]]$. These ideas apply also to the set $M[[s,t]]$ of power series in s and t.

DEFINITION 2. A Q-*module* is a Z_2-vector space M together with an additive map $Q_t : M \to M[[t]]$ called the *total square*, such that $Q_t \circ Q_s$ is symmetric in t and s, where Q_t is extended to a map $Q_t : M[[s]] \to M[[s,t]]$ by putting $Q_t(s) = s(s+t)$ and $Q_t(\sum_i a_i s^i) = \sum_i Q_t(a_i)Q_t(s)^i$. A Q-module is *graded* when M is Z-graded and $\dim(q_n(a)) = 2 \cdot \dim(a) + n$ for homogenous a.

The symmetry condition on $Q_t \circ Q_s$ is equivalent to a set of relations (called the Adem relations) on the q_n's. Let $\mathcal{K}$ denote the associative algebra generated by the q_n's subject to the Adem relations; we will call it the *Kudo-Araki algebra*. A Q-module is the same thing as a left $\mathcal{K}$-module; the free Q-module generated by a Z_2-vector space V is equal to $\mathcal{K} \otimes V$. Here is an explicit description of the Adem relations.

PROPOSITION 1. *A sequence of additive operations $q_n : M \to M$ defines a Q-module structure on M iff the q_n satisfy the following Adem relation for all $a \in M$:*

$$q_m(q_n(a)) = \sum_i \binom{i-n-1}{2i-m-n} q_{m+2n-2i}(q_i(a)).$$

PROOF. By definition we have

$$Q_t(Q_s(a)) = Q_t(\sum q_n(a)s^n) = \sum Q_t(q_n(a))Q_t(s)^n = \sum q_m(q_n(a))t^m[s(s+t)]^n.$$

The symmetry condition can be expressed as

$$\sum Q_t(q_n(a))[s(s+t)]^n = \sum q_j(q_i(a))s^j[t(t+s)]^i.$$

Thus we can compute $Q_t(q_n(a))$ by expanding the right hand side as a power series in $u = s(s+t)$. Observe that the change of parameter $u = st + s^2$ from s to u has a composition inverse if t is formally inverted (in the ring of Laurent series in t). We shall use residues to compute the coefficient of u^n in this expansion. Recall that if $f(u)$ is a Laurent series in u then the residue of $f(u)du$ is the coefficient of u^{-1} in $f(u)$; the coefficient of u^n is the residue of $f(u)u^{-n-1}du$. In our case $du = tds$ since we are working mod 2. Hence the coefficient of u^n in the right side of the equality is

$$\sum_{j,i\geq 0} q_j(q_i(a))\mathrm{Res}\left(\frac{s^j t^i (t+s)^i}{s^{n+1}(s+t)^{n+1}}tds\right) = \sum_{j,i\geq 0} q_j(q_i(a))t^{2i-2n+j}\binom{i-n-1}{n-j}.$$

Thus the symmetry condition can be expressed as

$$Q_t(q_n(a)) = \sum_{j,i\geq 0} q_j(q_i(a))t^{2i-2n+j}\binom{i-n-1}{n-j}.$$

The Adem relations are obtained by equating the coefficients of the two sides. QED

The next few propositions describe algebraic constructions of Q-modules and rings. We use the obvious product $M[[t]] \otimes N[[t]] \to (M \otimes N)[[t]]$; if the context is clear the product of $f(t) \in M[[t]]$ with $g(t) \in M[[t]]$ will be denoted by $f(t) \cdot g(t)$.

PROPOSITION 2. *If M and N are (graded) Q-modules (resp. Q-rings) then so is their tensor product $M \otimes N$ with $q_n(a \otimes b) = \sum_{i+j=n} q_i(a) \otimes q_j(b)$.*

PROOF. By definition we have $Q_t(a \otimes b) = Q_t(a) \cdot Q_t(b)$ with the notation introduced above. It is easy to see that $Q_t \circ Q_s(ab) = Q_t(Q_s(a)) \cdot Q_t(Q_s(b))$ and it follows that $Q_t \circ Q_s$ is symmetric in s, t. If M and N are graded and a and b are homogenous then

$$\dim(q_i(a) \otimes q_j(b)) = 2\dim(a) + i + 2\dim(b) + j = 2\dim(a \otimes b) + n.$$

Hence $M \otimes N$ is a graded Q-module. It is obvious that $Q_t : M \otimes N \to M \otimes N[[t]]$ is a ring homomorphism if M and N are Q-rings. Moreover, in this case $Q_0(a \otimes b) = Q_0(a) \otimes Q_0(b) = a^2 \otimes b^2 = (a \otimes b)^2$. QED

If R is a Q-ring then the product map $R \otimes R \to R$ is a Q-module map. If an R-module M is also a Q-module and the structure map $R \otimes M \to M$ is a Q-module map then we shall say that M is a QR-*module*. It is easy to see that $R \otimes \mathcal{K}$ is the free QR-module on one generator, hence it is an algebra. A QR-module is the same thing as a left module over $R \otimes \mathcal{K}$ (see Example 3 of appendix C).

PROPOSITION 3. *If R is a Q-ring then so are the polynomial ring $R[x_1,\dots,x_n]$ and the formal power series ring $R[[x_1,\dots,x_n]]$ if we put $Q_t(x_i) = x_i(x_i + t)$ for every $1 \leq i \leq n$. Moreover, these Q-rings are graded if R is graded and $\mathrm{codim}(x_i) = 1$. The same result is true for Q-modules instead of Q-rings.*

PROOF. Let us verify that $Z_2[x]$ is a Q-ring if we put $Q_t(x) = x(x+t)$. Obviously, $Q_0(x) = x^2$. For the symmetry we have

$$Q_t \circ Q_s(x) = Q_t(x(x+s)) = Q_t(x)(Q_t(x) + Q_t(s))$$
$$= x(x+t)(x(x+t) + s(s+t)) = x(x+t)(x+s)(x+s+t).$$

This proves that $Z_2[x]$ is a Q-ring. If $\operatorname{codim}(x) = 1$ then $x(x+t)$ is homogenous of codimension 2 when $\operatorname{codim}(t) = 1$. It follows that $Z_2[x]$ is a graded Q-ring. According to proposition 2, $R[x] = R \otimes Z_2[x]$ is a Q-ring. By continuity of Q_t it follows that $R[[x]]$ is also a Q-ring. It follows by induction on n that $R[x_1, \dots, x_n]$ and $R[[x_1, \dots, x_n]]$ are Q-rings. If M is a Q-module then according to proposition 2, $M[x_1, \dots, x_n] = M \otimes Z_2[x_1, \dots, x_n]$ is a Q-module. It follows by continuity of Q_t that $M[[x_1, \dots, x_n]]$ is a Q-module. The statement about gradings is easily checked. QED

Let us denote by $Z_2[\beta_*]$ the polynomial ring $Z_2[\beta_0, \beta_1, \dots]$.

PROPOSITION 4. *There is a unique Q-structure on the polynomial ring $Z_2[\beta_*]$ such that $Q_t(\beta)(x(x+t)) = \beta(x)\beta(x+t)$ where $\beta(x) = \sum_i \beta_i x^i$. This is a graded Q-ring if we take $\dim(\beta_i) = i$. We have the explicit formula*

$$q_r(\beta_n) = \sum_i \binom{r+i-1}{i} \beta_{n-i}\beta_{r+n+i}.$$

PROOF. For simplicity let us put $Z_2[\beta_*] = R$. The series $\beta(x)\beta(y)$ is symmetric in x and y. It follows that it can be expanded as a series in the elementary symmetric functions $t = x+y$ and $v = xy = x(x+t)$. This proves the existence and uniqueness of a power series $h(t,v)$ in $R[[t,v]]$ such that $h(t, x(x+t)) = \beta(x)\beta(x+t)$. Let $Q_t : R \to R[[t]]$ be the unique ring homomorphism such that $Q_t(\beta)(v) = h(t,v)$. Then $Q_t(\beta)(x(x+t)) = \beta(x)\beta(x+t)$. Putting $t = 0$ yields $Q_0(\beta)(x^2) = \beta(x)^2$; hence $Q_0(\beta_i) = \beta_i^2$ for every $i \geq 0$. It follows that $Q_t(x) = x^2$ for every $x \in R$ since the β_i's generate R as a ring. In order to prove the symmetry in s,t of $(Q_t \circ Q_s)(\beta)$ we first extend $Q_t : R \to R[[t]]$ to $R[[x]]$ by putting $Q_t(x) = x(x+t)$. Then we have $Q_t(\beta(x)) = Q_t(\beta)(x(x+t)) = \beta(x)\beta(x+t)$. Assuming that $Q_t(s) = s(s+t)$ we obtain $(Q_t \circ Q_s)(x) = x(x+t)(x+s)(x+s+t)$. Hence $(Q_t \circ Q_s)(x)$ is symmetric in s,t. Thus, we can prove the symmetry of $(Q_t \circ Q_s)(\beta)$ by proving the symmetry of $(Q_t \circ Q_s)(\beta(x))$. We have

$$Q_t(\beta(x+s)) = Q_t(\beta)(Q_t(x+s)) = Q_t(\beta)((x+s)(x+s+t))$$
$$= \beta(x+s)\beta(x+s+t).$$

Thus

$$(Q_t \circ Q_s)(\beta(x)) = Q_t(\beta(x)\beta(x+s)) = Q_t(\beta(x))Q_t(\beta(x+s))$$
$$= \beta(x)\beta(x+t)\beta(x+s)\beta(x+s+t)$$

is symmetric in s,t. This finishes the proof that $(Q_t \circ Q_s)(\beta)$ is symmetric. It follows that $(Q_t \circ Q_s)(y)$ is symmetric for every $y \in R$ since the coefficients of β generate R. This finishes the proof that $Z_2[\beta_*]$ is a Q-ring. We next compute $Q_t(\beta_n)$ by the method of residues. The identity $Q_t(\beta)(x(x+t)) = \beta(x)\beta(x+t)$ means that

$Q_t(\beta_n)$ is the coefficient of $v = x(x+t)$ in $\beta(x)\beta(x+t)$. We have $dv = tdx$ and $Q_t(\beta_n)$ is equal to

$$\mathrm{Res}\ \frac{\beta(x)\beta(x+t)}{x^{n+1}(x+t)^{n+1}}tdx = \sum_{i,j}\beta_i\beta_j\mathrm{Res}\ \frac{x^i(x+t)^j}{x^{n+1}(x+t)^{n+1}}tdx$$

$$= \sum_{i,j}\beta_i\beta_j\binom{j-n-1}{n-i}t^{i+j-2n}\ .$$

The formula for $q_r(\beta_n)$ is the coefficient of t^r in this series. The formula shows that $q_r(\beta_n)$ is homogenous of dimension $2n+r$ if $\dim(b_n) = n$. It then follows by the Cartan formula that $q_r(\beta_{n_1}\cdots\beta_{n_k})$ is homogenous of dimension $2(n_1+\cdots n_k)+r$. This proves that $Z_2[\beta_*]$ is a graded Q-ring. QED

Recall that for any ring R and any subset $S \subseteq R$ there is a fraction ring $S^{-1}R$ obtained by formally inverting the elements in S. Let $j : R \to S^{-1}R$ be the canonical map.

PROPOSITION 5. *If R is a Q-ring then so is the fraction ring $S^{-1}R$ and the canonical map $j : R \to S^{-1}R$ is a map of Q-rings.*

PROOF. For any $x \in R$ let us put $h(x) = \sum_n j(q_n(x))t^n$. This defines a ring homomorphism $h : R \to S^{-1}R[[t]]$. The constant term of the formal power series $h(x)$ is $j(q_0(x)) = j(x)^2$. Hence $h(x)$ has a multiplicative inverse when $x \in S$. It follows that there is a unique ring homomorphism $Q_t : S^{-1}R \to S^{-1}R[[t]]$ such that $Q_t(j(x)) = h(x)$ for $x \in R$. By definition, $Q_t(x/y) = Q_t(x)/Q_t(y)$ for any fraction $x/y \in S^{-1}R$. Hence $Q_0(x/y) = (x/y)^2$. The symmetry in s,t of $Q_t(Q_s(x/y))$ is obvious from the formula $Q_t(Q_s(x/y)) = Q_t(Q_s(x))/Q_t(Q_s(y))$. QED

The fraction ring $Z_2[b_0^{-1}, b_*] = \{b_0\}^{-1}Z_2[b_0, b_1, \ldots]$ supports a Hopf algebra structure called the *Faa di Bruno Hopf algebra*; we shall denote it by $\mathcal{B}$. By definition, the comultiplication is the map $\Delta : \mathcal{B} \to \mathcal{B} \otimes \mathcal{B}$ such that

$$\Delta(b) = (b \otimes 1) \circ (1 \otimes b)$$

where $b(x) = \sum_i b_i x^{i+1}$. The counit $\epsilon : \mathcal{B} \to Z_2$ is the map such that $\epsilon(b)(x) = x$. The power series $b(x)$ has a composition inverse $\bar{b}(x)$ since b_0 is invertible in $\mathcal{B}$. The antipode is the algebra map $\mathcal{B} \to \mathcal{B}$ such that $b(x) \mapsto \bar{b}(x)$.

COROLLARY. *The Faa di Bruno algebra $\mathcal{B}$ has a unique Q-ring structure such that $Q_t(b)(x(x+t)) = b(x)b(x+t)$ where $b(x) = \sum_i b_i x^{i+1}$. This is a graded Q-ring if we take $\dim(b_i) = i$.*

PROOF. Let $\beta(x) = \sum_i b_i x^i$. By proposition 4 there is a unique Q-structure on $Z_2[b_*] = Z_2[b_0, b_1, \ldots]$ such that $Q_t(\beta)(x(x+t)) = \beta(x)\beta(x+t)$. We have $b(x) = x\beta(x)$. Hence

$$Q_t(b)(x(x+t)) = x(x+t)\beta(x)\beta(x+t) = b(x)b(x+t).$$

The result then follows from proposition 5 since $\mathcal{B} = \{b_0\}^{-1}Z_2[b_*]$. QED

Recall that a formal power series $a(x)$ with coefficients in a Z_2-algebra R is *additive* if $a(x+y) = a(x)+a(y)$; this happens iff $a(x)$ has the form $a(x) = \sum a_i x^{2^i}$. If a_0 has an inverse then $a(x)$ has an additive composition inverse $\bar{a}(x)$. It follows that the algebra $Z_2[a_0^{\pm}, a_1, \dots] = \{a_0\}^{-1} Z_2[a_0, a_1, \dots]$ supports a Hopf algebra structure that we call the *extended Milnor Hopf algebra*; we shall denote it by $\mathcal{A}$. The comultiplication is the map $\Delta : \mathcal{A} \to \mathcal{A} \otimes \mathcal{A}$ such that

$$\Delta(a) = (a \otimes 1) \circ (1 \otimes a)$$

where $a(x) = \sum_i a_i x^{2^i}$.

PROPOSITION 6. *The extended Milnor algebra $\mathcal{A}$ has a unique Q-ring structure such that $Q_t(a)(x(x+t)) = a(x)a(x+t)$ where $a(x) = \sum_i a_i x^{2^i}$. This is a graded Q-ring if we take $\dim(a_n) = 2^n - 1$. We have the explicit formula*

$$Q_t(a_n) = a_n^2 + t^{-2^{n+1}}(\sum_{k=0}^{n} a_k t^{2^k})(\sum_{i=n+1}^{\infty} a_i t^{2^i}).$$

PROOF. Let us first prove the result for the ring $R = Z_2[a_0, a_1, \dots]$. The result for $\mathcal{A}$ will follow by inverting a_0. As in the proof of proposition 4 there is a power series $h(t, v)$ in $R[[t, v]]$ such that $h(t, x(x+t)) = a(x)a(x+t)$. Let us see that $h(t, x)$ is additive. The series $tx + x^2$ has a composition inverse in the ring $R[t^{-1}, t][[x]]$. Its inverse $r(x)$ is additive since $tx + x^2$ is additive. It follows that $h(t, x)$ is additive since $h(t, x(x+t)) = a(x)^2 + a(x)a(t)$ is additive and additive series are closed under composition. Hence there is a unique ring homomorphism $Q_t : R \to R[[t]]$ such that $Q_t(a)(v) = h(t, v)$. That this is a Q-structure is proved as in proposition 4. We justify the explicit formula as follows. We note that $Q_t(a)(x(x+t)) = a(x)a(x+t)$ iff

$$\sum Q_t(a_n)(x^2 + tx)^{2^n} = a(x)^2 + a(x)a(t).$$

By looking at the x and the $x^{2^{n+1}}$ terms here, we see that the identity holds iff

$$Q_t(a_0)t = a_0 a(t), \text{ and } Q_t(a_n) + Q_t(a_{n+1})t^{2^{n+1}} = a_n^2 + a_{n+1}a(t).$$

By multiplying both sides by $t^{2^{n+1}}$, we see that the second equation holds iff

$$Q_t(a_n)t^{2^{n+1}} + Q_t(a_{n+1})t^{2^{n+2}} = (a_n t^{2^n})^2 + (a_{n+1}t^{2^{n+1}})a(t).$$

This is essentially a recursive equation, and one can check directly that its (unique) solution is given by

$$Q_t(a_n)t^{2^{n+1}} = a_n^2 t^{2^{n+1}} + (\sum_{k=0}^{n} a_k t^{2^k})(\sum_{i=n+1}^{\infty} a_i t^{2^i})$$

as claimed. That $\mathcal{A}$ is graded if we take $\dim(a_n) = 2^n - 1$ is proved as in proposition 4. QED

REMARK. In particular we have

$$Q_t(a_0) = \sum_0^{\infty} a_0 a_i t^{2^i - 1}.$$

This shows that the Q-structure on $\mathcal{A}$ cannot survive if we insist that $a_0 = 1$.

Addendum to Section 1.

Nature supplies us with many examples of Q-rings through the cohomology of topological spaces and the homology of E_∞-spaces. Note that in this paper, homology and cohomology are always taken with coeficients in Z_2.

It is standard that the cohomology H^*X of a topological space X is a graded ring with natural Steenrod operations Sq^i for $i \geq 0$; see Steenrod, Epstein [1962], for instance. For $x \in H^n X$, we define $q_i(x) = Sq^{n-i}(x)$ for $0 \leq i \leq n$, and $q_i(x) = 0$ for $n < i$; then $R = H^*X$ is a graded Q-ring with elements of $H^n(X)$ having degree $-n$ (grading by codimension). For example, the Steenrod operations on $H^*RP^\infty = Z_2[x]$ are given by $Sq^0(x) = x$, $Sq^1(x) = x^2$ and $Sq^i(x) = 0$ for $i > 1$. Hence $Q_t(x) = x(x+t)$; thus if $X = RP^\infty \times \ldots RP^\infty$ (n copies) then $H^*X = Z_2[x_1, \ldots, x_n]$ is the Q-ring described in proposition 3.

It is also standard that if E is an E_∞-space then H_*E is a graded ring with Dyer-Lashof operations Q^j for $j \geq 0$. See May [1971], for instance, for a description of the properties of these operations. The operation $q_i : H_n E \to H_{2n+i}E$ is the Dyer-Lashof operation Q^{n+i}. In fact the operations q_i were explicitly used and studied in Kudo, Araki [1956]. The usual properties of Dyer-Lashof operations translate into the statement that H_*E is a graded Q-ring with $Q_t = \sum q_i t^i$. For example, the homology of the classifying space BO_* for real vector bundles is the polynomial ring $Z_2[\beta_*]$ where $\beta_0, \beta_1, \ldots$ is a basis of $H_*BO_1 = H_*RP^\infty$. It supports the generic power series $\beta(x) = \sum \beta_i x^i$ which arises in the theory of characteristic classes; see Milnor, Stasheff [1974], for instance. The Whitney sum of vector bundles gives BO_* the structure of an E_∞-space. The resulting Q-ring structure on $Z_2[\beta_*]$ is the one described in proposition 4; the explicit formula derived there is equivalent to one in Priddy [1975]. We will give another direct geometric proof of this result in a sequel to this paper.

For the standard Dyer-Lashof algebra (and some indications of its history) see May [1971], Madsen [1975], and May [1976], for instance.

The Faa di Bruno Hopf algebra $\mathcal{B} = Z_2[b_0^{-1}, b_0, b_1, \ldots]$ is studied in combinatorics; see Joni, Rota [1982] for instance. It is closely related to the dual of the Landweber-Novikov algebra; see Landweber [1967], Quillen [1971], Adams [1974], or Morava [1985], for instance.

The Adem relations are due to Adem [1957]. The idea of writing Adem relations via generating series occurs in Bisson [1977] and in Bullett, MacDonald [1982] (extended by Steiner [1984]). For an early application of the method of formal residues in algebraic toplogy, see the brief discussion on page 65 of Adams [1974].

2. The Kudo-Araki algebra $\mathcal{K}$ and its dual

Here we show that the coalgebra dual to $\mathcal{K}$ is exactly representing substitution of Ore polynomials. We consequently obtain two explicit basis of $\mathcal{K}$.

Recall that if M and N are Q-modules then so is $M \otimes N$, with the operation q_n defined by the Cartan formula:

$$q_n(x \otimes y) = \sum_{i+j=n} q_i(x) \otimes q_j(y).$$

Abstractly this defines a symmetric tensor product on the category of left $\mathcal{K}$-modules. The unit for this tensor product is Z_2 with its Q-ring structure. It follows from this that $\mathcal{K}$ is a cocommutative bialgebra. The comultiplication $\delta : \mathcal{K} \to \mathcal{K} \otimes \mathcal{K}$ can be defined as the unique map of Q-modules such that $\delta(1) = 1 \otimes 1$. It is an algebra map, and from the relation $\delta(q_n(1)) = q_n(\delta(1))$ it follows that

$$\delta(q_n) = \sum_{i+j=n} q_i \otimes q_j.$$

Similarly, the counit is the unique Q-module map $\epsilon : \mathcal{K} \to Z_2$ such that $\epsilon(1) = 1$. It is an algebra map, and from the relation $\epsilon(q_n(1)) = q_n(\epsilon(1))$ it follows that

$$\epsilon(q_n) = \begin{cases} 1 & \text{if } n = 0 \\ 0 & \text{otherwise.} \end{cases}$$

Thus $\mathcal{K}$ is a (cocommutative) bialgebra and we will give a description of its graded dual $\mathcal{K}'$. First, let us see that the coalgebra $\mathcal{K}$ splits as a direct sum of coalgebras. The Adem relations are homogenous of degree 2 since they only involve words $q_i q_j$ of length 2 on the generators. It follows that $\mathcal{K}$ is a graded ring, graded by "length of words", which we call the *exponent*. We shall denote by $\mathcal{K}(n)$ the homogenous component of exponent n. It is clear from the Cartan formula that δ is mapping $\mathcal{K}(n)$ into $\mathcal{K}(n) \otimes \mathcal{K}(n)$. Hence $\mathcal{K}(n)$ is a coalgebra and we have a decomposition of $\mathcal{K}$ as a direct sum of coalgebras:

$$\mathcal{K} = \bigoplus_n \mathcal{K}(n).$$

We shall describe the graded dual of each coalgebra $\mathcal{K}(n)$; for this we need the *dimension grading* on $\mathcal{K}(n)$. It is the unique grading which makes $\mathcal{K}$ a graded Q-module and has $dim(1) = 0$. In order to prove its existence we can associate to the generator q_i the affine transformation $\alpha_i : N \to N$ given by $\alpha_i(x) = 2x + i$; this is recording the behavior of q_i on the dimensions of elements in any graded Q-module. Then to each word $q_{i_1} \cdots q_{i_n}$ we associate the affine transformation $\alpha_{i_1} \circ \cdots \alpha_{i_n}$. The Adem relations in proposition 1 §1 express the word $q_m q_n$ in terms of the words $q_{m+2n-2i} q_i$. This is compatible with the identity $\alpha_m \circ \alpha_n = \alpha_{m+2n-2i} \circ \alpha_i$ satisfied by the affine transformations. It follows that $\mathcal{K}$ has a grading over the monoid of affine transformations $\{2^n x + d : n, d \geq 0\}$. If $y \in \mathcal{K}$ has "degree" $2^n x + d$ then we say that y is homogenous of *dimension* d and exponent n (or $y \in \mathcal{K}(n)$). In order to study $\mathcal{K}(n)$ we need a good basis and we shall obtain one by studying the n-fold iterated total square.

For a Q-module M, the *n-fold iterated total square* $Q^{(n)} = Q_{t_n, \ldots, t_1} : M \to M[[t_1, \ldots, t_n]]$ is defined inductively by

$$Q^{(n)}(x) = Q_{t_n}(Q^{(n-1)}(x)),$$

where Q_{t_n} is extended to $M[[t_1, \ldots, t_{n-1}]]$ by putting $Q_{t_n}(t_i) = t_i(t_i + t_n)$ for every $i < n$.

The 2-fold total square $Q_{t,s} = Q_t \circ Q_s : M \to M[[s, t]]$ is symmetric in s, t. It is also invariant under the automorphism σ of $M[[s, t]]$ defined by the substitution $s \mapsto s + t$ and $t \mapsto t$, since $\sigma(s(s+t)) = \sigma(s)(\sigma(s) + \sigma(t)) = (s+t)t$. It follows that $Q_{t,s}$ is invariant under the full group $GL(2, Z_2)$ of linear substitutions since this group is generated by σ and the transposition (s, t). In order to develop an expansion of the iterated total operations that takes the symmetry into account, we need to recall the theory of Dickson invariants. If R is a Z_2-algebra, we shall say that a polynomial $p(x) = \sum_{i=0}^{n} c_i\, x^{2^i}$ with coefficients in R is an *Ore polynomial of exponent* $\leq n$; see Ore [1933], Rota [1971]. From the identity $(x+y)^2 = x^2 + y^2$ it follows that $p(x+y) = p(x) + p(y)$; hence an Ore polynomial is additive. The set of roots of $p(x)$ is closed under addition and hence forms a Z_2-vector space. Generically, the appropriate symmetry group for the set of roots is the general linear group $GL(n) = GL(n, Z_2)$. Let us make this more precise. The group $GL(n)$ is acting by linear substitutions on the polynomial ring $Z_2[t_1, \dots, t_n]$; according to Dickson, the subring of invariants is the polynomial algebra $Z_2[t_1, \dots, t_n]^{GL(n)} = Z_2[w_0, \dots, w_{n-1}]$, where

$$W_n(x) = x^{2^n} + \sum_{i=0}^{n-1} w_i\, x^{2^i} = \prod_{v \in \langle t_1, \dots, t_n \rangle} (x+v)$$

and $\langle t_1, \dots, t_n \rangle$ is the Z_2-vector space generated by $t_1, \dots, t_n$. We shall denote the ring of Dickson invariants $Z_2[w_0, \dots, w_{n-1}]$ by $\mathcal{W}(n)$. Note that w_i is a homogenous polynomial of degree $2^n - 2^i$ (in $Z_2[t_1, \dots, t_n]$). For any Z_2-vector space M we have

$$M[t_1, \dots, t_n]^{GL(n)} = M[w_0, \dots, w_{n-1}] \quad \text{and}$$

$$M[[t_1, \dots, t_n]]^{GL(n)} = M[[w_0, \dots, w_{n-1}]].$$

These are graded by *codimension* where by definition $\mathrm{codim}(w_i) = 2^n - 2^i$.

LEMMA 1. *If M is a Q-module then $Q^{(n)} : M \to M[[t_1, \dots, t_n]]$ takes its values in $M[[w_0, \dots, w_{n-1}]]$. Hence there is an expansion*

$$Q^{(n)}(x) = \sum_{l(R)=n} q_R(x) w^R$$

with $q_R \in \mathcal{K}(n)$ and $w^R = w_0^{r_0} \cdots w_{n-1}^{r_{n-1}}$ for $R = (r_0, \dots, r_{n-1})$.

PROOF. It suffices to show that $Q_{t_n, \dots, t_1}$ is invariant under the action of $GL(n)$. But this group is generated by the transpositions $(t_i, t_{i+1}) \mapsto (t_{i+1}, t_i)$ $(1 \leq i < n)$ and the linear substitution σ such that $(t_{n-1}, t_n) \mapsto (t_{n-1} + t_n, t_n)$. If we apply the case $n = 2$ to the Q-module $M[[t_1, \dots, t_{n-2}]]$ with $s = t_{n-1}$ and $t = t_n$ we obtain that $Q_{t_n, \dots, t_1}$ is invariant under the action by σ and the transposition $(t_{n-1}, t_n) \mapsto (t_n, t_{n-1})$. But $Q_{t_n, \dots, t_1} = Q_{t_n} \circ Q_{t_{n-1}, \dots, t_1}$ is invariant under the transpositions $(t_i, t_{i+1}) \mapsto (t_{i+1}, t_i)$ for $1 \leq i < n-1$ since by the inductive hypothesis $Q_{t_{n-1}, \dots, t_1}$ is invariant under $GL(n-1)$. This proves the result. QED

Notice that if M is graded and $x \in M$ is homogenous then $Q^{(n)}(x)$ is homomgenous of dimension $2^n \dim(x)$, when $\mathrm{codim}(t_i) = 1$. Applying this observation to the case $M = \mathcal{K}$ and $x = 1$ we obtain that $\dim(q_R) = \mathrm{codim}(w^R)$. We shall prove that the q_R's form a basis of $\mathcal{K}(n)$. We first show that they linearly span $\mathcal{K}(n)$.

The polynomial $W_n(x)$ is related to the polynomial $W_{n-1}(x)$ by the formula $W_n(x) = W_{n-1}(x) W_{n-1}(x + t_n)$, or equivalently by $W_n(x) = W_{n-1}(x)(W_{n-1}(x) + W_{n-1}(t_n))$ since $W_{n-1}(x)$ is additive. This can be written as $W_n = V_n \circ W_{n-1}$ where $V_n(x) = x(x + u_n)$ and $u_n = W_{n-1}(t_n)$. Iterating, we obtain that $W_n = V_n \circ \cdots \circ V_1$ where $V_i(x) = x(x + u_i)$ and $u_i = W_{i-1}(t_i)$ for $1 \leq i \leq n$. We have

$$Z_2[w_0, \dots, w_{n-1}] \subseteq Z_2[u_1, \dots, u_n] \subseteq Z_2[t_1, \dots, t_n]$$

and the polynomials $u_1, \ldots, u_n$ are algebraically independent. Let us put $u^K = u_1^{k_1} \cdots u_n^{k_n}$ and $q^K = q_{k_1} \cdots q_{k_n}$ for $K = (k_1, \ldots, k_n) \in N^n$.

LEMMA 2. *The total square of order n has an expansion*

$$(*) \qquad Q^{(n)}(x) = \sum_{l(K)=n} q^K(x)u^K.$$

PROOF. Let us first verify by induction on n that if y is an element in a Q-ring such that $Q_t(y) = y(y+t)$ then $Q^{(n)}(y) = W_n(y)$. This is obvious for $n = 1$. Supposing that $Q^{(n-1)}(y) = W_{n-1}(y)$ we obtain

$$Q^{(n)}(y) = Q_{t_n}(Q^{(n-1)}(y)) = Q_{t_n}(W_{n-1}(y)) = \prod_{v \in \langle t_1, \ldots, t_{n-1} \rangle} Q_{t_n}(y+v)$$

$$= \prod_{v \in \langle t_1, \ldots, t_{n-1} \rangle} y(y+t_n) + v(v+t_n) = \prod_{v \in \langle t_1, \ldots, t_{n-1} \rangle} (y+v)(y+v+t_n)$$

$$= W_{n-1}(y)W_{n-1}(y+t_n) = W_n(y).$$

Let us now prove by induction on n that the expansion $(*)$ is valid for every n. By symmetry we have $Q^{(n)} = Q_{t_1, \ldots, t_n}$: thus $Q^{(n)}$ is the composite

$$M \xrightarrow{\;Q_{t_n}\;} M[[t_n]] \xrightarrow{\;Q^{(n-1)}\;} M[[t_1, \ldots, t_n]]$$

where the Q-structure is extended to $M[[t_n]]$ by putting $Q_t(t_n) = t_n(t_n + t)$. From the result just proved it follows that $Q^{(n-1)}(t_n) = W_{n-1}(t_n) = u_n$. Assuming that the expansion $(*)$ is valid for $n-1$ we obtain

$$Q^{(n)}(x) = Q^{(n-1)}\left(\sum_i q_i(x)t_n^i\right) = \sum_i Q^{(n-1)}(q_i(x))u_n^i$$

$$= \sum_i \sum_{l(K)=n-1} q^K(q_i(x))u^K u_n^i = \sum_{l(K)=n-1} \sum_i q^{K,i}(x)u^{K,i}$$

$$= \sum_{l(K)=n} q^K(x)u^K$$

where K, i denotes the concatenation $(k_1, \ldots, k_{n-1}, i)$ of $K = (k_1, \ldots, k_{n-1})$ and i. QED

The equality of the two expansions

$$\sum_{l(K)=n} q^K(x)u^K = \sum_{l(R)=n} q_R(x)w^R$$

of $Q^{(n)}(x)$ shows that each q^K is a linear combination of the q_R's. But the q^K's generate $\mathcal{K}(n)$ and it follows that the q_R's generate $\mathcal{K}(n)$. We shall prove that the q_R's are linearly independent by constructing a Q-module $\mathcal{W}'$ in which their images are linearly independent.

For any Z_2-algebra R let $D_n(R)$ denote the set of monic Ore polynomials of exponent n with coefficients in R. The functor $D_n : Alg \to Sets$ is represented by the algebra $\mathcal{W}(n)$, with $W_n(x)$ as the generic element; see appendix A for this terminology. Composition of Ore polynomials defines an operation $D_m \times D_n \to$

D_{m+n} that is represented by the ring homomorphism $\Delta : \mathcal{W}(m + n) \to \mathcal{W}(m) \otimes \mathcal{W}(n)$ such that

$$\Delta(W_{m+n}) = (W_m \otimes 1) \circ (1 \otimes W_n).$$

Collecting these maps together we obtain a coalgebra structure on $\mathcal{W} = \oplus_{n \geq 0} \mathcal{W}(n)$.

Recall that each $\mathcal{W}(n) = Z_2[w_0, \dots, w_{n-1}]$ is a ring graded by codimension, where $\operatorname{codim}(w_i) = 2^n - 2^i$. Let $\theta_R : \mathcal{W}(n) \to Z_2$ be the linear form picking the coefficient of w^R. The subspace $\mathcal{W}^i(n)$ with basis $(w^R : \operatorname{codim}(w^R) = i)$ is finite dimensional. Thus the *graded dual* $\mathcal{W}'(n)$ is the linear span of the θ_R's. It is graded by *dimension* where $\dim(\theta_R) = \operatorname{codim}(w^R)$. If $\mathcal{W}'_i(n)$ denotes its component of dimension i then we have $\mathcal{W}'_i(n) = [\mathcal{W}^i(n), Z_2]$. Each $\mathcal{W}'(n)$ has a coalgebra structure (δ, ϵ) obtained by dualising the algebra structure of $\mathcal{W}(n)$. From the relation $w^I w^J = w^{I+J}$ it follows by duality that

$$\delta(\theta_K) = \sum_{I+J=K} \theta_I \otimes \theta_J.$$

The counit $\epsilon : \mathcal{W}'(n) \to Z_2$ is the evaluation at $1 \in \mathcal{W}(n)$. In this way we obtain a coalgebra structure (δ, ϵ) on the direct sum $\mathcal{W}' = \bigoplus_n \mathcal{W}'(n)$.

PROPOSITION 1. *The graded dual $\mathcal{W}'$ is a bialgebra where the algebra structure is the convolution obtained from $\Delta : \mathcal{W} \to \mathcal{W} \otimes \mathcal{W}$ and the coalgebra structure is obtained by dualising the algebra structure on each $\mathcal{W}(n)$.*

PROOF. We saw that $\mathcal{W}'$ is a coalgebra with the maps (δ, ϵ) that dualise the algebra structure on each $\mathcal{W}(n)$. Let us see that $\mathcal{W}'$ is a subalgebra of the convolution algebra $[\mathcal{W}, Z_2]$ obtained from Δ. It is easy to verify that each $\Delta : \mathcal{W}(m+n) \to \mathcal{W}(m) \otimes \mathcal{W}(n)$ preserves codimension, if codimension on $\mathcal{W}(m) \otimes \mathcal{W}(n)$ is defined by $\operatorname{codim}(x \otimes y) = 2^n \operatorname{codim}(x) + \operatorname{codim}(y)$. It follows that if $f \in \mathcal{W}'_i(m)$ and $g \in \mathcal{W}'_j(n)$ then $f \star g \in \mathcal{W}'_k(m + n)$ where $\star$ denotes the convolution product and where $k = 2^n i + j$. It remains to verify that $\mathcal{W}'$ is a bialgebra. Let us show that the convolution product $\mathcal{W}' \otimes \mathcal{W}' \to \mathcal{W}'$ and the unit $Z_2 \to \mathcal{W}'$ are coalgebra maps. It suffices to prove that each convolution product $\mathcal{W}'(m) \otimes \mathcal{W}'(n) \to \mathcal{W}'(m + n)$ and each counit $\epsilon : \mathcal{W}'(n) \to Z_2$ is a coalgebra map. But this is true by duality since each convolution product is dualising the algebra map $\Delta : \mathcal{W}(m + n) \to \mathcal{W}(m) \otimes \mathcal{W}(n)$ and each ϵ is dualising the algebra map $Z_2 \to \mathcal{W}(n)$. QED

We next exhibit a Q-module structure on $\mathcal{W}'$. For $f \in \mathcal{W}'$ let $q_i(f) = f \star \theta_i$ where $\theta_i \in \mathcal{W}'(1)$ is the basis element. With this definition we have

$$Q_t(f) = \sum_i (f \star \theta_i) t^i = f \star \sum_i \theta_i t^i.$$

Let $v_t^\dagger : \mathcal{W}(1) \to Z_2[t]$ be the ring homomorphism (it is an isomorphism) representing the Ore polynomial $v_t(x) = x^2 + tx$ (see appendix A for the notation $(\ )^\dagger$). Then we have the expansions $v_t^\dagger = \sum_i \theta_i t^i$ and $Q_t(f) = f \star v_t^\dagger$.

LEMMA 3. *The operation $Q_t(f) = f \star v_t^\dagger$ defines a Q-module structure on $\mathcal{W}'$.*

PROOF. We need to prove that $Q_t \circ Q_s(f)$ is symmetric in s and t. Let $h^\dagger : \mathcal{W}(1) \to Z_2[s, t]$ be the ring homomorphism representing the Ore polynomial $h(x) = x^2 + s(s+t)x$. The polynomial $(h \circ v_t)(x) = x(x+t)(x+s)(x+s+t)$ is symmetric in s and t. Hence $(f \star h^\dagger) \star v_t^\dagger = f \star (h^\dagger \star v_t^\dagger) = f \star (h \circ v_t)^\dagger$ is symmetric in s and t. Thus

$$
\begin{aligned}
(f \star h^\dagger) \star v_t^\dagger &= \left(\sum_i q_i(f)(s^2 + st)^i \right) \star v_t^\dagger = \sum_i (q_i(f) \star v_t^\dagger)(s^2 + st)^i \\
&= \sum_i Q_t(q_i(f))(s^2 + st)^i = Q_t \sum_i q_i(f) s^i \\
&= Q_t \circ Q_s(f)
\end{aligned}
$$

is symmetric in s and t. QED

LEMMA 4. *(Generalised Cartan formulas) If M and N are Q-modules then*

$$
q^K(x \otimes y) = \sum_{I+J=K} q^I(x) \otimes q^J(y) \quad and \quad q_R(x \otimes y) = \sum_{I+J=R} q_I(x) \otimes q_J(y)
$$

for any $x \in M$ and $y \in N$.

PROOF. If R is a Q-ring then the n-fold total square is a ring homomorphism and we have $Q^{(n)}(xy) = Q^{(n)}(x)Q^{(n)}(y)$. This proves the two identities for Q-rings. For Q-modules M and N the argument can be carried out in the "enveloping" Q-ring $R = \tilde{\Lambda} M \otimes \tilde{\Lambda} N$ introduced in the next section. QED

Let $1 \in \mathcal{W}'(0)$ be the unit element for the convolution product. There is a unique map of Q-modules $i : \mathcal{K} \to \mathcal{W}'$ such that $i(1) = 1$ since $\mathcal{K}$ is free as Q-module.

THEOREM 1. *The map $i : \mathcal{K} \to \mathcal{W}'$ is an anti-isomorphism of algebras and an isomorphism of coalgebras. We have $i(q_R) = \theta_R$ for every R, and the operations q_R with $l(R) = n$ form a basis of $\mathcal{K}(n)$.*

PROOF. Let us first verify that i is an anti-homomorphism of algebras. It is preserving the unit element since $i(1) = 1$. The identity $i(xy) = i(y) \star i(x)$ will be proved if we show that it holds when x is a generator q_n of $\mathcal{K}$. We have $i(q_n \, y) = q_n(i(y)) = i(y) \star \theta_n$ since i is a map of Q-modules. It follows that $i(q_n \, y) = i(y) \star i(q_n)$ since $i(q_n) = q_n(1) = 1 \star \theta_n = \theta_n$. This finishes the proof that i is an anti-homomorphism of algebras. Let $V_k^\dagger : \mathcal{W}(1) \to Z_2[u_1, \ldots, u_n]$ be the ring homomorphism representing the Ore polynomial $V_k(x) = x(x + u_k)$. We have an expansion $V_k^\dagger = \sum_i \theta_i \, u_k^i$. If we apply i to the power series $Q^{(n)}(x)$ and use the expansions provided by Lemma 1 and 2 we obtain

$$
\sum_{l(R)=n} i(q_R) \, w^R = \sum_{l(K)=n} i(q^K) \, u^K
$$

where $q^K = q_{k_1} \cdots q_{k_n}$ and $u^K = u_1^{k_1} \cdots u_n^{k_n}$. But

$$
i(q^K) = i(q_{k_n}) \star \cdots \star i(q_{k_1}) = \theta_{k_n} \star \cdots \star \theta_{k_1}
$$

since $i(q_k) = \theta_k$ and i is anti-homomorphism. It follows that

$$
\sum_{l(K)=n} i(q^K) \, u^K = V_n^\dagger \star \cdots \star V_1^\dagger = (V_n \circ \cdots \circ V_1)^\dagger = W_n^\dagger.
$$

But

$$W_n^\dagger = \sum_{l(R)=n} \theta_R w^R$$

since W_n is the generic monic Ore polynomial and $W_n^\dagger : \mathcal{W}(n) \to \mathcal{W}(n)$ is the identity map. This proves that $i(q_R) = \theta_R$ for every R. It follows that the q_R's are linearly independent since the θ_R's are. Hence they form a basis and i is a linear isomorphism. It remains to show that i preserves the coalgebra structures. This follows from lemma 4 and the formula for $\delta(\theta_R)$ given above proposition 1. QED

PROPOSITION 2. *For each $n \geq 0$ the operations $q^K = q_{k_1} \cdots q_{k_n}$ with $k_1 \leq \cdots \leq k_n$ form a basis of $\mathcal{K}(n)$.*

PROOF. For every $n \geq 0$ let us order N^n lexicographicaly. Let $J_n \subseteq N^n$ be the set of increasing sequences $k_1 \leq \cdots \leq k_n$. If $m > r$ then the Adem relations in proposition 1 §1 give a rewrite rule that expresses $q_m q_r$ as a sum of terms $q_j q_i$ with $j < m$ on the right hand side. This is because $2i \geq m + r$, since the bottom of the binomial coefficients must be non-negative. Hence $j = m + 2r - 2i \leq r < m$. This shows that $q^K = q_{k_1} \cdots q_{k_n}$ can be rewritten as a linear combination of elements of smaller index when $K \notin J_n$. This proves that the elements q^K with $K \in J_n$ generate $\mathcal{K}(n)$. To prove they form a basis we use a counting argument for the dimension of $\mathcal{K}_m(n)$, the homogenous component of grade m of $\mathcal{K}(n)$ (here we say grade instead of dimension to avoid confusion). The elements q^K of grade m with $K \in J_n$ generate $\mathcal{K}_m(n)$. We have $\mathrm{grade}(q^K) = k_1 + k_2 2 + \cdots + k_n 2^n$. On the other hand, the elements q_R of grade m form a basis of $\mathcal{K}_m(n)$. We have $\mathrm{grade}(q_R) = \mathrm{codim}(w^R) = r_0(2^n - 2^0) + \cdots + r_{n-1}(2^n - 2^{n-1})$. But the transformation $r_i = k_{i+1} - k_i$ (with $r_0 = k_1$) defines a bijection between these two sets of elements. It follows that the q^K's of grade m with $K \in J_n$ form a basis of $\mathcal{K}_m(n)$. QED

Addendum to Section 2.

Historically, the Dyer-Lashof algebra was modelled upon the Steenrod algebra. As we discuss in section 4, the comultiplication for the Milnor coalgebra dual to the Steenrod algebra is representing composition of additive formal power series. The existence of a dual pairing between Dyer-Lashof operations and Dickson invariants is implicit in Madsen's thesis; see Madsen [1975]. There he describes a comultiplication which is dual to the composition of "upper-indexed" Dyer-Lashof operations. Wilkerson [1983] and Morava [1991] speculate on how to derive this coproduct directly from the algebra of the Dickson invariants.

If $V = Z_2^n$ then $H^*V = Z_2[t_1, \cdots, t_n]$ and $\mathcal{W}(n) = H^*V^{GL(n)}$ is the image of the ring homomorphism $H^*\Sigma_{2^n} \to H^*V$ induced by the natural "diagonal" inclusion $Z_2^n \subseteq \Sigma_{2^n}$. See Mui [1975], Madsen, Milgram [1979], or Mann, Milgram [1982] for more discussion of this point.

An expansion of iterated total operations in terms of Dickson invariants is discussed in Mui [1983] and in Lomonaco [1992].

Some good sources for algebraic background on the Dickson invariants are Mui [1975] and Wilkerson [1983]. The original reference given there is Dickson [1911]. See also Ore [1933] and Rota [1971].

3. Free Q-rings.

In this section we show that free Q-rings are polynomial rings. The result involves an explicit construction of the Q-ring freely generated by a Q-module; we show that it is the enveloping algebra of a predetermined squaring or *Frobenius* operation. We also introduce the concept of *rank* grading on a Q-ring.

We shall denote by $Q\langle V \rangle$ the Q-ring freely generated by a Z_2-vector space V. The free Q-ring on one generator will be denoted by $Q\langle x \rangle$. To construct $Q\langle V \rangle$ the first step is to take $\mathcal{K} \otimes V$ which is the free Q-module on V. The second step is to construct a Q-ring from $\mathcal{K} \otimes V$ or more generally from a Q-module. This is done with the *enveloping algebra* $\tilde{\Lambda}M = \tilde{\Lambda}(M, q_0)$ of a vector space M equiped with an endomorphism q_0. We have $Q\langle V \rangle = \tilde{\Lambda}(\mathcal{K} \otimes V)$ and in particular, $Q\langle x \rangle = \tilde{\Lambda}\mathcal{K}$.

Recall that the *Frobenius endomorphism* of a commutative Z_2-algebra is the map $x \mapsto x^2$. For a Q-ring this is the operation q_0. We shall sometimes say that a Z_2-vector space M equipped with an arbitrary endomorphism $q_0 : M \to M$ is a *Frobenius module*. It is *graded* when M is graded and $q_0(M_n) \subseteq M_{2n}$. The forgetful functor from commutative Z_2-algebras to Frobenius modules has a left adjoint $\tilde{\Lambda}(-)$ that we now describe. If M is a Frobenius module then the ring $\tilde{\Lambda}M$ is the quotient of the symmetric algebra $S(M)$ by the ideal generated by the differences $x^2 - q_0(x)$ for $x \in M$. We shall refer to $\tilde{\Lambda}M$ as the *enveloping algebra* of (M, q_0).

PROPOSITION 1. *If M is a Q-module, then the enveloping algebra $\tilde{\Lambda}M$ of (M, q_0) is a Q-ring. The functor $M \mapsto \tilde{\Lambda}(M)$ is left adjoint to the forgetful functor from Q-rings to Q-modules.*

PROOF. There is a unique ring homomorphism $Q_t : S(M) \to S(M)[[t]]$ extending the map $Q_t : M \to M[[t]]$. The ring map $Q_t \circ Q_s : S(M) \to S(M)[[s,t]]$ is symmetric in s and t since it is extending $Q_t \circ Q_s : M \to M[[s,t]]$ and the latter is symmetric. Let $\pi : S(M) \to \tilde{\Lambda}M$ be the canonical map and let us use the same notation for its extension $S(M)[[t]] \to \tilde{\Lambda}M[[t]]$. In particular, putting $t = 0$ we obtain $Q_0 \circ Q_s = Q_s \circ Q_0$ where $Q_0(s) = s^2$. We have $\pi Q_0(x) = \pi x^2$ for any $x \in S(M)$, since this relation is true for $x \in M$ and Q_0 and $(-)^2$ are ring homomorphisms. Then for any $x \in M$ we have

$$\pi Q_t(q_0(x)) = \pi Q_t Q_0(x) = \pi Q_0 Q_t(x) = \pi Q_t(x)^2 = \pi Q_t(x^2)$$

where $Q_0(t) = t^2$. It follows from this equality that there is a unique ring map $Q_t : \tilde{\Lambda}M \to \tilde{\Lambda}M[[t]]$ such that $Q_t(\pi(x)) = \pi(Q_t(x))$ for every $x \in S(M)$. From the relation $\pi Q_0(x) = \pi x^2$ for any $x \in S(M)$ it follows that $Q_0(x) = x^2$ for every $x \in \tilde{\Lambda}M$. The symmetry in s and t of $Q_t \circ Q_s : \tilde{\Lambda}M \to \tilde{\Lambda}M[[s,t]]$ follows from the symmetry of the corresponding map on $S(M)$. This proves that $\tilde{\Lambda}M$ is a Q-ring. Let us prove that $\tilde{\Lambda}(-)$ is left adjoint to the forgetful functor from Q-rings to Q-modules. We need to prove that if R is a Q-ring and $f : M \to R$ is a map of Q-modules then there is a unique map of Q-rings $\tilde{f} : \tilde{\Lambda}M \to R$ such that $\tilde{f}(x) = f(x)$ for $x \in M$. But we have $f(q_0(x)) = q_0(f(x)) = f(x)^2$ since f is a map of Q-modules. It follows that there is a unique map of rings $\tilde{f} : \tilde{\Lambda}M \to R$ such that $\tilde{f}(x) = f(x)$ for $x \in M$. It remains to verify that $Q_t(\tilde{f}(x)) = \tilde{f}(Q_t(x))$ for every $x \in \tilde{\Lambda}M$. But it suffices to verify this equality for $x \in M$ since both sides are ring homomorphisms and M generates $\tilde{\Lambda}M$ as a ring. For every $x \in M$

$Q_t(\tilde{f}(x)) = Q_t(f(x))$ and $\tilde{f}(Q_t(x)) = f(Q_t(x))$ and also $Q_t(f(x)) = f(Q_t(x))$ since f is a map of Q-modules. QED

COROLLARY. *The free Q-ring on a vector space V is $\tilde{\Lambda}(\mathcal{K} \otimes V)$. In particular $Q\langle x \rangle$, the free Q-ring on one generator, is $\tilde{\Lambda}\mathcal{K}$.*

PROOF. Each functor $\tilde{\Lambda}(-)$ and $\mathcal{K} \otimes (-)$ is left adjoint to the corresponding forgetful functor. The result follows from the fact that a composite of left adjoints is left adjoint to the composite; see MacLane [1971], for instance. QED

A Frobenius module is nothing but a module over the polynomial algebra $Z_2[q_0]$. Hence the free Frobenius module generated by a vector space V is $Z_2[q_0] \otimes V$. According to proposition 2 §2 the algebra $\mathcal{K}$ has a basis $q_{i_1} q_{i_2} \cdots q_{i_n}$ where $i_1 \leq i_2 \leq \cdots \leq i_n$. But each such $q_{i_1} q_{i_2} \cdots q_{i_n}$ can be written uniquely as $q_0^r q_{i_k} \cdots q_{i_n}$ where $0 < i_k \leq \ldots \leq i_n$. It follows that $\mathcal{K} = Z_2[q_0] \otimes \mathcal{K}^\flat$ where $\mathcal{K}^\flat$ is the linear span of the elements $q_{j_1} \cdots q_{j_n}$ such that $0 < j_1 \leq \cdots \leq j_n$.

PROPOSITION 2. *For any vector space V the free Q-ring $Q\langle V \rangle$ is isomorphic to the symmetric algebra $S(\mathcal{K}^\flat \otimes V)$. In particular, the free Q-ring on one generator $Q\langle x \rangle$ has a polynomial basis consisting of the elements $q_{i_1} \cdots q_{i_n}(x)$ where $0 < i_1 \leq \cdots \leq i_n$.*

PROOF. We saw above that $\mathcal{K} = Z_2[q_0] \otimes \mathcal{K}^\flat$. Hence for any vector space V, we have

$$Q\langle V \rangle = \tilde{\Lambda}(\mathcal{K} \otimes V) = \tilde{\Lambda}(Z_2[q_0] \otimes \mathcal{K}^\flat \otimes V).$$

The result then follows from the fact that a composite of left adjoints is left adjoint to the composite. More explicitly, the first functor $Z_2[q_0] \otimes (-)$ is left adjoint to the forgetful functor from Frobenius modules to Z_2-vector spaces; the second functor $\tilde{\Lambda}(-)$ is left adjoint to the forgetful functor from Z_2-algebras to Frobenius modules. Hence their composite $\tilde{\Lambda}(Z_2[q_0] \otimes (-))$ is left adjoint to the forgetful functor from algebras to vector spaces. Hence $\tilde{\Lambda}(Z_2[q_0] \otimes V)$ is the symmetric algebra $S(V)$ for any V. Thus $S(\mathcal{K}^\flat \otimes V) = \tilde{\Lambda}(Z_2[q_0] \otimes \mathcal{K}^\flat \otimes V) = Q\langle V \rangle$ for any V. QED

Let M be a Frobenius module with a basis $(u_i : i \in I)$. Consider the family of commutative monomials $u^S = u_{i_1} \cdots u_{i_n}$ where $S = \{i_1, \ldots, i_n\}$ runs through the finite subsets of I. By using the relation $u_i^2 = q_0(u_i)$ it is easy to see that the family (u^S) generates $\tilde{\Lambda}M$. We shall prove that the family (u^S) is a basis of $\tilde{\Lambda}M$. Consider the filtration $F_0 \subseteq F_1 \subseteq \cdots$ where F_n is the linear span of the monomials u^S with S of size at most n. The subgroup F_n does not depend on the basis $(u_i : i \in I)$ since it is the linear span of all the elements that can be written as a product of at most n elements in M. We have $1 \in F_0$ and $F_n F_m \subseteq F_{n+m}$. The associated graded algebra

$$\mathrm{gr}\tilde{\Lambda}M = \bigoplus_n F_n/F_{n-1}$$

is generated by F_1/F_0. Moreover, we have $x^2 = 0$ in F_2/F_1 for every $x \in F_1/F_0$ since $x^2 = q_0(x) \in F_1$. If ΛM is the exterior algebra on M then there is an algebra map $i : \Lambda M \to \mathrm{gr}\tilde{\Lambda}M$ extending the canonical map $M \to F_1/F_0$. The map i is surjective since F_1/F_0 generates $\mathrm{gr}\tilde{\Lambda}M$.

LEMMA 1. *The map* $i : \Lambda M \to \mathrm{gr}\tilde{\Lambda} M$ *is an isomorphism.*

PROOF. A Frobenius module is the same as a module over $Z_2[q_0]$. Any module is a directed colimit of finitely presented modules. Hence it suffices to prove the result for finitely presented modules over $Z_2[q_0]$ since the functors $M \mapsto \Lambda M$ and $M \mapsto \mathrm{gr}\tilde{\Lambda} M$ are preserving directed colimits. According to an elementary theorem of algebra, every finitely presented module over $Z_2[q_0]$ is a direct sum of cyclic modules. It suffices to prove the result in the case where M is a cyclic module, since for any Frobenius modules M and N we have a commutative square of canonical maps

$$
\begin{array}{ccc}
\Lambda(M) \otimes \Lambda(N) & \longrightarrow & \Lambda(M \oplus N) \\
\downarrow{\scriptstyle i \otimes i} & & \downarrow{\scriptstyle i} \\
\mathrm{gr}\tilde{\Lambda}(M) \otimes \mathrm{gr}\tilde{\Lambda}(N) & \longrightarrow & \mathrm{gr}\tilde{\Lambda}(M \oplus N)
\end{array}
$$

in which the horizontal arrows are isomorphisms. There are two cases to consider (depending on the type of M). If M is infinite cyclic then $M = Z_2[q_0]x$ is free over a generator x. It follows that $\tilde{\Lambda}(Z_2[q_0]x)$ is freely generated by x as a Z_2-algebra. Thus $\tilde{\Lambda} M = Z_2[x]$ and the canonical map $M \to \tilde{\Lambda} M$ is the linear map $Z_2[q_0]x \to Z_2[x]$ such that $q_0^n x \mapsto x^{2^n}$. But every natural number n can be expressed uniquely as a sum of distinct powers of 2. Hence every x^n can be written uniquely as a product $q_0^{i_1}(x) \ldots q_0^{i_r}(x)$ where $i_1 < \cdots < i_r$. Let us put $l(n) = r$. For any $r \geq 0$ let $B_r = \{x^n : l(n) = r\}$. Then B_1 is a basis of M and the set $B_0 \cup \cdots \cup B_r$ is a basis of F_r. It follows that B_r projects to a basis in F_r/F_{r-1} and the canonical map $\Lambda^r M \to F_r/F_{r-1}$ is bijective. This finishes the proof in the case where M is infinite cyclic. In the case where M is finite cyclic, we have $M = Z_2[q_0]x$ with x satisfying a defining relation $p(q_0)x = 0$ where

$$
p(q_0) = q_0^n + a_{n-1}q_0^{n-1} + \cdots + a_1 q_0 + a_0
$$

is a monic polynomial of degree n. But then ΛM has dimension 2^n since M has dimension n. We have observed that i is surjective. Thus the result will be proved if we show that $\mathrm{gr}\tilde{\Lambda} M$ has dimension 2^n. But $\mathrm{gr}\tilde{\Lambda} M$ and $\tilde{\Lambda} M$ have the same dimension since

$$
\sum_{i=0}^{n} \dim(F_i/F_{i-1}) = \sum_{i=0}^{n} \dim(F_i) - \dim(F_{i-1})
$$
$$
= \dim(F_n) - \dim(F_{-1}) = \dim \tilde{\Lambda} M
$$

Thus it suffices to prove that $\tilde{\Lambda} M$ has dimension 2^n. By definition, $\tilde{\Lambda} M$ is generated as a ring by an element x satisfying the relation $p(q_0)x = 0$ where $q_0 x = x^2$. Equivalently, it is generated by an element x satisfying the defining relation $q(x) = 0$ where $q(x)$ is the Ore polynomial

$$
q(x) = x^{2^n} + a_{n-1}x^{2^{n-1}} + \cdots + a_1 x^2 + a_0 x.
$$

It follows that $\tilde{\Lambda} M = Z_2[x]/(q(x))$ has dimension 2^n. QED

THEOREM 1. *(Basis theorem) Let M be a Frobenius module and let $(u_i : i \in I)$ be a basis of M. Then the monomials $u^S = u_{i_1} \cdots u_{i_n}$ where $S = \{i_1, \ldots, i_n\}$ runs through the finite subsets of I form a linear basis of $\tilde{\Lambda}M$.*

PROOF. For any $n \geq 0$ let B_n be the family of monomials u^S where S is of size n. It follows from the lemma that B_n gives a basis of F_n/F_{n-1}. An induction on n then shows that the disjoint union $B_0 \cup \cdots \cup B_n$ is a basis of F_n. The result is then obtained by letting $n \to \infty$. QED

COROLLARY. *If M is a Frobenius module the canonical map $i : M \to \tilde{\Lambda}M$ is injective. In particular, any Q-modules can be embedded in a Q-ring.*

We end this section by showing that free Q-rings are bigraded. Recall that a grading $(M_n : n \in Z)$ on a Q-module M is a *dimension grading* if $q_i : M_n \to M_{2n+i}$.

DEFINITION 1. A *rank grading* $(R[n] : n \in Z)$ on a Q-ring is a grading such that $1 \in R[0]$, $R[m]R[n] \subseteq R[n+m]$, and $q_i : R[n] \to R[2n]$. An *exponent grading* $(M(n) : n \geq 0)$ on a Q-module is a non-negative grading such that $q_i : M(n) \to M(n+1)$.

Two gradings $(V[i] : i \in I)$ and $(V_j : j \in J)$ on a vector space V are *compatible* if every homogenous component for one is graded with respect to the other. Then the two gradings have a common double refinement $(V[i]_j : (i,j) \in I \times J)$, so that

$$V[i] = \bigoplus_{j \in J} V[i]_j \quad \text{and} \quad V_j = \bigoplus_{i \in I} V[i]_j.$$

EXAMPLES. The Q-ring $Z_2[\beta_*]$ of proposition 4 §1 has compatible dimension and rank gradings with $\dim(\beta_i) = i$ and with $\mathrm{rank}(\beta_i) = 1$. The Q-module $\mathcal{K}$ has an exponent grading $(\mathcal{K}(n) : n \geq 0)$.

If V is a graded vector space then the Q-module $\mathcal{K} \otimes V$ has compatible rank and dimension gradings; we have $\dim(q \otimes x) = 2^n \dim(x) + i$ and $\exp(q \otimes x) = n$ for $q \in \mathcal{K}(n)_i$ and homogenous x.

PROPOSITION 3. *If a Q-module M has compatible exponent and dimension gradings then its enveloping algebra $\tilde{\Lambda}M$ has compatible rank and dimension gradings.*

PROOF. By construction $\tilde{\Lambda}M$ is the quotient of $S(M)$ by the ideal J generated by the elements $x^2 + q_0(x)$ where x runs through M. If $\dim(x) = n$ then $x^2 + q_0(x)$ is homogenous of dimension $2n$ since $\dim(q_0 x) = 2n + 0 = 2n$. If $x \in M$ is homogenous of exponent n let us say that it is of *rank 2^n*. Then $x^2 + q_0(x)$ is homogenous of rank 2^{n+1} since $\exp(q_0 x) = 1 + \exp(x)$. Thus the ring $\tilde{\Lambda}M = S(M)/J$ has two compatible gradings, one extending the dimension grading of M and the other extending its rank grading. It remains to prove that $\dim(q_i(x)) = 2 \dim(x) + i$ and $\mathrm{rank}(q_i(x)) = 2 \dim(x)$ for homogenous $x \in \tilde{\Lambda}M$. But this is true for homogenous $x \in M$ and the general case follows from the Cartan formula. QED

COROLLARY. *The free Q-ring $Q\langle V\rangle$ generated by a graded vector space V has compatible rank and dimension gradings.*

PROOF. The free Q-module on V is $\mathcal{K}\otimes V$. We saw above that it has compatible dimension and exponent gradings. The result then follows from the equality $Q\langle V\rangle = \tilde{\Lambda}(\mathcal{K}\otimes V)$. QED

Addendum to Section 3.

The results proved in this section are very close to results stated in section 2 of May [1976]. Our graded Q-modules correspond to the "allowable left modules" over the Dyer-Lashof algebra $\mathcal{R}$ used by May there.

The Q-ring $H_*(BO_*) = Z_2[\beta_*]$ is graded by dimension and rank as shown in the above example, and $H_m(BO_r)$ is the component with rank r and dimension m.

For any space X let $E_\infty(X)$ denote the free E_∞-space generated by X. From results in May [1976] (which extend the results in Dyer, Lashof [1962] for the case where X is a point) it follows that $H_*E_\infty(X) = Q\langle H_*X\rangle$, the free Q-ring generated by H_*X. Let $\Sigma_n X$ denote the space $E[n] \times_{\Sigma_n} X^n$, where $E[n]$ is a contractible space on which Σ_n acts freely. Then we have

$$E_\infty(X) = \sum_n E[n] \times_{\Sigma_n} X^n$$

and we have $H_*\Sigma_n X = Q\langle H_*X\rangle[n]$, the component of rank n. In particular, $H_*\Sigma_n = Q\langle x\rangle[n]$.

4. The extended Milnor coalgebra $\mathcal{A}$.

In order to discuss the Nishida relations in the next section we need to use the *extended Milnor Hopf algebra* $\mathcal{A} = Z_2[a_0^{-1}, a_0, a_1, \dots]$. We also introduce the *positive bialgebra* $\mathcal{A}^+$ and show that $\mathcal{K}$ is anti-isomorphic to a subalgebra of the convolution algebra $[\mathcal{A}^+, Z_2]$. A detailed discussion relating coactions by $\mathcal{A}$ and by the standard Milnor Hopf algebra $\mathcal{S}_*$ can be found in appendix B.

In this section Alg denotes the category of commutative Z_2-algebras, Mon the category of monoids and Grp the category of groups. Most bialgebras (resp. Hopf algebras) will be defined by specifying the functors $Alg \to Mon$ (resp. $Alg \to Grp$) that they represent (see appendix A).

We say that a formal power series $f(x)$ with coefficients in a Z_2-algebra R is *additive* if $f(x+y) = f(x)+f(y)$; this happens iff $f(x)$ has the form $f(x) = \sum f_i x^{2^i}$. The set of all additive series in $xR[[x]]$ is a monoid under composition that we shall denote by $A^+(R)$. If an additive f has a composition inverse then the inverse is also additive; this happens iff f_0 has an inverse. The set of all such additive invertible series forms a group $A(R) \subseteq A^+(R)$. The functor A^+ is represented by the polynomial ring $\mathcal{A}^+ = Z_2[a_0, a_1, \dots]$, with $a(x) = \sum_i a_i x^{2^i}$ as the generic element. The coproduct $\Delta : \mathcal{A}^+ \to \mathcal{A}^+ \otimes \mathcal{A}^+$ is given by

$$\Delta(a) = (a \otimes 1) \circ (1 \otimes a).$$

In this way, $\mathcal{A}^+$ is a bialgebra with no antipode. The functor A is represented by $\mathcal{A} = Z_2[a_0^{\pm}, a_1, \dots]$. It is a Hopf algebra that we call the *extended Milnor Hopf algebra*. The antipode $\chi : \mathcal{A} \to \mathcal{A}$ takes $a(x)$ to its composition inverse.

A right $\mathcal{A}$-comodule V comes equipped with a structure map $\psi : V \to V \otimes \mathcal{A}$ which we may refer to as an $\mathcal{A}$-*coaction* or as a *Milnor coaction*. A *ring coaction* is a Z_2-algebra R for which the coaction $\psi : R \to R \otimes \mathcal{A}$ is a ring homomorphism.

Suppose a coaction $\psi : V \to V \otimes \mathcal{A}$ actually lands in $V \otimes \mathcal{A}^+$ (so that no negative power of a_0 is involved). In this case we shall say that ψ is *positive*, so that a positive coaction is the same as a right $\mathcal{A}^+$-comodule.

EXAMPLE. A Milnor ring coaction is always representing an action by the functor A. For example, for any $R \in Alg$ let $F(R)$ denote the set of all power series with coefficients in R. The functor F is represented by the algebra $Z_2[\beta_*] = Z_2[\beta_0, \beta_1, \dots]$ with generic element $\beta(x) = \sum \beta_i x^i$. There is an action on the right $F(R) \times A(R) \to F(R)$ given by $(g(x), f(x)) \mapsto g(f(x))$, and it determines a (positive) Milnor ring coaction on $Z_2[\beta_*]$ with $(\psi\beta)(x) = \beta(a(x))$. More explicitly,

$$\sum_i \psi(\beta_i) \, x^i = \sum_i \beta_i \, a(x)^i.$$

Observe that this coaction is rank preserving, where $\mathrm{rank}(\beta_i) = 1$ for every $i \geq 0$. It follows that for each n, the homogenous component of rank n of $Z_2[\beta_*]$ is a comodule over $\mathcal{A}$.

The monomials $a^R = a_0^{r_0} a_1^{r_1} \cdots$ with $R = (r_0, r_1, \dots)$ form a basis of $\mathcal{A}$ (notice that $r_0 \in Z$). Any coaction $\psi : V \to V \otimes \mathcal{A}$ has a formal expansion

$$\psi(y) = \sum_R \psi_R(y) \otimes a^R$$

for $y \in V$. This expansion determines natural operations ψ_R on Milnor comodules. Recall that any linear form $f \in [\mathcal{A}, Z_2]$ gives a natural operation $y \mapsto f \bullet y$ (see appendix A). This is the left action of the convolution algebra $[\mathcal{A}, Z_2]$ on $\mathcal{A}$-comodules. For any linear form $f \in [\mathcal{A}, Z_2]$ we shall denote by $\langle f, r \rangle$ its value at $r \in \mathcal{A}$ and reserve the notation $f(y)$ for $f \bullet y$. For each monomial $a^R \in \mathcal{A}$ let ψ_R denote the linear form $\mathcal{A} \to Z_2$ which picks the coefficient of a^R. Then $\psi_R(y) = \psi_R \bullet y$.

For any Milnor coaction $\psi : V \to V \otimes \mathcal{A}$, the specialization $a(x) \mapsto v_t(x) = tx + x^2$ determines a total operation $Q_t^* : V \to V[t^{\pm}]$ taking values in Laurent polynomials with coefficients in V. The formal expansion

$$Q_t^*(y) = \sum_n q_n^*(y) t^n$$

for $y \in V$ determines a collection of linear operations $q_n^* : V \to V$, where n ranges over the *integers*.

Equivalently, if $v_t^\dagger : \mathcal{A} \to Z_2[t^{\pm}]$ is the ring homomorphism determined by the specialization $a(x) \mapsto v_t(x)$ then $Q_t^*(y) = v_t^\dagger \bullet y$, and the Laurent expansion $v_t^\dagger = \sum_n q_n^* t^n$ determines linear forms $q_n^* : \mathcal{A} \to Z_2$ with $q_n^*(y) = q_n^* \bullet y$ for $y \in V$.

PROPOSITION 1. *A coaction $\psi : V \to V \otimes \mathcal{A}$ is positive iff $Q_t^* : V \to V[t^{\pm}]$ actually lands in $V[t]$. In this case the operations $(x, q_n) \mapsto q_n^*(x)$ define a right $\mathcal{K}$-module structure on V.*

PROOF. It is obvious that Q_t^* lands in $V[t]$ when ψ is positive. Conversely, let us suppose that Q_t^* lands in $V[t]$. Recall the formula $Q_t^*(x) = v_t^\dagger \bullet x$ for $x \in V$ where $v_t^\dagger$ is the map $\mathcal{A}_* \to Z_2[t^{\pm}]$ representing the polynomial $v_t(x) = tx + x^2$. Let $W_n(x)$ be the Ore polynomial of exponent n used in §2. For any $y \in V$ let us put $Q^{*(n)}(y) = W_n^\dagger \bullet y$ where $W_n^\dagger : \mathcal{A} \to \mathcal{W}(n)[w_0^{-1}]$ is the map representing $W_n(x)$. Let us prove that $Q^{*(n)}$ actually lands in $V \otimes \mathcal{W}(n)$. Recall from lemma 2 §2 that $Z_2[w_0, \ldots, w_{n-1}] \subseteq Z_2[u_1, \ldots, u_n]$ where $W_n = V_n \circ \cdots \circ V_1$ and $V_i(x) = x(x + u_i)$. Let us put $R = Z_2[u_1^{\pm}, \ldots, u_n^{\pm}]$ and for every $1 \leq i \leq n$ let $V_i^\dagger$ be the map $\mathcal{A}_* \to R$ representing the polynomial $V_i(x)$. We have

$$Q^{*(n)}(y) = W_n^\dagger \bullet y = (V_n \circ \cdots \circ V_1)^\dagger \bullet y = V_n^\dagger \bullet \cdots \bullet V_1^\dagger \bullet y = Q_{u_n}^* \circ \cdots \circ Q_{u_1}^*(y)$$

and this shows that $Q^{*(n)}(y)$ lands in $V[u_1, \ldots, u_n]$. Observe now that the ring $Z_2[u_1, \ldots, u_n]$ is an integral extension of $Z_2[w_0, \ldots, w_{n-1}]$ since from §2 we have

$$Z_2[w_0, \ldots, w_{n-1}] \subseteq Z_2[u_1, \ldots, u_n] \subseteq Z_2[t_1, \ldots, t_n]$$

where $W_n(t_i) = 0$ and $u_i = W_{i-1}(t_i)$. Hence

$$Z_2[w_0, \ldots, w_{n-1}] = Z_2[w_0^{-1}, w_0, \ldots, w_{n-1}] \cap Z_2[u_1, \ldots, u_n]$$

since $Z_2[w_0, \ldots, w_{n-1}]$ is integrally closed. Thus

$$V[w_0, \ldots, w_{n-1}] = V[w_0^{-1}, w_0, \ldots, w_{n-1}] \cap V[u_1, \ldots, u_n]$$

and it follows that $Q^{*(n)}(y)$ lands in $V \otimes \mathcal{W}(n)$. It follows from this that $\psi(y)$ lands in $V \otimes \mathcal{A}$ since a monomial $a^R = a_0^{r_0} a_1^{r_1} \cdots$ with $R = (r_0, r_1, \ldots, r_{n-1}, 0, \ldots)$ belongs to $\mathcal{A}^+$ iff the monomial $W_n^\dagger(a^R) = w^R$ belongs to $\mathcal{W}(n)$. This finishes the proof of the first part. It remains to prove the last statement. If $h(x) = s(s + t)x + x^2$ then $h^\dagger \bullet y = \sum_n q_n^*(y))(s^2 + st)^n$. The polynomial $(h \circ v_t)(x) = x(x + t)(x + s)(x + s + t)$ is symmetric in s and t. Hence also $h^\dagger \bullet (v_t^\dagger \bullet y) = (h \circ v_t)^\dagger \bullet y$. Thus

$$h^\dagger \bullet (v_t^\dagger \bullet y) = h^\dagger \bullet \sum_n q_n^*(y) t^n = \sum_n (h^\dagger \bullet q_n^*(y)) t^n = \sum_n \sum_r q_r^*(q_n^*(y)) t^n (s^2 + st)^r$$

is symmetric in s and t. But this relation is formally the transpose of the Adem relations (proposition 1 §1) between the q_n's. QED

From any coaction $\psi : V \to V \otimes \mathcal{A}$ we extract a grading $(V_n : n \in Z)$ by specialising the generic additive series $a(x)$ to ux. More precisely, the grading coaction $\gamma : V \to V \otimes Z_2[u^{\pm}]$ is obtained by putting $a_0 = u$ and $a_i = 0$ for $i > 0$ in the expansion of ψ (see appendix B).

The *graded dual* V' of a graded vector space V is the direct sum of the dual spaces $V'^n = [V_n, Z_2]$ for $n \in Z$. If we think of V as graded by dimension, then we shall say that $z \in V'^n$ is of *codimension* n. For any coaction $\psi : V \to V \otimes \mathcal{A}$ we shall denote by $q_n : V' \to V'$ the transpose of the operation q_n^*, and define $Q_t : V' \to V'[t^{\pm}]$ by $Q_t = \sum q_n t^n$. Then $q_n(z) = z \star q_n^*$ and $Q_t(z) = z \star v_t^\dagger$ for $z \in V'$ where $v_t(x) = tx + x^2$. If $\psi : V \to V \otimes \mathcal{A}$ is positive then Q_t involves only positive powers of t.

COROLLARY. *If $\psi : V \to V \otimes \mathcal{A}$ is a positive coaction then the operation $Q_t : V' \to V'[t]$ defines a left $\mathcal{K}$-module structure on V'.*

PROOF. This is obtained by duality from the proposition above. It also follows directly by using the formula $Q_t(z) = z \star v_t^\dagger$. QED

REMARK. The Q-modules that appear as the graded dual of some positive Milnor comodule are rather special: they are negatively graded in dimension since they are positively graded in codimension; also the formal power series $Q_t(z)$ is a polynomial of degree $\leq n$ for $z \in V'^n$ and $q_n(z) = z$.

The operations Q_t^* and Q_t on a Milnor comodule and its graded dual correspond to *Steenrod operations*. Precisely, the *total Steenrod square* $Sq_t^* : V \to V[t]$ in a Milnor comodule V is given by $Sq_t^*(y) = S_t^\dagger \bullet y$ where S_t is the polynomial $x + tx^2$. There is an expansion $Sq_t(y) = \sum_{n \geq 0} Sq_*^n(y)t^n$ where $Sq_*^n : V \to V$ is homogenous of degree $-n$. The *total Steenrod square* $Sq_t : V' \to V'[t]$ on the graded dual V' is given by $Sq_t(z) = z \star S_t^\dagger$. There is an expansion $Sq_t(z) = \sum_{n \geq 0} Sq^n(z)t^n$ where $Sq^n : V' \to V'$ is homogenous of degree n. The operations Sq_*^n and Sq_i are transpose to each other. We have $\langle Sq^i(z), y \rangle = \langle z, Sq_i^*(y) \rangle$ for every z in V' and every y in V.

Here are some formulas for translating between the operations q_n^* and Sq_*^n in a Milnor comodule V and similarly for the operations q_n and Sq^n in the graded dual V'. Recall that $Q_t^* : V \to V[t^\pm]$ is given by $Q_t^*(y) = v_t^\dagger \bullet y$ for $y \in V$ where v_t is the polynomial $tx + x^2$. Also, $Q_t : V' \to V'[t^\pm]$ is given by $Q_t(z) = z \star v_t$ for $z \in V'$. From the identity $tx + x^2 = t(x + t^{-1}x^2)$ it follows that for and $y \in V_n$ and $z \in V'^n$ we have

$$Q_t^*(y) = \sum Sq_r^*(y)t^{n-2r} \quad \text{and} \quad Q_t(z) = \sum Sq^r(z)t^{n-r}.$$

Combined with proposition 1, these formulas show that a coaction is positive iff $Sq_r^* = 0$ on V_n when $2r > n$; equivalently, iff $Sq^r = 0$ on V'^n when $r > n$.

Every positive comodule has the structure of a left module over the convolution algebra $[\mathcal{A}^+, Z_2]$. The proposition above shows that every positive comodule has also a natural right $\mathcal{K}$-module structure. This indicates that the map $q_n \mapsto q_n^*$ gives an anti-homomorphism of algebras $j : \mathcal{K} \to [\mathcal{A}^+, Z_2]$. The first statement of the following proposition uses the duality $\mathcal{K}(n) \simeq \mathcal{W}'(n)$.

PROPOSITION 2. *The restriction of j to $\mathcal{K}(n)$ is dual to the map $W_n^\dagger : \mathcal{A}^+ \to \mathcal{W}(n)$ for which $a(x) \mapsto W_n(x)$. The map j is injective, so that j gives an anti-isomorphism between $\mathcal{K}$ and a subalgebra of $[\mathcal{A}^+, Z_2]$.*

PROOF. Let us prove the first statement. For each $n \geq 0$ the map $W_n^\dagger : \mathcal{A}^+ \to \mathcal{W}(n)$ represents the inclusion $D_n \subset \mathcal{A}^+$, where D_n is the functor which associates to each $R \in Alg$ the set $D_n(R)$ of monic Ore polynomials of exponent n in $R[x]$. Consider the following squares:

$$
\begin{array}{ccc}
D_m \times D_n & \longrightarrow & D_{m+n} \\
\cap \uparrow & & \cap \uparrow \\
& & \\
\mathcal{A}^+ \times \mathcal{A}^+ & \longrightarrow & \mathcal{A}^+
\end{array}
\qquad\qquad
\begin{array}{ccc}
\mathcal{W}(m) \otimes \mathcal{W}(n) & \xleftarrow{\Delta} & \mathcal{W}(n+m) \\
W_m^\dagger \otimes W_n^\dagger \uparrow & & \uparrow W_{n+m}^\dagger \\
& & \\
\mathcal{A}^+ \otimes \mathcal{A}^+ & \xleftarrow{\Delta} & \mathcal{A}^+.
\end{array}
$$

The first one commutes since the horizontal arrows represent the composition of polynomials or power series. Hence also the second. Let $h_n : \mathcal{W}'(n) \to [\mathcal{A}^+, Z_2]$ denote the transpose of $W_n^\dagger$. Taking duals we obtain a commutative square

$$
\begin{array}{ccc}
\mathcal{W}'(m) \otimes \mathcal{W}'(n) & \xrightarrow{\ \star\ } & \mathcal{W}'(n+m) \\
\downarrow{\scriptstyle h_m \otimes h_m} & & \downarrow{\scriptstyle h_{n+m}} \\
[\mathcal{A}^+, Z_2] \otimes [\mathcal{A}^+, Z_2] & \xrightarrow{\ \star\ } & [\mathcal{A}^+, Z_2].
\end{array}
$$

where $\star$ represents convolution. This shows that by collecting the maps h_n together we have an algebra map $h : \mathcal{W}' \to [\mathcal{A}^+, Z_2]$ (modulo the verification that h preserves the unit elements). Let us see that $j = hi$ where i is the anti-isomorphism $i : \mathcal{K} \simeq \mathcal{W}'$ described in Theorem 1 §2. For this it suffices to verify the equality $j(x) = hi(x)$ for elements $x \in \mathcal{K}(1)$, since $\mathcal{K}(1)$ generates $\mathcal{K}$ and the maps j and hi are (anti)-homomorphisms of algebras. We have $\sum_n h(\theta_n) w_0^n = W_1^\dagger$ since h is transpose to $W_1^\dagger$. According to Theorem 1 §2 we have $i(q_n) = \theta_n$ for every $n \geq 0$. But $W_1^\dagger = \sum_n q_n^* w_0^n$ since $W_1(x) = x^2 + w_0 x$. Hence $W_1^\dagger = \sum_n j(q_n) w_0^n$ since $j(q_n) = q_n^*$. Combining these equalities together we obtain that $h(i(q_n)) = j(q_n)$. This finishes the proof of the first statement. It remains to show that j is injective, or equivalently that h is injective. We shall use the basis (θ_K) of $\mathcal{W}'(n)$ dual to the monomial basis (w^K) (as discussed in §2). For every $K = (k_0, \ldots, k_{n-1})$ and $i \geq 0$ let us put $K, i = (k_0, \ldots, k_{n-1}, i, 0, \ldots)$. A direct computation shows that

$$
\langle h_n(\theta_K), w^R \rangle = \langle \theta_K, W_n^\dagger(w^R) \rangle = \begin{cases} 1 & \text{if } R = K, i \text{ for some } i \geq 0 \\ 0 & \text{otherwise .} \end{cases}
$$

Thus

$$
h_n(\theta_K) = \sum_i \psi_{K,i}.
$$

But the family $(\psi_{K,1} : n \geq 0, K \in N^n)$ is linearly independent. Hence also the family $(h_n(\theta_K) : n \geq 0, K \in N^n)$. This proves that h transforms a basis of $\mathcal{W}'$ into a linearly independent family. QED

Addendum to Section 4.

The idea of expressing the theory of Steenrod operations in terms of comodules over a Hopf algebra was introduced in Milnor [1958]. The method is explained in the setting of generalized cohomology theories in Adams [1974], for instance.

The extended Milnor Hopf algebra is implicit in Wilson [1980] (in connection with unstable operations and the Hopf ring of Eilenberg-MacLane spaces) and in in Kuhn [1994] (in connection with "generic modular representations"). There is a discussion of an extended Landweber-Novikov Hopf algebra in Morava [1985].

A Milnor coaction is positive (in our sense) iff its graded dual is an *unstable* module over the Steenrod algebra; see Steenrod, Epstein [1962], for the background. The homology of any topological space X is equipped with a natural (positive) Milnor coaction $\psi : H_* X \to H_* X \otimes \mathcal{A}$ which by duality determines the Steenrod operations on cohomology.

The total Steenrod operation Sq_t is used to great effect in Milnor, Stasheff [1974].

The coaction on $H_*(BO_*) = Z_2[\beta_*]$ is determined by $(\psi\beta)(x) = \beta(a(x))$ as in Example 1; see Adams [1974] or Switzer [1973], for instance. For each $n \geq 0$ the component of rank n in $Z_2[\beta_*]$ is a Milnor comodule isomorphic to $H_*(BO_n)$. For instance, $H_*(BO_1)$ is isomorphic to $Z_2\langle\beta_*\rangle$ and $H_*(BO_n)$ is isomorphic to the nth symmetric power of $Z_2\langle\beta_*\rangle$.

There have been years of work on the Steenrod algebra, including surprising recent quantum-leaps. We have not tried to encompass that work here. Some recent guides to the literature are Miller [1986], Lannes [1995], Schwartz [1994], and Wood [1995].

5. The Nishida relations

We give a description of the theory of Nishida relations based on the concept of Q-module and that of commutation operator between an algebra and a coalgebra, or equivalently between a monad and a comonad. A key ingredient is the Q-ring structure on the extended Milnor Hopf algebra $\mathcal{A}$. We show that a coaction $V \to V \otimes \mathcal{A}$ can be extended uniquely to a ring coaction $Q\langle V\rangle \to Q\langle V\rangle \otimes \mathcal{A}$ satisfying the Nishida relations.

DEFINITION 1. Suppose that (R, Q_t) is a Q-ring. A Q-parameter is an additive power series $f(x) \in R[[x]]$ such that $Q_t(f)(x(x+t)) = f(x)f(x+t)$. We shall sometimes say that f is a parameter for Q_t.

For any vector space V the vector space $(V \otimes R)[[t]]$ is a module over the ring $R[[t]]$. If the context is clear we shall denote by $f(t)g(t)$ the product between a power series $f(t)$ in $(V \otimes R)[[t]]$ and a power series $g(t)$ in $R[[t]]$.

PROPOSITION 1. Let $f(x)$ be a Q-parameter in a Q-ring R. If M is a Q-module then so is $M \otimes R$ with $Q'_t(x \otimes r) = Q_{f(t)}(x)Q_t(r)$ for $x \in M$ and $r \in R$.

PROOF. We shall show that $Q'_t Q'_s : M \otimes R \to M \otimes R[[s,t]]$ is symmetric in s and t. We have $Q'_t Q'_s(x \otimes r) = Q'_t(Q'_s(x))Q_t(Q_s(r))$ since $Q'_t(x \otimes r) = Q'_t(x)Q_t(r)$ for $x \in M$ and $r \in R$; hence it suffices to verify the symmetry of $Q'_t(Q'_s(x))$ for $x \in M$. From the relations $Q_t(f(s)) = Q_t(f)(Q_t(s)) = Q_t(f)(s(s+t)) = f(s)(f(s)+f(t))$ it follows that

$$Q'_t Q'_s(x) = Q'_t Q_{f(s)}(x) = Q'_t \sum_i q_i(x)f(s)^i = \sum_i Q'_t(q_i(x))Q_t(f(s))^i$$

$$= \sum_i Q_{f(t)}(q_i(x))[f(s)(f(s)+f(t))]^i.$$

But this last expression can be obtained by replacing s by $f(s)$ and t by $f(t)$ in the expression $\sum_i Q_t(q_i(x))(s(s+t))^i = Q_t Q_s(x)$. Hence the symmetry of $Q'_t Q'_s(x)$ follows from the symmetry of $Q_t Q_s(x)$. QED

Sometimes we shall denote by Q^f the Q-structure Q'_t obtained from a Q-parameter f. We shall often use the following principle whose proof is immediate: if $\alpha : R \to S$ is a map of Q-rings and $g = \alpha(f)$ then $M \otimes \alpha : M \otimes R \to M \otimes S$ is a map of Q-modules, where $M \otimes R$ and $M \otimes S$ are respectively equipped with Q^f and Q^g.

Recall from section 1 that the Q-ring structure on $\mathcal{A}$ is determined by the identity $Q_t(a)(x(x+t)) = a(x)a(x+t)$; hence $a(x)$ is a Q-parameter. Thus, if (M, Q_t) is a Q-module then $(M \otimes \mathcal{A}, Q'_t)$ is a Q-module.

DEFINITION 2. Let M have a Q-structure Q_t and a Milnor coaction $\psi : M \to M \otimes \mathcal{A}$; the *Nishida relations hold* for the pair (Q_t, ψ) if ψ is a Q-module map where $M \otimes \mathcal{A}$ is equipped with Q'_t.

We shall sometimes say that the Nishida relations *hold between* Q_t and ψ if they hold for the pair (Q_t, ψ).

EXAMPLE. The ring $Z_2[\beta_*] = Z_2[\beta_0, \beta_1, \dots]$ has a Q-structure determined by the identity $Q_t(\beta)(x(x+t)) = \beta(x)\beta(x+t)$ where $\beta(x) = \sum_i \beta_i x^i$. It has also a Milnor coaction $\psi : Z_2[\beta_*] \to Z_2[\beta_*] \otimes \mathcal{A}$ determined by the identity $\psi(\beta)(x) = \beta(a(x))$. Let us see that the Nishida relations hold. To prove that ψ is a map of Q-rings it suffices to show that $Q'_t(\beta')(x(x+t)) = \beta'(x)\beta'(x+t)$ where $\beta'(x) = \beta(a(x))$. We have

$$Q'_t(\beta')(x(x+t)) = Q'_t(\beta \circ a)(x(x+t)) = Q'_t(\beta) \circ Q_t(a)(x(x+t))$$
$$= Q'_t(\beta)(a(x)a(x+t)) = Q_{a(t)}(\beta)(a(x)(a(x)+a(t)))$$

But by substituting $t \mapsto a(t)$ and $x \mapsto a(x)$ in the relation $Q_t(\beta)(x(x+t)) = \beta(x)\beta(x+t)$ we obtain

$$Q_{a(t)}(\beta)(a(x)(a(x)+a(t))) = \beta(a(x))\beta(a(x)+a(t)) = \beta'(x)\beta'(x+t).$$

By combining these equalities we obtain the result:

PROPOSITION 2. *If the Nishida relations hold between Q_t and ψ then the following two conditions are satisfied:*
(i) *(M, Q_t) is a graded Q-module if M is given the grading from ψ*
(ii) *$\sum_j q_i^*(q_j(x))s^i t^j = \sum_{r,k} q_k(q_r^*(x))t^k s^r (s+t)^{k+r}$.*

PROOF. Let us suppose that ψ is a Q-module map. To prove (i) consider the map $u^\dagger : \mathcal{A} \to Z_2[u^\pm]$ such that $a(x) \mapsto ux$. There is a Q-structure on $Z_2[u^\pm]$ such that $Q_t(u) = u^2$. For this structure ux is a Q-parameter and it follows that $u^\dagger$ is a map of Q-rings. Hence $M \otimes Z_2[u^\pm]$ has a Q-structure Q'_t and $M \otimes u^\dagger : M \otimes \mathcal{A} \to M \otimes Z_2[u^\pm]$ preserves Q'_t. It follows that the grading coaction $\gamma = (M \otimes u^\dagger)\psi : M \to M \otimes Z_2[u^\pm]$ is a Q-module map. Thus $\gamma(Q_t(x)) = Q'_t(\gamma(x))$ for every $x \in M$. If $x \in M_n$ then $\gamma(x) = xu^n$ and $Q'_t(\gamma(x)) = Q'_t(xu^n) = Q_{ut}(x)u^{2n}$. The equality $\gamma(Q_t(x)) = Q'_t(\gamma(x))$ becomes

$$\sum_i \gamma(q_i(x))t^i = \sum_i \gamma(q_i(x))(ut)^i u^{2n}.$$

Equating the coefficients of t^i we obtain $\gamma(q_i(x)) = q_i(x)u^{2n+i}$, hence $q_i(x) \in M_{2n+i}$. This proves that M is a graded Q-module. Let us prove condition (ii). Let $v_s^\dagger$ be the map $\mathcal{A} \to Z_2[s^\pm]$ such that $a(x) \mapsto v_s(x) = sx + x^2$. There is a Q-structure on $Z_2[s^\pm]$ such that $Q_t(s) = s(s+t)$. For this structure $v_s(x) = x(x+s)$ is a Q-parameter and it follows that $v_s^\dagger : \mathcal{A} \to Z_2[s^\pm]$ is a map of Q-rings. Hence the composite $Q_s^* = (M \otimes v_s^\dagger)\psi : M \to M \otimes Z_2[s^\pm]$ is a Q-module map. Thus $Q_s^* Q_t(x) = Q'_t Q_s^*(x)$. But we have

$$Q_s^* Q_t(x) = Q_s^* \sum_j q_j(x)t^j = \sum_j Q_s^*(q_j(x))t^j = \sum_j q_i^*(q_j(x))s^i t^j \quad \text{and}$$

$$Q'_t Q^*_s(x) = Q'_t \sum_r q^*_r(x) s^r = \sum_r Q'_t(q^*_r(x))(Q_t(s))^r$$

$$= \sum_r Q_{v_s(t)}(q^*_r(x))(s(s+t))^r = \sum_{r,k} q_k(q^*_r(x))(t(t+s))^k(s(s+t))^r$$

$$= \sum_{r,k} q_k(q^*_r(x)) t^k s^r (s+t)^{k+r}.$$

QED

Later (in proposition 8) we shall prove a converse to this proposition.

REMARK. If we select the coefficient of $s^i t^j$ on each side of the equation $Q^*_s Q_t(x) = Q'_t Q^*_s(x)$ then we obtain the following relation (*):

$$q^*_i q_j(x) = \begin{cases} \sum_{k=0}^{j} \binom{\alpha}{j-k} q_k q^*_{\alpha-k}(x) & \text{if } i+j = 2\alpha \text{ is even} \\ 0 & \text{if } i+j \text{ is odd} \end{cases}$$

This is completely equivalent to the classical Nishida relations; see the addendum to this section.

There is a more general form of the Nishida relations which involves iterates of q_i's. We first compute the iterated total square $Q'^{(n)} : M \otimes R \to M \otimes R[[t_1, \dots, t_n]]$ of the Q-structure Q'_t on $M \otimes R$, when $f(x)$ is a Q-parameter in a Q-ring R. Let us denote by F the endomorphism of $R[[t_1, \dots, t_n]]$ that is obtained from the substitution $t_i \mapsto f(t_i)$. The endomorphism F commutes with the action of $GL(n)$ since f is additive. Hence F induces an endomorphism of $R[[w_0, \dots, w_{n-1}]]$, the subring of Dickson invariants. The image by F of the generic Ore polynomial $W_n(x)$ is given by

$$F(W_n)(x) = x^{2^n} + \sum_{i=0}^{n-1} F(w_i) x^{2^i} = \prod_{v \in \langle t_1, \dots, t_n \rangle} x + f(v) \quad .$$

For $x \in M$, let us put

$$Q_F^{(n)}(x) = \sum_{l(K)=n} q_K(x) F(w)^K$$

where $F(w)^K$ denotes $F(w^K)$.

LEMMA 1. We have $Q'^{(n)}(xr) = Q_F^{(n)}(x) Q^{(n)}(r)$ for $x \in M$ and $r \in R$.

PROOF. From the formula $Q'_t(xr) = Q'_t(x) Q_t(r)$ it follows by induction on n that $Q'^{(n)}(xr) = Q'^{(n)}(x) Q^{(n)}(r)$. Hence it suffices to prove that $Q'^{(n)}(x) = Q_F^{(n)}(x)$ for $x \in M$. Let $M \otimes F$ denote the endomorphism of $M \otimes R[[t_1, \dots, t_n]]$ that is obtained from the substitution $t_i \mapsto f(t_i)$ and let F_M denote its composite with the inclusion $\imath : M[[t_1, \dots, t_n]] \hookrightarrow M \otimes R[[t_1, \dots, t_n]]$. By definition we have $Q_F^{(n)} = F_M \circ Q^{(n)}$ where $Q^{(n)}$ is the n-fold iteration of Q_t on M. We shall prove by induction on n that $Q'^{(n)}(x) = F_M \circ Q^{(n)}(x)$ for every $x \in M$. This is clear for

$n = 0$ (and for $n = 1$). If $n > 0$ then consider the diagram

$$
\begin{array}{ccccc}
M & \xrightarrow{\;Q^{(n-1)}\;} & M[[t_1,\dots,t_{n-1}]] & \xrightarrow{\;Q_{t_n}\;} & M[[t_1,\dots,t_n]] \\
\downarrow{\scriptstyle i} & & \downarrow{\scriptstyle F_M} & & \downarrow{\scriptstyle F_M} \\
M \otimes R & \xrightarrow{\;Q'^{(n-1)}\;} & M \otimes R[[t_1,\dots,t_{n-1}]] & \xrightarrow{\;Q'_{t_n}\;} & M \otimes R[[t_1,\dots,t_n]].
\end{array}
$$

The composite of the top (resp. bottom) horizontal arrows is $Q^{(n)}$ (resp. $Q'^{(n)}$). Let us suppose as induction hypothesis that the left hand square commutes. The result will be proved if we show that the right hand square commutes. It suffices to verify the commutativity for $x \in M$, and for $x = t_i$ for every $i < n$. But if $x \in M$ then we have

$$
Q'_{t_n}(F_M(x)) = Q'_{t_n}(x) = Q_{f(t_n)}(x) = F_M(Q_{t_n}(x)).
$$

If $x = t_i$ and $i < n$ then we have

$$
Q'_{t_n}(F_M t_i) = Q'_{t_n}(f(t_i)) = Q_{t_n}(f)(t_i(t_i + t_n)) = f(t_i)(f(t_i) + f(t_n)).
$$

But this equals $F_M(Q_{t_n}(t_i))$ since $Q_{t_n}(t_i) = Q'_{t_n}(t_i) = t_i(t_i + t_n)$. QED

In the case where $R = \mathcal{A}$ and $f = a$ we obtain an endomorphism $F = A$ of the algebra $\mathcal{A}[[w_0,\dots,w_{n-1}]]$. Observe that $M \otimes \mathcal{A}[[w_0,\dots,w_{n-1}]]$ is a module over the ring $\mathcal{A}[[w_0,\dots,w_{n-1}]]$.

PROPOSITION 3. *If the Nishida relations hold between Q_t and ψ then for each $n \geq 0$ and $x \in M$ we have an equality of generating series*

$$
\sum_{l(K)=n}\sum_{R} \psi_R(q_K(x))\, a^R\, w^K = \sum_{l(K)=n}\sum_{R} q_K(\psi_R(x))\, Q^{(n)}(a)^R\, A(w)^K
$$

where $A(w)^K$ denotes $A(w^K)$ and $Q^{(n)}(a)^R$ denotes $Q^{(n)}(a^R)$.

PROOF. By hypothesis, $\psi : M \to M \otimes \mathcal{A}$ is a map of Q-module. Hence $\psi(Q^{(n)}(x)) = Q'^{(n)}(\psi(x))$ for every $n \geq 0$ and every $x \in M$. But we have

$$
\psi(Q^{(n)}(x)) = \sum_{l(K)=n} \psi(q_K(x))\, w^K = \sum_{l(K)=n}\sum_{R} \psi_R(q_K(x))\, a^R\, w^K
$$

and the lemma shows that

$$
Q'^{(n)}(\psi(x)) = Q'^{(n)} \sum_{R} \psi_R(x)\, a^R = \sum_{R} Q^{(n)}_A(\psi_R(x))\, Q^{(n)}(a^R)
$$

$$
= \sum_{l(K)=n}\sum_{R} q_K(\psi_R(x))\, A(w)^K\, Q^{(n)}(a)^R.
$$

QED

For any Z_2-vector space V the vector spaces $\mathcal{K} \otimes (V \otimes \mathcal{A})$ and $(\mathcal{K} \otimes V) \otimes \mathcal{A}$ are canonically isomorphic, but they support different Q-structures. The first is the free Q-module on $(V \otimes \mathcal{A})$ and the second has the Q-structure Q'_t. There is a unique Q-module map

$$\rho_V : \mathcal{K} \otimes (V \otimes \mathcal{A}) \to (\mathcal{K} \otimes V) \otimes \mathcal{A}$$

such that $\rho_V(1 \otimes x \otimes r) = 1 \otimes x \otimes r$, since the domain is free. This yields a natural transformation

$$\rho : KA \to AK$$

where K and A are the functors $K, A : Vect \to Vect$ given respectively by $K(V) = \mathcal{K} \otimes V$ and $A(V) = V \otimes \mathcal{A}$. We shall say that ρ is the *Nishida commutation operator*. The basic theory of commutation operators is given in appendix C.

PROPOSITION 4. *For each $n \geq 0$, $x \in V$ and $r \in \mathcal{A}$ there is an equality of generating series*

$$\sum_{l(K)=n} \rho_V(q_K \otimes x \otimes r) \, w^K = \sum_{l(K)=n} q_K \otimes x \otimes Q^{(n)}(r) \, A(w)^K.$$

In particular, the Nishida operator ρ_V is preserving the decomposition $\mathcal{K} \otimes V \otimes \mathcal{A} = \bigoplus_n \mathcal{K}(n) \otimes V \otimes \mathcal{A}$.

PROOF. The left hand side of this equality is an expansion of $\rho_V Q^{(n)}(1 \otimes x \otimes r)$. But $\rho_V Q^{(n)} = Q'^{(n)} \rho_V$ since ρ_V is a Q-module map. Moreover, $Q'^{(n)} \rho_V(1 \otimes x \otimes r) = Q'^{(n)}(1 \otimes x \otimes r) = Q_A^{(n)}(1 \otimes x) Q^{(n)}(r)$ by lemma 1. If we expand the last term of this equality we obtain the equality of generating series. The last statement is obvious from this equality. QED

The endofunctor K has a monad structure $m : KK \to K$, $u : I \to K$ obtained from the algebra structure on $\mathcal{K}$. An action $q : K(V) \to V$ by this monad is equivalent to a left $\mathcal{K}$-module structure on V, which is precisely a Q-module structure on V. Similarly, the functor A has a comonad structure $\Delta : A \to AA$, $\epsilon : A \to I$ obtained from the coalgebra structure on $\mathcal{A}$. A coaction $\psi : V \to A(V)$ by this comonad is equivalent to a right $\mathcal{A}$-comodule structure on V.

THEOREM 1. *The Nishida operator $\rho : KA \to AK$ respects the monad structure of K and the comonad structure of A. The Nishida relations hold for an action $q : K(M) \to M$ and a coaction and $\psi : M \to A(M)$ iff the pair (q, ψ) is ρ-commuting.*

REMARK. The theorem could be formulated in terms of a commutation operator between an algebra and a coalgebra rather than between a monad and a comonad. For this we need to use the operator

$$\tilde{\rho} : \mathcal{K} \otimes \mathcal{A} \to \mathcal{A} \otimes \mathcal{K}$$

obtained by postcomposing $\rho_{Z_2} : \mathcal{K} \otimes \mathcal{A} \to \mathcal{K} \otimes \mathcal{A}$ with the symmetry isomorphism $\mathcal{K} \otimes \mathcal{A} \simeq \mathcal{A} \otimes \mathcal{K}$. Note that the theorem would then say that $\tilde{\rho}$ is respecting the *opposite coalgebra structure* (Δ^o, ϵ) of $\mathcal{A}$ (and the algebra structure of $\mathcal{K}$). We have chosen the monadic formulation only for expository reasons. See the remark after Proposition 5 Appendix C.

PROOF. Let us first see that for any action $q : K(V) \to V$ the composite

$$KA(V) \xrightarrow{\rho_V} AK(V) \xrightarrow{A(q)} A(V)$$

is the action $q' : KA(V) \to A(V)$ defining the Q-structure Q'_t on $A(V)$. In the usual notation this composite is

$$\mathcal{K} \otimes (V \otimes \mathcal{A}) \xrightarrow{\rho_V} (\mathcal{K} \otimes V) \otimes \mathcal{A} \xrightarrow{q \otimes \mathcal{A}} V \otimes \mathcal{A}.$$

But ρ_V and $q \otimes \mathcal{A}$ are Q-module maps; hence so is their composite $(q \otimes \mathcal{A})\rho_V$. Moreover, $(q \otimes \mathcal{A})\rho_V(1 \otimes x \otimes r) = (q \otimes \mathcal{A})(1 \otimes x \otimes r) = x \otimes r$. This proves that $(q \otimes \mathcal{A})\rho_V = q'$. Applying proposition 2 of appendix C, we see that ρ commutes with the monad structure of K. Let us prove that ρ respects the comonad structure of A. For this we need the following lemma.

LEMMA 2. *If $f(x)$ and $g(x)$ are Q-parameters in Q-rings R and S respectively then so is $f \circ g$ in $R \otimes S$ for $Q'_t = Q^g$. Moreover, for any Q-module M we have $Q^{f \circ g} = (Q^f)^g$ in $M \otimes R \otimes S$.*

PROOF. From the relation $Q_t(f)(s(s+t)) = f(s)f(s+t)$ it follows by substituting $t \mapsto g(t)$ and $s \mapsto g(s)$ that $Q_{g(t)}(f)(g(s)(g(s) + g(t))) = f(g(s))f(g(s) + g(t))$. Hence

$$Q'_t(f \circ g)(s(s+t)) = Q_{g(t)}(f) \circ Q_t(g)(s(s+t)) = Q_{g(t)}(f)(g(s)(g(s) + g(t)))$$
$$= f(g(s))f(g(s+t))$$

and the first statement is proved. For the second statement we have

$$(Q^f)^g(x \otimes r \otimes s) = (Q^f)_{g(t)}(x \otimes r)Q_t(s) = Q_{f(g(t))}(x)Q_{g(t)}(r)Q_t(s)$$
$$= Q_{f \circ g(t)}(x)Q'_t(r \otimes s) = Q^{f \circ g}(x \otimes r \otimes s)$$

for any $x \in M$, $r \in R$ and $s \in S$. QED

We can now continue the proof of Theorem 1. If we apply proposition 3 of appendix C we see that it is enough to prove that the maps $V \otimes \Delta : V \otimes \mathcal{A} \to V \otimes \mathcal{A} \otimes \mathcal{A}$ and $M \otimes \epsilon : V \otimes \mathcal{A} \to V$ are Q-module maps for any Q-module V. Here $(V \otimes \mathcal{A}) \otimes \mathcal{A}$ has the Q-structure Q''_t obtained by successive applications of proposition 1. By definition $Q''_t = (Q^f)^g$, where $f(t)$ and $g(t)$ are respectively denoting the power series $(a \otimes 1)(t)$ and $(1 \otimes a)(t)$. We have $\Delta(a) = f \circ g$ and the lemma shows that $\Delta(a)$ is a change of parameter for the Q-structure Q'_t on $\mathcal{A} \otimes \mathcal{A}$. It follows that Δ is a map of Q-rings if $\mathcal{A} \otimes \mathcal{A}$ has the Q-structure Q'_t. Hence $M \otimes \Delta$ is a Q-module map if $(V \otimes \mathcal{A}) \otimes \mathcal{A}$ has the Q-structure $Q^{\Delta(a)}$. But the lemma shows that $Q^{\Delta(a)} = (Q^f)^g = Q''_t$, and this shows $M \otimes \Delta$ is a Q-module map if $(V \otimes \mathcal{A}) \otimes \mathcal{A}$ has the Q-structure Q''_t. The proof that $V \otimes \epsilon : V \otimes \mathcal{A} \to V$ is a Q-module map is immediate. Let us prove the last statement. Let $q : K(V) \to V$ be an action and $\psi : V \to A(V)$ a coaction. According to proposition 4 appendix C the pair (q, ψ) is ρ-commuting iff ψ is a $\mathcal{K}$-module map. QED

COROLLARY. *The unique Q-module map $\psi' : \mathcal{K} \otimes V \to (\mathcal{K} \otimes V) \otimes \mathcal{A}$ extending a coaction $\psi : V \to V \otimes \mathcal{A}$ is a coaction. If V is a Q-module with structure map $q : \mathcal{K} \otimes V \to V$ then the Nishida relations hold between q and ψ iff q is a comodule map where $\mathcal{K} \otimes V$ has the coaction ψ'.*

PROOF. The map ψ' is equal to the composite

$$\mathcal{K} \otimes V \xrightarrow{\mathcal{K} \otimes \psi} \mathcal{K} \otimes (V \otimes \mathcal{A}) \xrightarrow{\rho_V} (\mathcal{K} \otimes V) \otimes \mathcal{A}$$

since this composite is a Q-module map extending ψ. The Theorem shows that ρ respects the comonad structure of $\mathcal{A}$. Hence by the dual of proposition 2 appendix C the map ψ' is a coaction. Let us prove the last statement. According to the theorem 1 the Nishida relations hold for (q, ψ') iff the pair (q, ψ') is ρ-commuting. The result then follows from proposition 4 appendix C. QED

DEFINITION 3. The *natural extension* $\psi' : \mathcal{K} \otimes V \to (\mathcal{K} \otimes V) \otimes \mathcal{A}$ of a coaction $\psi : V \to V \otimes \mathcal{A}$ is the unique Q-module map extending ψ.

Let us calculate the grading that is obtained from the coaction ψ' on $\mathcal{K} \otimes V$. Observe first that the inclusion $V \subset \mathcal{K} \otimes V$ is homogenous of degree 0 since ψ' extends ψ. Hence $\dim(1 \otimes x) = \dim(x)$ for any homogenous $x \in V$. Observe also that the Nishida relations hold since ψ' is a Q-module map. Hence, according to proposition 2, $\mathcal{K} \otimes V$ is a graded Q-module with the grading induced from ψ'. This shows in particular that $\dim(q_n \otimes x) = 2 \dim(1 \otimes x) + n = 2 \dim(x) + n$ for any homogenous $x \in V$. More information about ψ' is contained in the following proposition.

PROPOSITION 5. *Let* $\psi' : \mathcal{K} \otimes V \to (\mathcal{K} \otimes V) \otimes \mathcal{A}$ *be the natural extension of a coaction. Then for every* $n \geq 0$ *and* $x \in V$ *we have an equality of generating series*

$$\sum_{l(K)=n} \psi'(q_K \otimes x)\, w^K = \sum_i \sum_{l(K)=n} q_K \otimes x_i \otimes Q^{(n)}(r_i)\, A(w)^K$$

where $\psi(x) = \sum_i x_i \otimes r_i$. *In particular,* ψ' *preserves the decomposition* $\mathcal{K} \otimes V = \bigoplus_n \mathcal{K}(n) \otimes V$, *and* ψ' *is positive when* ψ *is.*

PROOF. If $x \in V$ and $\psi(x) = \sum_i x_i \otimes r_i$ then

$$\sum_{l(K)=n} \psi'(q_K \otimes x)\, w^K = \sum_{l(K)=n} \rho_V(q_K \otimes x_i \otimes r_i)\, w^K$$

since $\psi' = \rho_V(\mathcal{K} \otimes \psi)$. The formula then follows from proposition 4. The second statement is clear from the formula. If ψ is positive then $\psi(x) = \sum_i x_i \otimes r_i$ where $r_i \in \mathcal{A}^+$. But then $Q^{(n)}(r_i) \in \mathcal{A}^+$ since $\mathcal{A}^+$ is a sub-Q-ring of $\mathcal{A}$. The formula then shows that ψ' is positive. QED

The natural extension of the trivial coaction on Z_2 is the unique Q-module map $\psi' : \mathcal{K} \to \mathcal{K} \otimes \mathcal{A}$ such that $\psi'(1) = 1$. We shall say that this is the *natural coaction* on $\mathcal{K}$ and denote it by ψ.

COROLLARY. *The natural coaction* $\psi : \mathcal{K} \to \mathcal{K} \otimes \mathcal{A}$ *is positive and preserves the decomposition* $\mathcal{K} = \bigoplus_n \mathcal{K}(n)$. *For each* $n \geq 0$ *we have an equality of generating series*

$$\sum_{l(K)=n} \psi(q_K) w^K = \sum_{l(K)=n} q_K\, A(w)^K.$$

T. P. BISSON AND A. JOYAL

PROOF. It suffices to specialise the formula of the proposition in the case where $x = 1$ and $\psi(1) = 1 \otimes 1$. QED

It follows from this corollary that each natural coaction $\psi : \mathcal{K}(n) \to \mathcal{K}(n) \otimes \mathcal{A}$ is positive. By the corollary to proposition 1 §4 there is a corresponding Q-structure on the graded dual $\mathcal{K}'(n) = \mathcal{W}(n)$.

PROPOSITION 6. *The Q-structure on $\mathcal{W}(n)$ is a Q-ring structure. It is determined by the identity $Q_t(W_n)(x(x+t)) = W_n(x)W_n(x+t)$ where $W_n(x)$ is the generic Ore polynomial. $\mathcal{W}(n)$ is a graded Q-ring with $codim(w_i) = 2^n - 2^i$.*

PROOF. The family of elements $(q_K : l(K) = n)$ forms a basis of $\mathcal{K}(n)$ which is dual to the monomial basis $(w^K : l(K) = n)$ of $\mathcal{W}(n)$. By definition, the operation Q_t on $\mathcal{W}(n)$ is obtained by transposing the operation Q_t^* on $\mathcal{K}(n)$. Hence

$$\sum_{l(K)=n} q_K Q_t(w^K) = \sum_{l(K)=n} Q_t^*(q_K) w^K.$$

By definition, Q_t^* is obtained by specialising ψ with $a(x) \mapsto xt + x^2$. This means that we have

$$\sum_{l(K)=n} Q_t^*(q_K) w^K = \sum_{l(K)=n} q_K A(w)^K$$

where the endomorphism A is defined by $a(x) = xt + x^2$. Combining these equalities, we have that $Q_t(w^R) = A(w)^R$ for every R and this proves that $Q_t = A$ if $a(x) = xt + x^2$. By definition of the endomorphism A we have

$$A(W_n)(x(x+t)) = \prod_{v \in \langle t_1, \ldots, t_n \rangle} x(x+t) + a(v) = \prod_{v \in \langle t_1, \ldots, t_n \rangle} x(x+t) + v(v+t)$$

$$= \prod_{v \in \langle t_1, \ldots, t_n \rangle} (x+v)(x+v+t) = W_n(x)W_n(x+t).$$

In particular $q_0(W_n) = Q_0(W_n) = W_n^2$, and it follows that q_0 is the Frobenius endomorphism of $\mathcal{W}(n)$. This finishes the proof that $\mathcal{W}(n)$ is a Q-ring. The grading on $\mathcal{W}(n)$ is exactly dual to the dimension grading of $\mathcal{K}(n)$. QED

Let Vect^K denote the category of Q-modules and let Vect^A denote the category of Milnor comodules. We shall denote by Vect^{KA} the category of vector spaces equipped with a pair (Q_t, ψ) for which the Nishida relations hold.

PROPOSITION 7.
(i) *The functor $M \mapsto M \otimes \mathcal{A}$ is right adjoint to the forgetful functor $\mathrm{Vect}^{KA} \to \mathrm{Vect}^K$, where $M \otimes \mathcal{A}$ has the Q-structure Q_t'.*
(ii) *The functor $V \mapsto \mathcal{K} \otimes V$ is left adjoint to the forgetful functor $\mathrm{Vect}^{KA} \to \mathrm{Vect}^A$, where $\mathcal{K} \otimes V$ has Milnor coaction ψ'.*

PROOF. This follows directly from proposition 5 of appendix C. QED

We shall now prove a converse to proposition 2. We need the following result.

LEMMA 3. *Let $f : U \to V$ be a linear map homogenous of degree 0 between two $\mathcal{A}$-comodules. Then f is a comodule map iff f commutes with Q_t^*.*

PROOF. The implication $\Rightarrow$ is trivial. Let us suppose that $f \circ q_n^* = q_n^* \circ f$ for every $n \in Z$. It follows from the discussion above proposition 2 §4 that $f(Sq_*^n(x)) = Sq_*^n(f(x))$ for every $n \geq 0$ and $x \in U$. Hence f is a map of $\mathcal{S}_*$-comodules, since it is a classical result that the Sq_*^n's generate; see Milnor [1958], for instance. Thus f is a map of $\mathcal{A}$-comodules by proposition 2, appendix B. QED

PROPOSITION 8. *If conditions (i) and (ii) of proposition 2 are satisfied for (M, Q_t, ψ) then the Nishida relations hold.*

PROOF. We must show that $\psi : M \to M \otimes \mathcal{A}$ is a Q-module map. We shall use the fact that the following diagram commutes

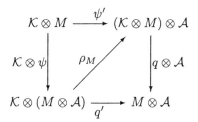

where q is the structure map of the Q-module M, q' is the structure map of $\mathcal{K} \otimes M$ equipped with Q_t' and ψ' is the natural extension of ψ (the commutativity is clear from the proofs of theorem 1 and its corollary; see also proposition 4 appendix C). Let us now suppose that conditions (i) and (ii) are satisfied for (M, Q_t, ψ). According to proposition 5, the map ψ' induces a coaction $\psi'(1)$ on $\mathcal{K}(1) \otimes M$. We shall begin by proving that the restriction $q(1)$ of q to $\mathcal{K}(1) \otimes M$ is a comodule map. For this we shall use the lemma above with $f = q(1)$. Let us first see that f is homogenous of degree 0. We saw above proposition 5 that $\dim(q_n \otimes x) = 2 \dim(x) + n$ for homogenous $x \in M$. But $\dim(q_n(x)) = 2 \dim(x) + n$ by condition (i). Thus f is homogenous of degree 0 since $f(q_n \otimes x) = q_n(x)$. It remains to verify that f preserves the operation Q_t^*. We shall show that $f(Q_s'^*(y)) = Q_s^*(f(y))$ for every $y = q_n \otimes x \in \mathcal{K}(1) \otimes M$. If we specialise the formula of proposition 5 with $n = 1$ and $a(x) \mapsto v_s(x) = x(x + s)$ we obtain that

$$\sum_n Q_s'^*(q_n \otimes x)t^n = \sum_i \sum_n q_n \otimes q_i^*(x)(s^2 + st)^i(t^2 + st)^n$$

for every $x \in M$ since $Q_s^*(x) = \sum_i q_i^*(x)s^i$ and $Q_t(s) = s^2 + st$. Hence

$$\sum_n f(Q_s'^*(q_n \otimes x))t^n = \sum_i \sum_n f(q_n \otimes q_i^*(x))(s^2 + st)^i(t^2 + st)^n$$
$$= \sum_i \sum_n q_n(q_i^*(x))(s^2 + st)^i(t^2 + st)^n$$

for every $x \in M$. On the other hand

$$\sum_n Q_s^*(f(q_n \otimes x))t^n = \sum_n Q_s^*(q_n(x))t^n$$

and this shows that the equality $f(Q'^*_s(q_n \otimes x)) = Q^*_s(f(q_n \otimes x))$ follows from condition (ii). We have proved that $q(1)$ is a comodule map. From the observation made at the beginning of the proof, the diagram

$$
\begin{array}{ccc}
\mathcal{K}(1) \otimes M & \xrightarrow{\psi'(1)} & (\mathcal{K}(1) \otimes M) \otimes \mathcal{A} \\
\mathcal{K}(1) \otimes \psi \downarrow & \rho_M(1) \nearrow & \downarrow q(1) \otimes \mathcal{A} \\
\mathcal{K}(1) \otimes (M \otimes \mathcal{A}) & \xrightarrow[q'(1)]{} & M \otimes \mathcal{A}
\end{array}
$$

commutes where $q'(1)$ denotes the restriction of q' and $\rho_M(1)$ denotes the restriction of $\rho_M(1)$ (which exists by Proposition 4). Hence if one of the following two squares commutes then so does the other:

$$
\begin{array}{ccc}
\mathcal{K}(1) \otimes M \xrightarrow{\mathcal{K}(1) \otimes \psi} \mathcal{K}(1) \otimes (M \otimes \mathcal{A}) & \qquad & \mathcal{K}(1) \otimes M \xrightarrow{\psi'(1)} (\mathcal{K}(1) \otimes M) \otimes \mathcal{A} \\
q(1) \downarrow \qquad\qquad \downarrow q'(1) & \qquad & q(1) \downarrow \qquad\qquad \downarrow q(1) \otimes \mathcal{A} \\
M \xrightarrow[\psi]{} M \otimes \mathcal{A} & \qquad & M \xrightarrow[\psi]{} M \otimes \mathcal{A}.
\end{array}
$$

But the second square commutes since we have shown that $q(1)$ is a comodule map. Hence also the first. But this means that ψ is a Q-module map since $\mathcal{K}(1)$ generates $\mathcal{K}$. QED

PROPOSITION 9. *If M and N are Q-modules with Milnor coactions for which the Nishida relations hold then so is $M \otimes N$.*

PROOF. According to proposition 8, it suffices to show that the conditions (i) and (ii) of proposition 2 are satisfied. This is clear for (i) since the grading on $M \otimes N$ is compatible with the Q-structure. For condition (ii) we need to verify that

$$
Q^*_s Q_t(x \otimes y) = \sum_{r,k} q_k(q^*_r(x \otimes y))(t^2 + st)^k (s^2 + st)^r
$$

for $x \otimes y \in M \otimes N$. But this follows directly from the Cartan formula for the q_k's and the q^*_j's. QED

We end this section with a few propositions about extending Milnor coactions from modules to rings so as to satisfy the Nishida relations.

If M is a Frobenius module (see §2 for this concept) then so is $M \otimes \mathcal{A}$ with $q_0(x \otimes r) = q_0(x) \otimes r^2$. We shall say that a coaction $\psi : M \to M \otimes \mathcal{A}$ *commutes* with a Frobenius endomorphism $q_0 : M \to M$ if the relation $\psi(q_0(x)) = q_0(\psi(x))$ holds. This is certainly the case when q_0 is obtained from a Q-structure Q_t and the Nishida relations hold. Recall that a Q-structure on M extends to a Q-ring structure on $\tilde{\Lambda} M$ (proposition 1 §3).

PROPOSITION 10. *If a coaction $\psi : M \to M \otimes \mathcal{A}$ commutes with a Frobenius endomorphism $q_0 : M \to M$ then it has a unique extension $\psi : \tilde{\Lambda} M \to \tilde{\Lambda} M \otimes \mathcal{A}$ as* *a ring coaction. Moreover, if q_0 is obtained from a Q-structure Q_t and the Nishida relations hold for (Q_t, ψ) then they also hold for their ring extension to $\tilde{\Lambda}M$.*

PROOF. Let us suppose that ψ commutes with q_0. Then

$$\psi(q_0(x)) = q_0(\psi(x)) = \sum_R q_0(\psi_R(x)) \otimes a^{2R}$$

for any $x \in M$. Let $f : M \to \tilde{\Lambda}M \otimes \mathcal{A}$ be the composite $(i \otimes \mathcal{A})\psi$ where $i : M \to \tilde{\Lambda}M$ is the inclusion. Then $i(q_0(x)) = i(x)^2$ and it follows from the equality above that $f(q_0(x)) = f(x)^2$. Hence by the universal property of $\tilde{\Lambda}M$ there is a unique ring homomorphism $\overline{\psi} : \tilde{\Lambda}M \to \tilde{\Lambda}M \otimes \mathcal{A}$ extending f. The coassociativity of $\overline{\psi}$ follows from the coassociativity of ψ together with the fact that M generates $\tilde{\Lambda}M$ as a ring. For the counit the proof is similar. This finishes the proof of the first statement. Let us now suppose that q_0 is obtained from a Q-structure Q_t and that the Nishida relations hold for (Q_t, ψ). Then f is a Q-module map since ψ and i are. Hence $\overline{\psi}$ is a Q-ring map since $\tilde{\Lambda}M$ is freely generated by M as a Q-ring (proposition 1 §3). This proves that the Nishida relations hold for $(\overline{Q}_t, \overline{\psi})$; where $\overline{Q}_t$ denotes the ring extension of Q_t to $\tilde{\Lambda}M$. QED

COROLLARY. *Let $Q\langle V \rangle$ be the Q-ring freely generated by a Milnor comodule (V, ψ). Then the coaction $\psi : V \to V \otimes \mathcal{A}$ has a unique extension $\psi' : Q\langle V \rangle \to V\langle M \rangle \otimes \mathcal{A}$ as ring coaction for which the Nishida hold.*

PROOF. The uniqueness of ψ' is clear since $Q\langle V \rangle$ is free over V as a Q-ring and since ψ' is a map of Q-rings if it is a ring coaction satisfying the Nishida relations. To prove its existence it suffices to take for ψ' the Q-module map extending ψ. We then have to prove that ψ' is a coaction. This can be proved directly or can be seen by applying the proposition to the natural extension $\psi' : \mathcal{K} \otimes V \to \mathcal{K} \otimes V \otimes \mathcal{A}$ since $Q\langle V \rangle = \tilde{\Lambda}(\mathcal{K} \otimes V)$. QED

Addendum to Section 5.

Suppose that X is an E_∞-space. The *classical Nishida relations* describe how the operations Sq_*^i commute with the Dyer-Lashof operations. The relations are expressed as an equality

$$Sq_*^a Q^b(x) = \sum_c \binom{b-a}{a-2c} Q^{b+c-a} Sq_*^c(x)$$

where Q^b is the Dyer-Lashof operation; see Nishida [1968] or May [1971], for instance. Let us show that these are equivalent to the relation (*) in the remark following proposition 2. By definition, $Q^b(x) = q_{b-n}(x)$ if $x \in H_n(X)$ and $b \geq n$, and $Q^b(x) = 0$ otherwise. Recall that if $x \in H_n(X)$ then we have $Sq_*^c(x) = q_{n-2c}^*(x)$ and $q_i^*(x) = 0$ unless $i + n$ is even. Supposing that $x \in H_n(X)$ we have $Sq_*^a Q^b(x) = q_i^* q_j(x)$ where $i = b + n - 2a$ and $j = b - n$. The relation (*) then gives $Sq_*^a Q^b(x) = \sum_k \binom{b-a}{b-n-k} q_k q_{b-a-k}^*(x)$. Half of the terms in the sum are null since $q_{b-a-k}^*(x) = 0$ when $b - a - k + n$ is odd. If we write $k = b + 2c - a - n$ then we have $q_k q_{b-a-k}^*(x) = Q^{b+c-a} Sq_*^c(x)$ and $b - n - k = a - 2c$. This proves the claim. It follows that the classical Nishida relations are equivalent to condition (ii) of Proposition 2. Hence, by proposition 8, our Nishida relations hold for $H_* X$.

In a sequel to this paper, we will use cobordism theory to give a direct geometric proof that our Nishida relations hold for the homology of an E_∞-space.

6. Q-coalgebras.

In this section we study some natural coalgebra structures on Q-modules and rings.

DEFINITION 1. A Q-*coalgebra* is a Q-module with coalgebra structure (M, δ, ϵ) such that $\delta : M \to M \otimes M$ and $\epsilon : M \to Z_2$ are Q-module maps. If in addition M is a Q-ring and a bialgebra we shall say that it is a Q-*bialgebra*.

The Kudo-Araki algebra $\mathcal{K}$ is an example of a Q-coalgebra. Many of the usual constructions with coalgebras and algebras can be done with Q-coalgebras and Q-rings. For example, if C is a Q-coalgebra and R is a Q-ring then the set of Q-module maps $C \to R$ is a subalgebra of the convolution algebra $[C, R]$. If C is a Q-bialgebra then the set of Q-ring maps $C \to R$ is a monoid for the convolution product. Conversely, if a functor from Q-rings to monoids is representable by a Q-ring then this Q-ring has a Q-bialgebra structure. If a Q-bialgebra has an antipode then this antipode is a map of Q-rings.

EXAMPLE 1. Consider the functor $R \mapsto E(R)$ which associates to a Q-ring R the set $E(R)$ of elements $u \in R$ such that $Q_t(u) = u(u + t)$. This functor is representable by the Q-ring $Z_2[x]$ with Q-structure given by $Q_t(x) = x(x + t)$. The functor E has a group structure since $E(R)$ is closed under addition. We obtain a Q-bialgebra structure on $Z_2[x]$ with $\delta : Z_2[x] \to Z_2[x] \otimes Z_2[x]$ given by $\delta(x) = x \otimes 1 + 1 \otimes x$.

EXAMPLE 2. Consider the functor $R \mapsto F(R)$ which associates to a Q-ring R the set of formal power series $f(x) \in R[[x]]$ satisfying $Q_t(f)(x(x+t)) = f(x)f(x+t)$. This functor is representable by the Q-ring $Z_2[\beta_*] = Z_2[\beta_0, \beta_1, \ldots]$ of proposition 4 §1. The generic formal power series is $\beta(x) = \sum_i \beta_i x^i$. If $f \in F(R)$ and $g \in F(R)$ then

$$Q_t(fg)(x(x + t)) = Q_t(f)(x(x + t))Q_t(g)(x(x + t))$$
$$= f(x)f(x + t)g(x)g(x + t) = (fg)(x)(fg)(x + t).$$

It follows that $F(R)$ has a monoid structure. The map $\delta : Z_2[\beta_*] \to Z_2[\beta_*] \otimes Z_2[\beta_*]$ such that $\delta(\beta) = \beta \otimes \beta$ is representing the product of the monoid F. It follows that δ is a map of Q-rings. Hence we have described a Q-bialgebra structure on $Z_2[\beta_*]$.

EXAMPLE 3. The Q-ring $Q\langle x \rangle$ is a Q-bialgebra in two ways. Since $Q\langle x \rangle$ is free, a Q-ring map $Q\langle x \rangle \to R$ is determined by its values at x. Thus $Q\langle x \rangle$ is representing the forgetful functor U from Q-rings to Sets. The functor U has also an algebra structure since $U(R) = R$ is an algebra. It follows that U has two monoid structures, one obtained from the multiplication and the other from the addition. The multiplication is represented by the map of Q-rings $\delta : Q\langle x \rangle \to Q\langle x \rangle \otimes Q\langle x \rangle$ such that $\delta(x) = x \otimes x$. The addition is represented by the map of Q-rings $\sigma : Q\langle x \rangle \to Q\langle x \rangle \otimes Q\langle x \rangle$ such that $\sigma(x) = x \otimes 1 + 1 \otimes x$. Put together, these two comultiplications are giving $Q\langle x \rangle$ the structure of a (dual) Hopf ring (see the remark below).

EXAMPLE 4. The Q-ring $Q\langle x, x^{-1} \rangle = \{x\}^{-1}Q\langle x \rangle$ is representing the functor which associates to a Q-ring R its group $\mu(R)$ of invertible elements. Hence $Q\langle x, x^{-1} \rangle$ is an Hopf algebra.

Let $C = (C, \delta, \epsilon)$ be a coalgebra. Let us denote by $\delta : Q\langle C \rangle \to Q\langle C \rangle \otimes Q\langle C \rangle$ and $\epsilon : Q\langle C \rangle \to Z_2$ the Q-ring maps respectively extending $\delta : C \to C \otimes C$ and $\epsilon : C \to Z_2$.

PROPOSITION 1. *If C is a coalgebra then $Q\langle C \rangle$ is a Q-bialgebra with δ and ϵ.*

PROOF. For any Q-ring R the set $F(R)$ of Q-ring maps $Q\langle C \rangle \to R$ is in natural one to one correspondance with the set $[C, R]$ of linear maps $C \to R$ since $Q\langle C \rangle$ is free on C. But $[C, R]$ is a monoid for the convolution product defined from the coalgebra structure on C. Hence $F(R)$ has a natural monoid structure which must correspond to a coalgebra structure on $Q\langle C \rangle$. The verification that this coalgebra structure is represented by δ and ϵ is straightforward. QED

REMARK. The proof of the proposition shows that if C is a coalgebra then $Q\langle C \rangle$ is a (dual) *Hopf ring* (a ring object in the category opposite to algebras). This is because the functor $R \mapsto [C, R]$ which it represents has an algebra structure (the convolution algebra). The multiplicative structure is represented by δ; the additive structure is represented by the map σ studied in proposition 2 below.

COROLLARY. *The coalgebra structure on $Q\langle x \rangle$ defined by the Q-ring maps such that $\delta(x) = x \otimes x$ and $\epsilon(x) = 1$ restricts to induce the the coalgebra structure on $\mathcal{K}$.*

PROOF. We have to verify that the square

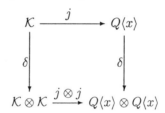

commutes. But the four sides are Q-module maps and the square commutes on the generator $1 \in \mathcal{K}$ since $j(1) = x$, $\delta(1) = 1 \otimes 1$ and $\delta(x) = x \otimes x$. QED

For any Frobenius module M let $\sigma : \tilde{\Lambda}M \to \tilde{\Lambda}M \otimes \tilde{\Lambda}M$ and $\nu : \tilde{\Lambda}M \to Z_2$ be the ring maps such that $\sigma(x) = x \otimes 1 + 1 \otimes x$ and $\epsilon(x) = 0$ for $x \in V$. Recall that an element x of a bialgebra is *primitive* if $\delta(x) = x \otimes 1 + 1 \otimes x$.

LEMMA 1. *For any Frobenius module M the maps σ and ν define a commutative Hopf algebra structure on $\tilde{\Lambda}M$. The subgroup of primitive elements of $\tilde{\Lambda}M$ is $M \subset \Lambda M$.*

PROOF. The first statement is easily proved using representable functors. By definition, $\sigma(x) = x \otimes 1 + 1 \otimes x$ for every $x \in M$ and this means that every element of M is primitive. Conversely, let us prove that every primitive element of $\tilde{\Lambda}M$ belongs to M. If $q_0 = 0$ then $\tilde{\Lambda}M$ is the exterior algebra ΛM and this is a classical result. We shall reduce the general problem to this case. Let $F_0 \subseteq F_1 \subseteq \cdots$ be the filtration of $\tilde{\Lambda}M$ used in Lemma 1 §3. According to this lemma $\mathrm{gr}\tilde{\Lambda}M = \Lambda M$. Suppose now that $x \in \tilde{\Lambda}M$ is a non-zero primitive element and let $n \geq 0$ be the smallest integer such that $x \in F_n$. Then the corresponding element $\bar{x}$ in $F_n/F_{n-1} = \Lambda^n M$ is primitive and it follows that $n = 1$. This shows that $x \in F_1$. It remains to prove that $x \in M$. Using the basis theorem 1 §3 it is easy to see that an element $x \in F_1$

belongs to M iff $\epsilon(x) = 0$. But $\epsilon(x) = 0$ since ϵ preserves primitive elements and 0 is the only primitive element of Z_2. This proves that $x \in M$. QED

PROPOSITION 2. *For any vector space V let $\sigma : Q\langle V \rangle \to Q\langle V \rangle \otimes Q\langle V \rangle$ and $\nu : Q\langle V \rangle \to Z_2$ be the Q-ring maps such that $\sigma(x) = x \otimes 1 + 1 \otimes x$ and $\epsilon(x) = 0$ for $x \in V$. Then σ and ν are defining a commutative Hopf algebra structure on $Q\langle V \rangle$. For this structure, the subgroup of primitive elements is $\mathcal{K} \otimes V \subset Q\langle V \rangle$.*

PROOF. To prove that $Q\langle V \rangle$ is a commutative Hopf algebra it suffices to show that the functor F it represents has a natural abelian group structure. But for any Q-ring R the set $F(R)$ is in natural bijection with the set $[V, R]$ of linear maps $V \to R$. Addition gives this set an abelian group structure. The verification that the structure is represented by σ and ν is straightforward. It remains to prove the second part. But this is a direct consequence of the preceding lemma applied to the case where $M = \mathcal{K} \otimes V$. QED

COROLLARY. *The Kudo-Araki algebra $\mathcal{K}$ is the subgroup of primitive elements of $Q\langle x \rangle$ with the Hopf algebra structure defined by σ and ν.*

REMARK. Here is another proof of the corollary. According to proposition 2 §3 we have $Q\langle x \rangle = S(\mathcal{K}^\flat)$. For any vector space U the symmetric algebra $S(U)$ has a Hopf algebra structure obtained from the addition operation on U. It is easy to see that the subgroup of primitive elements of $S(U)$ is the closure cl U of $U \subset S(U)$ under the Frobenius automorphism of $S(U)$. If $U = \mathcal{K}^\flat$ then the set of primitive element of $S(\mathcal{K}^\flat)$ is cl $\mathcal{K}^\flat = \mathcal{K}$.

Addendum to Section 6.

The homology of any space X has a coalgebra structure $\delta : H_*X \to H_*X \otimes H_*X$ obtained from the diagonal $X \to X \times X$. If X is an E_∞-space then H_*X is a Q-bialgebra; it is a Hopf algebra if X is group-like. For example the homology of BO_* (which is the classifying space for real vector bundles) is the Q-bialgebra $Z_2[\beta_*]$ considered in example 2. As another example, the homology of $\Omega^\infty S^\infty$ (which is the classifying space of cohomotopy) is the Hopf algebra $Q\langle x, x^{-1} \rangle$ considered in example 4.

If $E_\infty(X)$ is the free E_∞-space generated by a space X then $H_*E_\infty(X)$ is the free Q-ring generated by H_*X. Our proposition 1 relates the diagonal structure on $H_*E_\infty(X) = Q\langle H_*X \rangle$ with the diagonal structure on H_*X. In particular, the diagonal $H_*\Sigma_* \to H_*\Sigma_* \otimes H_*\Sigma_*$ is given by the map of Q-rings $\delta : Q\langle x \rangle \to Q\langle x \rangle \otimes Q\langle x \rangle$ such that $\delta(x) = x \otimes x$.

The algebra $Q\langle x \rangle$ becomes a Hopf algebra when equipped with the map $\sigma : Q\langle x \rangle \to Q\langle x \rangle \otimes Q\langle x \rangle$ such that $\delta(x) = x \otimes 1 + 1 \otimes x$. Here is the "geometric" origin of this map. The ring structure on $H_*\Sigma = H_*B\Sigma_*$ is derived from the map $\alpha : B\Sigma_* \times B\Sigma_* \to B\Sigma_*$ induced by the subgroup inclusions $\Sigma_m \times \Sigma_n \to \Sigma_{n+m}$. It follows that α has the homotopy type of a finite covering space. By transfer we obtain a map $\alpha^! : H_*\Sigma_* \to H_*\Sigma_* \otimes H_*\Sigma_*$. This is σ.

A good source for coalgebras is Sweedler [1969]. For background on Hopf rings, see Ravenel, Wilson [1977], for instance. For some Hopf ring calculations directly related to our work here, see Turner [1996].

Appendix A

We recall a few basic algebraic facts about representable functors, Hopf algebras and convolution products.

Let Alg be the category of (commutative) Z_2-algebras. Recall that a set valued functor $F : Alg \to Sets$ is *represented* by a pair (a, A), where $A \in Alg$ and $a \in F(A)$, if for any Z_2-algebra R and any $b \in F(R)$ there is a unique algebra map $f : A \to R$ such that $F(f)(a) = b$. The element $a \in F(A)$ is said to be *universal* or *generic* in F. An element $b \in F(R)$ is then a *specialisation* of a; if $F(f)(a) = b$ we say that f is representing b and we shall write $f = b^\dagger$. Thus $b = F(b^\dagger)(a)$ and $f = F(f)(a)^\dagger$. Notice that for any $b \in F(R)$ and any map $g : R \to S$ we have $F(g)(b)^\dagger = g \circ b^\dagger$. Notice also that $a^\dagger$ is the identity map $A \to A$.

If G is another functor represented by (b, B) we say that a natural transformation $t : F \to G$ is *represented* by a map $f : B \to A$ if $G(f)(b) = t_A(a)$. The product functor $F \times G$ is then represented by $((a \otimes 1, 1 \otimes b), A \otimes B)$, where $a \otimes 1 \in F(A \otimes B)$ denotes the image of a along the canonical map $A \to A \otimes B$ and similarly for $1 \otimes b$.

For any pair U and V of Z_2-vector spaces we shall denote by $[U, V]$ the vector space of linear maps $U \to V$. Recall that if $\mathcal{C}$ is a coalgebra and R is an algebra then $[\mathcal{C}, R]$ is an associative algebra for the convolution product. More generally, let $\phi : M \to M \otimes \mathcal{C}$ be a right comodule over $\mathcal{C}$. The *convolution* of $f \in [M, R]$ with $g \in [\mathcal{C}, R]$ is the element $f \star g \in [M, R]$ obtained by composing

$$M \xrightarrow{\ \phi\ } M \otimes \mathcal{C} \xrightarrow{\ f \otimes g\ } R \otimes R \xrightarrow{\ m\ } R$$

where m is multiplication. The convolution product gives $[M, R]$ the structure of a right module over $[\mathcal{C}, R]$.

Recall that for any $g \in [\mathcal{C}, R]$ the *dot product* $g \bullet (-) : M \otimes R \to M \otimes R$ is the R-linear extension of the composite

$$M \xrightarrow{\ \phi\ } M \otimes \mathcal{C} \xrightarrow{\ M \otimes g\ } M \otimes R.$$

The dot product gives $M \otimes R$ the structure of a left module over $[\mathcal{C}, R]$. For any $f \in [M, R]$, $g \in [\mathcal{C}, R]$ and $x \in M \otimes R$ we have the identity

$$\langle f, g \bullet x \rangle = \langle f \star g, x \rangle$$

where $\langle -, - \rangle$ denotes the natural pairing $[M, R] \otimes (M \otimes R) \to R$. If $\mathcal{C}$ is a bialgebra representing a functor G and $\alpha \in G(\mathcal{C})$ is the generic element then $\alpha \bullet x = \phi(x)$ for every $x \in M$.

If a pair $(a, \mathcal{G})$ represents a monoid-valued functor G then $\mathcal{G}$ is a bialgebra with coproduct $\Delta : \mathcal{G} \to \mathcal{G} \otimes \mathcal{G}$ representing the multiplication $G \times G \to G$, and with the augmentation $\epsilon : \mathcal{G} \to Z_2$ representing the unit element $1 \to G$. If $u, v \in G(R)$ are represented by algebra maps $f, g : \mathcal{G} \to R$, then the convolution product $f \star g$ represents the product $uv \in G(R)$. This can be expressed by the identity $(uv)^\dagger = u^\dagger \star v^\dagger$. If G is group-valued then $\mathcal{G}$ is a Hopf algebra with the antipode $\chi : \mathcal{G} \to \mathcal{G}$ representing the inverse operation $G \to G$. If $u \in G(R)$ is represented by $f : \mathcal{G} \to R$ then u^{-1} is represented by $f \circ \chi : \mathcal{G} \to R$.

A good source for representable functors is MacLane [1971]. For Hopf algebras and convolution algebras see Sweedler [1969] or Joyal, Street [1991c], for instance.

Appendix B

In this appendix we relate comodules over the extended Milnor coalgebra $\mathcal{A}$ to graded comodules over the standard Milnor coalgebra $\mathcal{S}_*$.

The vector space $Z_2\langle I\rangle$ freely generated by a set I has a coalgebra structure with comultiplication obtained from the diagonal $I \to I \times I$ and counit obtained from the map $I \to 1$. We shall say that it is a *diagonal coalgebra*. A coaction $\gamma : V \to V \otimes Z_2\langle I\rangle$ is equivalent to a grading $(V_i : i \in I)$ on V. More precisely, from a grading $(V_i : i \in I)$ we obtain a coaction $\gamma(x) = \sum_i \gamma_i(x) \otimes i$ where $\gamma_i : V \to V_i$ is the projection operator; conversely, from a coaction γ we obtain a grading with $x \in V_i$ iff $\gamma(x) = x \otimes i$.

For any Z_2-algebra R let $\mu(R)$ be the group of invertible elements of R. The functor $\mu : Alg \to Grp$ is represented by the algebra $Z_2[u^\pm] = Z_2[u^{-1}, u]$ of Laurent polynomials, with u as the generic element and with coproduct given by $\delta(u) = u \otimes u$. It is a diagonal coalgebra since $\delta(u^n) = u^n \otimes u^n$ for every $n \in Z$. It follows that a coaction $\gamma : V \to V \otimes Z_2[u^\pm]$ is equivalent to a Z-grading on V. We have $x \in V_n$ iff $\gamma(x) = x \otimes u^n$.

Recall the discussion of the extended Milnor coalgebra $\mathcal{A} = Z_2[a_0^\pm, a_1, \dots]$ in §4. The coalgebra $\mathcal{A}$ is representing the functor $A : Alg \to Grp$ which associates to each R the group $A(R)$ of invertible additive power series in $xR[[x]]$. Let us introduce two gradings on $\mathcal{A}$, each obtained from an action of the multiplicative group μ on A. The action by conjugation $(f(x), r) \mapsto r^{-1}f(rx)$ gives the *dimension grading*, for which $\dim(a_i) = 2^i - 1$. The action on the left $(r, f(x)) \mapsto rf(x)$ gives the *rank grading*, for which $\mathrm{rank}(a_i) = 1$. The monomials $a^R = a_0^{r_0}a_1^{r_1}\cdots$ with $R = (r_0, r_1, \dots)$ form a basis of $\mathcal{A}$ (notice that $r_0 \in Z$). For any $R = (r_0, r_1, \dots)$ (with $r_0 \in Z$) we have $\dim(a^R) = r_1(2^1 - 1) + r_2(2^2 - 1) + \cdots$ and $\mathrm{rank}(a^R) = r_0 + r_1 + \cdots$. We shall write $\dim(a^R) = \dim(R)$ and $\mathrm{rank}(a^R) = \mathrm{rank}(R)$.

Recall that a coaction $\psi : V \to V \otimes \mathcal{A}$ has a formal expansion

$$\psi(y) = \sum_R \psi_R(y) \otimes a^R$$

for $y \in V$; this determines natural operations ψ_R on Milnor comodules.

From any coaction $\psi : V \to V \otimes \mathcal{A}$ we can extract a grading $(V_n : n \in Z)$ by specialising the generic additive series $a(x)$ to ux. More precisely, the grading coaction $\gamma : V \to V \otimes Z_2[u^\pm]$ is obtained by putting $a_0 = u$ and $a_i = 0$ for $i > 0$ in the expansion of ψ. Hence

$$\gamma(y) = \sum_n \psi_{(n,0,\dots)}(y)u^n$$

for $y \in V$, and this means that $\psi_{(n,0,\dots)}$ is the projection operator $V \to V_n$. Notice that $\gamma(y) = u^\dagger \bullet y$ where $u^\dagger$ is the ring homomorphism $\mathcal{A} \to Z_2[u^\pm]$ such that $a(x) \mapsto ux$.

PROPOSITION 1. *Each ψ_R is an operation $\psi_R : V_{n+r} \to V_r$, where $n = dim(R)$ and $r = rank(R)$; $\psi_R = 0$ on V_m for $m \neq n + r$. In particular, $\psi : V \to V \otimes \mathcal{A}$ is preserving the (total) dimension.*

PROOF. Notice the identity $(ua)^R = a^R u^{\text{rank} R}$ where $(ua)(x) = u(a(x))$. From the associativity of ψ we obtain that

$$\sum \gamma(\psi_R(y))\, a^R = \sum \psi_R(y)\, (ua)^R = \sum \psi_R(y)\, a^R\, u^{\text{rank} R}.$$

Hence $\gamma(\psi_R(y)) = \psi_R(y)u^{\text{rank} R}$ and this proves that $\psi_R(y)$ is of dimension rank(R). Let us prove that ψ_R is 0 on V_m except when $m = \dim(R) + \text{rank}(R)$. Notice the identity $(au)^R = a^R u^{\dim R + \text{rank} R}$ where $(au)(x) = a(ux)$. From the associativity of ψ we obtain that

$$\sum_m \psi(\gamma_m(y))u^m = \sum \psi_R(y)\, (au)^R = \sum \psi_R(y)\, a^R\, u^{\dim R + \text{rank} R}.$$

By equating the coefficient of u^m on each side of this equality we obtain the result. This proves also that ψ_R is lowering dimension by an amount equal to $\dim(R) = \dim(a^R)$. Hence ψ is homogenous of degree 0. QED

The classical Milnor coalgebra $\mathcal{S}_*$ is representing the functor $S : Alg \to Grp$ which associates to $R \in Alg$ the group of additive power series $f(x) = \sum f_i x^{2^i}$ with $f_0 = 1$ (it is a group for the operation of substitution). We have $\mathcal{S}_* = Z_2[\xi_1, \xi_2, \dots]$, where $\xi(x) = x + \sum_{i>0} \xi_i\, x^{2^i}$ is the generic series. The coproduct $\Delta : \mathcal{S}_* \to \mathcal{S}_* \otimes \mathcal{S}_*$ is given by

$$\Delta(\xi) = (\xi \otimes 1) \circ (1 \otimes \xi).$$

The monomials $\xi^K = \xi_1^{k_1} \xi_2^{k_2} \cdots$ with $K = (k_1, k_2, \dots)$ form a basis of $\mathcal{S}_*$. An $\mathcal{S}_*$-coaction $\phi : V \to V \otimes \mathcal{S}_*$ has a formal expansion

$$\phi(x) = \sum_K \phi_K(x) \otimes \xi^K.$$

The *dimension grading* on $\mathcal{S}_*$ is given by $\dim(\xi_i) = 2^i - 1$. We say that a comodule $\phi : V \to V \otimes \mathcal{S}_*$ is *graded* if V is graded by dimension and ϕ preserves dimension. Suppose now that $\psi : V \to V \otimes \mathcal{A}$ is a coaction. Then V has a grading obtained from the specialisation $a(x) \mapsto ux$. Also, V has a coaction $\phi : V \to V \otimes \mathcal{S}_*$ obtained from the specialisation $a(x) \mapsto \xi(x)$. The coaction ϕ is preserving dimension since ψ and the specialisation $a(x) \mapsto \xi(x)$ are both dimension preserving. We have

$$\phi(x) = \sum_K \phi_K(x)\xi^K = \sum_K \sum_n \psi_{n,K}(x)\xi^K$$

where $n, K = (n, k_1, k_2, \dots)$ for $n \in Z$ and $K = (k_1, k_2, \dots)$. More precisely, for any $R = (r_0, r_1, \dots)$ we have

$$\psi_R(x) = \begin{cases} \phi_K(x) & \text{if } x \in V_n \\ 0 & \text{otherwise .} \end{cases}$$

where $K = (r_1, r_2, \dots)$ and $n = \dim(R) + \text{rank}(R)$. The following result is easy to prove:

PROPOSITION 2. *The specialisations $a(x) \mapsto \xi(x)$ and $a(x) \mapsto ux$ induce a one-to-one correspondence $\psi \mapsto \phi$ between the $\mathcal{A}$-coactions and the graded $\mathcal{S}_*$-coactions.*

REMARK. The above result depends on the fact that $\mathcal{A}$ is the semi-direct (tensor) product $Z_2[u^{\pm}] \otimes \mathcal{S}_*$ (see examples 1 and 3 of appendix C). The functorial group A is a semi-direct product $\mu \times S$ since the projection $f \mapsto f_0$ is a split homomorphism $A(R) \to \mu(R)$ with kernel $S(R)$.

WARNING. The traditional Milnor Hopf algebra is actually the opposite Hopf algebra $(\mathcal{S}_*, \Delta^o, \epsilon)$, with comultiplication Δ^o given by $\Delta^o(\xi) = (1 \otimes \xi) \circ (\xi \otimes 1)$. The comultiplication Δ^o is therefore representing composition of power series in the reverse order. The antipode is an isomorphism between $(\mathcal{S}_*, \Delta, \epsilon)$. and $(\mathcal{S}_*, \Delta^o, \epsilon)$. Right comodules over one may be regarded as left comodules over the other. Following Milnor [1958] most authors work with the corresponding *left* coaction over the Hopf algebra $(\mathcal{S}_*, \Delta^o, \epsilon)$. There is an intricate discussion of questions relating to left and right coactions in Boardman [1982].

Appendix C

The purpose of this appendix is to sketch an abstract theory of commutation operators.

Recall that a *monoidal or tensor category* is a category $\mathcal{C}$ equipped with a (tensor) product functor $\otimes : \mathcal{C} \times \mathcal{C} \to \mathcal{C}$ and a unit object $I \in \mathcal{C}$ together with natural associativity and unit isomorphisms, $A \otimes (B \otimes C) \simeq (A \otimes B) \otimes C$ and $A \otimes I \simeq A \simeq I \otimes A$, satisfying standard coherence conditions (see MacLane [1971]). A tensor category is *strict* if the associativity and unit isomorphisms are identity maps. We shall use MacLanes's coherence theorem. For all practical purpose it says that we can safely identify all the objects obtained by multiplying a given sequence of objects according to various bracketing patterns. One version of the theorem says that any tensor category is equivalent to a strict one. The tensor power $A^{\otimes n} = A^n$ of an object is defined inductively: $A^0 = I$ and $A^{n+1} = A \otimes A^n = A^n \otimes A$.

DEFINITION 1. Let $\mathcal{C}$ be a tensor category. A *commutation operator* between two objects A, B in $\mathcal{C}$ is an arrow $\rho : A \otimes B \to B \otimes A$.

The *powers* $^q\rho^p : A^p \otimes B^q \to B^q \otimes A^p$ of a commutation operator are easily defined with symbolic planar representations (Joyal, Street [1991b]). The diagram defining $^q\rho^p$ is a $p \times q$ diamond grid. For example, in the 3×4 case it is:

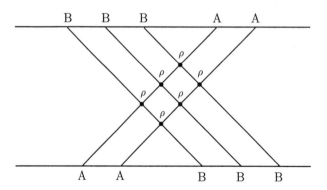

The operator $^3\rho^2 : A^2 \otimes B^3 \longrightarrow B^3 \otimes A^2$

The powers ρ^p and $^q\rho$ can be defined by induction: $\rho^0 : B \to B$ is the identity map and ρ^{p+1} is the composite

$$A^{p+1} \otimes B \xrightarrow{A \otimes \rho^p} A \otimes B \otimes A^p \xrightarrow{\rho \otimes A^p} B \otimes A^{p+1};$$

and similarly for $^q\rho$. We have $^q\rho^p = {}^q(\rho^p) = ({}^q\rho)^p$.

DEFINITION 2. Let $\mathcal{C}$ be a tensor category. We shall say that a commutation operator $\rho : A \otimes B \to B \otimes A$ *respects* an arrow $f : A^p \to A^q$, or that f *is (left) compatible with* ρ, if the square

$$
\begin{array}{ccc}
A^p \otimes B & \xrightarrow{f \otimes B} & A^q \otimes B \\
\downarrow{\scriptstyle \rho^p} & & \downarrow{\scriptstyle \rho^q} \\
B \otimes A^p & \xrightarrow{B \otimes f} & B \otimes A^q
\end{array}
$$

commutes. There is a similar concept of (right) compatibility for an arrow $g : B^p \to B^q$.

EXAMPLE. A Yang-Baxter operator $R : A \otimes A \to A \otimes A$ is a self-respecting commutation operator (Joyal, Street [1991a]).

PROPOSITION 1. *If $\rho : A \otimes B \to B \otimes A$ respects $f : A^p \to A^q$ and $g : A^k \to A^r$ the it respects their tensor product $f \otimes g : A^{p+k} \to A^{q+r}$, and their composite $gf : A^p \to A^r$ if $q = r$. If $f : A^p \to A^q$ is respected by ρ then it is respected by $^r\rho : A \otimes B^r \to B^r \otimes A$ for any $r \geq 0$.*

PROOF. These results are obvious with planar diagram representations. We leave their verification to the reader. QED

Recall that a *monoid* in a tensor category $\mathcal{C}$ is a triple $K = (K, m, u)$ where $m : K \otimes K \to K$ and the $u : I \to K$, and such that the following associativity and unit diagrams commute:

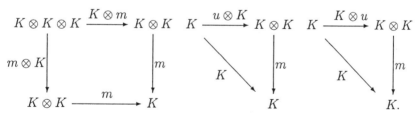

As for ordinary monoids, the unit for a given m is unique when it exists. A monoid structure (m, u) is thus determined by m. Recall that a (left)*action* of K on an object S is an arrow $q : K \otimes S \to S$ such that the following associativity and unit diagrams commute:

$$
\begin{array}{ccc}
K \otimes K \otimes S & \xrightarrow{K \otimes q} & K \otimes S \\
\downarrow{\scriptstyle m \otimes S} & & \downarrow{\scriptstyle q} \\
K \otimes S & \xrightarrow{q} & S
\end{array}
\qquad
\begin{array}{ccc}
S & \xrightarrow{u \otimes S} & K \otimes S \\
& \searrow{\scriptstyle S} & \downarrow{\scriptstyle q} \\
& & S.
\end{array}
$$

We also say that $S = (S, q)$ is an K-*object*. If S and T are K-objects we say that a map $f : S \to T$ is an K-*map* if it respects the actions.

Recall that a *comonoid* $A = (A, \Delta, \epsilon)$ in a tensor category $\mathcal{C}$ is a monoid in the opposite category. The *comultiplication* $\Delta : A \to A \otimes A$ and the *counit* satisfy dual conditions expressed by the coassociativity and counit diagrams. A *(left) coaction* $\psi : A \to A \otimes S$ of A on S is defined dualy. We shall say that $S = (S, \psi)$ is an A-*object*. There is also the concept of A-maps between A-objects.

DEFINITION 3. We shall say that a commutation operator $\rho : K \otimes A \to A \otimes K$ respects a monoid structure (K, m, u) if it respects m and u. A commutation operator ρ can similarly respect a (co)monoid structure on A, or on K.

Suppose that a commutation operator $\rho : K \otimes A \to A \otimes K$ respects the monoid structures (K, m, u) and (A, n, v). Then the composite

$$A \otimes K \otimes A \otimes K \xrightarrow{A \otimes \rho \otimes K} A \otimes A \otimes K \otimes K \xrightarrow{n \otimes m} A \otimes K.$$

is a monoid structure on $A \otimes K$. We shall sometime denote this monoid by $A \otimes_\rho K$. An action by $A \otimes_\rho K$ can be analysed in terms of a pair of ρ-*commuting* actions by A and K. Two actions $q : K \otimes S \to S$ and $r : A \otimes S \to S$ are said to ρ-*commute* if the following diagram commutes:

$$
\begin{array}{ccc}
K \otimes A \otimes S & \xrightarrow{\rho \otimes S} & A \otimes K \otimes S \\
\downarrow{\scriptstyle K \otimes r} & & \downarrow{\scriptstyle A \otimes q} \\
K \otimes S \xrightarrow{\ q\ } & S \xleftarrow{\ r\ } & A \otimes S;
\end{array}
$$

In this case the composite

$$A \otimes K \otimes S \xrightarrow{A \otimes q} A \otimes S \xrightarrow{\ r\ } S$$

is an action by $A \otimes_\rho K$. This defines a one to one correspondence between the actions by $A \otimes_\rho K$ and the pairs (q, r) of ρ-commuting actions.

EXAMPLE 1. Let G be a group and $H, K \subseteq G$ be subgroups. Let us suppose that the map $H \times K \to G$ given by $(x, y) \to xy$ is bijective. Then for any $(x, y) \in K \times H$ there is a unique pair $(u, v) \in H \times K$ such that $xy = uv$. If we put $\rho(x, y) = (u, v)$ then we obtain a commutation operator $\rho : K \times H \to H \times K$ (in the category *Sets* of sets). It is easy to see that ρ respects the group structures of H and K. A simple example of such a pair (H, K) is given by a semidirect decomposition $G = H \times K$. We then have $\rho(x, y) = (xyx^{-1}, x)$ if K is the normal subgroup. By duality we obtain examples with Hopf algebras. From appendix B we know that the extended coalgebra $\mathcal{A}$ is a semidirect (tensor) product $Z_2[u^\pm] \otimes \mathcal{S}_*$ since the functorial group A which it represents is a semi-direct product $\mu \times S$.

EXAMPLE 2. For any vector space V (over a field k) let $T(V)$ be the tensor algebra of V. If S is a vector space then a linear map $\alpha : V \otimes S \to S \otimes T(V)$ has a unique extension $\rho : T(V) \otimes S \to S \otimes T(V)$ as an operator respecting the algebra structure of $T(V)$. In particular, any linear map $\alpha : S \to S \otimes k[x]$ has a unique extension $\rho : k[x] \otimes S \to S \otimes k[x]$ as an operator respecting the algebra structure of $k[x]$. Suppose now that S is an algebra and that $\alpha(s) = \alpha_0(s) \otimes 1 + \alpha_1(s) \otimes x$

for some linear maps $\alpha_0, \alpha_1 : S \to S$. In this case ρ is respecting the algebra structure of S iff α_0 is an algebra endomorphism and α_1 is an α_0-derivation (that is $\alpha_1(st) = \alpha_1(s)t + \alpha_0(s)\alpha_1(t)$ for any $s, t \in S$). We then obtain an algebra structure on $S[x] = S \otimes k[x] = S \otimes_\rho k[x]$ that is called an *Ore extension* of S. For example, if $k = Z_2$ and S is commutative we can take $\alpha_0(s) = s^2$ and $\alpha_1 = 0$. Then the Ore extension $S[x]$ is isomorphic to the algebra of Ore polynomials with coefficients in S, with the operation of substitution as multiplication.

EXAMPLE 3. If R is a Q-ring then $R \otimes \mathcal{K}$ is a Q-module and there is a unique Q-module map $\rho : \mathcal{K} \otimes R \to R \otimes \mathcal{K}$ such that $\rho(1 \otimes x) = x \otimes 1$. The operator ρ respects the algebra structures of R and $\mathcal{K}$. It follows that $R \otimes \mathcal{K}$ has an algebra structure obtained from ρ. A module over this algebra is the same as a QR-module. This example is a special case of the general concept of semi-direct product between a bialgebra and an algebra. If B is a bialgebra then the tensor product of two (left) B-modules has a B-module structure derived from the comultiplication of B. If N is a B-module then $N \otimes B$ is a B-module and there is a unique B-module map $\rho : B \otimes N \to N \otimes B$ such that $\rho(1 \otimes x) = x \otimes 1$. It is easy to see that ρ respects the algebra structure of B. If N has an algebra structure for which multiplication $N \otimes N \to N$ and units $I \to B$ are B-modules maps then this structure is respected by ρ. In this case there is a semi-direct product algebra $N \otimes B$. A module V over $N \otimes B$ is the same thing as a module V over N in the category of B-modules (this means that V is a B-module and that the structure map $N \otimes M \to M$ is a B-module map). There is a dual result with comodules and coalgebras. For example, if $B = Z_2[u^\pm]$ then a comodule over $Z_2[u^\pm]$ is a graded vector spaces. The coalgebra $\mathcal{S}_*$ is graded, it is thus a coalgebra in the category of comodules over $Z_2[u^\pm]$. Hence there is a commutation operator $\rho : \mathcal{S}_* \otimes Z_2[u^\pm] \to Z_2[u^\pm] \otimes \mathcal{S}_*$ and a corresponding semi-direct product $\mathcal{A} = Z_2[u^\pm] \otimes \mathcal{S}_*$. It follows that a comodule over $\mathcal{A}$ is the same as a graded comodule over $\mathcal{S}_*$ (see proposition 2 appendix B).

EXAMPLE 4. The category $[\mathcal{C}, \mathcal{C}]$ of endofunctors of a category $\mathcal{C}$ is monoidal with composition for the tensor product. A monoid (K, m, u) in this category is called a *monad* (MacLane [1971]). The concept of commutation operator $\rho : KM \to MK$ respecting two monads is due to Jon Beck [1969] and it is called a *Beck distributive law* by category theorists.

PROPOSITION 2. *Let $K = (K, m, u)$ be a monoid and let $\rho : K \otimes A \to A \otimes K$ be a commutation operator. Then the following two conditions are equivalent:*
(i) *ρ respects (K, m, u);*
(ii) *for any action $q : K \otimes S \to S$ the composite*

$$K \otimes A \otimes S \xrightarrow{\rho \otimes S} A \otimes K \otimes S \xrightarrow{A \otimes q} A \otimes S$$

is an action $q' : K \otimes A \otimes S \to A \otimes S$.

PROOF. The implication (i) $\Rightarrow$ (ii) is left to the reader. We sketch the proof of (ii) $\Rightarrow$ (i). If we take $S = K$ and $q = m$ the map q' is the composite

$$K \otimes A \otimes K \xrightarrow{\rho \otimes K} A \otimes K \otimes K \xrightarrow{A \otimes m} A \otimes K.$$

By hypothesis q' is a coaction. If we precompose the associativity diagram for q' with the arrow $K \otimes K \otimes A \otimes u : K \otimes K \otimes A \to K \otimes K \otimes A \otimes K$ and simplify we obtain the compatibility diagram of m with ρ. The compatibility diagram of u with ρ is obtained similarly but with the unit diagram of q' instead. QED

PROPOSITION 3. *Let* $K = (K, m, u)$ *be a monoid and* $A = (A, \Delta, \epsilon)$ *be a comonoid. If a commutation operator* $\rho : K \otimes A \rightarrow A \otimes K$ *respects* (K, m, u) *then the following two conditions are equivalent:*
(i) ρ *respects* (A, Δ, ϵ);
(ii) *for any action* $q : K \otimes S \rightarrow S$ *the maps* $\epsilon \otimes S : A \otimes S \rightarrow S$ *and* $\Delta \otimes S : A \otimes S \rightarrow A \otimes A \otimes S$ *are* K-*maps.*

PROOF. The implication (i) $\Rightarrow$ (ii) is left to the reader. We sketch the proof of (ii) $\Rightarrow$ (i). If $S = K$ and $q = m$ the arrow q' is the composite

$$K \otimes A \otimes K \xrightarrow{\rho \otimes K} A \otimes K \otimes K \xrightarrow{A \otimes m} A \otimes K$$

and q'' is the composite

$$K \otimes A \otimes A \otimes K \xrightarrow{^2\rho \otimes K} A \otimes A \otimes K \otimes K \xrightarrow{A \otimes m} A \otimes A \otimes K.$$

If we precompose with $K \otimes A \otimes u$ the diagram expressing that $\Delta \otimes K$ is an K-map and simplify we obtain the compatibility diagram of Δ with ρ. The compatibility of ϵ with ρ is obtained similarly but by using $\epsilon \otimes K$ instead of $\Delta \otimes K$. QED

Suppose now that $\rho : K \otimes A \rightarrow A \otimes K$ respects the monoid structure $K = (K, m, u)$ and the comonoid structure $A = (A, \Delta, \epsilon)$. We shall say that an action $q : K \otimes S \rightarrow S$ and a coaction $\psi : S \rightarrow A \otimes S$ ρ-*commute* if the following diagram commutes:

$$
\begin{array}{ccccc}
K \otimes S & \xrightarrow{\ q\ } & S & \xrightarrow{\ \psi\ } & A \otimes S \\
{\scriptstyle K \otimes \psi}\big\downarrow & & & & \big\uparrow{\scriptstyle A \otimes q} \\
K \otimes A \otimes S & & \xrightarrow{\ \rho \otimes S\ } & & A \otimes K \otimes S.
\end{array}
$$

DEFINITION 4. A ρ-*biaction* is a ρ-commuting pair (q, ψ) of an action $q : K \otimes S \rightarrow S$ and a coaction $\psi : S \rightarrow A \otimes S$.

EXAMPLE. Let $Z_2[q_0]$ be the polynomial ring and let $Z_2[u^{\pm}]$ be the ring of Laurent polynomials with its bialgebra structure given by $\delta(u) = u \otimes u$. Then the commutation operator $\rho : Z_2[q_0] \otimes Z_2[u^{\pm}] \rightarrow Z_2[u^{\pm}] \otimes Z_2[q_0]$ given by $\rho(q_0^r \otimes u^n) = u^{2^r n} \otimes q_0$ respects the algebra structure of $Z_2[q_0]$ and the coalgebra structure of $Z_2[u^{\pm}]$. A ρ-biaction is a graded module M equipped with a Frobenius operator $q_0 : M \rightarrow M$ doubling the dimension. Perhaps a more interesting example is the commutation operator $\rho : \mathcal{K} \otimes Z_2[u^{\pm}] \rightarrow Z_2[u^{\pm}] \otimes \mathcal{K}$ defined as follow: consider the Q-structure on $Z_2[u^{\pm}] \otimes \mathcal{K}$ that is given by

$$Q_t(u^n \otimes x) = u^{2n}Q_{ut}(x) = u^{2n}\sum_i u^i \otimes q_n(x)t^i.$$

The vector space $\mathcal{K} \otimes Z_2[u^{\pm}]$ is free as a Q-module over $Z_2[u^{\pm}]$. By definition, ρ is the unique Q-module map such that $\rho(1 \otimes u^n) = u^n \otimes 1$. A ρ-biaction is graded

Q-module M. These two examples are special case of a more general commutation operator $\tilde{\rho} : \mathcal{K} \otimes \mathcal{A} \to \mathcal{A} \otimes \mathcal{K}$ that is playing a central role in our theory of the Nishida relations in §5.

By the dual of proposition 2 for any coaction $\psi : S \to A \otimes S$ the composite

$$K \otimes S \xrightarrow{K \otimes \psi} K \otimes A \otimes S \xrightarrow{\rho} A \otimes K \otimes S$$

is a coaction $\psi' : K \otimes S \to A \otimes K \otimes S$.

PROPOSITION 4. *Suppose that $\rho : K \otimes A \to A \otimes K$ respects the monoid structure of K and the comonoid structure of A. If $q : K \otimes S \to S$ is an action and $\psi : S \to A \otimes A$ is a coaction then the following three conditions are equivalent:*
(i) *the pair (q, ψ) ρ-commutes;*
(ii) *ψ is an K-map;*
(iii) *q is an A-map.*

PROOF. By definition of q' and ψ' the following diagram commutes:

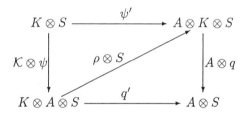

This shows that the pair (q, ψ) is ρ-commuting iff one of the following two squares commute

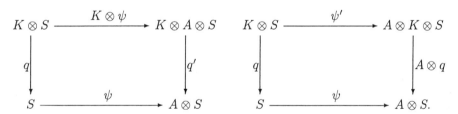

But the left hand side square commutes iff ψ is an K-map and the right hand square commutes iff q is an A-map. QED

Let $\mathcal{C}^K$ and $\mathcal{C}^A$ be the categories of K-actions and A-coactions respectively. Let $\mathcal{C}^{KA}$ be the category of ρ-biactions (with maps respecting both action and coaction). We have the following proposition whose verification is left to the reader.

PROPOSITION 5. *Suppose that $\rho : K \otimes A \to A \otimes K$ respects the monoid structure $K = (K, m, u)$ and the comonoid structure $A = (A, \Delta, \epsilon)$.*
(i) *For any coaction $\psi : S \to A \otimes S$ the pair $(m \otimes S, \psi')$ is a ρ-biaction on $K \otimes S$. The functor $S \mapsto K \otimes S$ is left adjoint to the forgetful functor $\mathcal{C}^{KA} \to \mathcal{C}^A$.*
(ii) *For any action $q : K \otimes S \to S$ the pair $(q', \Delta \otimes S)$ is a ρ-biaction on $A \otimes S$. The functor $S \mapsto A \otimes S$ is right adjoint to the forgetful functor $\mathcal{C}^{KA} \to \mathcal{C}^K$.*

REMARK. The theory of commutation operator presented here must be slightly reformulated in the case of the the monoidal category $[\mathcal{C}, \mathcal{C}]$ of endofunctor of a category $\mathcal{C}$. In this case the K-objects of a monoid (i.e. of a monad on $\mathcal{C}$) should be taken in $\mathcal{C}$ rather than in $[\mathcal{C}, \mathcal{C}]$. Recall that an action by K on an object $S \in \mathcal{C}$ is a map $q : K(S) \to S$ for which the following associativity and unit diagrams commute:

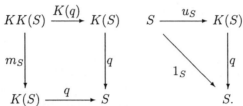

These K-objects are called K-algebras by category theorists (MacLane [1971]). Similarly, the A-objects of a comonad are called A-coalgebras. Let $\rho : KA \to AK$ be a commutation operator respecting the monad structure of K and the comonad structure of A. We say that an action $q : K(S) \to S$ and a coaction $\psi : S \to A(S)$ ρ-commute if the following diagram commutes:

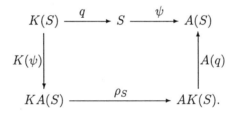

We say that (S, q, ψ) is a ρ-bialgebra. All the results of this appendix are valid when properly formulated for K-algebras, A-coalgebras in $\mathcal{C}$ instead of K-objects and A-objects in $[\mathcal{C}, \mathcal{C}]$.

References

[1974]   J.F. Adams, *Stable homotopy and generalized homology*, University of Chicago Press, 1974.

[1978]   J.F. Adams, *Infinite loop spaces*, vol. 90, Ann. of Math. Studies, Princeton, 1978.

[1957]   J. Adem, *The relations on Steenrod powers of cohomology classes*, Algebraic Geometry and Topology, Princeton University Press, 1957, pp. 191-238.

[1956]   S. Araki & T. Kudo, *Topology of H_n-spaces and H_n-squaring operations*, Mem. Fac. Sci. Kyusyu Univ. Ser. A **10** (1956), 85-120.

[1966]   M.F. Atiyah, *Power operations in K-theory*, Quart. J. Math. Oxford (2) **17** (1966), 165-193, reprinted in Atiyah, *K-theory*, Benjamin, 1967.

[1971]   M. Barratt, D. Kahn & S.Priddy, *On $\Omega^\infty S^\infty$ and the infinite symmetric group*, Proc. of Sympos. Pure Math. **22** (1971), A.M.S.

[1969]   J. Beck, *Distributive laws*, Seminar on Triple and Categorical Homology Theory, Lecture Notes in Mathematics, vol. 80, Springer Verlag, 1969.

[1977]   T.P. Bisson, *Divided sequences and bialgebras of homology operations*, Ph.D. Thesis, Duke, 1977.

[1995a] T.P. Bisson & A. Joyal, *The Dyer-Lashof algebra in bordism*, extended abstract, C.R. Math. Rep. Acad. Sci. Canada **17** (1995), 135-140.

[1995b] _____, *Nishida relations in bordism and homology*, extended abstract, C.R. Math. Rep. Acad. Sci. Canada **17** (1995), 141-146.

[1982]   J.M. Boardman, *The eightfold way to BP operations*, Can. Math. Soc. Conf. Proc. **2** (1982), no. 1.

[1960]   W. Browder, *Homology operations and loop spaces*, Illinois J. Math. (1960), 347-357.

[1982]   S.R. Bullett & I.G. Macdonald, *On the Adem relations*, Topology **21** (1982), 329-332.

[1911]   L.E. Dickson, *A fundamental system of invariants of the general modular linear group with a solution of the form problem*, Trans. A.M.S. **12** (1911), 75-98.

[1958]   A. Dold, *Homology of the symmetric products and other functors of complexes*, Ann. of Math. **68** (1958), 54-80.

[1962]   E. Dyer & R.K. Lashof, *Homology of iterated loop spaces*, Amer. J. Math. **84** (1962), 35-88.

[1958]   A. Grothendieck, *La théorie des classes de Chern*, Bull. Soc. Math. France **86** (1958), 137-154.

[1979]   P. Hoffman, *τ-Rings and Wreath Product Representations*, Lect. Notes in Math., vol. 746, Springer-Verlag, 1979.

[1982]   S.A. Joni & G.-C. Rota, *Coalgebras and bialgebras in combinatorics*, Umbral Calculus and Hopf Algebra, Contemporary Mathematics, vol. 6, A.M.S., 1982, pp. 1-47.

[1991a] A. Joyal & R. Street, *Tortile Yang Baxter operators in tensor categories*, J. of Pure and Applied Algebra **71** (1991), 43-51.

[1991b] ———, *The geometry of tensor calculus I*, Advances in Mathematics **88** (1991), no. 1, 55-112.

[1991c] ———, *An introduction to Tannaka duality and quantum groups*, Category Theory Proceedings: Como 1990, Lecture Notes in Math., vol. 1488, Springer-Verlag, 1991, pp. 411-492.

[1973]   D. Knutson, *λ-Rings and the Representation Theory of the Symmetric Group*, Lecture Notes in Mathematics, vol. 308, Springer Verlag, 1973.

[1994]   N. Kuhn, *Generic representations of the finite general linear groups and the Steenrod algebra*, Am. J. Math. **116** (1994), 327-360.

[1967]   P. Landweber, *Cobordism operations and Hopf algebras*, Trans. A.M.S. **27** (1967), 94-110.

[1995]   J. Lannes, *Applications dont la source est un classifiant*, Proceedings of the International Congress of Mathematicans: Zurich, 1994, pp. p566-573.

[1992]   L. Lomonaco, *Normalized operations in cohomology*, 1990 Barcelona Conference on Algebraic Topology (J. Aguadé, M. Castellet & F.R. Cohen, eds.), Lecture Notes in Mathematics, vol. 1509, Springer-Verlag, 1992, pp. p240-249.

[1971]   S. MacLane, *Categories for the Working Mathematician*, Springer Verlag, 1971.

[1979]   I.G. Macdonald, *Symmetric Functions and Hall Polynomials*, Oxford Univ. Press, Oxford, 1979.

[1982]   B.M. Mann & R.J. Milgram, *On the Chern classes of the regular representations of some finite groups*, Proc. of the Edinburgh Math. Soc. **25** (1982), 259-268.

[1975]   I. Madsen, *On the action of the Dyer-Lashof algebra in $H_*(G)$*, Pacific J. Math. **60** (1975), 235-275.

[1979]   I. Madsen & R. J. Milgram, *The classifying spaces for surgery and cobordism of manifolds*, Ann. of Math. Studies, vol. 92, Princeton 1979.

[1970]   J.P. May, *A general algebraic approach to Steenrod operations*, Lecture Notes in Math., vol. 168, Springer-Verlag, 1970, pp. 153-231.

[1971]   J.P. May, *Homology operations on infinite loop spaces*, Proc. Symp. Pure Math. (A. Liulevicius, ed.), vol. XXII, A.M.S., 1971, pp. 171-185.

[1976]   J.P. May, *The homology of E_∞-spaces*, The homology of iterated loop spaces (F.R. Cohen, T. J. Lada & J.P. May, eds.), Lecture Notes in Math., vol. 533, Springer, 1976, pp. 1-68.

[1965]   R.J. Milgram, *Iterated Loop spaces*, Ann. of Math. **84** (1966), 386-403.

[1987]   H. Miller, *The Sullivan conjecture and homotopical representation theory*, Proc. Int. Cong. of Math. 1986: Berkeley, Calif. (A.M. Gleason, ed.), A.M.S., 1987, pp. 580-589.

[1958]   J.W. Milnor, *The Steenrod algebra and its dual*, Ann. Math. **67** (1958), 150-171.

[1974]   J.W. Milnor & J.D. Stasheff, *Characteristic Classes*, Princeton University Press, 1974.

[1985]   J. Morava, *Noetherian localization of categories of cobordism comodules*, Ann. Math. **121** (1985), 1-39.

[1993]   J. Morava, *Some examples of Hopf algebras and Tannakian categories*, Contemp. Math. (M.C. Tangora., ed.), Algebraic Topology, Oaxtepec 1991, vol. 146, A.M.S., 1993, pp. 349-359.

[1975] H. Mui, *Modular invariant theory and the cohomology algebras of symmetric spaces*, J. Fac. Sci. Univ. Tokyo **22** (1975), 319-369.

[1983] ———, *Dickson invariants and Milnor basis of the Steenrod algebra*, Eger International Colloquium in Topology, 1983.

[1984] ———, *Homology operations derived from modular coinvariants*, Algebraic topology, Göttingen (L Smith, ed.), Lecture Notes in Math., vol. 1172, 1984, pp. 85-115.

[1957] M. Nakaoka, *Cohomology of symmetric products*, J. Inst. Polyt. **7** (1957), Osaka City Univ., 121-144.

[1960] ———, *Decomposition theorem for homology groups of symmetric groups*, Ann. of Math. **71** (1960), no. 1, 16-42.

[1961] ———, *Homology of the infinite symmetric group*, Ann. of Math. **73** (1961), no. 2, 229-257.

[1982] W. Nichols & M.E. Sweedler, *Hopf algebras and combinatorics*, Umbral Calculus and Hopf Algebra, Contemporary Mathematics, vol. 6, A.M.S., 1982.

[1968] G. Nishida, *Cohomology operations in iterated loopspaces*, Proc. Japan Acad. **44** (1968), 104-109.

[1933] O. Ore, *On a special class of polynomials*, Trans. A.M.S. **35** (1933), 559-584; Correction Trans. A.M.S. **36** (1934), p. 275.

[1972] S. Priddy, *Transfer, symmetric groups, and stable homotopy theory*, Algebraic K-theory I (H. Bass, ed.), Lecture Notes in Math., vol. 341, Springer-Verlag, 1972, pp. 244-255.

[1975] ———, *Dyer-Lashof operations for the classifying spaces of certain matrix groups*, Quart. J. Math. Oxford (3) **26** (1975), 179-193.

[1971] D.G. Quillen, *Elementary proofs of some results of cobordism theory using Steenrod operations*, Adv. in Math. **7** (1971), 29-56.

[1977] D.C. Ravenel & W.S. Wilson, *The Hopf ring for complex cobordism*, J. of Pure and Applied Algebra **9** (1977), 241-280.

[1971] G.C. Rota, *Combinatorial Theory and Invariant Theory*, Notes by L. Guibas, (Summer 1971), Bowdoin College, Maine.

[1994] L. Schwartz, *Unstable modules over the Steenrod algebra and Sullivan's fixed point set conjecture*, U. of Chicago Press, 1994.

[1974] G. Segal, *Categories and cohomology theories*, Topology **13** (1974), 293-312.

[1972] N.E. Steenrod, *Cohomology operations and obstruction to extending continuous functions*, (lectures given in 1957), Adv. in Math. **8** (1972).

[1953] ———, *Homology groups of symmetric groups and reduced power operations*, Proc. Nat. Can. Sci. **39** (1953).

[1957] ———, *Cohomology operations derived from the symmetric groups*, Comment. Math. Helv. **31** (1957), 195-218.

[1962] N.E. Steenrod & D.B.A. Epstein, *Cohomology Operations*, Ann. of Math. Studies, vol. 50, Princeton University Press, 1962.

[1984] R. Steiner, *Homology operations and power series*, Glasgow Math. J. **24** (1984), 161-168.

[1969] M.E. Sweedler, *Hopf Algebras*, W.A. Benjamin, New York, 1969.

[1973] R.M. Switzer, *Homology comodules*, Invent. Math. **20** (1973), 97-102.

[1996] P.R. Turner, *Dickson Coinvariants and the homology of QS^0*, Math. Zeit. (to appear).

[1983] C. Wilkerson, *A primer on the Dickson invariants*, Contemporary Math. (H.R.Miller & S.B.Priddy, eds.), Proc. of the Northwestern homotopy theory conference: 1982, vol. 19, 1983; a revised version of this paper is available on WWW at hopf.math.purdue.edu.

[1980] W.S. Wilson, *Brown-Peterson homology-an introduction and sampler*, A.M.S. Regional Conference Series in Mathematics, vol. 48, 1982.

[1995] R.W. Wood, *Differential operators and the Steenrod algebra*, submitted to London Math. Soc.; See also: *An introduction to the Steenrod algebra through differential operators*, preprint 1995, available at www.LeHigh.edu, the archive of the Lehigh Algebraic Topology Discussion List.

CANISIUS COLLEGE, BUFFALO, N.Y. (U.S.A).
E-mail address: bisson@canisius.edu

DÉPARTEMENT DE MATHÉMATIQUES, UQAM, MONTRÉAL, QUÉBEC H3C 3P8
E-mail address: joyal@math.uqam.ca

Contemporary Mathematics
Volume **202**, 1997

Operadic tensor products and smash products

J. P. MAY

ABSTRACT. Let k be a commutative ring. E_∞ k-algebras are associative
and commutative k-algebras up to homotopy, as codified in the action of
an E_∞ operad; A_∞ k-algebras are obtained by ignoring permutations. Using a particularly well-behaved E_∞ algebra, we explain an associative and
commutative operadic tensor product $\boxtimes$ that effectively hides the operad:
an A_∞ algebra or E_∞ algebra A is defined in terms of maps $k \longrightarrow A$ and
$A \boxtimes A \longrightarrow A$ such that the obvious diagrams commute, and similarly for
modules over A. This makes it little more difficult to study these algebraic
objects than it is to study their classical counterparts. We also explain
a topological analogue of the theory. This gives a symmetric monoidal
category of modules over the sphere spectrum S whose derived category
is equivalent to the classical stable homotopy category. The existence of
this category allows the wholesale importation of algebraic techniques into
stable homotopy theory.

There will not be time to go into this, but the algebraic theory has applications to mixed Tate motives in algebraic geometry and the topological
theory has applications to the construction and study of MU-module spectra, the construction of generalized Künneth and universal coefficient spectral sequences, a construction of the algebraic K-theory of S-algebras that
includes Quillen's algebraic K-theory of discrete rings and Waldhausen's
algebraic K-theory of spaces, a construction of the topological Hochschild
homology of an S-algebra that generalizes Bökstedt's THH, and a completion theorem for equivariant complex cobordism and any of its modules
analogous to the Atiyah-Segal completion theorem in equivariant K-theory.

1. The category of $\mathbb{C}$-modules and the product $\boxtimes$

Let $\mathscr{C}$ be an operad in a cocomplete symmetric monoidal category $\mathscr{S}$ with
product $\otimes$ and unit κ. We are thinking of the category of differential graded
modules over a commutative ring k and will restrict to it shortly. In general,
$\mathbb{C} = \mathscr{C}(1)$ is a monoid in $\mathscr{S}$. In our algebraic context, this means that $\mathbb{C}$ is a

1991 *Mathematics Subject Classification*. Primary 18-02, 55-02; Secondary 18C99, 55P42.

This talk is largely based on Part V of [**7**] and on [**5**], to which the interested reader is
referred for details and more complete references.

The author was supported in part by NSF Grant #DMS-9423300.

DGA. We call left $\mathbb{C}$-objects $\mathbb{C}$-modules in any case. In the algebraic situation, if $\mathscr{C}$ is unital and the augmentation $\epsilon : \mathbb{C} \to k$ is a quasi-isomorphism, then the derived categories $\mathscr{D}_k$ and $\mathscr{D}_\mathbb{C}$ are equivalent.

Via instances of the structural maps γ, we have a left action of $\mathbb{C}$ and a right action of $\mathbb{C} \otimes \mathbb{C}$ on $\mathscr{C}(2)$, and these actions commute with each other. Thus we have a bimodule structure on $\mathscr{C}(2)$. Let M and N be left $\mathbb{C}$-modules. Clearly $M \otimes N$ is a left $\mathbb{C} \otimes \mathbb{C}$-module via the given actions. This makes sense of the following definition of the "operadic tensor product $\boxtimes$".

DEFINITION 1.1. For $\mathbb{C}$-modules M and N, define $M \boxtimes N$ to be the $\mathbb{C}$-module

$$M \boxtimes N = \mathscr{C}(2) \otimes_{\mathbb{C}\otimes\mathbb{C}} M \otimes N.$$

This definition deserves more study than it has been given. In special cases, it has led to interesting algebraic results, which are joint work with Igor Kriz and which I shall describe. These are largely motivated by applications to mixed Tate motives, but I will not go into that. An analogous operadic smash product has led to really rather spectacular results in topology, which are joint work with Elmendorf, Kriz, and Mandell [5]. I will give an introduction to the parallel topological theory at the end.

We concentrate on algebra. Here we have an operadic Hom functor $\mathrm{Hom}^{\boxtimes}$ on $\mathbb{C}$-modules to go with the operadic tensor product. Since its precise definition is dictated by the adjunction, I won't bother writing it down.

LEMMA 1.2. *There is a natural adjunction isomorphism*

$$\mathscr{M}_\mathbb{C}(M \boxtimes N, P) \cong \mathscr{M}_\mathbb{C}(M, \mathrm{Hom}^{\boxtimes}(N, P)).$$

The product $\boxtimes$ is always commutative.

LEMMA 1.3. *There is a canonical commutativity isomorphism of $\mathbb{C}$-modules*

$$\tau : M \boxtimes N \longrightarrow N \boxtimes M.$$

PROOF. Use the action of the transposition $\sigma \in \Sigma_2$ on $\mathscr{C}(2)$ together with the transposition isomorphisms $\mathbb{C} \otimes \mathbb{C} \to \mathbb{C} \otimes \mathbb{C}$ and $M \otimes N \to N \otimes M$. $\square$

Associativity is more subtle and requires an exceptionally well-behaved operad. Although we are working algebraically, the following basic result comes from topology.

THEOREM 1.4. *There is an E_∞ operad $\mathscr{C} = C_*(\mathscr{L})$, where $\mathscr{L}$ is the "linear isometries operad", for which there is a canonical associativity isomorphism of $\mathbb{C}$-modules*

$$(L \boxtimes M) \boxtimes N \cong L \boxtimes (M \boxtimes N).$$

In fact, for any j-tuple $M_1, \ldots, M_j$ of $\mathbb{C}$-modules, there is a canonical isomorphism

$$M_1 \boxtimes \cdots \boxtimes M_j \cong \mathscr{C}(j) \otimes_{\mathbb{C}^j} (M_1 \otimes \cdots M_j),$$

where the iterated product on the left is associated in any fashion. For $j \geq 2$, the
j-fold $\boxtimes$-power $\mathbb{C}^{\boxtimes j}$ is isomorphic to $\mathscr{C}(j)$ as a $(\mathbb{C}, \mathbb{C}^j)$-bimodule.

It is also true that $\mathscr{C}(j)$ is isomorphic to $\mathbb{C}$ as a left $\mathbb{C}$-module. These properties
are quite miraculous. I will explain why they are true later, after describing some
of the implications. We restrict attention to this particular operad $\mathscr{C}$ from now
on. As long as we are considering E_∞ operads, there is no loss of generality:
algebras and their modules over other E_∞ operads can be converted functorially
to algebras and their modules over $\mathscr{C}$.

Note that k is a $\mathbb{C}$-module via the augmentation $\mathbb{C} \to k$. The degeneracy map
$\sigma_1 : \mathscr{C}(2) \to \mathscr{C}(1) = \mathbb{C}$ induces a natural unit map.

LEMMA 1.5. *There is a natural map of $\mathbb{C}$-modules $\lambda : k \boxtimes N \to N$.*

This map is not an isomorphism, but another special property of the linear
isometries operad implies that it is usually a quasi-isomorphism.

LEMMA 1.6. *$\sigma_1 : \mathscr{C}(2) \to \mathbb{C}$ is a homotopy equivalence of right $\mathbb{C}$-modules.*

There is not time to describe it today, but there is a theory of cell modules over
DGA's that is just like the topological theory of CW complexes or CW spectra,
except much simpler. Briefly, free modules on suspensions of k serve as analogs
of spheres and thus as the domains of attaching maps for algebraic cones. We
pass to derived categories by approximating general modules by quasi-isomorphic
cell modules. Via such cell approximations, the functors $\boxtimes$ and $\text{Hom}^{\boxtimes}$ induce
a derived tensor product $\overset{L}{\boxtimes}$ and a derived Hom functor $R\,\text{Hom}^{\boxtimes}$ on $\mathscr{D}_{\mathbb{C}}$. These
functors are nicely related to the derived tensor product and Hom functors on
k-modules.

PROPOSITION 1.7. *If N is a cell module, then $\lambda : k \boxtimes N \to N$ is a quasi-*
isomorphism, the functor $M \boxtimes N$ of M preserves exact sequences and quasi-
isomorphisms, and the k-module $M \boxtimes N$ is naturally quasi-isomorphic to $M \otimes N$.
Therefore the equivalence of derived categories $\mathscr{D}_{\mathbb{C}} \to \mathscr{D}_k$ induced by the forgetful
functor from $\mathbb{C}$-modules to k-modules carries $\overset{L}{\boxtimes}$ to $\overset{L}{\otimes}$.

PROPOSITION 1.8. *If N is any $\mathbb{C}$-module, then the adjoint $N \longrightarrow \text{Hom}^{\boxtimes}(k, N)$*
of λ is a quasi-isomorphism and the functor $\text{Hom}^{\boxtimes}(M, N)$ of M preserves ex-
act sequences of cell $\mathbb{C}$-modules. If M is a cell $\mathbb{C}$-module, then the functor
$\text{Hom}^{\boxtimes}(M, N)$ of N preserves exact sequences and quasi-isomorphisms, and the
k-module $\text{Hom}^{\boxtimes}(M, N)$ is quasi-isomorphic to $\text{Hom}(M, N)$. Therefore the equiv-
alence of derived categories $\mathscr{D}_{\mathbb{C}} \to \mathscr{D}_k$ induced by the forgetful functor carries
$R\,\text{Hom}^{\boxtimes}$ to $R\,\text{Hom}$.

Remark 1.9. It would be of interest to construct an E_∞ operad with the properties that we have quoted by purely algebraic methods. One further property

is out of reach. We would like $\lambda : k \boxtimes k \to k$ to be an isomorphism. The indecomposable quotient $\mathscr{C}(2) \otimes_{\mathscr{C}(1) \otimes \mathscr{C}(1)} k$ would then be k. We have required that $\mathscr{C}(j)_n = 0$ for $n < 0$, and it is natural to require further that $\mathscr{C}(1)_0 = k$. This property would then imply that $\mathscr{C}(2)_0 = k$, contradicting the requirement that $\mathscr{C}(2)$ be Σ_2-free.

2. A new description of A_∞ and E_∞ algebras and modules

We consider $\mathscr{C}$-algebras and their modules, where $\mathscr{C} = C_*(\mathscr{L})$. We call these A_∞ algebras when we ignore permutations and E_∞ algebras when we retain them.

Restricting the action to $j = 1$, we see that an A_∞ algebra is a $\mathbb{C}$-module with additional structure. From $j = 0$ we obtain a unit $\eta : k \longrightarrow A$ and from $j = 2$ we obtain a product $\phi : A \boxtimes A \to A$. The rest of the operad action is actually determined by this portion of it. The following result is the analog of a theorem first discovered in a much deeper topological context.

THEOREM 2.1. *An A_∞ algebra A determines and is determined by a $\mathbb{C}$-module with a unit map $\eta : k \longrightarrow A$ and a product map $\phi : A \boxtimes A \to A$ such that the following diagrams commute:*

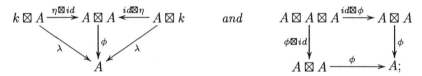

A is an E_∞ algebra if the following diagram also commutes:

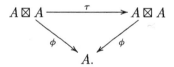

Although I will not have time to go into detail, there is an even simpler and more convenient way to package these reinterpretations of A_∞ and E_∞ algebras.

PROPOSITION 2.2. *The category of $\mathbb{C}$-modules under k admits a variant $\square$ of the product $\boxtimes$ with respect to which it is symmetric monoidal with unit k. An A_∞ or E_∞ algebra is precisely a monoid or commutative monoid in this category.*

To illustrate the force of this result, I will give a formal consequence. Recall that the tensor product of commutative DGA's is their coproduct in the category of commutative DGA's. The proof consists of categorical diagram chases that now carry over to our more general context.

COROLLARY 2.3. *Let A and B be A_∞ algebras. Then $A \square B$ is an A_∞ algebra. If M is an A-module and N is a B-module, then $M \boxtimes N$ is an $A \square B$-module.*

If A and B are E_∞ algebras, then $A \, \square \, B$ is an E_∞ algebra and is the coproduct of A and B in the category of E_∞ algebras.

We have a similar reinterpretation of the notion of an A-module.

THEOREM 2.4. *Let A be an A_∞ or E_∞ algebra. An A-module is a $\mathbb{C}$-module M together with a map $\mu : A \boxtimes M \to M$ such that the following diagrams commute:*

$$
\begin{array}{ccc}
k \boxtimes M \xrightarrow{\eta \boxtimes id} A \boxtimes M & \quad and \quad & A \boxtimes A \boxtimes M \xrightarrow{id \boxtimes \mu} A \boxtimes M \\
\searrow^{\lambda} \quad \downarrow^{\mu} & & \quad \downarrow^{\phi \boxtimes id} \qquad \qquad \downarrow^{\mu} \\
M & & A \boxtimes M \xrightarrow{\quad \mu \quad} M.
\end{array}
$$

When A is an E_∞ algebra, this implies that we obtain the same A-modules for A regarded as an E_∞ algebra as for A regarded by neglect of structure as an A_∞ algebra. This is far from obvious with the original operadic definitions.

3. $U(A)$ and the derived category of A-modules

Fix an A_∞ algebra A. The category of A-modules is isomorphic to the category of $U(A)$-modules. It therefore enjoys all of the formal properties that are familiar from the context of modules over a non-commutative DGA. Here we are thinking of A-modules as defined in the ground category of k-modules. However, the ideas above dictate that we sometimes change our point of view and regard A-modules as defined in the category of $\mathbb{C}$-modules. Since $U(k) = \mathbb{C}$, this means that we are thinking of the ground category as the category of E_∞ k-modules rather than the category of k-modules.

From the first point of view we see that $M \otimes K$ and $\mathrm{Hom}(K, M)$ are A-modules if M is an A-module and K is a k-module. From the second point of view, we see that $M \boxtimes L$ and $\mathrm{Hom}^{\boxtimes}(L, M)$ are A-modules if M is an A-module and L is a $\mathbb{C}$-module.

To develop the cell theory of A-modules, we need a free functor from k-modules to A-modules, and we already have a free functor from k-modules to $\mathbb{C}$-modules, namely $K \to \mathbb{C} \otimes K$. The following observation gives a free functor $L \to A \lhd L$ from $\mathbb{C}$-modules to A-modules.

LEMMA 3.1. *For a $\mathbb{C}$-module L, define $A \lhd L$ to be the pushout displayed in the diagram*

$$
\begin{array}{ccc}
k \boxtimes L \xrightarrow{\eta \boxtimes id} A \boxtimes L \\
\downarrow^{\lambda} \qquad \qquad \downarrow \\
L \xrightarrow{\quad\quad} A \lhd L.
\end{array}
$$

Then, for A-modules M,

$$
\mathscr{M}_A(A \lhd L, M) \cong \mathscr{M}_{\mathbb{C}}(L, M).
$$

LEMMA 3.2. *For a k-module K, define an A-module FK by*

$$FK = A \lhd (\mathbb{C} \otimes K).$$

Then, for A-modules M,

$$\mathscr{M}_A(FK, M) \cong \mathscr{M}_k(K, M).$$

Now recall that we have already constructed the free A-module functor for an algebra A over any operad: since the category of A-modules is isomorphic to the category of $U(A)$-modules, the free A-module generated by a k-module K must be $U(A) \otimes K$. We are entitled to the following consequence, which is special to the linear isometries operad. Note that the unit of $U(A)$ determines a natural k-map $K \to U(A) \otimes K$.

PROPOSITION 3.3. *For k-modules K, the natural map $FK \to U(A) \otimes K$ is an isomorphism of A-modules.*

In particular, Fk is isomorphic as an A-module to $U(A)$. It simplifies matters to assume from now on that A is augmented. This assumption makes it fairly easy to prove the following result, which gives homological control of free A-modules, something that is rather difficult to achieve for algebras over general operads.

PROPOSITION 3.4. *If K is a cell k-module, then the A-map $\alpha : FK \to A \otimes K$ induced by the canonical k-map $K \to A \otimes K$ is a quasi-isomorphism. If K is a free k-module with zero differential, then $H_*(FK)$ is the free $H_*(A)$-module generated by K.*

We can now parrot the development of derived categories via cell modules over DGA's in our more general context of E_∞ algebras. For an actual DGA A, we have two categories of A-modules in sight, namely ordinary ones and A_∞ ones. The latter are the same as $U(A)$-modules, and we have the following consistency statement.

PROPOSITION 3.5. *If A is a DGA, then the quasi-isomorphism $\alpha : U(A) \cong Fk \to A$ is a map of DGA's. Therefore it induces an equivalence of categories from the derived category $\mathscr{D}_A$ to the derived category $\mathscr{D}_{U(A)}$.*

4. The tensor product of A-modules

We have not yet defined tensor products of modules over A_∞ algebras. We can mimic classical algebra.

DEFINITION 4.1. Let A be an A_∞ algebra and let M be a right and N be a left A-module. Define $M \boxtimes_A N$ to be the coequalizer displayed in the following diagram of $\mathbb{C}$-modules:

$$M \boxtimes A \boxtimes N \underset{\mathrm{id} \boxtimes \nu}{\overset{\mu \boxtimes \mathrm{id}}{\rightrightarrows}} M \boxtimes N \longrightarrow M \boxtimes_A N,$$

where μ and ν are the given actions of A on M and N.

We construct the derived tensor product $\overset{L}{\boxtimes}_A$ by approximating one of the variables by a cell A-module.

When $A = k$, our new $M \boxtimes_k N$ coincides with $M \boxtimes N$. We have used the notation $\boxtimes_A$ to avoid confusion with $\otimes_A$ in the case of a DGA A regarded as an A_∞ algebra.

PROPOSITION 4.2. *If A is a DGA and M and N are A-modules, then the derived tensor product $M \overset{L}{\boxtimes}_A N$ is isomorphic in the derived category $\mathscr{D}_k$ to the classical derived tensor product $M \overset{L}{\otimes}_A N$.*

The new tensor product enjoys the same formal properties as the classical tensor product over DGA's. We give some examples.

LEMMA 4.3. *For a right A-module M and left A-module N,*

$$M \boxtimes_A N \cong N \boxtimes_{A^{op}} M.$$

LEMMA 4.4. *Let L be an (A, B)-bimodule, M be a (B, C)-bimodule, and N be a (C, D)-bimodule. Then $L \boxtimes_B M$ is an (A, C)-bimodule and*

$$(L \boxtimes_B M) \boxtimes_C N \cong L \boxtimes_B (M \boxtimes_C N)$$

as (A, D)-bimodules.

LEMMA 4.5. *The action $\nu : A \boxtimes N \longrightarrow N$ of a left A-module N factors through a map of A-modules $\lambda : A \boxtimes_A N \longrightarrow N$.*

The homological behavior of the functor $\boxtimes_A$ is unclear from its definition as a coequalizer, but analysis of its behavior on free modules leads to the following result.

PROPOSITION 4.6. *Let N be a cell A-module. Then $\lambda : A \boxtimes_A N \to N$ is a quasi-isomorphism and the functor $M \boxtimes_A N$ of M preserves exact sequences and quasi-isomorphisms.*

We have a Hom functor $\mathrm{Hom}_A^{\boxtimes}$ to go with our new tensor product. It is defined as a suitable equalizer, as in the context of DGA's.

LEMMA 4.7. *For $\mathbb{C}$-modules L and left A-modules M and N, there is a natural adjunction isomorphism*

$$\mathscr{M}_A(L \boxtimes M, N) \cong \mathscr{M}_\mathbb{C}(L, \mathrm{Hom}_A^{\boxtimes}(M, N)).$$

Analyis of $\mathrm{Hom}_A^{\boxtimes}(M, N)$ on free A-modules M gives homological control.

LEMMA 4.8. *Let M be a cell A-module. Then the functor $\mathrm{Hom}_A^{\boxtimes}(M, N)$ preserves exact sequences and quasi-isomorphisms in the variable N. It also preserves exact sequences of cell A-modules in the variable M.*

We construct the derived functor $R\operatorname{Hom}_A^{\boxtimes}$ by approximating the contravariant variable by a cell A-module.

5. Generalized Eilenberg-Moore spectral sequences

Fix an A_∞ algebra A. Since our derived tensor product and Hom functors generalize those of DGA's, the following definition is reasonable. We are grading by subscripts, with differentials lowering degrees.

DEFINITION 5.1. Define

$$\operatorname{Tor}_*^A(M,N) = H_*(M \overset{L}{\boxtimes}_A N) \quad \text{and} \quad \operatorname{Ext}_A^*(M,N) = H_{-*}(R\operatorname{Hom}_A^{\boxtimes}(M,N)).$$

These functors enjoy the same general properties as in the case of DGA's: exact triangles in either variable induce long exact sequences on passage to Tor or Ext, Tor preserves direct sums in either variable, and Ext converts direct sums in M to direct products and preserves direct products in N. The behavior on free modules is

$$\operatorname{Tor}_*^A(M,FK) \cong H_*(M \otimes K) \quad \text{and} \quad \operatorname{Ext}_A^*(FK,N) \cong H_{-*}(\operatorname{Hom}(K,N)).$$

The crucial point of this generalized definition of Tor and Ext is that we have Eilenberg-Moore spectral sequences for their calculation, just as for DGA's. Following the usual grading convention, write $H_*(M) = H^{-*}(M)$.

THEOREM 5.2. *There are natural spectral sequences of the form*

$$E_{p,q}^2 = \operatorname{Tor}_{p,q}^{H_*(A)}(H_*(M),H_*(N)) \Longrightarrow \operatorname{Tor}_{p+q}^A(M,N)$$

and

$$E_2^{p,q} = \operatorname{Ext}_{H^*(A)}^{p,q}(H^*(M),H^*(N)) \Longrightarrow \operatorname{Ext}_A^{p+q}(M,N).$$

6. E_∞ algebras and duality

Let A be an E_∞ algebra. The study of E_∞ modules works exactly the same way as the study of modules over commutative DGA's.

THEOREM 6.1. *If M and N are A-modules, then $M \boxtimes_A N$ and $\operatorname{Hom}_A^{\boxtimes}(M,N)$ have canonical A-module structures deduced from the A-module structure of M or, equivalently, N. The tensor product over A is associative and commutative, and the unit maps $A \boxtimes_A M \to M$ and $A \to \operatorname{Hom}_A^{\boxtimes}(A,N)$ are maps of A-modules. There is a natural adjunction isomorphism*

$$\mathscr{M}_A(L \boxtimes_A M, N) \cong \mathscr{M}_A(L, \operatorname{Hom}_A^{\boxtimes}(M,N)).$$

The derived category $\mathscr{D}_A$ is symmetric monoidal under $\overset{L}{\boxtimes}_A$, and the adjunction passes to the derived category.

PROPOSITION 6.2. *If M and M' are cell A-modules, then so is $M \otimes_A M'$.*

Again, write $H_*(A) = H^{-*}(A)$; it is an associative and (graded) commutative algebra.

COROLLARY 6.3. $\mathrm{Tor}_*^A(M, N)$ and $\mathrm{Ext}_A^*(M, N)$ are $H_*(A)$-modules, and there are natural commutativity and associativity isomorphisms of $H_*(A)$-modules

$$\mathrm{Tor}_*^A(M, N) \cong \mathrm{Tor}_*^A(N, M)$$

and

$$\mathrm{Tor}_*^A(L \boxtimes_A M, N) \cong \mathrm{Tor}_*^A(L, M \boxtimes_A N).$$

The spectral sequences above are spectral sequences of differential $H_*(A)$-modules.

The formal properties we have stated imply many others, just as for classical commutative DGA's and their modules. For example,

$$\mathrm{Hom}_A^{\boxtimes}(M \boxtimes_A L, N) \cong \mathrm{Hom}_A^{\boxtimes}(M, \mathrm{Hom}_A^{\boxtimes}(L, N))$$

because the two sides represent isomorphic functors on modules. Using this, a formal argument shows that we have a composition pairing

$$\mathrm{Hom}_A^{\boxtimes}(M, N) \boxtimes_A \mathrm{Hom}_A^{\boxtimes}(L, M) \longrightarrow \mathrm{Hom}_A^{\boxtimes}(L, N).$$

This pairing is associative and commutative. It induces a Yoneda product on Ext.

PROPOSITION 6.4. There is a natural, associative, and unital system of pairings

$$\mathrm{Ext}_A^*(M, N) \otimes_{H_*(A)} \mathrm{Ext}_A^*(L, M) \longrightarrow \mathrm{Ext}_A^*(L, N).$$

Moreover, there is an induced pairing of Ext spectral sequences that coincides with the algebraic Yoneda pairing on the E_2-level and converges to this pairing of Ext groups.

Formal duality theory, as developed by Dold-Puppe [2] and others [8] in algebraic topology and by Deligne [1] and others in algebraic geometry, applies verbatim to the present context. We define $M^\vee = \mathrm{Hom}_A^{\boxtimes}(M, A)$, and we say that a cell A-module M is "strongly dualizable" if it has a coevaluation map $\bar\eta : A \longrightarrow M \boxtimes M^\vee$ such that the following diagram commutes in $\mathscr{D}_A$:

$$
\begin{array}{ccc}
A & \xrightarrow{\ \bar\eta\ } & M \otimes_A M^\vee \\
{\scriptstyle \eta}\downarrow & & \downarrow{\scriptstyle \tau} \\
\mathrm{Hom}_A(M, M) & \xleftarrow[\ \nu\ ]{} & M^\vee \otimes_A M.
\end{array}
$$

When M is strongly dualizable, various natural maps such as

$$\rho : M \longrightarrow M^{\vee\vee}$$

and

$$\nu : M^\vee \boxtimes_A N \longrightarrow \mathrm{Hom}_A^{\boxtimes}(M, N)$$

induce isomorphisms in $\mathscr{D}_A$, exactly as if A were a classical k-algebra, without differential, and M were a finitely generated projective A-module. The last isomorphism has the following implication, which is an algebraic version of Spanier-Whitehead duality in topology.

PROPOSITION 6.5. *For a strongly dualizable A-module M and any A-module N,*

$$\text{Tor}_n^A(M^\vee, N) \cong \text{Ext}_A^n(M, N).$$

The following observation, which is due to Greenlees, makes clear that finite cell A-modules and their direct summands should be viewed as analogues of finitely generated free and projective modules in ordinary commutative algebra.

THEOREM 6.6. *A cell A-module is strongly dualizable if and only if it is a direct summand up to homotopy of a finite cell A-module.*

7. The linear isometries operad

The algebraic discussion above was all predicated on the good properties of the topological linear isometries E_∞ operad. We proceed to define that operad, which was already implicit in Boardman's approach [15] to the stable homotopy category, before operads were invented.

Let $U \cong \mathbb{R}^\infty$ be a countably infinite dimensional real inner product space, topologized as the union of its finite dimensional subspaces. Let U^j be the direct sum of j copies of U. Define $\mathscr{L}(j)$ to be the set of linear isometries $U^j \to U$ with the function space topology. Note that a linear isometry is an injection but not necessarily an isomorphism. The space $\mathscr{L}(0)$ is the point i, $i : 0 \to U$, and $\mathscr{L}(1)$ contains the identity $1 : U \to U$. The left action of Σ_j on U^j by permutations induces a free right action of Σ_j on $\mathscr{L}(j)$. The structure maps

$$\gamma : \mathscr{L}(k) \times \mathscr{L}(j_1) \times \cdots \times \mathscr{L}(j_k) \longrightarrow \mathscr{L}(j_1 + \cdots + j_k)$$

are defined by

$$\gamma(g; f_1, \ldots, f_k) = g \circ (f_1 \oplus \cdots \oplus f_k).$$

The crucial associativity property of $\mathscr{C}$ stems from an associativity property of $\mathscr{L}$ that was first observed by Mike Hopkins. We need a categorical definition in order to state it properly.

DEFINITION 7.1. Working in an arbitrary category, suppose given a diagram

$$A \overset{e}{\underset{f}{\rightrightarrows}} B \overset{g}{\longrightarrow} C$$

in which $ge = gf$. The diagram is called a split coequalizer if there are maps

$$h : C \to B \quad \text{and} \quad k : B \to A$$

such that $gh = \text{id}_C$, $fk = \text{id}_B$, and $ek = hg$.

Observe that $\mathscr{L}(1)$ acts from the left on any $\mathscr{L}(i)$, via γ, hence $\mathscr{L}(1) \times \mathscr{L}(1)$ acts from the left on $\mathscr{L}(i) \times \mathscr{L}(j)$. Note too that $\mathscr{L}(1) \times \mathscr{L}(1)$ acts from the right on $\mathscr{L}(2)$. Let us denote these actions by ν and μ, respectively.

LEMMA 7.2 (HOPKINS). *For $i \geq 1$ and $j \geq 1$, the diagram*

$$\mathscr{L}(2) \times \mathscr{L}(1) \times \mathscr{L}(1) \times \mathscr{L}(i) \times \mathscr{L}(j) \underset{\mathrm{id} \times \nu}{\overset{\mu \times \mathrm{id}}{\rightrightarrows}} \mathscr{L}(2) \times \mathscr{L}(i) \times \mathscr{L}(j) \overset{\gamma}{\longrightarrow} \mathscr{L}(i+j)$$

is a split coequalizer of spaces.

PROOF. Choose isomorphisms $s : U^i \to U$ and $t : U^j \to U$ and define

$$h(f) = (f \circ (s \oplus t)^{-1}, s, t)$$

and

$$k(f; g, g') = (f; g \circ s^{-1}, g' \circ t^{-1}; s, t).$$

It is trivial to check the required identities. $\square$

Observe that, while covariant functors need not preserve coequalizers in general, they clearly do preserve split coequalizers. This applies to the singular chain complex functor and leads to the following result.

PROPOSITION 7.3. *Let $i \geq 1$ and $j \geq 1$. Then the structural map γ of the operad $\mathscr{C} = C_*(\mathscr{L})$ induces an isomorphism*

$$\mathscr{C}(2) \otimes_{\mathbb{C} \otimes \mathbb{C}} \mathscr{C}(i) \otimes \mathscr{C}(j) \longrightarrow \mathscr{C}(i+j).$$

We use this to construct the promised natural associativity isomorphism

$$(L \boxtimes M) \boxtimes N \cong L \boxtimes (M \boxtimes N),$$

and we claim that both sides are naturally isomorphic to

$$\mathscr{C}(3) \otimes_{\mathbb{C}^3} L \otimes M \otimes N.$$

Note that $N \cong \mathbb{C} \otimes_{\mathbb{C}} N$. We have the isomorphisms

$$(L \boxtimes M) \boxtimes N \cong \mathscr{C}(2) \otimes_{\mathbb{C}^2} (\mathscr{C}(2) \otimes_{\mathbb{C}^2} L \otimes M) \otimes (\mathbb{C} \otimes_{\mathbb{C}} N)$$
$$\cong (\mathscr{C}(2) \otimes_{\mathbb{C}^2} \mathscr{C}(2) \otimes \mathscr{C}(1)) \otimes_{\mathbb{C}^3} (L \otimes M \otimes N)$$
$$\cong \mathscr{C}(3) \otimes_{\mathbb{C}^3} (L \otimes M \otimes N).$$

The symmetric argument shows that this is also isomorphic to $L \boxtimes (M \boxtimes N)$.

In view of the generality of the previous proposition, the argument iterates to prove that all j-fold iterated $\boxtimes$ products are canonically isomorphic to

$$\mathscr{C}(j) \otimes_{\mathbb{C}^j} M_1 \otimes \cdots \otimes M_j.$$

When all $M_i = \mathbb{C}$, this gives an isomorphism $\mathbb{C}^{\boxtimes j} \cong \mathscr{C}(j)$ of $(\mathbb{C}, \mathbb{C}^j)$-bimodules. We also claimed an isomorphism of left $\mathbb{C}$-modules between $\mathscr{C}(j)$ and $\mathbb{C}$. This arises from the evident fact that if $t : U^j \longrightarrow U$ is an isomorphism, then composition with t and t^{-1} give inverse homeomorphisms of left $\mathcal{L}(1)$-spaces between $\mathcal{L}(j)$ and $\mathcal{L}(1)$.

We also used the following observation about $\mathcal{L}$.

LEMMA 7.4. *The degeneracy map* $\sigma_1 : \mathcal{L}(2) \to \mathcal{L}(1)$ *is an* $\mathcal{L}(1)$-*equivariant homotopy equivalence.*

This does not exhaust the remarkable properties of this truly miraculous operad. The following property is not at all obvious and is not inherited in algebra.

LEMMA 7.5. *The orbit* $\mathcal{L}(2)/\mathcal{L}(1) \times \mathcal{L}(1)$ *is a point.*

8. Applications

The singular cochains of topological spaces can be given E_∞ algebra structures. The singular chains of E_∞ spaces and thus of infinite loop spaces have evident E_∞ algebra structures. Bloch's Chow complexes in algebraic geometry can be transformed into quasi-isomorphic E_∞ algebras. The last example leads to one version of a theory of mixed Tate motives [6, 7]. The first example has been studied by Smirnov [12, 13]. In both cases, much more remains to be done. We turn from algebra to topology to explain fully worked out applications of a parallel theory that we have developed there [5].

9. The stable homotopy category

Since the early 1960's, serious work in stable algebraic topology has taken place in a suitable category of spectra, which plays the role in stable algebraic topology that the category of spaces plays in unstable algebraic topology. Until recently, algebraic topologists always worked, not on the point-set level, but in the stable homotopy category of spectra. Thus associativity, commutativity, and unity were understood as properties that held only "up to homotopy". The reason for this crudity of structure was that we believed that it was impossible to construct a good category of spectra with a smash product that was associative, commutative, and unital on the point-set level, so that it made no sense to ask for precise point-set level algebraic structure.

The topological analog of the algebraic theory that I have described gives such a seemingly impossible category. This theory allows point-set definitions of ring, module, and algebra spectra and thus gives rise to a new subject of stable topological algebra. Most of the important examples of ring spectra in the homotopical sense arise from ring spectra in the new sense: the sphere spectrum, the Eilenberg-MacLane spectra, the spectra of algebraic and topological K-theory, the spectra of cobordism theory, and so on. Many of the examples come from multiplicative infinite loop space theory, which was itself by far the deepest of the earlier applications of topological operads.

For a new style commutative ring spectrum R and R-module M, it makes sense to construct the localization $M[T^{-1}]$ of M at a multiplicatively closed subset $T \subset \pi_*(R)$ and the quotient M/IM and completion $M\hat{_I}$ of M at an ideal $I \subset \pi_*(R)$. There are new torsion products and Ext groups that include the Tor and Ext groups of classical algebra and both classical and new homology and cohomology theories in topology as special cases. The new theory has already had many applications [**3, 5**] and is rapidly becoming a standard tool. The applications include:

- Simple new constructions of important spectra that previously were constructed by indirect means and with little algebraic structure.
- Simple constructions of a host of useful spectral sequences — universal coefficient, Künneth, generalized Eilenberg-Moore, etc.
- New and compatible constructions of Quillen's algebraic K-theory of rings and Waldhausen's algebraic K-theory of spaces.
- A new construction and generalization of topological Hochschild homology and of spectral sequences for its calculation.
- A completion theorem for equivariant complex cobordism and its modules analogous to the Atiyah-Segal completion theorem in equivariant K-theory.

10. A sketch of the definitions

Let me give a thumbnail sketch of the modern foundations of stable homotopy theory [**4**]. Let $U = \mathbb{R}^\infty$ be the sum of countably many copies of $\mathbb{R}$, with its standard inner product. A prespectrum T consists of based spaces $T(V)$ for each finite dimensional inner product space $V \subset U$ together with maps $\sigma : T(V) \wedge S^{W-V} \longrightarrow T(W)$ when $V \subset W$, where $W - V$ is the orthogonal complement of V in W and S^{W-V} is its one-point compactification. A map $f : T \longrightarrow T'$ of prespectra is a collection of maps $fV : TV \longrightarrow T'V$ that are strictly compatible with the structure maps.

A spectrum E is a prespectrum whose adjoint structure maps $\tilde{\sigma} : E(V) \longrightarrow \Omega^{W-V}E(W)$ are homeomorphisms. A map $f : E \longrightarrow E'$ of spectra is a map of underlying prespectra. The forgetful functor $\ell : \mathscr{S} \longrightarrow \mathscr{P}$ from the category of spectra to the category of prespectra has a left adjoint $L : \mathscr{P} \longrightarrow \mathscr{S}$. For a based space X, define $QX = \cup \Omega^q \Sigma^q X$, where the inclusions are given by suspension of maps. The suspension spectrum of X is

$$\Sigma^\infty X = L\{\Sigma^n X\} = \{Q\Sigma^n X\}.$$

In particular, $S = \Sigma^\infty S^0$ is the sphere, or zero sphere, spectrum.

The zeroth space E_0 of a spectrum is denoted $\Omega^\infty E$; such spaces are called infinite loop spaces. The functors Σ^∞ and Ω^∞ are left and right adjoint. More generally, there is a shift desuspension functor Σ_V^∞ that is left adjoint to the Vth space functor. We define sphere spectra for integers n by

$$S^n = \Sigma^\infty S^n \text{ if } n \geq 0 \quad \text{and} \quad S^{-n} = \Sigma^\infty_{\mathbb{R}^n} S^0 \text{ if } n > 0.$$

We define the smash product of a prespectrum T and a based space X by

$$(T \wedge X)(V) = TV \wedge X.$$

We then define $E \wedge X = L(\ell E \wedge X)$. Taking $X = I_+$, this gives us a notion of homotopy between maps of spectra. We define homotopy groups of spectra by

$$\pi_n(E) = [S^n, E].$$

We say that a map of spectra is a weak equivalence if it induces an isomorphism of homotopy groups. We have a homotopy category $h\mathscr{S}$, in which homotopic maps are identified. The stable homotopy category $\bar{h}\mathscr{S}$ is obtained from $h\mathscr{S}$ by adjoining formal inverses to the weak equivalences.

The essential point is to define the smash product of spectra. On the prespectrum level, we define

$$(T \wedge T')(V \oplus V') = T(V) \wedge T'(V').$$

We lift the construction to spectra by use of the left adjoint L. The key fact is that $E \wedge E'$ is now a spectrum indexed on inner product spaces in $U \oplus U$ rather than in U. We call this the external smash product.

Given a linear isometry $f : U \oplus U \longrightarrow U$, we can construct a functor f_* from spectra indexed on $U \oplus U$ to spectra indexed on U. The composite $f_*(E \wedge E')$ is an internal smash product. Different choices of f give rise to equivalent functors when we pass to the stable homotopy category, and this is what implies the associativity, commutativity, and unity of the internal smash product that we use when defining ring spectra and modules in the homotopical sense.

This smash product is crying out to be reinterpreted in terms of the operad $\mathscr{L}$. There is a "twisted half-smash product" that allows us to glue together all of the j-fold internal smash products into a single j-fold smash product

$$\mathscr{L}(j) \ltimes (E_1 \wedge \cdots \wedge E_j).$$

It is equivalent to each of the j-fold internal smash products determined by a choice of isometry $U^j \longrightarrow U$. This smash product, although canonical, is not associative.

We can define an action of $\mathbb{L} = \mathscr{L}(1)$ on a spectrum E by means of a suitable map $\mathscr{L}(1) \ltimes E \longrightarrow E$. This gives us the notion of an $\mathbb{L}$-spectrum. Examples include all suspension spectra. For $\mathbb{L}$-spectra E and E', we can define an operadic smash product

$$E \wedge_{\mathscr{L}} E' = \mathscr{L}(2) \ltimes_{\mathscr{L}(1) \times \mathscr{L}(1)} E \wedge E'$$

that is again an $\mathbb{L}$-spectrum. This smash product is commutative, and the split coequalizer property of the linear isometries operad implies that it is also associative. It is not unital, but there is a natural map $\lambda : S \wedge_{\mathscr{L}} E \longrightarrow E$

of $\mathbb{L}$-spectra that is always a weak equivalence of spectra. It is not usually an isomorphism, but our last surprising property (Lemma 7.5) of the linear isometries operad implies that it is an isomorphism for $E = S$ and for $E = S \wedge_{\mathscr{L}} E'$ for any $\mathbb{L}$-spectrum E'.

We define an S-module to be an $\mathbb{L}$-spectrum M such that $\lambda : S \wedge_{\mathscr{L}} M \longrightarrow M$ is an isomorphism. For two S-modules M and M', we define

$$M \wedge_S M' = M \wedge_{\mathscr{L}} M'.$$

The category of S-modules is symmetric monoidal under $\wedge_S$, with unit S, and there is even a function S-module functor F_S such that

$$\mathscr{M}_S(M \wedge_S M', M'') \cong \mathscr{M}_S(M, F_S(M, M')),$$

where $\mathscr{M}_S$ is the category of S-modules. We define the derived category $\mathscr{D}_S$ by adjoining formal equivalences to the weak equivalences in the homotopy category $h\mathscr{D}_S$. This category is equivalent to the stable homotopy category $\bar{h}\mathscr{S}$, and the equivalence preserves smash products and function spectra.

From here, the topological theory is precisely parallel to the algebraic theory. Note the startling fact that, in view of the strict unity property, the topological theory is actually better behaved algebraically than the algebraic theory!

Given the category of S-modules, we define an S-algebra R by requiring a unit $S \longrightarrow R$ and product $R \wedge_S R \longrightarrow R$ such that the evident unit and associativity diagrams commute. We say that R is a commutative S-algebra if the evident commutativity diagram also commutes. We define a left R-module similarly, by requiring a map $R \wedge_S M \longrightarrow M$ such that the evident unit and associativity diagrams commute.

For a right R-module M and left R-module N, we define $M \wedge_R N$ by a coequalizer diagram

$$M \wedge_S R \wedge_S N \underset{\mathrm{Id}\wedge_S \nu}{\overset{\mu\wedge_S\mathrm{Id}}{\rightrightarrows}} M \wedge_S N \longrightarrow M \wedge_R N,$$

where μ and ν are the given actions of R on M and N. If R is commutative, then the smash product of R-modules is an R-module, the category $\mathscr{M}_R$ of R-modules is symmetric monoidal with unit R, and there is a function R-module functor $F_R(M, N)$ with the usual adjunction. We define R-algebras exactly as we defined S-algebras, via unit and product maps $R \longrightarrow A$ and $A \wedge_R A \longrightarrow A$. All of the standard formal properties of modules, rings, and algebras go over directly to the new subject of stable topological algebra. For example, the smash product $A \wedge_R A'$ of commutative R-algebras A and A' is their coproduct in the category of commutative R-algebras.

Thinking homotopically, $\mathscr{M}_R$ has a derived category $\mathscr{D}_R$ that is obtained by inverting the maps of R-modules that are weak equivalences of underlying spectra. For an S-algebra R, the category $\mathscr{D}_R$ is not just a tool for the the study of classical algebraic topology, but an interesting new subject of study in its own right. When $R = Hk$ for a discrete ring k, $\mathscr{D}_R$ is equivalent to $\mathscr{D}_k$.

What about examples? There is a notion of an A_∞ ring spectrum E that Quinn, Ray, and I defined in 1972 [**10**]. It is specified by an action of the operad $\mathscr{L}$ on a spectrum E. Such an action is given by maps

$$\theta_j : \mathscr{L}(j) \ltimes E^j \longrightarrow E$$

such that the evident associativity and unity diagrams commute. If the θ_j are Σ_j-equivariant, then E is said to be an E_∞ ring spectrum. It turns out that if E is an A_∞ ring spectrum, then $S \wedge_{\mathscr{L}} E$ is an S-algebra; if E is an E_∞ ring spectrum, then $S \wedge_{\mathscr{L}} E$ is a commutative S-algebra. Thus the earlier definitions are essentially equivalent to the new ones, and the earlier work gives a plenitude of examples.

Observe that the operad $\mathscr{L}$ has been exploited in several essentially different ways in this theory.

Descriptively: Thom spectra arise in nature with an action of $\mathscr{L}$ [**10**]. This observation was the starting point of everything I talked about today.

Constructively: $\mathscr{L}$ acts on Steiner's analog [**14**] of the infinite little cubes operad; this fact is the starting point of multiplicative infinite loop space theory [**11**], in which one constructs E_∞ ring spectra from E_∞ ring spaces.

Foundationally: Properties of $\mathscr{L}$ discovered long after the applications just cited led to our construction of the symmetric monoidal category of S-modules.

As this conference has shown, this theory is just one of many in which operads play a central mathematical role. For me, after enjoying their company for 25 years, I still find operads remarkable and delightful creatures, with a knack for springing surprises. Thanks for joining the fun.

BIBLIOGRAPHY

1. P. Deligne. Catégories tannakiennes. In *The Grothendieck Festschrift*, Volume 2, 111–195. Birkhauser, 1990.
2. A. Dold and D. Puppe. Duality, trace, and transfer. In *Proc. International Conference on Geometric Toplogy*, 81–102. PWN – Polish Scient. Publishers, 1980.
3. A. Elmendorf, J. P. C. Greenlees, I. Kriz, and J. P. May. Commutative algebra in stable homotopy theory and a completion theorem. *Mathematical Research Letters* 1(1994), 225-239.
4. A. D. Elmendorf, I. Kriz, M. A. Mandell, and J. P. May. Modern foundations for stable homotopy theory. In "A handbook of algebraic topology", edited by I. M. James. Elsevier Science. 1995.
5. A. Elmendorf, I. Kriz, A. M . Mandell, and J. P. May. Rings, modules, and algebras in stable homotopy theory. Amer. Math. Soc. Surveys and Monographs. To appear.
6. I. Kriz and J. P. May. Derived categories and motives. *Mathematical Research Letters*, 1(1994), 87–94.
7. I. Kriz and J. P. May. Operads, Algebras, Modules, and Motives. Astérisque Vol 133. 1995.
8. L. G. Lewis, Jr., J. P. May, and M. Steinberger (with contributions by J. E. McClure). *Equivariant stable homotopy theory.Springer Lecture Notes in Mathematics* Vol. 1213. Springer, 1986.

9. J. P. May. *The Geometry of Iterated Loop Spaces. Springer Lecture Notes in Mathematics* Vol. 271. Springer, 1972.

10. J. P. May. E_∞ *ring spaces and* E_∞ *ring spectra. Springer Lecture Notes in Mathematics* Vol. 577. Springer, 1977.

11. J. P. May. Multiplicative ininite loop space theory. *J. Pure and Applied Algebra*, 26(1982), 1–69.

12. V. A. Smirnov. On the cochain complex of topological spaces. Math. USSR Sbornik 43(1982), 133-144.

13. V. A. Smirnov. Homotopy theory of coalgebras. Math. USSR Izvestiya 27(1986). 575-592.

14. R. Steiner. A canonical operad pair. *Math. Proc. Camb. Phil. Soc.*, 86(1979), 443–449.

15. R. Vogt. Boardman's stable homotopy category. Aarhus University Lecture Notes Series, 21, 1970.

THE UNIVERSITY OF CHICAGO, CHICAGO, IL 60637
E-mail address: may@math.uchicago.edu

Contemporary Mathematics
Volume **202**, 1997

Homotopy Gerstenhaber algebras and topological field theory

Takashi Kimura, Alexander A. Voronov, and Gregg J. Zuckerman

ABSTRACT. We prove that the BRST complex of a topological conformal field theory is a homotopy Gerstenhaber algebra, as conjectured by Lian and Zuckerman in 1992. We also suggest a refinement of the original conjecture for topological vertex operator algebras. We illustrate the usefulness of our main tools, operads and "string vertices" by obtaining new results on Vassiliev invariants of knots and double loop spaces.

Two-dimensional topological quantum field theory (TQFT) at its most elementary level is the theory of $\mathbb{Z}$-graded commutative associative algebras (with some additional structure) [**38**]. Thus, it came as something of a surprise when several groups of mathematicians realized that the physical state space of a 2D TQFT has the structure of a $\mathbb{Z}$-graded Lie algebra, relative to a new grading equal to the old grading minus one. Moreover, the commutative and Lie products fit together nicely to give the structure of a Gerstenhaber algebra (G-algebra), a $\mathbb{Z}$-graded Poisson algebra for which the Poisson bracket has degree -1 (see Section 1). This G-algebra structure is best understood in the framework of 2D topological conformal field theories (TCFTs) (see Section 5.2) wherein operads of moduli spaces of Riemann surfaces play a fundamental role.

G-algebras arose explicitly in M. Gerstenhaber's work on the Hochschild cohomology theory for associative algebras (see Section 1 for this and several other contexts for the theory of G-algebras). Operads arose

1991 *Mathematics Subject Classification.* Primary 55S20, 81T40, 57M25, 17B69 Secondary 57R19, 53Z05, 14H10.

Research of the first author was supported in part by an NSF postdoctoral research fellowship.

Research of the second author was supported in part by NSF grant DMS-9402076.

Research of the third author was supported in part by NSF grant DMS-9307086.

in the work of J. Stasheff, Gerstenhaber and later work of P. May on
the recognition problem for iterated loop spaces. Eventually, F. Cohen discovered that the homology of a double loop space is naturally
a G-algebra, see Section 1; in fact, a double loop space is naturally
an algebra over the little disks operad employed by Cohen and also
Boardman and Vogt, see Section 6.2. (The reader should consult the
article [33] by May in these proceedings.)

In joint work, B. Lian and G. Zuckerman [30] (see also the joint
work of M. Penkava and A. Schwarz [35]) discovered the above-men-
tioned G-algebra structure in the context of topological vertex operator algebras (TVOAs) (see Section 2), which are a powerful algebraic
starting point for the construction of 2D topological conformal field
theories. Lian and Zuckerman also gave a number of concrete
constructions of various examples of TVOAs, TCFTs and G-algebras.

In an attempt to understand Lian-Zuckerman's work geometrically,
E. Getzler [16] found a G-algebra structure in the physical state space
of an abstract TCFT (as well as in a topological "massive" quantum
field theory). Getzler's ongoing work with J. Jones was already dealing
with G-algebras as the $n = 2$ case of n-algebras. Segal's ideas, see [37],
played an essential role in Getzler's discovery. In particular, Segal had
already developed an extension to TCFTs of his geometric category
approach [36] to conformal field theory.

Later on Y.-Z. Huang [22] found a third approach which combined
the ideas of Lian, Zuckerman and Getzler. In particular, Huang took
steps towards the construction of a TCFT from a TVOA; this work
was based on Huang's earlier demonstration of how to construct a
(tree-level) holomorphic conformal field theory (CFT) from a vertex
operator algebra (VOA). Such a connection between Segal's geometric
approach to conformal field theory and Borcherds' algebraic definition
[6] of a VOA (see also the book of I. Frenkel, J. Lepowsky and A.
Meurman [11]) had already been suggested in some public lectures by
I. Frenkel [10].

From the very beginning of the above development, it was understood that the physical state space of a TCFT is merely the cohomology of a much more enormous object, the BRST complex of the TCFT.
Thus there arose the question: does the G-algebra structure on the vector space of physical states come from a higher homotopy G-algebra
structure on the BRST complex itself? This question was explicitly
raised by Lian and Zuckerman in their work on TVOAs and associated
TCFTs. They found that in the BRST complex of a TVOA, all of the
identities of a G-algebra fail to hold on the nose, but they continue
to hold up to homotopy. They then asked whether these homotopies

could be continued to an infinite hierarchy of higher homotopies, such as those found in the work of Stasheff on A_∞-algebras and the more recent work of Stasheff and T. Lada on L_∞-algebras.

The inspiration for the search for higher homotopy algebras in topological conformal field theory arose in the related context of closed string field theory, see Stasheff [41]. The explicit discussion of higher homotopy algebras in string field theory appeared in work of Stasheff [40], M. Kontsevich [27], and E. Witten and B. Zwiebach [46, 48]. The later joint papers of T. Kimura, A. Voronov and Stasheff [25, 26] constructed L_∞ and C_∞ structures using the operadic approach. These papers also include a conceptual explanation of the relationship between string field theory and TCFTs.

However, research on higher homotopies suffered from a lack of a proper definition of a higher homotopy Gerstenhaber algebra (homotopy G-algebra). Recently, various definitions have been put forward, in particular in work of V. Ginzburg and M. Kapranov [19], Getzler and Jones [17], and Gerstenhaber and Voronov [13]. In the current paper, we use the term G_∞-algebra to refer to a particular definition of a homotopy G-algebra appearing in the work of Getzler and Jones (Definition 4.1). This definition is based on pioneering work of R. Fox and L. Neuwirth [9]. G_∞-algebras are governed by what we call the G_∞-operad.

The main new result of the current paper is the proof that the BRST complex of a TCFT is indeed a G_∞-algebra. In fact, a (tree-level) TCFT itself is defined to be an algebra over a particular topological operad. Thus, in this paper, both the "classical" theory of topological operads as well as the recent theory of linear operads play essential roles.

The original question of Lian and Zuckerman can now be formulated precisely (see our Conjecture 2.3): does a TVOA carry a natural G_∞ structure? Since we have answered the analogous question for a TCFT, a crucial step still remains in the program to answer the original question: the completion of Huang's work on the construction of a TCFT from a TVOA. Such a construction should identify the BRST complex of the TCFT with an appropriate topological completion of the BRST complex of the TVOA. We look forward to the successful conclusion of this program.

One of the essential tools of our paper is M. Wolf and Zwiebach's "string vertices", which make a bridge between a topological operad of punctured Riemann spheres and the infinite dimensional topological operad responsible for CFTs. Amazingly, the latter operad plays a key role in the subjects of Vassiliev invariants and double loop spaces. In particular, string vertices combined with the approach of our paper yield Vassiliev invariants of knots in Section 6.1 and the structure of a homotopy G-algebra on the singular chain complex of a double loop space in Section 6.2.

ACKNOWLEDGMENT . We are very grateful to T. Q. T. Le, B. Lian, A. S. Schwarz, and J. Stasheff for helpful discussions. T.K. and A.A.V. express their sincere gratitude to J.-L. Loday and J. Stasheff for their hospitality at the wonderful conference in Luminy. T.K. and G.J.Z. would like to thank A.A.V. for inviting them to the terrific conference at Hartford. A.A.V. also thanks IHES for offering him excellent conditions for work on the project in June of 1995.

1. Gerstenhaber algebras

A *Gerstenhaber algebra* or a *G-algebra* is a graded vector space H with a dot product xy defining the structure of a graded commutative algebra and with a bracket $[x, y]$ of degree -1 defining the structure of a graded Lie algebra, such that the bracket with an element is a derivation of the dot product:

$$[x, yz] = [x, y]z + (-1)^{(\deg x - 1) \deg y} y[x, z],$$

where $\deg x$ denotes the degree of an element x. In other words, a G-algebra is a specific graded version of a Poisson algebra.

This structure arises naturally in a number of contexts, such as the following.

EXAMPLE 1.1. Let A be an associative algebra and $C^n(A, A) = \operatorname{Hom}(A^{\otimes n}, A)$ be its Hochschild complex. Then the dot product defined as the usual cup product up to a sign

$$(1) \quad (x \cdot y)(a_1, \ldots, a_{k+l}) = (-1)^{kl}(x \cup y)(a_1, \ldots, a_{k+l})$$
$$= (-1)^{kl} x(a_1, \ldots, a_k) y(a_{k+1}, \ldots, a_{k+l}),$$

where x and y are k- and l-cochains and $a_i \in A$, and a G-bracket $[x, y]$ define the structure of a G-algebra on the Hochschild cohomology $H^n(A, A)$. The bracket was introduced by Gerstenhaber [12] in order to describe the obstruction for extending a first order deformation of the algebra A to the second order. The following definition of the bracket is due to Stasheff [39]. Considering the tensor coalgebra $T(A) = \bigoplus_{n=0}^\infty A^{\otimes n}$ with the comultiplication $\Delta(a_1 \otimes \cdots \otimes a_n) = \sum_{k=0}^n (a_1 \otimes \cdots \otimes a_k) \otimes (a_{k+1} \otimes \cdots \otimes a_n)$, we can identify the Hochschild cochains $\operatorname{Hom}(A^{\otimes n}, A)$ with the coderivations $\operatorname{Coder} T(A)$ of the tensor coalgebra $T(A)$. Then the G-bracket $[x, y]$ is defined as the (graded) commutator of coderivations. In fact, the Hochschild complex $C^\bullet(A, A)$ is a differential graded Lie algebra with respect to this bracket.

EXAMPLE 1.2. Let $A_n^\bullet$ be the $\mathbb{Z}$-graded commutative algebra generated by n variables $x_1, x_2, ..., x_n$, of degree zero, and n more variables $\partial_{x_1}, ..., \partial_{x_n}$ of degree one. We refer to an element of this algebra as a polyvector field. The elements of degree zero are interpreted as functions, the elements of degree one as vector fields, those of degree two as bivector fields, and so on. The dot product is simply the graded commutative multiplication of polyvector fields.

Long ago, Schouten and Nijenhuis [34] defined a bracket operation on polyvector fields (they thought of such fields as antisymmetric contravariant tensor fields). The Schouten-Nijenhuis bracket $[P, Q]$ is characterized by the following:

1. For any two functions f and g, $[f, g] = 0$.
2. If f is a function and X is a vector field, $[X, f] = -[f, X] = Xf$.
3. If X and Y are vector fields, then $[X, Y]$ is the standard bracket of the vector fields.
4. Together, the dot product and the Schouten-Nijenhuis bracket endow $A_n^\bullet$ with the structure of a Gerstenhaber algebra.

Let C_n be the polynomial algebra in n variables. It is known [21] that $H^\bullet(C_n, C_n)$ is canonically isomorphic as a G-algebra to the algebra $A_n^\bullet$.

EXAMPLE 1.3. Let $\mathfrak{g}$ be any Lie algebra, and let $\Lambda^\bullet\mathfrak{g}$ be the Grassmann algebra generated by $\mathfrak{g}$. Define a bracket $[X, Y]$ on $\Lambda^\bullet\mathfrak{g}$ by requiring the following:

1. If a and b are scalars, $[a, b] = 0$.
2. If X is in $\mathfrak{g}$ and a is a scalar, then $[X, a] = 0$.
3. If X and Y are in $\mathfrak{g}$, then $[X, Y]$ is the Lie bracket in $\mathfrak{g}$.
4. The wedge product together with the bracket product endow $\Lambda^\bullet\mathfrak{g}$ with the structure of a Gerstenhaber algebra.

EXAMPLE 1.4. Let M be a manifold (differentiable, complex, algebraic, etc.) Let $F(M)$ be the commutative algebra of functions (of the appropriate type–differentiable, holomorphic, regular, etc.) on M. Let $V^\bullet(M)$ be the algebra of polyvector fields on M, with the operations of wedge product and the Schouten-Nijenhuis bracket, defined by analogy with the bracket in $A_n^\bullet$. Then $V^\bullet(M)$ is a Gerstenhaber algebra. We can regard $V^\bullet(M)$ as the commutative superalgebra of functions on ΠT^*M, the cotangent bundle of M with the fibers made into odd supervector spaces. The G-bracket in $V^\bullet(M)$ is the odd Poisson bracket associated to the canonical odd symplectic two-form on ΠT^*M.

Let P be a bivector field on M. We can always construct a bracket operation on the algebra $F(M)$ by the formula

$$\{f, g\}_P = \iota(P)(df \wedge dg) = (df \wedge dg)(P),$$

where $\iota(P)$ denotes contraction of P against a two-form. The bracket $\{,\}_P$ satisfies the Jacobi identity if and only if the Schouten-Nijenhuis bracket $[P, P]$ is zero. In this case, the algebra $F(M)$ becomes what is known as a Poisson algebra; moreover, the derivation $\sigma_P = [P, -]$ has square zero and turns $V^\bullet(M)$ into a differential graded G-algebra, whose cohomology G-algebra is known as the Poisson cohomology of M relative to P.

EXAMPLE 1.5. Let X be a topological space and let $\Omega^2 X$ be the two-fold loop space of X. Let $A^\bullet(X)$ denote the homology of $\Omega^2 X$ with rational coefficients. We endow $A^\bullet(X)$ with the structure of a $\mathbb{Z}$-graded commutative algebra, via the Pontrjagin product. $A^\bullet(X)$ is a Hopf algebra which is freely generated as a graded commutative algebra by its subspace P of primitive elements. Moreover, P is isomorphic to the rational homotopy of $\Omega^2 X$. P is $\mathbb{Z}$-graded, and we have $P_n = \pi_{n+2}(X) \otimes_{\mathbb{Z}} \mathbb{Q}$, for n nonnegative, and 0 otherwise.

The rational homotopy of the ordinary loop space of X is a $\mathbb{Z}$-graded Lie algebra, which we denote by L. The bracket is called the Samelson product. It is known that $P_n = L_{n+1}$. Thus, the algebra $A^\bullet(X)$ is isomorphic to the graded exterior algebra $\Lambda^\bullet L$. In particular, $A^\bullet(X)$ is a G-algebra of a type generalizing Example 1.3 above.

1.1. Operads in action. For a primer on operads, algebras over operads and the little disks operad, see P. May's paper [32] in this volume. Here we recall briefly the definition of the little disks operad in relation to G-algebras. The *little disks operad* is the collection $\{D(n), n \geq 1\}$ of topological spaces $D(n)$ with an action of the permutation group and operad compositions. The space $D(n)$ consists of embeddings of n little disks in the unit disk via dilation and translation. The operad composition $\circ_i : D(m) \times D(n) \to D(m + n - 1)$ is defined by contracting the unit disk with m little disks inside it to fit into the ith little disk in the other unit disk and erasing the seam. Since $D(n)$ is naturally an open subset in $\mathbb{R}^{3n}$, it is a topological operad and one can naturally obtain an operad of graded vector spaces from it, applying the functor of homology. This operad $H_\bullet(D(n))$ is called the *G-operad* in view of the following theorem.

THEOREM 1.1 (F. Cohen [**7, 8**]). *The structure of a G-algebra on a $\mathbb{Z}$-graded vector space is equivalent to the structure of an algebra over the homology little disks operad $H_\bullet(D(n))$.*

2. Topological vertex operator algebras

In this section, we give a brief introduction to VOAs, in order to present the Lian-Zuckerman conjecture in an updated form. VOAs will not show up in the rest of the paper, but we intend to get back to them in a subsequent paper. See [**30**] for more discussion.

DEFINITION 2.1 (Quantum operators). Consider the integrally bigraded complex vector space $V^\bullet[\cdot] = \bigoplus_{g \in \mathbb{Z}, \Delta \in \mathbb{Z}} V^g[\Delta]$; if v is in $V^g[\Delta]$ we will write $|v| = g =$ the ghost number of v and $||v|| = \Delta =$ the weight of v. Let z be a formal variable with degrees $|z| = 0$ and $||z|| = -1$. Then, it makes sense to speak of a homogeneous *bi-infinite* formal power series

$$\phi(z) = \sum_{n \in \mathbb{Z}} \phi(n) z^{-n-1}$$

of degrees $|\phi(z)|$, $||\phi(z)||$, where the coefficients $\phi(n)$ are homogeneous linear maps in $V^\bullet[\cdot]$ of degrees $|\phi(n)| = |\phi(z)|$, $||\phi(n)|| = -n - 1 + ||\phi(z)||$. Note then that the terms $\phi(n) z^{-n-1}$ indeed have the same degrees $|\phi(z)|$, $||\phi(z)||$ for all n. We call a finite sum of such series a *quantum operator* on $V^\bullet[\cdot]$, and denote the bigraded linear space of quantum operators as $\mathrm{QO}(V^\bullet[\cdot])$. We will denote the special operator $\phi(0)$ by the symbol $\mathrm{Res}_z \phi(z)$.

DEFINITION 2.2. A *vertex operator graded algebra* consists of the following ingredients:

1. An integrally bigraded complex vector space $V^\bullet[\cdot]$.
2. A linear map $Y : V^\bullet[\cdot] \to \mathrm{QO}(V^\bullet[\cdot])$ such that Y has bidegree (0,0). We call Y the *vertex map*, and if v is in $V^\bullet[\cdot]$, we let $Y(v, z) = Y(v)(z)$ denote the *vertex operator* associated to v. The map Y is subject to the following axioms:
   (a) Let v and v' be elements of $V^\bullet[\cdot]$: then $\mathrm{Res}_z(z^m Y(v, z) v')$ vanishes for m sufficiently positive.
   (b) (Cauchy-Jacobi identity) Let v and v' be elements of $V^\bullet[\cdot]$ and let $f(z, w)$ be a Laurent polynomial in z, w, and $z - w$. Then we have the identity

$$\mathrm{Res}_w \mathrm{Res}_{z-w} Y(Y(v, z - w)v', w) f(z, w)$$
$$= \mathrm{Res}_z \mathrm{Res}_w Y(v, z) Y(v', w) f(z, w)$$
$$- (-1)^{|v||v'|} \mathrm{Res}_w \mathrm{Res}_z Y(v', w) Y(v, z) f(z, w).$$

   (c) There exists a distinguished element 1 in $V^0[0]$ such that $Y(1, z)$ is the identity operator and such that for any v in

$V^\bullet[\cdot]$, the result $Y(v, z)1$ is a power series in z and

$$\lim_{z \to 0} Y(v, z)1 = v.$$

3. A distinguished element F in $V^0[1]$ such that $F_0 = \text{Res}_z\, Y(F, z)$ defines the ghost number grading: if v is in $V^g[\Delta]$, $F_0 v = gv$.
4. A distinguished element L in $V^0[2]$ such that if we define an operator $L_n = \text{Res}_z(z^{n+1} Y(L, z))$ for every integer n, we have the following:
   (a) For v in $V^g[\Delta]$, $L_0 v = \Delta v$.
   (b) For any v we have $Y(L_{-1}v, z) = \partial Y(v, z)$.
   (c) For some fixed complex number c (the central charge), we have

$$[L_m, L_n] = (m - n)L_{m+n} + \frac{c}{12}(m^3 - m)\delta_{m,-n}1.$$

DEFINITION 2.3. A *topological vertex operator algebra* (*TVOA*) is a vertex operator graded algebra $(V^\bullet[\cdot], Y, 1, F, L)$ equipped with two additional distinguished elements J in $V^1[1]$ and G in $V^{-1}[2]$ such that the following axioms hold:

1. The operator $Q = \text{Res}_z\, Y(J, z)$ satisfies $Q^2 = 0$. Q is called the BRST charge, or BRST coboundary operator. Q has bidegree $(1, 0)$: $|Q| = 1$, $\|Q\| = 0$.
2. $[Q, Y(G, z)] = Y(L, z)$.

DEFINITION 2.4. Let $(V^\bullet[\cdot], Y, 1, F, L, J, G)$ be a TVOA. Let v and v' be elements of $V^\bullet[\cdot]$.

1. The *dot product* of v and v' is the element

$$v \cdot v' = \text{Res}_z(z^{-1} Y(v, z)v').$$

2. The *bracket product* of v and v' is the element

$$[v, v'] = (-1)^{|v|} \text{Res}_w \text{Res}_{z-w} Y(Y(G, z - w)v, w)v'.$$

LEMMA 2.1. *Let $V^\bullet[\cdot]$ be a TVOA.*

1. *The BRST operator Q is a derivation of both the dot and bracket products, which therefore induce dot and bracket products respectively on the BRST cohomology $H^\bullet(V, Q)$ of $V^\bullet[\cdot]$ relative to Q.*
2. *Every BRST cohomology class is represented by a BRST-cocycle of weight 0. Thus, the weight grading induces the trivial grading in BRST cohomology.*
3. *The BRST cohomology is trivial unless the central charge $c = 0$.*

THEOREM 2.2. *Let $V^\bullet[\cdot]$ be a TVOA. With respect to to the induced dot and bracket products, the BRST cohomology $H^\bullet(V, Q)$ of $V^\bullet[\cdot]$ relative to Q is a Gerstenhaber algebra.*

For an interesting study of the identities satisfied by the dot and bracket products in the TVOA $V^\bullet[\cdot]$ itself, see F. Akman [1]. In particular, Akman finds that the space $V^\bullet[\cdot]$ endowed with the bracket is a $\mathbb{Z}$-graded Leibniz algebra [31].

CONJECTURE 2.3. *Let $V^\bullet[\cdot]$ be a TVOA. Then the dot product and the skew-symmetrization of the bracket defined above can be extended to the structure of a G_∞-algebra (see below) on $V^\bullet[\cdot]$.*

This conjecture makes precise sense in the light of our current work: it refines the question posed by Lian and Zuckerman, who expected A_∞ and L_∞ structures to mix together.

3. Classical story: homotopy associative and homotopy Lie algebras

Inasmuch as G-algebras combine properties of commutative associative and Lie algebras, homotopy G-algebras make a similar combination of homotopy associative and homotopy Lie algebras; see Section 4.3 below. (The commutativity is not completely lost, either: homotopy G-algebras will also provide a homotopy for the commutativity of the dot product). Before discussing homotopy G-algebras, let us recall definitions of the more traditional homotopy associative and homotopy Lie algebras.

DEFINITION 3.1 (Homotopy associative (A_∞-) algebras). A *homotopy associative algebra* is a complex $V = \sum_{i \in \mathbb{Z}} V_i$ with a differential d, $d^2 = 0$, of degree 1 and a collection of n-ary products M_n:

$$M_n(v_1, \ldots, v_n) \in V, \qquad v_1, \ldots, v_n \in V, \ n \geq 2,$$

which are homogeneous of degree $2 - n$ and satisfy the relations

$$dM_n(v_1, \ldots, v_n) + \sum_{i=1}^{n} \epsilon(i) M_n(v_1, \ldots, dv_i, \ldots, v_n)$$

$$= \sum_{\substack{k+l=n+1 \\ k,l \geq 2}} \sum_{i=0}^{l-1} \epsilon(k,i) M_l(v_1, \ldots, v_i, M_k(v_{i+1}, \ldots, v_{i+k}), v_{i+k+1}, \ldots, v_n),$$

where $\epsilon(i) = (-1)^{\deg v_1 + \cdots + \deg v_{i-1}}$ is the sign picked up by taking d through $v_1, \ldots, v_{i-1}$, $\epsilon(k,i) = (-1)^{k(\deg v_1 + \cdots + \deg v_i)}$ is the sign picked up by M_k passing through $v_1, \ldots, v_i$.

For $n = 3$, the above identity shows that the binary product M_2 is associative up to a homotopy, provided by the ternary product M_3.

DEFINITION 3.2 (Homotopy Lie (L_∞-) algebras). A *homotopy Lie algebra* is a complex $V = \sum_{i \in \mathbb{Z}} V_i$ with a differential d, $d^2 = 0$, of degree 1 and a collection of n-ary brackets:

$$[v_1, \ldots, v_n] \in V, \qquad v_1, \ldots, v_n \in V, \ n \geq 2,$$

which are homogeneous of degree $3 - 2n$ and super (or graded) symmetric:

$$[v_1, \ldots, v_i, v_{i+1}, \ldots, v_n] = (-1)^{|v_i||v_{i+1}|}[v_1, \ldots, v_{i+1}, v_i, \ldots, v_n],$$

$\deg v$ denoting the degree of $v \in V$, and satisfy the relations

$$d[v_1, \ldots, v_n] + \sum_{i=1}^{n} \epsilon(i)[v_1, \ldots, dv_i, \ldots, v_n]$$

$$= \sum_{\substack{k+l=n+1 \\ k,l \geq 2}} \sum_{\substack{\text{unshuffles } \sigma: \\ \{1,2,\ldots,n\}=I_1 \cup I_2, \\ I_1=\{i_1,\ldots,i_k\}, \ I_2=\{j_1,\ldots,j_{l-1}\}}} \epsilon(\sigma)[[v_{i_1}, \ldots, v_{i_k}], v_{j_1}, \ldots, v_{j_{l-1}}],$$

where $\epsilon(i) = (-1)^{\deg v_1 + \cdots + \deg v_{i-1}}$ is the sign picked up by taking d through $v_1, \ldots, v_{i-1}$, $\epsilon(\sigma)$ is the sign picked up by the elements v_i passing through the v_j's during the unshuffle of $v_1, \ldots, v_n$, as usual in superalgebra.

For $n = 3$, the above identity shows that the binary bracket $[v_1, v_2]$ satisfies the Jacobi identity up to a homotopy, provided by the next bracket $[v_1, v_2, v_3]$.

4. Homotopy G-algebras

Due to Fred Cohen's Theorem [8], the structure of a G-algebra on a vector space V is equivalent to the structure on V of an algebra over the homology operad $H_\bullet(D(n))$, $n \geq 1$, of the little disks operad $D(n)$. In most general terms, a homotopy Gerstenhaber algebra or homotopy G-algebra is an algebra over a "resolution" of this G-operad $G_\bullet(n) = H_\bullet(D(n))$, $n \geq 1$, i.e., an operad $hG_\bullet(n)$ of complexes whose cohomology is identified with $G_\bullet(n)$. Different resolutions lead to different notions of homotopy G-algebras. A minimal resolution to answer the question of Deligne that the Hochschild complex is a homotopy G-algebra was introduced in [13], two free resolutions were constructed in [17]. The obvious singular-chain resolution $C_\bullet(D(n))$ is too large for our purposes: it produces too many higher operations (homotopies).

EXAMPLE 4.1 (Hochschild complex of an associative algebra). Let A be an associative algebra and $C^n(A, A) = \text{Hom}(A^{\otimes n}, A)$ its Hochschild cochain complex. Define the following collection of multilinear operations, called *braces*, cf. [15], on $C^\bullet(A, A)$:

$$\{x\}\{x_1, \ldots, x_n\}(a_1, \ldots, a_m) :=$$

$$\sum (-1)^\varepsilon x(a_1, \ldots, a_{i_1}, x_1(a_{i_1+1}, \ldots), \ldots, a_{i_n}, x_n(a_{i_n+1}, \ldots), \ldots, a_m)$$

for $x, x_1, \ldots, x_n \in C^\bullet(A, A)$, $a_1, \ldots, a_m \in A$, where the summation runs over all possible substitutions of $x_1, \ldots, x_n$ into x in the prescribed order and $\varepsilon := \sum_{p=1}^n (\deg x_p - 1) i_p$. The braces $\{x\}\{x_1, \ldots, x_n\}$ are homogeneous of degree $-n$, i.e., $\deg\{x\}\{x_1, \ldots, x_n\} = \deg x + \deg x_1 + \cdots + \deg x_n - n$. We will also adopt the following convention:

$$x \circ y := \{x\}\{y\}.$$

In addition, the usual cup product (altered by a sign) defined by (1) and the differential

$$(dx)(a_1, \ldots, a_{n+1})$$

$$:= (-1)^{\deg x} a_1 x(a_2, \ldots, a_{n+1})$$

$$+ (-1)^{\deg x} \sum_{i=1}^n (-1)^i x(a_1, \ldots, a_{i-1}, a_i a_{i+1}, a_{i+2}, \ldots, a_{n+1})$$

$$- x(a_1, \ldots, a_n) a_{n+1}$$

define the structure of a differential graded (DG) associative algebra on $C^\bullet(A, A)$. In their turn the braces satisfy the following identities.

$$(2) \quad \{\{x\}\{x_1, \ldots, x_m\}\}\{y_1, \ldots, y_n\}$$

$$= \sum_{0 \le i_1 \le \cdots \le i_m \le n} (-1)^\varepsilon \{x\}\{y_1, \ldots, y_{i_1}, \{x_1\}\{y_{i_1+1}, \ldots\}, \ldots,$$

$$y_{i_m}, \{x_m\}\{y_{i_m+1}, \ldots\}, \ldots, y_n\},$$

where $\varepsilon := \sum_{p=1}^m (\deg x_p - 1) \sum_{q=1}^{i_p} (\deg y_q - 1)$, i.e., the sign is picked up by the $\{x_i\}$'s passing through the $\{y_j\}$'s in the shuffle.

(3)

$$\{x_1 \cdot x_2\}\{y_1, \ldots, y_n\} = \sum_{k=0}^n (-1)^\varepsilon \{x_1\}\{y_1, \ldots, y_k\} \cdot \{x_2\}\{y_{k+1}, \ldots, y_n\},$$

where $\varepsilon = (\deg x_2) \sum_{p=1}^{k} (\deg y_p - 1)$.

(4)
$$d(\{x\}\{x_1, \ldots, x_{n+1}\}) - \{dx\}\{x_1, \ldots, x_{n+1}\}$$
$$- (-1)^{\deg x - 1} \sum_{i=1}^{n+1} (-1)^{\deg x_1 + \cdots + \deg x_{i-1} - i - 1} \{x\}\{x_1, \ldots, dx_i, \ldots, x_{n+1}\}$$
$$= - (-1)^{\deg x (\deg x_1 - 1)} x_1 \cdot \{x\}\{x_2, \ldots, x_{n+1}\}$$
$$- (-1)^{\deg x} \sum_{i=1}^{n} (-1)^{\deg x_1 + \cdots + \deg x_i - i} \{x\}\{x_1, \ldots, x_i \cdot x_{i+1}, \ldots, x_{n+1}\}$$
$$+ (-1)^{\deg x + \deg x_1 + \cdots + \deg x_n - n} \{x\}\{x_1, \ldots, x_n\} \cdot x_{n+1}$$

The structure of braces $\{x\}\{x_1, \ldots, x_n\}$, $n \geq 1$, and a dot product xy satisfying the above identities on a complex $V = \bigoplus_n V^n$ of vector spaces was called in [13] a *homotopy G-algebra*. Thus the above braces and dot product provide the Hochschild complex $C^\bullet(A, A)$ of an associative algebra A with with the structure of a homotopy G-algebra. From the definition of a homotopy G-algebra, it is easy to find the definition of a homotopy G-operad: it is an operad of complexes with braces $\{x\}\{x_1, \ldots, x_n\}$ and a dot product xy as generators and the associativity and the above identities for the braces and the dot product as defining relations.

REMARK 1. A more general notion of a homotopy G-algebra is that of B_∞-algebra; see Getzler and Jones [17]. A $\mathbb{Z}$-graded vector space $V^\bullet$ is a B_∞-algebra if and only if the bar coalgebra $BV = \bigoplus_{n=0}^{\infty} (V[-1])^{\otimes n}$, has the structure of a DG bialgebra, that is to say, a bialgebra with a differential which is simultaneously a derivation and a coderivation. The differential gives rise to a differential on V, as well as the dot product on V and higher A_∞ products. The multiplication, compatible with the coproduct, on BV gives rise to braces on V and some higher braces. If the higher products and the higher braces vanish identically, we retrieve the homotopy G-algebra structure of the previous example.

In this paper we will need the following still more general geometric definition of a homotopy G-algebra of Getzler and Jones [17]; see also [14]. Consider Fox-Neuwirth's cellular partition of the configuration spaces $F(n, \mathbb{C}) = \mathbb{C}^n \setminus \Delta$, where Δ is the weak diagonal $\cup_{i \neq j} \{x_i = x_j\}$, of n distinct points on the complex plane $\mathbb{C}$: cells are labeled by ordered partitions of the set $\{1, \ldots, n\}$ into subsets with arbitrary reorderings within each subset. This reflects grouping points lying on common vertical lines on the plane and ordering the points lexicographically.

For each n, take the quotient cell complex $K_\bullet\mathcal{M}(n)$ by the action of translations $\mathbb{R}^2$ and dilations $\mathbb{R}_+^*$. These quotient spaces do not form an operad, but one can glue lower $K_\bullet\mathcal{M}(n)$'s to the boundaries of higher $K_\bullet\mathcal{M}(n)$'s to form a cellular operad $K_\bullet\mathcal{M} = \{K_\bullet\mathcal{M}(n) \mid n \geq 2\}$. In fact, the underlying spaces $\underline{\mathcal{M}}(n)$ are manifolds with corners compactifying $\mathcal{M}(n)$.

The resulting space $\underline{\mathcal{M}}(n)$ is an S^1-bundle over the real compactification $\underline{\mathcal{M}}_{0,n+1}$ of the moduli space $\mathcal{M}_{0,n+1}$ of $n+1$-punctured curves of genus zero, see [14, 17]. The space $\underline{\mathcal{M}}(n)$ can be also interpreted as a "decorated" moduli space, see [14]. Indeed, it can be identified with the moduli space of data $(C; p_1, \ldots, p_{n+1}; \tau_1, \ldots, \tau_m, \tau_\infty)$, where C is a stable algebraic curve with $n+1$ punctures $p_1, \ldots, p_{n+1}$ and m double points. For each i, $1 \leq i \leq m$, τ_i is the choice of a tangent direction at the ith double point to the irreducible component that is furthest away from the "root", i.e., from the component of C containing the puncture $\infty := p_{n+1}$, while τ_∞ is a tangent direction at ∞. The stability of a curve is understood in the sense of Mumford's geometric invariant theory: each irreducible component of C must be stable, i.e., admit no infinitesimal automorphisms. The operad composition is given by attaching the ∞ puncture on a curve to one of the other punctures on another curve, remembering the tangent direction at each new double point.

Cells in this cellular operad $K_\bullet\mathcal{M}$ are enumerated by pairs (T, p), where T is a directed rooted tree with n labeled initial vertices and one terminal vertex, labeling a component of the boundary of $\underline{\mathcal{M}}(n)$, and p is an ordered partition, as above, of the set $\mathrm{in}(v)$ of incoming edges for each vertex v of the tree T.

In [17], it is shown that a complex V is a homotopy G-algebra in the sense of the example above iff it is an algebra over the operad $K_\bullet\mathcal{M}$ satisfying the following condition. The structure mappings

$$K_\bullet\mathcal{M}(n) \to \mathrm{Hom}(V^{\otimes n}, V),$$

for the algebra V over the operad $K_\bullet\mathcal{M}$ send all cells in $K_\bullet\mathcal{M}(n)$ to zero, except cells of two kinds:

1. $(\delta_n; i_1|i_2 \ldots i_n)$, where δ_n is the corolla, the tree with one root and n edges, connecting it to the remaining n vertices, corresponding to the configuration where the points $i_2, \ldots, i_n$ sit on a vertical line, the i_kth point being below the i_{k+1}st, and the i_1st point is in the half-plane to the left of the line;

2. $(\delta_2; i_1 i_2)$, corresponding to the configuration where all the points sit on a single vertical line, the i_1st point being below the i_2nd.

Cells of the first kind give rise to the braces $\{x_{i_1}\}\{x_{i_2}, \ldots, x_{i_n}\}$, $n \geq 2$, and cells of the second kind give rise to the dot products $x_{i_1} x_{i_2}$. The relations (3) and (4), and the associativity of the dot product, follow from the combinatorial structure of the cell complex. The relation (2) does not necessarily follow from the cell complex. Thus the homotopy G-operad of Example 4.1 is a quotient operad of $K_\bullet \mathcal{M}$. In particular, a homotopy G-algebra of Example 2 is an algebra over $K_\bullet \mathcal{M}$. In fact, the B_∞-operad corresponding to B_∞-algebras of Remark 1 is an intermediate operad between $K_\bullet \mathcal{M}$ and the homotopy G-operad of Example 4.1. Thus, every B_∞-algebra is a $K_\bullet \mathcal{M}$-algebra with certain operations also set to zero, and every homotopy G-algebra of Example 4.1 is a B_∞-algebra with some further operations vanishing.

The main purpose of this paper is to give a nontrivial example of an algebra over the operad $K_\bullet \mathcal{M}$ and thus answer the question of Lian and Zuckerman [30] about the homotopy structure of a TCFT. We will adopt the following definition of a homotopy G-operad and a homotopy G-algebra, introduced in [17] under the names of a homotopy 2-algebra and a braid algebra. We think it is appropriate to call them the G_∞-operad and a G_∞-algebra, respectively.

DEFINITION 4.1. The G_∞-*operad* is the operad $K_\bullet \mathcal{M}$ of complexes. An algebra over it is called a G_∞-*algebra*.

4.1. Some lower operations. Let V be a G_∞-algebra, as in Definition 4.1. This structure assumes a collection of n-ary operations $\mu_{T,p}$ on V associated with each cell of $K_\bullet \mathcal{M}(n)$, i.e., with each pair (T, p), where T is an n-tree labeling a component of the boundary of $\underline{\mathcal{M}}(n)$ and p is an ordered partition of incoming vertices for each vertex of T, see above. Here we would like to illustrate this complicated algebraic structure with an explicit description of the operations and relations for $n = 2$ and 3.

For $n = 2$, $\underline{\mathcal{M}}(2) = \mathcal{M}(2) \cong S^1$ and there are just two types of cells: $(\delta_2; 1|2)$ and $(\delta_2; 12)$, dividing the circle into two intervals and two points. Denote the corresponding operations on V by $v_1 \circ v_2 = \{v_1\}\{v_2\}$ and $v_1 v_2$, where v_1 and $v_2 \in V$. The cells $(\delta_2; 2|1)$ and $(\delta_2; 21)$ are obtained by applying the transposition $\tau_{12} \in S_2$ to the above two, and therefore the corresponding operations are just $v_2 \circ v_1$ and $v_2 v_1$.

We need also to introduce the following bracket, which, unlike the circle product, induces an operation on the d-cohomology:

(5) $$[v_1, v_2] = v_1 \circ v_2 - (-1)^{(\deg v_1 - 1)(\deg v_2 - 1)} v_2 \circ v_1.$$

For $n = 3$, $\mathcal{M}(3) \subset \underline{\mathcal{M}}(3)$ contains cells of the following types: $(\delta_3; 1|2|3)$, $(\delta_3; 1|23)$, $(\delta_3; 12|3)$, and $(\delta_3; 123)$, corresponding to operations which we will denote by $\{v_1\}\{v_2\}\{v_3\}$, $\{v_1\}\{v_2, v_3\}$, $\{v_1, v_2\}\{v_3\}$, and $M_3(v_1, v_2, v_3)$, respectively.

4.2. Some lower identities. There are no identities between compositions of operations, because the operad $\underline{\mathcal{M}}$ is free as an operad, and so is $K_\bullet\underline{\mathcal{M}}$. But there are identities involving the differential d, because the boundary of a cell in $K_\bullet\underline{\mathcal{M}}(n)$ is a linear combination of other cells — $K_\bullet\underline{\mathcal{M}}$ is a cellular operad after all. The sign rule we use for the boundaries of cells in the sequel is as follows. We introduce an orientation on cells in the configuration space $\underline{\mathcal{M}}(n)$, ordering their coordinates according to the rule: first, going from left to right, the x coordinates of the lines on which the points group, then, going lexicographically from left to right and from top to bottom, the y coordinates of the points in each group.

The boundary of the 1-cell $(\delta_2; 1|2) \in K_\bullet\underline{\mathcal{M}}(2)$ is

$$\partial(\delta_2; 1|2) = (\delta_2; 12) - (\delta_2; 21).$$

This incidence relation between cells implies the following *homotopy commutativity* relation for the dot product:

$$(6) \quad d(v_1 \circ v_2) - dv_1 \circ v_2 - (-1)^{\deg v_1 - 1} v_1 \circ dv_2$$
$$= v_1 v_2 - (-1)^{(\deg v_1)(\deg v_2)} v_2 v_1.$$

We also have $\partial(\delta_2; 12) = 0$, which yields the following *derivation property* of d with respect to the dot product:

$$d(v_1 v_2) - dv_1 v_2 - (-1)^{\deg v_1} v_1 dv_2 = 0.$$

For the top cell in $K_\bullet\underline{\mathcal{M}}(3)$, we have

$$\partial(\delta_3; 1|2|3) = -(\delta_3; 1|23) + (\delta_3; 1|32)$$
$$+ (\delta_3; 12|3) - (\delta_3; 21|3)$$
$$+ (\delta_2 \circ_1 \delta_2, (1|2) \circ_1 (1|2)) - (\delta_2 \circ_2 \delta_2; (1|2) \circ_2 (1|2)),$$

the circles on the figure meaning "magnifying glasses", here showing that a pair of points on the sphere bubbled off onto another "microscopic" sphere attached at a double point. This happens in much the same way that a sphere pinches off to form a genus zero curve with a node as one goes to the compactification divisor in the moduli space of genus zero curves. This implies

$$(7) \quad d(\{v_1\}\{v_2\}\{v_3\}) - \{dv_1\}\{v_2\}\{v_3\} - (-1)^{\deg v_1 - 1}\{v_1\}\{dv_2\}\{v_3\}$$
$$- (-1)^{\deg v_1 + \deg v_2}\{v_1\}\{v_2\}\{dv_3\}$$
$$= -\{v_1\}\{v_2, v_3\} + (-1)^{(\deg v_2 - 1)(\deg v_3 - 1)}\{v_1\}\{v_3, v_2\}$$
$$+ \{v_1, v_2\}\{v_3\} - (-1)^{(\deg v_1 - 1)(\deg v_2 - 1)}\{v_2, v_1\}\{v_3\}$$
$$+ (v_1 \circ v_2) \circ v_3 - v_1 \circ (v_2 \circ v_3).$$

The incidence relations

$$\partial(\delta_3; 1|23)$$
$$= -(\delta_3; 123) + (\delta_3; 213) - (\delta_3; 231)$$
$$+ (\delta_2 \circ_2 \delta_2; (1|2) \circ_2 (12))$$
$$- \tau_{12}(\delta_2 \circ_2 \delta_2; (12) \circ_2 (1|2)) - (\delta_2 \circ_1 \delta_2; (12) \circ_1 (1|2))$$

and similarly

$$\partial(\delta_3; 12|3)$$
$$= -(\delta_3; 123) + (\delta_3; 132) - (\delta_3; 312)$$
$$- (\delta_2 \circ_1 \delta_2; (1|2) \circ_1 (12))$$
$$+ (\delta_2 \circ_2 \delta_2; (12) \circ_2 (1|2)) + \tau_{12}(\delta_2 \circ_1 \delta_2; (12) \circ_1 (1|2)),$$

where τ_{12} is the transposition exchanging 1 and 2, imply

$$(8) \quad d(\{v_1\}\{v_2, v_3\}) - \{dv_1\}\{v_2, v_3\})$$
$$- (-1)^{\deg v_1 - 1}\{v_1\}\{dv_2, v_3\} - (-1)^{\deg v_1 + \deg v_2}\{v_1\}\{v_2, dv_3\}$$
$$= -M_3(v_1, v_2, v_3) + (-1)^{(\deg v_1)(\deg v_2)} M_3(v_2, v_1, v_3)$$
$$- (-1)^{\deg v_1 (\deg v_2 + \deg v_3)} M_3(v_2, v_3, v_1) + v_1 \circ (v_2 \cdot v_3)$$
$$- (-1)^{(\deg v_1 - 1)\deg v_2} v_2 \cdot (v_1 \circ v_3) - (v_1 \circ v_2) \cdot v_3$$

and

$$(9) \quad d(\{v_1, v_2\}\{v_3\}) - \{dv_1, v_2\}\{v_3\})$$
$$- (-1)^{\deg v_1 - 1}\{v_1, dv_2\}\{v_3\} - (-1)^{\deg v_1 + \deg v_2}\{v_1, v_2\}\{dv_3\}$$
$$= -M_3(v_1, v_2, v_3) + (-1)^{(\deg v_2)(\deg v_3)} M_3(v_1, v_3, v_2)$$
$$- (-1)^{\deg v_3 (\deg v_1 + \deg v_2)} M_3(v_3, v_1, v_2) - (v_1 \cdot v_2) \circ v_3$$
$$+ v_1 \cdot (v_2 \circ v_3) + (-1)^{(\deg v_3 - 1)\deg v_2}(v_1 \circ v_3) \cdot v_2,$$

which can be regarded as the *homotopy left and right Leibniz rules* for the circle product with respect to the dot product, altered by the trilinear A_∞ product M_3.

Finally, for the 1-cell $(\delta_3; 123)$, we have

$$\partial(\delta_3; 123) = (\delta_2 \circ_1 \delta_2; (12) \circ_1 (12)) - (\delta_2 \circ_2 \delta_2; (12) \circ_2 (12)),$$

whence

$$(10) \quad d(M_3(v_1, v_2, v_3)) - M_3(dv_1, v_2, v_3) - (-1)^{\deg v_1} M_3(v_1, dv_2, v_3)$$
$$- (-1)^{\deg v_1 + \deg v_2} M_3(v_1, v_2, dv_3)$$
$$= (v_1 v_2) v_3 - v_1 (v_2 v_3),$$

so that the dot product is *homotopy associative* with M_3 being the *associator* for it. In fact, the M_n's form an A_∞-algebra, which is clearly seen at the operad level: the A_∞-operad, which is the natural cellular model of the configuration operad of the real line, embeds into $K_\bullet \underline{\mathcal{M}}$ as a real part.

Note that the commutator $[\cdot, \cdot]$ of (5) satisfies a graded *homotopy Leibniz rule*:

$$(11) \quad [v_1, v_2 v_3] - [v_1, v_2]v_3 - (-1)^{(\deg v_1 - 1)\deg v_2} v_2 [v_1, v_3]$$
$$= -d(\{v_1\}\{v_2, v_3\}) + \{dv_1\}\{v_2, v_3\} + (-1)^{\deg v_1 - 1}\{v_1\}\{dv_2, v_3\}$$
$$+ (-1)^{\deg v_1 + \deg v_2}\{v_1\}\{v_2, dv_3\} + (-1)^{\deg v_1(\deg v_2 + \deg v_3)}(d(\{v_2, v_3\}\{v_1\}$$
$$- (-1)^{\deg v_2 + \deg v_3}\{v_2, v_3\}\{dv_1\} - \{dv_2, v_3\}\{v_1\}$$
$$- (-1)^{\deg v_2 - 1}\{v_2, dv_3\}\{v_1\}),$$

which is a consequence of (8) and (9). The *homotopy Jacobi identity*

$$(12) \quad [[v_1, v_2], v_3] + (-1)^{(\deg v_1 - 1)(\deg v_2 + \deg v_3)}[[v_2, v_3], v_1]$$
$$+ (-1)^{(\deg v_3 - 1)(\deg v_1 + \deg v_2)}[[v_3, v_1], v_2]$$
$$= \{\text{the differential of the sum of all permutations of } \{v_1\}\{v_2\}\{v_3\}\}$$

for the bracket $[,]$ follows from (7). More generally, the graded symmetrizations of the operations $\{v_1\}\{v_2\}\ldots\{v_n\}$ form a homotopy Lie (or L_∞-) algebra: the symmetrizations are just the fundamental cycles of $\underline{\mathcal{M}}(n)$, which freely generate an operad with the differential coming from contraction of edges of trees labeling a basis in this free operad; see Beilinson-Ginzburg [4]. As shown by Hinich-Schechtman [20], this is nothing but the homotopy Lie operad. In addition, the circle product and the higher braces are a sort of universal enveloping pre-Lie algebra, cf. [12], for the L_∞-algebra.

REMARK 2. The dot product $v_1 v_2$ and the bracket $[v_1, v_2]$ descend to the cohomology of a G_∞-algebra and endow the cohomology with a G-algebra structure; see Equations (6), (10), (11), (12).

4.3. Homotopy Gerstenhaber, associative and Lie algebras and operads.

While G-algebras combine commutative associative and Lie algebras, the G-operad describing the class of G-algebras is a natural combination of the commutative and the Lie operads $\mathcal{C}omm$ and $\mathcal{L}ie$, describing the classes of commutative and Lie algebras, respectively. Indeed, we can identify the latter operads with $H_0(\underline{\mathcal{M}}, \mathbb{C})$ and $H_{n-1}(\underline{\mathcal{M}}, \mathbb{C})$. The first identification follows from connectedness of the spaces $\underline{\mathcal{M}}(n)$: $H_0(\underline{\mathcal{M}}(n), \mathbb{C}) \cong \mathbb{C} \cong \mathcal{C}omm(n)$, the second follows Fred

Cohen's Theorem [8]: $H_{n-1}(\underline{\mathcal{M}}(n), \mathbb{C}) \cong H_{n-1}(F(n, \mathbb{C}), \mathbb{C}) \cong \mathcal{L}ie(n)$. The G-operad is just $H_\bullet(\underline{\mathcal{M}}(n), \mathbb{C})$, according to another part of Cohen's Theorem, thereby neatly interpolating the commutative and the Lie operads.

The associative operad $\mathcal{A}ssoc$ can also be seen in this moduli space picture: $H_0(\underline{\mathcal{M}}_r(n), \mathbb{C}) \cong H_0(F(n, \mathbb{R}), \mathbb{C}) \cong \mathbb{C}[S_n] \cong \mathcal{A}ssoc(n)$, where $F(n, \mathbb{R})$ is the configuration space of n points on the real line, and $\underline{\mathcal{M}}_r(n)$ is the quotient of it by translations and dilations, compactified similarly to $\underline{\mathcal{M}}(n)$ above. The natural embedding $\mathbb{R} \to \mathbb{C}$, say, as the y axis, induces a morphism $\mathcal{A}ssoc = H_0(\underline{\mathcal{M}}_r, \mathbb{C}) \to H_\bullet(\underline{\mathcal{M}}, \mathbb{C})$ of the associative operad to the G-operad, of course through the commutative one.

Similarly, as we have seen in the previous section, G_∞-algebras combine the structures of A_∞- and L_∞-algebras. Accordingly, the G_∞-operad neatly incorporates the A_∞- and the L_∞-operads. The A_∞-operad is the operad consisting of all cells of the real moduli space operad $\underline{\mathcal{M}}_r$. It embeds as a cellular operad into $K_\bullet\underline{\mathcal{M}}$ if we identify $\mathbb{R}$ with the y axis on the complex plane $\mathbb{C}$ and just consider the cells which are formed by all points grouping on a vertical line. On the other hand, the L_∞-operad is the suboperad of the G_∞-operad $K_\bullet\underline{\mathcal{M}}$ generated by the sums of all top cells of $\underline{\mathcal{M}}(n)$'s, according to Beilinson-Ginzburg-Hinich-Schechtman; see the end of the previous section.

5. Topological Field Theories

The topological field theories described here are sometimes called *cohomological field theories* and are algebras over the operad of smooth singular chains on moduli spaces of punctured Riemann spheres with decorations.

5.1. Reduced Conformal Field Theories.
Let V be a topological vector space endowed with a collection of smooth maps $\Psi : F(n, \mathbb{C}) \to \mathrm{Hom}(V^{\otimes n}, V)$, one for each $n \geq 1$, taking $(z_1, z_2, \dots, z_n) \mapsto \Psi_{(z_1, z_2, \dots, z_n)}$ satisfying the following axioms:

1. The map Ψ should be S_n-equivariant, where the permutation group S_n acts on both $F(n, \mathbb{C})$ and $\mathrm{Hom}(V^{\otimes n}, V)$ in the obvious ways.
2. $\Psi_{(az_1+b, \dots, az_n+b)} = \Psi_{(z_1, \dots, z_n)}$ for all $a \in \mathbb{R}^+$ and $b \in \mathbb{C}$.
3. As the configuration of points $(z_1, \dots, z_n)$ approaches a composition of configurations $(t_1, \dots, t_p) \circ_i (w_1, \dots, w_{n-p+1})$ for some $i =$

$1, \ldots, p$ then we have

$$\Psi_{(z_1,\ldots,z_n)}(v_1, \ldots, v_n) \longrightarrow$$

$$\Psi_{(t_1,\ldots,t_p)}(v_1, \ldots, v_{i-1}, \Psi_{(w_1,\ldots,w_{n-p+1})}(v_i, \ldots, v_{i+n-p}), v_{i+n-p+1}, \ldots, v_n),$$

where the compositions are understood to be in $\underline{\mathcal{M}}$. This algebraic structure is said to be a *reduced conformal field theory (reduced CFT)*.

The last axiom replaces the usual associativity axiom for VOAs and encodes the various ways in which points can come together in the plane. The operators $\Psi_{(z,0)}(v_1, v_2)$ are analogous to vertex operators $Y(v_1, z)v_2$ of a VOA. It is clear that we can redescribe this structure in the following way.

PROPOSITION 5.1. *V is a reduced conformal field theory if and only if V is an $\underline{\mathcal{M}}$-algebra.*

A reduced topological conformal field theory (reduced TCFT) can be understood as an algebra over the smooth singular chain operad $C_\bullet(\underline{\mathcal{M}})$ of the configuration operad $\underline{\mathcal{M}}$ by analogy with TCFTs of [**45**]. For technical reasons — smooth singular chains, i.e., finite sums of smooth mappings from simplices to the manifold, do not naturally contain cells of $K_\bullet \underline{\mathcal{M}}$ among them — we prefer to use the following definition, imitating Segal's definition of a (full) TCFT [**38**], see also the next section.

DEFINITION 5.1. *A reduced topological conformal field theory (reduced TCFT)* is a complex (V, d) of vector spaces, called a *BRST complex or a state space*, and a collection of operator-valued differential forms $\Omega_n \in \Omega^\bullet(\underline{\mathcal{M}}(n), \operatorname{Hom}(V^{\otimes n}, V))$, one for each $n \geq 1$, such that

1.
$$\pi \Omega_n = \pi^* \Omega_n \qquad \text{for each } \pi \in S_n,$$

where π on the left-hand side acts naturally on $\operatorname{Hom}(V^{\otimes n}, V)$ and π^* is the geometric action of π by relabeling the punctures,

2.
$$d_{\mathrm{DR}} \Omega_n = d_{\mathrm{BRST}} \Omega_n,$$

where d_{DR} is the de Rham differential and d_{BRST} is the natural differential on the space $\operatorname{Hom}(V^{\otimes n}, V)$ of n-linear operators on V,

3.
$$\circ_i^* (\Omega_{m+n-1}) = \Omega_m \otimes \Omega_n \qquad \text{for each } i = 1, \ldots, m,$$

where $\circ_i : \underline{\mathcal{M}}(m) \times \underline{\mathcal{M}}(n) \to \underline{\mathcal{M}}(m+n-1)$ is the operad law.

THEOREM 5.2. *A reduced TCFT implies a natural G_∞-algebra structure on the BRST complex V.*

PROOF. Given a reduced TCFT, we obtain the structure of an algebra over the G_∞-operad

$$K_\bullet \underline{\mathcal{M}}(n) \to \operatorname{Hom}(V^{\otimes n}, V), \qquad n \geq 1,$$

by integrating the forms Ω_n over the cells:

$$C \mapsto \int_C \Omega_n.$$

$\square$

5.2. Conformal Field Theories. Let $\mathcal{P}(n)$ be the moduli space of Riemann spheres with $n + 1$ distinct, ordered, holomorphically embedded unit disks which do not overlap except, possibly, along their boundaries. The permutation group on n elements S_n acts on $\mathcal{P}(n)$ by permuting the ordering of the first n punctures. The collection $\mathcal{P} = \{ \mathcal{P}(n) \}$ forms an operad of complex manifolds where for all Σ in $\mathcal{P}(n)$ and Σ' in $\mathcal{P}(n')$, $\Sigma \circ_i \Sigma'$ in $\mathcal{P}(n + n' - 1)$ is obtained by cutting out the $n' + 1$st unit disk of Σ', the ith disk of Σ and sewing along their boundaries using the identification $z \mapsto \frac{1}{z}$. A *(tree level, $c = 0$) conformal field theory (CFT)* is an algebra over $\mathcal{P}$. "Tree level" means that only genus 0 Riemann surfaces appear, while $c = 0$ means that V is a representation of $\mathcal{P}$ rather than a projective representation (see Segal [38] for details). We shall restrict to such CFTs for simplicity. An important class of examples of holomorphic CFTs, that is, CFTs where the algebra maps are holomorphic, come from vertex operator algebras through the work of Huang-Lepowsky [23].

DEFINITION 5.2 (Segal [38], Getzler [16]). A *topological conformal field theory (TCFT)* is a complex (V, d) and a collection of forms $\Omega_n \in \Omega^\bullet(\mathcal{P}(n), \operatorname{Hom}(V^{\otimes n}, V))$ satisfying the same axioms (1)–(3) of Definition 5.1. The complex (V, d) is sometimes called the *BRST complex or the state space* of the theory in the physics literature. Also, the cohomology of the complex is called the *space of physical states*.

Again, we shall restrict to tree level, $c = 0$ TCFTs, although physical examples of TCFT's do not always have $c = 0$. The forms Ω_n can be constructed using the operator formalism from algebraic data known as a *string background* (see [2, 25] for details). However, we shall not need this fact.

PROPOSITION 5.3. *Let (V, d) be a TCFT. Then (V, d) is a reduced TCFT.*

PROOF. This result follows from the fact that there exists a morphism of operads (called *string vertices*) $s : \underline{\mathcal{M}} \to \mathcal{P}$ — indeed, there exists a family of such morphisms. Consequently, any TCFT is a reduced TCFT. A similar result was used by Kimura, Stasheff, and Voronov in [25] to prove the existence of a homotopy Lie algebra structure on a natural subcomplex of a TCFT.

The construction of s uses a geometric result due to Wolf and Zwiebach [47], who showed that every conformal class of Riemann spheres with at least three punctures can be endowed with a unique metric compatible with its complex structure which solves a minimal area problem for metrics on the punctured Riemann sphere subject to the constraint that the length of any homotopically nontrivial closed curve on the punctured sphere is greater than or equal to 2π. This minimal area metric decomposes the punctured sphere into flat cylinders foliated by closed geodesics with circumference 2π. A morphism s is obtained for each pair (l, f_l) where l is a real number greater than π and $f_l : [2\pi, \infty) \to \mathbb{R}$ is a smooth monotonically increasing function such that $f_l(2\pi) = 2\pi$ and $\lim_{x \to +\infty} f_l(x) = 2l$.

Let Σ represent a point in $\underline{\mathcal{M}}(n)$; since each irreducible component of Σ can be identified with the configuration of, say m, points on $\mathbb{C}$ quotiented by dilations and translations, assign to each point a tangent direction which points along the positive real axis in $\mathbb{C}$ and then assign the minimal area metric to each irreducible component of Σ minus its punctures and double points. Shrink all internal flat cylinders in each irreducible component with a height h greater than 2π to a flat cylinder with height $f_l(h)$. Furthermore, the minimal area metric and the tangent directions at each puncture and on each side of every double point gives rise to a holomorphically embedded unit disk centered there. $s(\Sigma)$ in $\mathcal{P}(n)$ is obtained by sewing together the irreducible components on each side of every double point. The remaining curve has no remaining double points and has a holomorphically embedded unit disk around each puncture. □

Combining the previous proposition with Theorem 5.2, we conclude the following.

COROLLARY 5.4. *Let (V, d) be a TCFT. Then (V, d) admits the structure of a G_∞-algebra.*

The existence of an A_∞-algebra structure on the state space of a TCFT was observed in [24], cf. the homotopy associative structure of the HIKKO open string-field theory group noticed by Stasheff [40].

REMARK 3. In particular, we see (Section 4.2) the structure of a homotopy commutative A_∞-algebra extending a dot product and an L_∞-algebra extending a bracket, naturally merged into one structure. This new L_∞ structure is not totally independent of the one observed by Witten and Zwiebach [46, 48] which was explained operadically in [25]. The L_∞ structure of the present paper extends the one studied before. At the operadic level, in the latter case, one projects along the phases (that is why semirelative BRST cochains are needed), while in the former one takes a section of this projection, conveniently provided by the moduli space $\underline{\mathcal{M}}$ of punctures on the sphere with an arrow at the ∞ puncture.

6. Applications of string vertices

6.1. Vassiliev knot invariants. Let $\widehat{\mathcal{M}}_{g,n}$ be the moduli space of stable curves of genus g and n ordered, distinct punctures whose double points are decorated with tangent directions, one on each irreducible component on either side of each double point, quotiented by the diagonal group of rotations by $U(1)$. $\widehat{\mathcal{M}}_{g,n}$ is a $6g - 6 + 2n$ dimensional compact, oriented orbifold with corners which can be constructed by making real blowups along the irreducible components of the divisor of the moduli space of stable curves of genus g and n punctures $\overline{\mathcal{M}}_{g,n}$. The collection of spaces $\widehat{\mathcal{M}} = \{ \widehat{\mathcal{M}}_{g,n} \}$ does not have natural composition maps between them, unlike $\overline{\mathcal{M}} = \{ \overline{\mathcal{M}}_{g,n} \}$, because for any two punctures which are to be attached together, there is no natural way to choose tangent directions at the double points. However, the space of (smooth) singular chains $C_\bullet(\widehat{\mathcal{M}}) = \{ C_\bullet(\widehat{\mathcal{M}}_{g,n}) \}$ does have natural composition maps between them by using the transfer which comes from attaching two curves together at two punctures and then averaging over the entire S^1 of tangent directions at the double points. $C_\bullet(\widehat{\mathcal{M}})$ forms a generalization of an operad called a *modular operad*, a notion due to Getzler and Kapranov [18], which generalizes operads in two ways. The first is that there is no natural "outgoing" puncture in this case — any two punctures can be attached including two on a single stable curve — while the second is that higher genus stable curves are allowed. *Throughout the remainder of this section, all operads will be assumed to be modular operads unless otherwise stated.*

$\widehat{\mathcal{M}}_{g,n}$ is a stratified space whose strata are indexed by *stable n-graphs*. These are (connected) graphs with n external legs whose vertices are decorated with a nonnegative integer. One associates a stable graph to each point in $\widehat{\mathcal{M}}_{g,n}$ by associating to each irreducible component, a corolla with an external leg associated to each puncture and double point on that component, an integer assigned to that vertex corresponding to the genus of the irreducible component, and then attaching the corollas together whenever the irreducible components share a double point. A stratum associated to a stable n-graph consists of all points in $\widehat{\mathcal{M}}_{g,n}$ whose associated graph is the given one. The stratification of $\widehat{\mathcal{M}}_{g,n}$ is obtained by pulling up the stratification of $\overline{\mathcal{M}}_{g,n}$ via the canonical projection map $\widehat{\mathcal{M}}_{g,n} \to \overline{\mathcal{M}}_{g,n}$ which forgets the tangent directions at each double point. The union of strata gives rise to a canonical filtration of $\widehat{\mathcal{M}}$ whose associated homology spectral sequence has an E^1-term $E^1 = \{ E^1_{g,n} \}$ which forms an operad of chain complexes. E^1 contains a suboperad $\mathcal{G} = \{ \mathcal{G}_{g,n} \}$ called *the graph complex* of Kontsevich [29] which is nothing more than the Feynman transform [18], a generalization of the operadic bar construction to modular operads, of the commutative operad.

Kontsevich [28] showed that the homology of the graph complex $H_\bullet(\mathcal{G}_{g,n})$ has a special significance in the theory of knots. He observed that the space of primitive chord diagrams of order n is isomorphic to $\oplus_{g=0}^{n} H_0(\mathcal{G}_{g,n-g+1})_{S_{n-g+1}}$. The space of chord diagrams $\mathcal{A}$ then is nothing more than the symmetric algebra over the space of primitive chord diagrams, since $\mathcal{A}$ is a commutative, cocommutative Hopf algebra. A *weight system* is an element in $\mathcal{A}^*$. A key theorem of Kontsevich [28] and Bar-Natan [3] states that weight systems are in one to one correspondence with finite type knot invariants due to Vassiliev [43]. Such knot invariants as the Jones, Alexander-Conway polynomials and their generalizations are constructed from Vassiliev invariants. Most familiar examples of weight systems are constructed from simple Lie algebras with an invariant metric (see [3]) and their representations, although it is has recently been discovered that not all weight systems are of this kind [44]. The following theorem was very likely known to Kontsevich, who used homotopy Lie algebras with an invariant inner product and graph complexes to construct Vassiliev knot invariants.

THEOREM 6.1. *Let (V, d) be an algebra over $\mathcal{G}$, Kontsevich's graph complex; then there is an associated family of weight systems.*

PROOF. The algebra structure morphism $\mathcal{G} \to \mathcal{E}nd_V$ induces the morphism $m : H_\bullet(\mathcal{G}) \to \mathcal{E}nd_{H_\bullet(V)}$ which can be extended to a homomorphism of (ungraded) commutative, associative algebras $m : \mathcal{A} \to \mathcal{S}W$ where $\mathcal{S}W$ denotes the (ungraded) symmetric algebra over the vector space W. Here W is a $\mathbb{Z}$-graded vector space whose degree

n subspace is isomorphic to $\oplus_{g=0}^{n} \left(\mathcal{E}nd_{H_{\bullet}(V),(g,n-g+1)} \right)_{S_{n-g+1}}$. Therefore, given a vector in the symmetric algebra over $H_{\bullet}(V)$, one obtains a weight system by contracting with a suitable inner product on $H_{\bullet}(V)$. $\qquad\square$

Examples of such $\mathcal{G}$-algebras come from TCFT's modulo a slight technicality. Let (V, d) be a $(c = 0)$ TCFT. That is, let $\mathcal{P}_{g,n}$ be the moduli space of genus g Riemann surfaces with n holomorphically embedded unit disks which do not intersect, except possibly along the boundaries in $\mathcal{P}_{0,2}$. They assemble into the operad $\mathcal{P} = \{\mathcal{P}_{g,n}\}$. A $(c = 0)$ TCFT is a collection of endomorphism-valued differential forms $\Omega_{g,n}$, $n \geq 1$, on $\mathcal{P}_{g,n}$ satisfying the natural modular operad generalization of the axioms of Definition 5.1 of Section 5.2.

Let $U(1)$ be the subgroup of $\mathcal{P}_{0,2}$ which consists of the set of Riemann spheres with the standard chart about 0 and the standard one around ∞ rotated by multiplication by a phase. If (V, d) is a TCFT then let $\Delta : V \to V$ be the unary operation associated to $U(1)$ which is regarded as a 1-cocycle in $\mathcal{P}_{0,2}$. The kernel V_r of Δ forms a subcomplex of (V, d) called the *semi-relative BRST complex*.

Let $\overline{\mathcal{N}}_{g,n}$ be the moduli space of stable curves of genus g with n distinct, ordered punctures which has the same decorations as points in $\widehat{\mathcal{M}}_{g,n}$ but which have, in addition, tangent directions at each puncture. The collection $\overline{\mathcal{N}} = \{\overline{\mathcal{N}}_{g,n}\}$ forms an operad. The string vertices introduced in the previous section can also be described as a morphism of (nonmodular) operads $s : \overline{\mathcal{N}} \to \mathcal{P}$ (string vertices) which are provided by minimal area metrics. These minimal area metrics are proven to exist in genus zero but are only conjectured to exist in higher genus [47]. We shall assume that they exist in what follows.

Using the string vertices, we can pull back the forms $\Omega_{g,n}$ to $\overline{\mathcal{N}}_{g,n}$ and then push them forward to forms in $\Omega^{\bullet}(\widehat{\mathcal{M}}_{g,n}, \mathrm{Hom}(V_r^{\otimes n}, V_r))$, as in [25].

THEOREM 6.2. *Let (V, d) be a TCFT; then (V_r, d) is an algebra over $\mathcal{G}$, thereby giving rise to a family of weight systems.*

PROOF. By integration of the forms $\Omega_{g,n}$, (V_r, d) becomes an algebra over Kontsevich's graph complex $\mathcal{G}$ which appears in the top row in the E^1 term in the homology spectral sequence of $\widehat{\mathcal{M}}$ associated to the canonical filtration. Now apply the previous theorem. $\qquad\square$

It is interesting that any two given TCFT's which are homotopic through the space of TCFT's will give rise to isomorphic weight systems. Therefore, there will be families of weight systems associated to each component of the moduli space of TCFT's.

6.2. Double loop spaces. The following theorem, a byproduct of *string vertices*, generalizes Stasheff's characterization of loop spaces as A_∞-spaces, i.e., algebras over his polyhedra operad, to double loop spaces. It is also a refinement of the Boardman-Vogt-May-Fadell-Neuwirth characterization of double loop spaces considered up to homotopy as algebras over the little disks operad $D(n)$, $n \geq 1$.

THEOREM 6.3. *Any double loop space is an algebra over the operad $\underline{M}$. In particular, the singular chain complex of a double loop space is a homotopy G-algebra, more precisely, an algebra over the singular chain operad $C_\bullet(\underline{M})$.*

PROOF. The standard construction of Boardman and Vogt [5] provides a double loop space with the natural structure of an algebra over the little disks operad $D(n)$, $n \geq 1$. The same argument gives the structure of an algebra over the operad $\mathcal{P}$ of Riemann spheres with holomorphic holes. String vertices deliver an operad morphism $\underline{M} \to \mathcal{P}$, which yields a morphism $C_\bullet(\underline{M}) \to C_\bullet(\mathcal{P})$. $\square$

If we found a singular chain representative of each cell in $K_\bullet(\underline{M})$ compatible with the operad structure, i.e., a morphism $K_\bullet(\underline{M}) \to C_\bullet(\underline{M})$ of operads, we would be able to answer the following question. (In the case of Stasheff polyhedra, see more below, this morphism does exist).

QUESTION 6.4. *Is the singular chain complex of a double loop space a G_∞-algebra?*

Note that string vertices offer an alternative approach to the study of loop spaces compared to that given by May's approximation theory, see for example J. Stasheff's contribution [42] to this volume. Ideally, approximation theory would provide a construction of a space $K\underline{M}X$ homotopy equivalent to the double loop space $\Omega^2\Sigma^2 X$ of the double suspension of a given topological space X, such that $K\underline{M}X$ is an algebra over the cellular operad $K_\bullet\underline{M}$. Another approach is the content of a promised theorem of Getzler and Jones [17, Introduction].

It would be interesting to see whether string vertices exist in the case of the little intervals operad. Here the collection of string vertices should be nothing but a morphism of operads from the compactified spaces of configurations of points on the real line to the little intervals operad. Since this configuration operad is isomorphic to Stasheff's polyhedra operad, see Kontsevich [29], it may yield a simpler "quantum" proof of Stasheff's famous theorem, saying that any loop space is an algebra over the Stasheff polyhedra operad, and in particular, the singular chain complex of a loop space is a homotopy associative (A_∞-) algebra.

References

1. F. Akman, *On some generalizations of Batalin-Vilkovisky algebras*, Preprint, Cornell University, 1995, `q-alg/9506027`.
2. L. Alvarez-Gaume, C. Gomez, G. Moore, and C. Vafa, *Strings in the operator formalism*, Nuclear Phys. B **303** (1988), 455–521.
3. D. Bar-Natan, *On Vassiliev knot invariants*, Topology **34** (1995), no. 2, 423–472.
4. A. Beilinson and V. Ginzburg, *Infinitesimal structure of moduli spaces of G-bundles*, Internat. Math. Research Notices (1992), no. 4, 63–74.
5. J. M. Boardman and R. M. Vogt, *Homotopy invariant algebraic structures on topological spaces*, Lecture Notes in Math., vol. 347, Springer-Verlag, 1973.
6. R. E. Borcherds, *Vertex operator algebras, Kac-Moody algebras and the Monster*, Proc. Natl. Acad. Sci. USA **83** (1986), 3068–3070.
7. F. R. Cohen, *The homology of C_{n+1}-spaces, $n \geq 0$*, The homology of iterated loop spaces, Lecture Notes in Math., vol. 533, Springer-Verlag, 1976, pp. 207–351.
8. _____, *Artin's braid groups, classical homotopy theory and sundry other curiosities*, Contemp. Math. **78** (1988), 167–206.
9. T. Fox and L. Neuwirth, *The braid groups*, Math. Scand. **10** (1962), 119–126.
10. I. B. Frenkel, Lectures at the Institute for Advanced Study, January 1988.
11. I. B. Frenkel, J. Lepowsky, and A. Meurman, *Vertex operator algebras and the Monster*, Academic Press, New York, 1988.
12. M. Gerstenhaber, *The cohomology structure of an associative ring*, Ann. of Math. **78** (1963), 267–288.
13. M. Gerstenhaber and A. A. Voronov, *Higher order operations on Hochschild complex*, Functional Anal. Appl. **29** (1995), no. 1, 1–6.
14. _____, *Homotopy G-algebras and moduli space operad*, Internat. Math. Research Notices (1995), 141–153.
15. E. Getzler, *Cartan homotopy formulas and the Gauss-Manin connection in cyclic homology*, Israel Math. Conf. Proc. **7** (1993), 65–78.
16. _____, *Batalin-Vilkovisky algebras and two-dimensional topological field theories*, Commun. Math. Phys. **159** (1994), 265–285, `hep-th/9212043`.
17. E. Getzler and J. D. S. Jones, *Operads, homotopy algebra and iterated integrals for double loop spaces*, Preprint, Department of Mathematics, MIT; Department of Mathematics Northwestern University, March 1994, `hep-th/9403055`.
18. E. Getzler and M. Kapranov, *Modular operads*, Preprint, Department of Mathematics, MIT, August 1994, dg-ga/9408003.
19. V. Ginzburg and M. Kapranov, *Koszul duality for operads*, Duke Math. J. **76** (1994), 203–272.
20. V. Hinich and V. Schechtman, *Homotopy Lie algebras*, Adv. Studies Sov. Math. **16** (1993), 1–18.
21. G. Hochschild, B. Kostant, and A. Rosenberg, *Differential forms on regular affine algebras*, Trans. Amer. Math. Soc. **102** (1962), 383–408.

22. Y.-Z. Huang, *Operadic formulation of topological vertex algebras and Gersten-haber or Batalin-Vilkovisky algebras*, Commun. Math. Phys. **164** (1994), 105–144, `hep-th/9306021`.

23. Y.-Z. Huang and J. Lepowsky, *Vertex operator algebras and operads*, The Gelfand Mathematics Seminars, 1990–1992 (Boston), Birkhäuser, 1993, `hep-th/9301009`, pp. 145–161.

24. T. Kimura, *Operads of moduli spaces and algebraic structures in topological conformal field theory*, Moonshine, the Monster, and Related Topics (Providence) (C. Dong and G. Mason, eds.), Contemporary Math., vol. 193, Amer. Math. Soc., 1996, pp. 159–190.

25. T. Kimura, J. Stasheff, and A. A. Voronov, *On operad structures of moduli spaces and string theory*, Commun. Math. Phys. **171** (1995), 1–25, `hep-th/9307114`.

26. _____, *Homology of moduli spaces of curves and commutative homotopy algebras*, The Gelfand Mathematics Seminars, 1993–1994 (J. Lepowsky and M. M. Smirnov, eds.), Birkhäuser, 1996, to appear.

27. M. Kontsevich, *Formal (non)-commutative symplectic geometry*, The Gelfand Mathematics Seminars, 1990-1992 (L. Corwin, I. Gelfand, and J. Lepowsky, eds.), Birkhäuser, 1993, pp. 173–187.

28. _____, *Vassiliev's knot invariants*, Adv. Sov. Math. **16** (1993), no. 2, 137 – 150.

29. _____, *Feynman diagrams and low-dimensional topology*, First European Congress of Mathematics, Vol. II (Paris, 1992) (Basel), Progr. Math., vol. 120, Birkhäuser, 1994, pp. 97–121.

30. B. H. Lian and G. J. Zuckerman, *New perspectives on the BRST-algebraic structure of string theory*, Commun. Math. Phys. **154** (1993), 613–646, `hep-th/9211072`.

31. J.-L. Loday, *Une version non commutative des algebres de Lie: les algebres de Leibniz*, Enseign. Math. (2) **39** (1993), 269–293.

32. J. P. May, *Definitions: Operads, algebras and modules*, Operads: Proceedings of Renaissance Conferences (J.-L. Loday, J. Stasheff, and A. A. Voronov, eds.), Amer. Math. Soc., 1996, in this volume.

33. _____, *Operads, algebras and modules*, Operads: Proceedings of Renaissance Conferences (J.-L. Loday, J. Stasheff, and A. A. Voronov, eds.), Amer. Math. Soc., 1996, in this volume.

34. A. Nijenhuis, *Jacobi-type identities for bilinear differential concomitants of certain tensor fields*, Indag. Math. **17** (1955), 390–403.

35. M. Penkava and A. Schwarz, *On some algebraic structures arising in string theory*, Perspectives in mathematical physics (Cambridge, MA), Conf. Proc. Lecture Notes Math. Phys., vol. III, Internat. Press, 1994, `hep-th/9212072`, pp. 219–227.

36. G. Segal, *Two-dimensional conformal field theories and modular functors*, IXth Int. Congr. on Mathematical Physics (Bristol; Philadelphia) (B. Simon, A. Truman, and I. M. Davies, eds.), IOP Publishing Ltd, 1989, pp. 22–37.

37. _____, *Lectures at Cambridge University*, summer 1992.

38. _____, *Topology from the point of view of Q.F.T.*, Lectures at Yale University, March 1993.

39. J. Stasheff, *On the homotopy associativity of H-spaces, II*, Trans. Amer. Math. Soc. **108** (1963), 293–312.

40. _____, *Higher homotopy algebras: string field theory and Drinfeld's quasi-Hopf algebras*, Proceedings of the XXth International Conference on Differential Geometric Methods in Theoretical Physics (New York, 1991) (River Edge, NJ), vol. 1, 1992, pp. 408–425.

41. _____, *Closed string field theory, strong homotopy Lie algebras and the operad actions of moduli space*, Perspectives on Mathematics and Physics (R.C. Penner and S.T. Yau, eds.), International Press, 1994, hep-th/9304061, pp. 265–288.

42. _____, *From operads to 'phisically' inspired theories*, Operads: Proceedings of Renaissance Conferences (J.-L. Loday, J. Stasheff, and A. A. Voronov, eds.), Amer. Math. Soc., 1996, in this volume.

43. V. A. Vassiliev, *Cohomology of knot spaces*, Theory of singularities and its applications (V. I. Arnold, ed.), Amer. Math. Soc., 1990, pp. 23–69.

44. P. Vogel, *Algebraic structures on modules of diagrams*, Preprint, Université Paris VII, August 1995.

45. A. A. Voronov, *Topological field theories, string backgrounds and homotopy algebras*, Proceedings of the XXIInd International Conference on Differential Geometric Methods in Theoretical Physics, Ixtapa-Zihuatanejo, México (J. Keller and Z. Oziewicz, eds.), vol. 4, Advances in Applied Clifford Algebras (Proc. Suppl.), no. S1, 1994, pp. 167–178.

46. E. Witten and B. Zwiebach, *Algebraic structures and differential geometry in two-dimensional string theory*, Nucl. Phys. B **377** (1992), 55–112.

47. M. Wolf and B. Zwiebach, *The plumbing of minimal area surfaces*, Jour. Geom. Phys. **15** (1994), 23–56.

48. B. Zwiebach, *Closed string field theory: Quantum action and the Batalin-Vilkovisky master equation*, Nucl. Phys. B **390** (1993), 33–152.

DEPARTMENT OF MATHEMATICS, BOSTON UNIVERSITY, BOSTON, MA 02215
E-mail address: kimura@math.bu.edu

DEPARTMENT OF MATHEMATICS, UNIVERSITY OF PENNSYLVANIA, PHILADELPHIA, PA 19104-6395
E-mail address: voronov@math.upenn.edu

DEPARTMENT OF MATHEMATICS, YALE UNIVERSITY, NEW HAVEN, CT 06520
E-mail address: zuckerman@math.yale.edu

Contemporary Mathematics
Volume **202**, 1997

Intertwining Operator Algebras, Genus-Zero Modular Functors and Genus-Zero Conformal Field Theories

YI-ZHI HUANG

ABSTRACT. We describe the construction of the genus-zero parts of conformal field theories in the sense of G. Segal from representations of vertex operator algebras satisfying certain conditions. The construction is divided into four steps and each step gives a mathematical structure of independent interest. These mathematical structures are intertwining operator algebras, genus-zero modular functors, genus-zero holomorphic weakly conformal field theories, and genus-zero conformal field theories.

1. Introduction

Conformal field theory (see for example, [**BPZ**], [**Wi**] and [**MS**]) is a physical theory related to many branches of mathematics, e.g., infinite-dimensional Lie algebras and Lie groups, sporadic finite simple groups, modular forms and modular functions, elliptic genera and elliptic cohomology, Calabi-Yau manifolds and mirror symmetry, and quantum groups and 3-manifold invariants. Recently there have been efforts by mathematicians to develop conformal field theory as a rigorous mathematical theory so that it can be used seriously in the near future to solve mathematical problems. A mathematically precise definition of conformal field theory was first given by Segal [**S**] several years ago. Segal's definition is geometric and is motivated by the path integral formulation of some conformal field theories. Though this definition is conceptually simple, it is difficult to see from it some very subtle but important properties of conformal field theories. Because of this, it is even more difficult to construct directly conformal field theories satisfying Segal's geometric definition.

1991 *Mathematics Subject Classification*. Primary 17B69; Secondary 17B67, 17B68, 81R10, 81T40.

Supported in part by NSF grants DMS-9301020 and DMS 9596101 and by DIMACS, an NSF Science and Technology Center funded under contract STC-88-09648.

In practice, physicists and mathematicians have studied concrete models of conformal field theories for many years and the methods used are mostly algebraic. These studies can also be summarized to give algebraic formulations of conformal field theory. The theory of vertex operator algebras is such an algebraic formulation. Many concrete examples of conformal field theories are formulated and studied in algebraic formulations and very detailed calculations can be carried out in these formulations. But, on the other hand, algebraic formulations have a disadvantage that the higher-genus parts of conformal field theories cannot even be formulated, and thus in algebraic formulations it is not easy to see the geometric and topological applications of conformal field theory.

Intuitively, it is expected that algebraic and geometric formulations are equivalent. But this equivalence, especially the construction of conformal field theories satisfying the geometric definition from conformal field theories satisfying the algebraic definition, is a highly nontrivial mathematical problem. In the present paper, we describe the construction of the genus-zero parts of conformal field theories from representations of vertex operator algebras satisfying certain conditions.

This construction can actually be divided into four steps and each step gives us a mathematical structure of independent interest. The first step is to construct an "intertwining operator algebra" from the irreducible representations of a vertex operator algebra satisfying certain conditions. The second step is to construct a "genus-zero modular functor" (a partial operad of a certain type) from an intertwining operator algebra. The third step is to construct a "genus-zero holomorphic weakly conformal field theory" (an algebra over the partial operad of the genus-zero modular functor satisfying certain additional properties) from the intertwining operator algebra. The last step is to construct a "genus-zero conformal field theory" from a genus-zero holomorphic weakly conformal field theory when the genus-zero holomorphic weakly conformal field theory is unitary in a certain sense. The first and the second steps are described in Sections 3 and 4, respectively. The third and the fourth are both described in Section 5.

We introduce the notions of intertwining operator algebra and genus-zero modular functor in Section 3 and 4, respectively. The notions of genus-zero holomorphic weakly conformal field theory and genus-zero conformal field theory are introduced in Section 5. The notion of intertwining operator algebra is new. The other three notions are modifications of the corresponding notions introduced by Segal [S] in the genus-zero case. We also give a brief description of the operadic formulation of the notion of vertex operator algebra in Section 2.

The construction described in this paper depends on the solutions of two problems. The first is the precise geometric description of the central charge of a vertex operator algebra. This problem is solved completely in [H4] (see also [H1] and [H2]). The second is the associativity of intertwining operators for a vertex operator algebra satisfying certain conditions. This is proved in

[**H3**] using the tensor product theory for modules for a vertex operator algebra (see [**HL3**]–[**HL6**], [**H3**]). All the results described in the present paper are consequences of the results obtained in [**H1**], [**H4**], [**HL3**], [**HL4**], [**HL6**], [**H3**] and [**H5**].

Acknowledgment. I would like to thank J.-L. Loday, J. Stasheff and A. A. Voronov for inviting me to attend these two conferences.

Notations:

$\mathbb{C}$: the (structured set of) complex numbers.

$\mathbb{C}^{\times}$: the nonzero complex numbers.

$\mathbb{R}$: the real numbers.

$\mathbb{Z}$: the integers.

$\mathbb{Z}_{+}$: the positive integers.

$\mathbb{N}$: the nonnegative integers.

2. Operadic formulation of the notion of vertex operator algebra

We give a brief description of the operadic formulation of the notion of vertex operator algebra in this section. We assume that the reader is familiar with operads and algebras over operads. See, for example, [**M**] for an introduction to these notions. The material in this section is taken from [**H1**], [**HL1**], [**HL2**] and [**H4**]. Since these references contain detailed and precise definitions and theorems, here we only present the main ideas. The main purpose of this section is to convince those readers unfamiliar with vertex operator algebras that vertex operator algebras are in fact very natural mathematical objects, and to introduce informally some basic concepts and notations in preparation for later sections. For details, see the references above. Note that though the operadic formulation of the notion of vertex operator algebra described below is very natural from the viewpoint of operads and geometry, historically vertex operator algebras occurred in mathematics and physics for completely different reasons. See the introduction of [**FLM**] and the introductory material in the references above for detailed historical discussions. Here we only mention that the geometric meanings of vertex operators and their duality properties (the associativity and the commutativity) are due to Frenkel [**F**], and following the ideas of Frenkel, Tsukada constructed in [**T**] vertex operator algebras associated to lattices (or tori) using a rigorous version of string path integral. For the algebraic formulation of the notion of vertex (operator) algebra, see [**B**], [**FLM**] and [**FHL**]. Recently, Beilinson and Drinfeld introduced a notion of chiral algebra based on algebraic $\mathcal{D}$-modules. This notion is essentially equivalent to the notion of vertex algebra without vacuum. See [**HL7**] for details.

As a motivation, we begin with associative algebras . Let $C(j)$, $j \in \mathbb{N}$, be the moduli space of circles (i.e., compact connected smooth one-dimensional manifolds) with $j + 1$ ordered points (called *punctures*), the 0-th negatively oriented, the others positively oriented, and with smooth local coordinates vanishing at these punctures. Then it is easy to see that $C(j)$ can be identified naturally with the set of permutations $(\sigma(1), \ldots, \sigma(j))$ of $(1, \ldots, j)$. Since this set can in turn be identified in the obvious way with the symmetric group S_j, the moduli space $C(j)$ can also be naturally identified with S_j, with the group S_j acting on $C(j)$ according to the left multiplication action on S_j. That is, S_j permutes the orderings of the positively oriented punctures. Given any two circles with punctures and local coordinates vanishing at the punctures, we can define the notion of sewing them together at any positively oriented puncture on the first circle and the negatively oriented puncture on the second circle by cutting out an open interval of length $2r$ centered at the positively oriented puncture on the first circle (using the local coordinate) and cutting out an open interval of length $2/r$ around the negatively oriented puncture on the second circle in the same way, and then by identifying the boundaries of the remaining parts using the two local coordinates and the map $t \longmapsto 1/t$; we assume that the corresponding closed intervals contain no other punctures. The ordering of the punctures of the sewn circle is given by "inserting" the ordering for the second circle into that for the first. Note that in general not every two circles with punctures and local coordinates can be sewn together at a given positively oriented puncture on the first circle. But it is clear that such a pair of circles with punctures and local coordinates is equivalent to a pair which can be sewn together. Also, the equivalence class of the sewn circle with punctures and local coordinates depends only on the two equivalence classes. Thus we obtain a sewing operation on the moduli space of circles with punctures and local coordinates. Given an element of $C(k)$ and an element of $C(j_s)$ for each j_s, $s = 1, \ldots, k$, we define an element of $C(j_1 + \cdots + j_k)$ by sewing the element of $C(k)$ at its s-th positively oriented puncture with the element of $C(j_s)$ at its negatively oriented puncture, for $s = 1, \ldots, k$, and by "inserting" the orderings for the elements of the $C(j_s)$ into the ordering for the element of $C(k)$. The identity is the unique element of $C(1)$. It is straightforward to verify that $C = \{C(j)\}_{j \in \mathbb{N}}$ is indeed an operad. We call C the *circle operad*.

Since in one dimension smooth structures and conformal structures are the same, the moduli space $C(j)$ can also be thought of as the moduli space of circles with conformal structures, with $j + 1$ ordered punctures with the 0-th negatively oriented and the others positively oriented, and with local conformal coordinates vanishing at these punctures. In fact, we have just defined three operads, namely, C, the corresponding conformal moduli space and $\{S_j\}_{j \in \mathbb{N}}$, and these three operads are isomorphic.

It is well known that the category of algebras over the operad $\{S_j\}_{j \in \mathbb{N}}$ is isomorphic to the category of associative algebras. Thus the category of algebras over C (or the category of algebras over the operad of the corresponding conformal moduli space) is isomorphic to the category of associative algebras.

We have just seen that associative algebras are algebras over an operad obtained from one-dimensional objects. It is very natural to consider algebras over operads obtained from two-dimensional objects. The simplest two-dimensional objects are topological spheres, i.e., genus-zero compact connected smooth two-dimensional manifolds. If we consider the operad of the moduli spaces of genus-zero compact connected smooth two-dimensional manifolds with punctures and local coordinates vanishing at these punctures, then the category of algebras over this operad is isomorphic to the category of commutative associative algebras. We do not obtain new algebras. On the other hand, in dimension two, conformal structures are much richer than smooth structures. Since conformal structures in two real dimensions are equivalent to complex structures in one complex dimension, it is natural to consider operads constructed from genus-zero compact connected one-dimensional complex manifolds with punctures and local coordinates vanishing at these punctures.

Roughly speaking, vertex operator algebras are algebras over a certain operad constructed from genus-zero compact connected one-dimensional complex manifolds with punctures and local coordinates. But the operad in this case has an analytic structure and this analytic structure becomes algebraic in a certain sense when the operad is extended to a partial operad. Vertex operator algebras have properties which are not only the reflections of the operad structure above, but also the reflections of the analytic structure of the extended partial operad. These analyticity properties of vertex operator algebras are very subtle and important features of the theory of vertex operator algebras. They are essential to the constructions and many applications of conformal field theories.

We now begin to describe the notion of vertex operator algebra using the language of operads. A *sphere with* $1 + n$ *tubes* ($n \in \mathbb{N}$) is a genus-zero compact connected one-dimensional complex manifold S with $n + 1$ distinct, ordered points $p_0, \ldots, p_n$ (called *punctures*) on S with p_0 negatively oriented and the other punctures positively oriented, and with local analytic coordinates $(U_0, \varphi_0), \ldots, (U_n, \varphi_n)$ vanishing at the punctures $p_0, \ldots, p_n$, respectively, where for $i = 0, \ldots, n$, U_i is a local coordinate neighborhood at p_i (i.e., an open set containing p_i) and $\varphi_i : U_i \to \mathbb{C}$, satisfying $\varphi_i(p_i) = 0$, is a local analytic coordinate map vanishing at p_i. Let S_1 and S_2 be spheres with $1 + m$ and $1 + n$ tubes, respectively. Let $p_0, \ldots, p_m$ be the punctures of S_1, $q_0, \ldots, q_n$ the punctures of S_2, (U_i, φ_i) the local coordinate at p_i for some fixed i, $0 < i \leq m$, and (V_0, ψ_0) the local coordinate at q_0. Assume that there exists a positive number r such that $\varphi_i(U_i)$ contains the closed disc $\bar{B}_0^r$ centered at 0 with radius r and $\psi_0(V_0)$ contains the closed disc $\bar{B}_0^{1/r}$ centered at 0 with radius $1/r$. Assume also that p_i and q_0 are the only punctures in $\varphi_i^{-1}(\bar{B}_0^r)$ and $\psi_0^{-1}(\bar{B}_0^{1/r})$, respectively. In this case we say that *the i-th tube of S_1 can be sewn with the 0-th tube of S_2*. From S_1 and S_2, we obtain a sphere with $1 + (m + n - 1)$ tubes by cutting $\varphi_i^{-1}(B_0^r)$ and $\psi_0^{-1}(B_0^{1/r})$ from S_1 and S_2, respectively, and then identifying the boundaries of the resulting surfaces using the map $\varphi_i^{-1} \circ J \circ \psi_0$ where J is the map from $\mathbb{C}^\times$ to itself given by $J(z) = 1/z$. The negatively oriented puncture of this sphere with tubes is p_0 and the positively oriented punctures (with ordering) of this sphere with tubes are $p_0, \ldots, p_{i-1}, q_1, \ldots, q_n, p_{i+1}, \ldots, p_m$. The local coordinates vanishing at these punctures are given in the obvious way. Thus we have a partial operation—the *sewing operation*—in the collection of spheres with tubes. We define the notion of conformal equivalence between two spheres with tubes in the obvious way except that two spheres with tubes are also said to be conformally equivalent if the only differences between them are local coordinate neighborhoods at the punctures. For any $n \in \mathbb{N}$, the space of equivalence classes of spheres with $1 + n$ tubes is called the *moduli space of spheres with $1 + n$ tubes*. For $n \in \mathbb{Z}_+$, the moduli space of spheres with $1 + n$ tubes can be identified with $K(n) = M^{n-1} \times H \times (\mathbb{C}^\times \times H)^n$ where M^{n-1} is the set of elements of $\mathbb{C}^{n-2}$ with nonzero and distinct components and H is the set of all sequences A of complex numbers such that $(\exp(\sum_{j \in \mathbb{Z}_+} A_j x^{j+1} \frac{d}{dx}))x$ is a convergent power series in some neighborhood of 0. We think of each element of $K(n)$, $n \in \mathbb{Z}_+$, as the sphere $\mathbb{C} \cup \{\infty\}$ equipped with negatively oriented puncture ∞ and positively oriented ordered punctures $z_1, \ldots, z_{n-2}, 0$, with an element of H specifying the local coordinate at ∞ and with n elements of $\mathbb{C}^\times \times H$ specifying the local coordinates at the other punctures. Analogously, the moduli space of spheres with $1 + 0$ tube can be identified with $K(0) = \{B \in H \mid B_1 = 0\}$. From now on we shall refer to $K(n)$ as the moduli space of spheres with $1 + n$ tubes, $n \in \mathbb{N}$. The sewing operation for spheres with tubes induces a partial operation on the $\cup_{n \in \mathbb{N}} K(n)$. It is still called the sewing operation and is denoted $_i\infty_0$.

Let $K = \{K(n)\}_{n \in \mathbb{N}}$. For any $k \in \mathbb{Z}_+$, $j_i, \ldots, j_k \in \mathbb{N}$, and any elements of $K(k)$, $K(j_1), \ldots, K(j_k)$, by successively sewing the 0-th tube of elements of $K(j_i)$, $i = 1, \ldots, k$, with the i-th tube of the element of $K(k)$, we obtain an element of $K(j_1 + \cdots + j_k)$ if the conditions to perform the sewing operation are satisfied. Thus we obtain a partial map γ_K from $K(k) \times K(j_1) \times \cdots K(j_k)$ to $K(j_1 + \cdots + j_k)$. Let $I \in K(1)$ be equivalence class of the standard sphere $\mathbb{C} \cup \{\infty\}$ with ∞ the negatively oriented puncture, 0 the only positively oriented puncture, and with standard local coordinates vanishing at ∞ and 0. For any $j \in \mathbb{N}$, $\sigma \in S_j$ and $Q \in K(j)$, $\sigma(Q)$ is defined to be the conformal equivalence class of spheres with $1 + j$ tubes obtained from members of the class Q by permuting the orderings of their positively oriented punctures using σ. Thus S_j acts on $K(j)$. It is easy to see that K together with γ_K, I and the actions of S_j on $K(j)$, $j \in \mathbb{N}$ satisfies all the axioms for an operad except that the substitution (or composition) map γ_K are only partially defined. Thus K is an partial operad. We call K the *sphere partial operad* or *vertex partial operad*.

For any two elements of the moduli space of spheres with tubes, in general the i-th tube of the first element might not be able to be sewn with the 0-th tube of the second element. But from the definition of sewing operation, we see that after rescaling the local coordinate at the i-th puncture of the first element or the local coordinate at the 0-th puncture of the second element, the i-th tube of the first element can always be sewn with the 0-th tube of the second element.

So we see that though K is partial, it is *rescalable*, that is, after rescaling, the substitution map is always defined. Since all rescalings of a local coordinate form a group isomorphic to $\mathbb{C}^\times$, K is an example of a $\mathbb{C}^\times$-*rescalable partial operad* and $\mathbb{C}^\times$ is the *rescaling group* of K.

For any $n \in \mathbb{N}$, $K(n)$ is an infinite-dimensional complex manifold. To be more precise, $K(n)$ is a complex (LB)-manifold, that is, a manifold modeled on an (LB)-space over $\mathbb{C}$ (the strict inductive limit of an increasing sequence of Banach spaces over $\mathbb{C}$) such that the transition maps are complex analytic (see, for example, [**K**] for the definition of (LB)-space and Appendix C of [**H4**] for the precise definition of complex (LB)-manifold). It is proved in [**H1**] and [**H4**] that the sewing operation and consequently the substitution map are analytic and even algebraic in a certain sense. So K is an analytic $\mathbb{C}^\times$-rescalable partial operad.

Note that in the operadic formulation of the notion of associative algebra, the nontrivial element σ_{12} of S_2 generates the operad $\{S_j\}_{j \in \mathbb{N}}$ for associative algebras. In fact, the associativity of the product in an associative algebra is the reflection of a property of σ_{12}. This element σ_{12} is called an *associative element* of the operad $\{S_j\}_{j \in \mathbb{N}}$ and $\{S_j\}_{j \in \mathbb{N}}$ is an example of an *associative operad*.

For any $z \in \mathbb{C}^\times$, let $P(z) \in K(2)$ be the conformal equivalence class of the sphere $\mathbb{C} \cup \{\infty\}$ with ∞ the negatively oriented puncture, z and 0 the first and second positively oriented punctures, respectively, and with the standard local coordinates vanishing at these punctures. Then $P(z)$ together with $K(0)$ and $K(1)$ also generates the partial operad K and $P(z)$ also has a property similar to that of σ_{12} (see [**H4**]). We call $P(z)$ an *associative element* of K and K is an example of an *associative analytic $\mathbb{C}^\times$-rescalable partial operad*.

In general, vertex operator algebras have nonzero central charges. To incorporate central charges, we need determinant line bundles and their complex powers. For any $n \in \mathbb{N}$, there is a holomorphic line bundle $\mathrm{Det}(n)$ over $K(n)$ called the *determinant line bundle*. The family $\mathrm{Det} = \{\mathrm{Det}(n)\}_{n \in \mathbb{N}}$ is an associative analytic $\mathbb{C}^\times$-rescalable partial operad such that the projections from $\mathrm{Det}(n)$ to $K(n)$ for all $n \in \mathbb{N}$ give a morphism of associative analytic $\mathbb{C}^\times$-rescalable partial operads and such that when restricted to the fibers, the substitution maps are linear maps. For any $c \in \mathbb{C}$ and $n \in \mathbb{N}$, the complex power $\mathrm{Det}^c = \{\mathrm{Det}^c(n)\}_{n \in \mathbb{N}}$ of Det is a well-defined associative analytic $\mathbb{C}^\times$-rescalable partial operad such that for any $n \in \mathbb{N}$, $\mathrm{Det}^c(n)$ is a holomorphic line bundle over $K(n)$ equal to $\mathrm{Det}(n)$, such that the projections from $\mathrm{Det}^c(n)$ to $K(n)$ for all $n \in \mathbb{N}$ give a morphism of associative analytic $\mathbb{C}^\times$-rescalable partial operads and such that when restricted to the fibers, the substitution maps are linear maps. For detailed descriptions of determinant line bundles and their complex powers, see [**H4**]. We denote $\mathrm{Det}^{1/2}(n)$ by $\tilde{K}(n)$ for $n \in \mathbb{N}$ and the partial operad $\mathrm{Det}^{1/2}$ by $\tilde{K}$. Then for any $c \in \mathbb{C}$, $\mathrm{Det}^{c/2} = \tilde{K}^c$.

To define algebras over a $\mathbb{C}^\times$-rescalable partial operad, we need to define a $\mathbb{C}^\times$-rescalable partial operad constructed from a vector space. Actually, we

have to introduce the notion of "partial pseudo-operad." Let $V = \coprod_{n \in \mathbb{R}} V_{(n)}$
be an $\mathbb{R}$-graded vector space such that $\dim V_{(n)} < \infty$ for all $n \in \mathbb{R}$ and $V' = \coprod_{n \in \mathbb{R}} V_{(n)}$. For any $n \in \mathbb{N}$, let $\mathcal{H}_V(n)$ be the space of linear maps from $V^{\otimes n}$
to $\overline{V} = \prod_{n \in \mathbb{R}} V_{(n)} = V'^*$, where $'$ and $*$ denote the functors of taking restricted
duals and duals of $\mathbb{R}$-graded vector spaces. Since images of elements of $\mathcal{H}_V(n)$
are in general in $\overline{V}$ but not in V, the substitutions are not defined in general.
For example, for $f \in \mathcal{H}_V(2)$, $g_1, g_2 \in \mathcal{H}_V(1)$ and $v_1, v_2 \in V$, $f(g_1(v_1), g_2(v_2))$ is
not defined in general. For any $n \in \mathbb{C}$, let P_n be the projection from V to $V_{(n)}$.
Then for any $m, n \in \mathbb{C}$, $f(P_m(g_1(v_1)), P_n(g_2(v_2)))$ is an element of $\overline{V} = V'^*$. If
for any $v_1, v_2 \in V$ and $v' \in V'$, the series $\langle v', \sum_{m,n \in \mathbb{C}} f(P_m(g_1(v_1)), P_n(g_2(v_2))) \rangle$
is absolutely convergent, then we obtain an element of $\mathcal{H}_V(2)$. In this way, we
obtain a partial map from $\mathcal{H}_V(2) \times \mathcal{H}_V(1) \times \mathcal{H}_V(1)$ to $\mathcal{H}_V(2)$. In general we can
define partial substitution maps for the family $\{\mathcal{H}_V(n)\}_{n \in \mathbb{N}}$ in this way. There
is also an identity $I_V \in \mathcal{H}_V(1)$ which is the identity map from V to $V \subset \overline{V}$.
The symmetric group S_n obviously acts on $\mathcal{H}_V(n)$. In general, however, the
substitution maps do not satisfy associativity even when both sides exist. We
call a structure like $\mathcal{H}_V$ a *partial pseudo-operad*. In particular, partial operads are
partial pseudo-operads. *Morphisms between partial pseudo-operads* are defined
in the obvious way.

For any $c \in \mathbb{C}$, a pseudo-algebra over $\tilde{K}^c$ is a $\mathbb{Z}$-graded vector space V and a
morphism of partial pseudo-operads from $\tilde{K}^c$ to $\mathcal{H}_V$. An algebra over $\tilde{K}^c$ is a
pseudo-algebra V over $\tilde{K}^c$ such that the image of the morphism from $\tilde{K}^c$ to $\mathcal{H}_V$
is a partial operad.

A *vertex associative algebra of central charge* c is an algebra V over $\tilde{K}^c$ such
that $V_{(n)} = 0$ for n sufficiently small and the morphism from $\tilde{K}^c$ to $\mathcal{H}_V$ is the
morphism is linear on the fibers of $\tilde{K}^c(n)$, $n \in \mathbb{N}$, and is meromorphic in a certain
sense (see [HL1], [HL2] and [H4] for the precise definition).

The following theorem announced in [HL1]–[HL2] and proved in [H4] is the
main result of the operadic formulation of the notion of vertex operator algebra:

THEOREM 2.1. *The category of vertex operator algebras of central charge c is
isomorphic to the category of vertex associative algebras of central charge c.*

3. Intertwining operator algebras

In this section, the properties of the algebra of intertwining operators for a
vertex operator algebra satisfying certain conditions are summarized to formulate the notion of intertwining operator algebra. The axioms in the definition
can be relaxed or modified to give many variants of this notion. Since we are
interested in constructing genus-zero modular functors and weakly holomorphic
conformal field theories from intertwining operator algebras, we shall only discuss
the version introduced below in this paper.

Let A be an n-dimensional commutative associative algebra over $\mathbb{C}$. Then for
any basis $\mathcal{A}$ of A, there are structure constants $\mathcal{N}_{a_1 a_2}^{a_3} \in \mathbb{C}$, $a_1, a_2, a_3 \in \mathcal{A}$, such

that

$$a_1 a_2 = \sum_{a_3 \in \mathcal{A}} \mathcal{N}_{a_1 a_2}^{a_3} a_3$$

for any $a_1, a_2 \in \mathcal{A}$. Assume that A has a basis $\mathcal{A} \subset A$ containing the identity $e \in A$ such that all the structure constants $\mathcal{N}_{a_1 a_2}^{a_3}$, $a_1, a_2, a_3 \in \mathcal{A}$, are in $\mathbb{N}$. Note that in this case, for any $a_1, a_2 \in \mathcal{A}$,

$$\mathcal{N}_{e a_1}^{a_2} = \delta_{a_1, a_2} = \begin{cases} 1 & a_1 = a_2, \\ 0 & a_1 \neq a_2. \end{cases}$$

The commutativity and the associativity of A give the following identities:

$$\mathcal{N}_{a_1 a_2}^{a_3} = \mathcal{N}_{a_2 a_1}^{a_3}$$

$$\sum_{a \in \mathcal{A}} \mathcal{N}_{a_1 a_2}^{a} \mathcal{N}_{a a_3}^{a_4} = \sum_{a \in \mathcal{A}} \mathcal{N}_{a_1 a}^{a^4} \mathcal{N}_{a_2 a_3}^{a},$$

for $a_1, a_2, a_3, a_4 \in \mathcal{A}$.

For a vector space $W = \coprod_{a \in \mathcal{A}, n \in \mathbb{R}} W_{(n)}^a$ doubly graded by $\mathbb{R}$ and $\mathcal{A}$, let

$$W_{(n)} = \coprod_{a \in \mathcal{A}} W_{(n)}^a$$

$$W^a = \coprod_{n \in \mathbb{R}} W_{(n)}^a.$$

Then

$$W = \coprod_{n \in \mathbb{R}} W_{(n)} = \coprod_{a \in \mathcal{A}} W^a.$$

The *graded dual of* W is the vector space $W' = \coprod_{a \in \mathcal{A}, n \in \mathbb{R}} (W_{(n)}^a)^*$ doubly graded by $\mathbb{R}$ and $\mathcal{A}$, where for any vector space V, V^* denotes the dual space of V. We shall denote the canonical pairing between W' and W by $\langle \cdot, \cdot \rangle_W$.

For a vector space W and a formal variable x, we shall in this paper denote the space of all formal sums of the form $\sum_{n \in \mathbb{R}} w_n x^n$ by $W\{x\}$. Note that we only allow real powers of x. For any $z \in \mathbb{C}$, we shall always choose $\log z$ so that

$$\log z = \log |z| + i \arg z \text{ with } 0 \leq \arg z < 2\pi.$$

DEFINITION 3.1. An *intertwining operator algebra of central charge* $c \in \mathbb{C}$ consists of the following data:

(i) A finite-dimensional commutative associative algebra A and a basis $\mathcal{A}$ of A containing the identity $e \in A$ such that all the structure constants $\mathcal{N}_{a_1 a_2}^{a_3}$, $a_1, a_2, a_3 \in \mathcal{A}$, are in $\mathbb{N}$.

(ii) A vector space

$$W = \coprod_{a \in \mathcal{A}, n \in \mathbb{R}} W_{(n)}^a, \text{ for } w \in W_{(n)}^a, \ n = \text{wt } w, \ a = \text{clr } w$$

doubly graded by $\mathbb{R}$ and $\mathcal{A}$ (graded by *weight* and by *color*, respectively).

(iii) For each triple $(a_1, a_2, a_3) \in \mathcal{A} \times \mathcal{A} \times \mathcal{A}$, an $\mathcal{N}_{a_1 a_2}^{a_3}$-dimensional subspace $\mathcal{V}_{a_1 a_2}^{a_3}$ of the vector space of all linear maps $W_{a_1} \otimes W_{a_2} \to W_{a_3}\{x\}$, or equivalently, an $\mathcal{N}_{a_1 a_2}^{a_3}$-dimensional vector space $\mathcal{V}_{a_1 a_2}^{a_3}$ whose elements are linear maps

$$
\begin{aligned}
\mathcal{Y} : W_{a_1} &\to \mathrm{Hom}(W_{a_2}, W_{a_3})\{x\} \\
w &\mapsto \mathcal{Y}(w, x) = \sum_{n \in \mathbb{R}} \mathcal{Y}_n(w) x^{-n-1} \text{ (where } \mathcal{Y}_n(w) \in \mathrm{End}\, W\text{)}.
\end{aligned}
$$

(iv) Two distinguished vectors $\mathbf{1} \in W^e$ (*the vacuum*) and $\omega \in W^e$ (*the Virasoro element*).

These data satisfy the following conditions for $a_1, a_2, a_3, a_4, a_5, a_6 \in \mathcal{A}$, $w_i \in W_{a_i}$, $i = 1, 2, 3$, and $w' \in W'_{a_4}$:

(i) The *grading-restriction conditions*:

$$
\dim W_{(n)}^a < \infty \text{ for } n \in \mathbb{Z}, a \in \mathcal{A},
$$
$$
W_{(n)}^a = 0 \text{ for } n \text{ sufficiently small and for all } a \in \mathcal{A} ,
$$

(ii) The *single-valuedness condition*: for any $\mathcal{Y} \in \mathcal{V}_{e a_1}^{a_1}$,

$$
\mathcal{Y}(w_1, x) \in \mathrm{Hom}(W_{a_1}, W_{a_1})[[x, x^{-1}]].
$$

(iii) The *lower-truncation property for vertex operators*: for any $\mathcal{Y} \in \mathcal{V}_{a_1 a_2}^{a_3}$, $\mathcal{Y}_n(w_1) w_2 = 0$ for n sufficiently large.

(iv) The *identity property*: for any $\mathcal{Y} \in \mathcal{V}_{e a_1}^{a_1}$, there is $\lambda_{\mathcal{Y}} \in \mathbb{C}$ such that $\mathcal{Y}(\mathbf{1}, x) = \lambda_{\mathcal{Y}} I_{W_{a_1}}$, where $I_{W_{a_1}}$ on the right is the identity operator on W_{a_1}.

(v) The *creation property*: for any $\mathcal{Y} \in \mathcal{V}_{a_1 e}^{a_1}$, there is $\mu_{\mathcal{Y}} \in \mathbb{C}$ such that $\mathcal{Y}(w_1, x)\mathbf{1} \in W[[x]]$ and $\lim_{x \to 0} \mathcal{Y}(w_1, x)\mathbf{1} = \mu_{\mathcal{Y}} w_1$ (that is, $\mathcal{Y}(w_1, x)\mathbf{1}$ involves only nonnegative integral powers of x and the constant term is $\mu_{\mathcal{Y}} w_1$).

(vi) The *convergence properties*: for any $m \in \mathbb{Z}_+$, $a_i, b_i, c_i \in \mathcal{A}$, $w_i \in W^{a_i}$, $\mathcal{Y}_i \in \mathcal{V}_{a_i \, b_{i+1}}^{c_i}$, $i = 1, \ldots, m$, $w' \in (W^{c_1})'$ and $w \in W^{b_m}$, the series

$$
\langle w', \mathcal{Y}_1(w_1, x_1) \cdots \mathcal{Y}_m(w_m, x_m) w \rangle_{W^{c_1}} \big|_{x_i^n = e^{n \log z_i}, i=1,\ldots,m, n \in \mathbb{R}}
$$

is absolutely convergent when $|z_1| > \cdots > |z_m| > 0$, and for any $\mathcal{Y}_1 \in \mathcal{V}_{a_1 a_2}^{a_5}$ and $\mathcal{Y}_2 \in \mathcal{V}_{a_5 a_3}^{a_4}$, the series

$$
\langle w', \mathcal{Y}_2(\mathcal{Y}_1(w_1, x_0) w_2, x_2) w_3 \rangle_W \big|_{x_0^n = e^{n \log z_1 - z_2}, x_2^n = e^{n \log z_2}, n \in \mathbb{R}}
$$

is absolutely convergent when $|z_2| > |z_1 - z_2| > 0$.

(vii) The *associativity*: for any $\mathcal{Y}_1 \in \mathcal{V}_{a_1 a_5}^{a_4}$ and $\mathcal{V}_{a_2 a_3}^{a_5}$, there exist $\mathcal{Y}_3 \in \mathcal{V}_{a_1 a_2}^{a}$ and $\mathcal{Y}_4^a \in \mathcal{V}_{a a_3}^{a_4}$ for all $a \in \mathcal{A}$ such that the (multi-valued) analytic function

$$
\langle w', \mathcal{Y}_1(w_1, x_1) \mathcal{Y}_2(w_2, x_2) w_3 \rangle_W \big|_{x_1 = z_1, x_2 = z_2}
$$

on $\{(z_1, z_2) \in \mathbb{C} \times \mathbb{C} \mid |z_1| > |z_2| > 0\}$ and the (multi-valued) analytic function

$$\sum_{a \in \mathcal{A}} \langle w', \mathcal{Y}_4^a(\mathcal{Y}_3^a(w_1, x_0)w_2, x_2)w_3\rangle_W\big|_{x_0=z_1-z_2, x_2=z_2}$$

on $\{(z_1, z_2) \in \mathbb{C} \times \mathbb{C} \mid |z_2| > |z_1 - z_2| > 0\}$ are equal on $\{(z_1, z_2) \in \mathbb{C} \times \mathbb{C} \mid |z_1| > |z_2| > |z_1 - z_2| > 0\}$.

(viii) The *Virasoro algebra relations*: Let Y be the element of $\mathcal{V}_{ea_1}^{a_1}$ such that $Y(\mathbf{1}, x) = I_{W_{a_1}}$ and let $Y(\omega, x) = \sum_{n \in \mathbb{Z}} L(n)x^{-n-2}$. Then

$$[L(m), L(n)] = (m-n)L(m+n) + \frac{c}{12}(m^3 - m)\delta_{m+n,0}$$

for $m, n \in \mathbb{Z}$.

(ix) The *$L(0)$-grading property*: $L(0)w = nw = (\mathrm{wt}\ w)w$ for $n \in \mathbb{R}$ and $w \in W_{(n)}$.

(x) The *$L(-1)$-derivative property*: For any $\mathcal{Y} \in \mathcal{V}_{a_1a_2}^{a_3}$,

$$\frac{d}{dx}\mathcal{Y}(w_1, x) = \mathcal{Y}(L(-1)w_1, x).$$

(xi) The *skew-symmetry*: There is a linear map Ω from $\mathcal{V}_{a_1a_2}^{a_3}$ to $\mathcal{V}_{a_2a_1}^{a_3}$ such that for any $\mathcal{Y} \in \mathcal{V}_{a_1a_2}^{a_3}$,

$$\mathcal{Y}(w_1, x)w_2 = e^{xL(-1)}(\Omega(\mathcal{Y}))(w_2, y)w_1\big|_{y^n=e^{in\pi}x^n}.$$

We shall denote the intertwining operator algebra defined above by

$$(W, \mathcal{A}, \{\mathcal{V}_{a_1a_2}^{a_3}\}, \mathbf{1}, \omega)$$

or simply by W. The commutative associative algebra A is called the *Verlinde algebra* or the *fusion algebra of W*. The linear maps in $\mathcal{V}_{a_1a_2}^{a_3}$ are called *intertwining operators of type $\binom{a_3}{a_1a_2}$*.

Remark 3.2. In the definition above, the second convergence property in axiom (vi) can be derived from the first one using the skew-symmetry. We include the second convergence property in the definition since we would like to state the associativity without proving this fact. The associativity or the skew-symmetry can be replaced by the following *commutativity*: for any $\mathcal{Y}_1 \in \mathcal{V}_{a_1a_5}^{a_4}$ and $\mathcal{Y}_2 \in \mathcal{V}_{a_2a_3}^{a_5}$, there exist $\mathcal{Y}_5^a \in \mathcal{V}_{a_2a}^{a_4}$ and $\mathcal{Y}_6^a \in \mathcal{V}_{a_1a_3}^a$ for all $a \in \mathcal{A}$ such that the (multi-valued) analytic function

$$\langle w', \mathcal{Y}_1(w_1, x_1)\mathcal{Y}_2(w_2, x_2)w_3\rangle_W\bigg|_{x_1=z_1, x_2=z_2}$$

on $\{(z_1, z_2) \in \mathbb{C} \times \mathbb{C} \mid |z_1| > |z_2| > 0\}$ and the (multi-valued) analytic function

$$\sum_{a \in \mathcal{A}} \langle w', \mathcal{Y}_5^a(w_2, x_1)\mathcal{Y}_6^a(w_1, x_2)w_3\rangle_W\bigg|_{x_1=z_1, x_2=z_2}$$

on $\{(z_1, z_2) \in \mathbb{C} \times \mathbb{C} \mid |z_2| > |z_1| > 0\}$ are analytic extensions of each other.

Remark 3.3. For any $a_1, a_2, a_3, a_4, a_5 \in \mathcal{A}$, the associativity gives a linear map from $\mathcal{V}_{a_1 a_5}^{a_4} \otimes \mathcal{V}_{a_2 a_3}^{a_5}$ to $\oplus_{a \in \mathcal{A}} \mathcal{V}_{a_1 a_2}^{a} \otimes \mathcal{V}_{a a_3}^{a_4}$. Thus for any $a_1, a_2, a_3, a_4 \in \mathcal{A}$, we obtain a linear map from $\oplus_{a \in \mathcal{A}} \mathcal{V}_{a_1 a}^{a_4} \otimes \mathcal{V}_{a_2 a_3}^{a}$ to $\oplus_{a \in \mathcal{A}} \mathcal{V}_{a_1 a_2}^{a} \otimes \mathcal{V}_{a a_3}^{a_4}$. It is easy to show that these linear maps are in fact linear isomorphisms. These linear isomorphisms are called the *fusing isomorphisms* and the associated matrices under any basis are called *fusing matrices*. Similarly, the commutativity in Remark 3.2 gives a linear isomorphism from $\oplus_{a \in \mathcal{A}} \mathcal{V}_{a_1 a}^{a_4} \otimes \mathcal{V}_{a_2 a_3}^{a}$ to $\oplus_{a \in \mathcal{A}} \mathcal{V}_{a_2 a}^{a_4} \otimes \mathcal{V}_{a_1 a_3}^{a}$ for any $a_1, a_2, a_3, a_4 \in \mathcal{A}$. These linear isomorphisms are called the *braiding isomorphisms* or the associated matrices under any basis are called *braiding matrices*.

In the rest of this section, we assume that the reader is familiar with the notions of abelian intertwining algebra, module for a vertex operator algebra, and intertwining operator among three modules for a vertex operator algebra. We also assume that the reader knows the basic concepts and results in the representation theory of vertex operator algebras, for example, the rationality of vertex operator algebras and the conditions to use the tensor product theory for modules for a vertex operator algebra. See [**DL**], [**FLM**], [**FHL**], [**HL3**]–[**HL6**] and [**H3**].

It is easy to verify the following:

PROPOSITION 3.4. *An abelian intertwining algebra in the sense of Dong and Lepowsky [**DL**] satisfying in addition the grading-restriction conditions is an intertwining operator algebra whose Verlinde algebra is the group algebra of the abelian group associated with the abelian intertwining algebra. In particular, vertex operator algebras are intertwining operator algebras.*

The results in [**FHL**], [**HL3**], [**HL4**], [**HL6**], [**H3**] imply the following result:

THEOREM 3.5. *Let V be a rational vertex operator algebra satisfying the conditions to use the tensor product theory for V-modules, $\mathcal{A} = \{a_1, \ldots, a_m\}$ the set of all equivalence classes of irreducible V-modules, and $W^{a_1}, \ldots, W^{a_m}$ representatives of $a_1, \ldots, a_m$, respectively. Then there is a natural commutative associative algebra structure on the vector space A spanned by $\mathcal{A}$ such that $W = \coprod_{i=1}^{m} W^{a_i}$ has a natural structure of intertwining operator algebra whose Verlinde algebra is A.*

We know that minimal Virasoro vertex operator algebras are rational (see [**DMZ**] and [**Wa**]). In [**H5**], it is proved that any vertex operator algebra containing a vertex operator subalgebra isomorphic to a tensor product algebra of minimal Virasoro vertex operator algebras satisfies the conditions to use the tensor product theory. The proof uses the representation theory of the Virasoro algebra and the Belavin-Polyakov-Zamolodchikov equations (see [**BPZ**]) for minimal models. We also know that the vertex operator algebras associated to Wess-Zumino-Novikov-Witten models (WZNW models) are rational (see [**FZ**]). In [**HL8**], the representation theory of the Virasoro algebra and the Belavin-Polyakov-Zamolodchikov equations for minimal models are replaced by the representation theory of affine Lie algebras and the Knizhnik-Zamolodchikov equations (see [**KZ**]) for WZNW models, respectively, to show that any vertex operator algebra containing subalgebras isomorphic to a tensor product algebra of vertex operator algebras associated to WZNW models also satisfies the conditions to use the tensor product theory. So we have:

COROLLARY 3.6. *Let V be a rational vertex operator algebra containing a vertex operator subalgebra isomorphic to a tensor product algebra of minimal Virasoro vertex operator algebras or to a tensor product algebra of vertex operator algebras associated to WZNW models, $\mathcal{A} = \{a_1, \ldots, a_m\}$ the set of all equivalence classes of irreducible V-modules, and $W^{a_1}, \ldots, W^{a_m}$ representatives of $a_1, \ldots, a_m$, respectively. Then there is a natural commutative associative algebra structure on the vector space A spanned by $\mathcal{A}$ such that $W = \coprod_{i=1}^{m} W^{a_i}$ has a natural structure of intertwining algebra whose Verlinde algebra is A.*

It is easy to see that in general the Verlinde algebras for intertwining operator algebras obtained from these vertex operator algebras are not group algebras. Thus these intertwining operator algebras are not abelian intertwining algebras. So we do have many examples of intertwining operator algebras which are not abelian intertwining algebras.

4. Genus-zero modular functors

In the definition of intertwining operator algebra, one of the data is the collections of the vector spaces $\mathcal{V}_{a_1 a_2}^{a_3}$, $a_1, a_2, a_3 \in \mathcal{A}$. From these vector spaces, without using many of the properties of the operators in them, we can construct a geometric object. The properties of these geometric objects are in fact the axioms in the notions of genus-zero modular functor and rational genus-zero modular functor.

We first give the definition of genus-zero modular functor.

DEFINITION 4.1. A *genus-zero modular functor* is an analytic $\mathbb{C}^{\times}$-rescalable partial operad $\mathcal{M}$ together with a morphism $\pi : \mathcal{M} \to K$ of $\mathbb{C}^{\times}$-rescalable partial operads satisfying the following axioms:

(i) For any $j \in \mathbb{N}$ the triple $(\mathcal{M}(j), K(j), \pi)$ is a finite-rank holomorphic vector bundle over $K(j)$.

(ii) For any $\mathcal{Q} \in \mathcal{M}(k)$, $\mathcal{Q}_1 \in \mathcal{M}(j_1), \ldots, \mathcal{Q}_k \in \mathcal{M}(j_k)$, $k, j_1, \ldots, j_k \in \mathbb{N}$, the substitution $\gamma_{\mathcal{M}}(\mathcal{Q}; \mathcal{Q}_1, \ldots, \mathcal{Q}_k)$ in $\mathcal{M}$ exists if (and only if)

$$\gamma_K(\pi(\mathcal{Q}); \pi(\mathcal{Q}_1), \ldots, \pi(\mathcal{Q}_k))$$

exists.

(iii) Let $Q \in K(k)$, $Q_1 \in K(j_1), \ldots, Q_k \in K(j_k)$, $k, j_1, \ldots, j_k \in \mathbb{N}$, such that $\gamma(Q; Q_1, \ldots, Q_k)$ exists. The map from the Cartesian product of the fibers over $Q, Q_1, \ldots, Q_k$ to the fiber over $\gamma_K(Q; Q_1, \ldots, Q_k)$ induced from the substitution map of $\mathcal{M}$ is multilinear and gives an isomorphism from the tensor product of the fibers over $Q, Q_1, \ldots, Q_k$ to the fiber over $\gamma_K(Q; Q_1, \ldots, Q_k)$.

(iv) The actions of the symmetric groups on $\mathcal{M}$ are isomorphisms of holomorphic vector bundles covering the actions of the symmetric groups on K.

Homomorphisms and *isomorphisms* of genus-zero modular functors are defined in the obvious way.

Let $\mathcal{M}$ be a genus-zero modular functor. For any $k \in \mathbb{Z}_+$, $j_1, \ldots, j_k \in \mathbb{N}$, let $\mathcal{M}_i(k), \mathcal{M}_i(j_1), \ldots, \mathcal{M}_i(j_k)$ be vector subbundles of $\mathcal{M}(k), \mathcal{M}(j_1), \ldots, \mathcal{M}(j_k)$, respectively, for $i = 1, 2$, and $P_i(k), P_i(j_1), \ldots, P_i(j_k)$ the projections from $\mathcal{M}(k)$, $\mathcal{M}(j_1), \ldots, \mathcal{M}(j_k)$ to $\mathcal{M}_i(k), \mathcal{M}_i(j_1), \ldots, \mathcal{M}_i(j_k)$, respectively, for $i = 1, 2$. Then $\gamma_\mathcal{M} \circ (P_1(k) \times P_1(j_1) \cdots \times P_1(j_k))$ and $\gamma_\mathcal{M} \circ (P_2(k) \times P_2(j_1) \cdots \times P_2(j_k))$ are both homomorphisms of vector bundles from $\mathcal{M}(k) \times \mathcal{M}(j_1) \times \cdots \times \mathcal{M}(j_k)$ to $\mathcal{M}(j_1 + \cdots + j_k)$. So the sum of $\gamma_\mathcal{M} \circ (P_1(k) \times P_1(j_1) \cdots \times P_1(j_k))$ and $\gamma_\mathcal{M} \circ (P_2(k) \times P_2(j_1) \cdots \times P_2(j_k))$ is well-defined. Similarly, sums of more than two such homomorphisms are also well-defined.

We next give the definition of rational genus-zero modular functor:

DEFINITION 4.2. A *rational genus-zero modular functor* is a genus-zero modular functor $\mathcal{M}$ and a finite set $\mathcal{A}$ satisfying the following axioms:

(v) For any $n \in \mathbb{N}$ and $a_0, a_1, \ldots, a_n \in \mathcal{A}$, there are finite-rank holomorphic vector bundles $\mathcal{M}_{a_1 \cdots a_n}^{a_0}(n)$ over $K(n)$ such that

$$\mathcal{M}(n) = \oplus_{a_0, a_1, \ldots, a_n \in \mathcal{A}} \mathcal{M}_{a_1 \cdots a_n}^{a_0}(n)$$

where $\oplus$ denotes the direct sum operation for vector bundles.

(vi) For any $n \in \mathbb{N}$ and $b_0, b_1, \ldots, b_n \in \mathcal{A}$, let $P^{b_0}(n)$ be the projection from $\mathcal{M}(n)$ to $\oplus_{a_1, \ldots, a_n \in \mathcal{A}} \mathcal{M}_{a_1 \cdots a_n}^{b_0}(n)$ and $P_{b_1 \cdots b_n}(n)$ the projection from $\mathcal{M}(n)$ to $\oplus_{a_0 \in \mathcal{A}} \mathcal{M}_{b_1 \cdots b_n}^{a_0}(n)$. Then for any $k \in \mathbb{Z}_+$, $j_1, \ldots, j_k \in \mathbb{N}$,

$$\gamma_\mathcal{M} = \sum_{b_1, \ldots, b_k \in \mathcal{A}} \gamma_\mathcal{M} \circ (P_{b_1 \cdots b_k}(k) \times P^{b_1}(j_1) \times \cdots P^{b_k}(j_k)).$$

We shall denote the rational genus-zero modular functor just defined by $(\mathcal{M}, \mathcal{A})$.

The simplest examples of rational genus-zero modular functors are $\tilde{K}^c$, $c \in \mathbb{C}$.

Using the methods developed in [**H4**] and using intertwining operators instead of vertex operators for vertex operator algebras, we have:

THEOREM 4.3. *Let* $(W, \mathcal{A}, \{\mathcal{V}_{a_1 a_2}^{a_3}\}, \mathbf{1}, \omega)$ *be an intertwining operator algebra of central charge* c. *Then there is a canonical rational genus-zero modular functor* $(\mathcal{M}_W, \mathcal{A})$ *such that for any* $a_1, a_2, a_3 \in \mathcal{A}$, $\mathcal{M}^{a_1}(0) = \tilde{K}^c(0)$ *and the fiber of* $\mathcal{M}_{a_1 a_2}^{a_3}(2)$ *is isomorphic to* $\mathcal{V}_{a_1 a_2}^{a_3}$.

In particular, by Theorem 4.3 and Corollary 3.6, we have:

COROLLARY 4.4. *Let V be a rational vertex operator algebra containing a vertex operator subalgebra isomorphic to a tensor product algebra of minimal Virasoro vertex operator algebras or to a tensor product algebra of vertex operator algebras associated to WZNW models, $\mathcal{A} = \{a_1, \ldots, a_m\}$ the set of all equivalence classes of irreducible V-modules, and $W^{a_1}, \ldots, W^{a_m}$ representatives of $a_1, \ldots, a_m$, respectively. Then there is a canonical rational genus-zero modular functor $(\mathcal{M}_V, \mathcal{A})$ such that for any $a_1, a_2, a_3 \in \mathcal{A}$, $\mathcal{M}^{a_1}(0) = \tilde{K}^c(0)$ and the fiber of $\mathcal{M}^{a_3}_{a_1 a_2}(2)$ is isomorphic to $\mathcal{V}^{a_3}_{a_1 a_2}$, where $c \in \mathbb{C}$ is the central charge of V and $\mathcal{V}^{a_3}_{a_1 a_2}$ is the space of all intertwining operators for V of type $\binom{W^{a_3}}{W^{a_1} W^{a_2}}$.*

Genus-zero modular functors contain almost all of the topological information one can obtain from an intertwining operator algebra. To see this, we need a result of Segal [**S**]:

PROPOSITION 4.5. *Let $\mathcal{M}$ be a genus-zero modular functor. Then there are canonical projectively flat connections on the vector bundles $\mathcal{M}(n)$, $n \in \mathbb{N}$, such that they are compatible with the substitution maps for the partial operad $\mathcal{M}$.*

For any $n \in \mathbb{Z}_+$, we can embed the configuration space

$$F(n) = \{(z_1, \ldots, z_n) \in \mathbb{C}^n \mid z_i \neq z_j \text{ for } i \neq j\}$$

into $K(n)$ in the obvious way. Thus given any genus-zero modular functor $\mathcal{M}$, for any $n \in \mathbb{Z}_+$, $\mathcal{M}(n)$ is pulled back to a vector bundle over $F(n)$ and the canonical connection on $\mathcal{M}(n)$ is pulled back to a connection on the pull-back vector bundle over $F(n)$. Since the actions of symmetric groups on $\mathcal{M}$ are isomorphisms of holomorphic vector bundles covering the actions of the symmetric actions on K, the pull-back vector bundle and the pull-back connection over $F(n)$ induce a vector bundle over $F(n)/S_n$ and a connection over this vector bundle for any $n \in \mathbb{Z}_+$. It is not difficult to prove the following:

PROPOSITION 4.6. *Let $\mathcal{M}$ be a genus-zero modular functor. For any $n \in \mathbb{Z}_+$, the connections on the pull-back vector bundle over $F(n)$ and on the induced vector bundle over $F(n)/S_n$ are flat.*

We know that a flat connection on a vector bundle over a connected manifold gives a structure of a representation of the fundamental group of the manifold on the fiber of the vector bundle at any point on the manifold. We also know that the braid group B_n on n strings is by definition the fundamental group of $F(n)/S_n$. So we obtain:

THEOREM 4.7. *Let $\mathcal{M}$ be a genus-zero modular functor, $n \in \mathbb{Z}_+$ and $\mathcal{V}$ the fiber at any point in $F(n) \subset K(n)$ of the vector bundle $\mathcal{M}(n)$. Then $\mathcal{V}$ has a natural structure of a representation of the braid group B_n on n strings.*

In particular, from a rational vertex operator algebra containing a vertex operator subalgebra isomorphic to a tensor product algebra of minimal Virasoro vertex operator algebras or to a tensor product algebra of vertex operator algebras associated to WZNW models, we obtain representations of braid groups. In the case that the vertex operator algebras are associated to WZNW models based on the Lie group $SU(2)$, the corresponding representations of the braid groups are the same as those obtained by Tsuchiya and Kanie [**TK**] and the corresponding knot invariants are the Jones polynomials for knots [**J**]. (In fact, from a genus-zero modular functor, we obtain not only representations of the braid groups, but also representations of braid groups with twists.)

Let $\mathcal{M} = \{\mathcal{M}(n)\}_{n \in \mathbb{N}}$ be a genus-zero modular functor and $\overline{\mathcal{M}}(n)$, $n \in \mathbb{N}$, the complex conjugates of the holomorphic vector bundles $\mathcal{M}(n)$. We define the *complex conjugate partial operad of* $\mathcal{M}$ to be the sequence $\overline{\mathcal{M}} = \{\overline{\mathcal{M}}(n)\}_{n \in \mathbb{N}}$ with the obvious partial operad structure induced from that of $\mathcal{M}$. Let

$$\mathcal{M} \otimes \overline{\mathcal{M}} = \{\mathcal{M}(n) \otimes \overline{\mathcal{M}}(n)\}_{n \in \mathbb{N}}$$

where, on the right-hand side, $\otimes$ is the tensor product operation for vector bundles over $K(n)$, $n \in \mathbb{N}$. Then the partial operad structures on $\mathcal{M}$ and $\overline{\mathcal{M}}$ induce a partial operad structure on $\mathcal{M} \otimes \overline{\mathcal{M}}$. In particular, for any $c \in \mathbb{C}$, since $\tilde{K}^c$ is a genus-zero modular functor, we obtain partial operads $\overline{\tilde{K}^c}$ and $\tilde{K}^c \otimes \overline{\tilde{K}^c}$. Using these partial operads, we have the following notion:

DEFINITION 4.8. A genus-zero modular functor $\mathcal{M}$ is *unitary* if there is a complex number $c \in \mathbb{C}$ and a morphism of analytic partial operads from $\mathcal{M} \otimes \overline{\mathcal{M}}$ to $\tilde{K}^c \otimes \overline{\tilde{K}^c}$ satisfying the following conditions:

(i) For any $n \in \mathbb{N}$ and $Q \in K(n)$, the morphism from $\mathcal{M} \otimes \overline{\mathcal{M}}$ to $\tilde{K}^c \otimes \overline{\tilde{K}^c}$ maps the fiber of $\mathcal{M}(n) \otimes \overline{\mathcal{M}}(n)$ at Q linearly to the fiber of $\tilde{K}^c(n) \otimes \overline{\tilde{K}^c}(n)$ at Q.

(ii) The linear map above from the fiber of $\mathcal{M}(n) \otimes \overline{\mathcal{M}}(n)$ at Q to the fiber of $\tilde{K}^c(n) \otimes \overline{\tilde{K}^c}(n) = \mathbb{C}$ at Q when viewed as a linear map from the fiber of $\mathcal{M}(n) \otimes \overline{\mathcal{M}}(n)$ at Q to $\mathbb{C}$ gives a positive-definite Hermitian form on the fiber of $\mathcal{M}(n)$ at Q.

The complex number c is called the *central charge* of the unitary genus-zero modular functor $\mathcal{M}$.

Let $(W, \mathcal{A}, \{\mathcal{V}^{a_3}_{a_1 a_2}\}, \mathbf{1}, \omega)$ be an intertwining algebra of central charge c. Assume that there are positive-definite Hermitian forms on $\mathcal{V}^{a_3}_{a_1 a_2}$ for all $a_1, a_2, a_3 \in \mathcal{A}$. These positive-definite Hermitian forms induce positive-definite Hermitian forms on $\oplus_{a \in \mathcal{A}} \mathcal{V}^{a_4}_{a_1 a} \otimes \mathcal{V}^a_{a_2 a_3}$ and $\oplus_{a \in \mathcal{A}} \mathcal{V}^a_{a_1 a_2} \otimes \mathcal{V}^{a_4}_{a a_3}$ for any $a_1, a_2, a_3, a_4 \in \mathcal{A}$. We have:

PROPOSITION 4.9. *Let* $(W, \mathcal{A}, \{\mathcal{V}^{a_3}_{a_1 a_2}\}, \mathbf{1}, \omega)$ *be an intertwining operator algebra of central charge* c. *If there are positive-definite Hermitian forms on* $\mathcal{V}^{a_3}_{a_1 a_2}$ *for all* $a_1, a_2, a_3 \in \mathcal{A}$ *such that the fusing isomorphism from* $\oplus_{a \in \mathcal{A}} \mathcal{V}^{a_4}_{a_1 a} \otimes$

$\mathcal{V}^a_{a_2 a_3}$ to $\oplus_{a \in \mathcal{A}} \mathcal{V}^a_{a_1 a_2} \otimes \mathcal{V}^{a_4}_{a a_3}$ pulls the induced positive-definite Hermitian form on $\oplus_{a \in \mathcal{A}} \mathcal{V}^a_{a_1 a_2} \otimes \mathcal{V}^{a_4}_{a a_3}$ back to that on $\oplus_{a \in \mathcal{A}} \mathcal{V}^{a_4}_{a_1 a} \otimes \mathcal{V}^a_{a_2 a_3}$ for any $a_1, a_2, a_3, a_4 \in \mathcal{A}$, then these positive-definite Hermitian forms give a unitary structure to the rational genus-zero modular functor $(\mathcal{M}_W, \mathcal{A})$.

In particular, we have:

PROPOSITION 4.10. *The genus-zero modular functors obtained from minimal Virasoro vertex operator algebras and from vertex operator algebras associated to WZNW models are unitary.*

5. Genus-zero conformal field theories

Modular functors only reflect the geometric objects constructed from the vector spaces $\mathcal{V}^{a_3}_{a_1 a_2}$, $a_1, a_2, a_3 \in \mathcal{A}$, without using the properties of the operators in these spaces. Also, an intertwining operator algebra has the vacuum and the Virasoro element which are not reflected in the definition of (rational) genus-zero modular functor. These data and the properties they satisfy are reflected in the following definition of genus-zero weakly holomorphic conformal field theory:

DEFINITION 5.1. Let $\mathcal{M}$ be a genus-zero modular functor. A *genus-zero weakly holomorphic conformal field theory over* $\mathcal{M}$ is an $\mathbb{R}$-graded vector space W and a morphism of partial pseudo-operads from $\mathcal{M}$ to the endomorphism partial pseudo-operad $\mathcal{H}_W$ (see Section 2) satisfying the following additional axioms:

(i) The image of the morphism from $\mathcal{M}$ to $\mathcal{H}_W$ is a partial operad.
(ii) The morphism from $\mathcal{M}$ to the endomorphism partial pseudo-operad $\mathcal{H}_W$ is linear on the fibers of the bundles $\mathcal{M}(n)$, $n \in \mathbb{N}$.
(iii) This morphism is holomorphic.

A genus-zero weakly holomorphic conformal field theory over a rational genus-zero modular functor is called a *rational genus-zero holomorphic weakly conformal field theory.*

Using the method developed in [**H4**], we have:

THEOREM 5.2. *Let $(W, \mathcal{A}, \{\mathcal{V}^{a_3}_{a_1 a_2}\}, \mathbf{1}, \omega)$ be an intertwining operator algebra and $(\mathcal{M}_W, \mathcal{A})$ the corresponding rational genus-zero modular functor. Then W has a canonical structure of a rational holomorphic genus-zero weakly conformal field theory.*

In particular, by Theorem 5.2 and Corollary 3.6, we have:

COROLLARY 5.3. *Let V be a rational vertex operator algebra containing a vertex operator subalgebra isomorphic to a tensor product algebra of minimal Virasoro vertex operator algebras or a vertex operator algebras associated to a WZNW model, $\mathcal{A} = \{a_1, \ldots, a_m\}$ the set of all equivalence classes of irreducible V-modules, and $W^{a_1}, \ldots, W^{a_m}$ representatives of $a_1, \ldots, a_m$, respectively. Then* *there is a canonical rational genus-zero weakly holomorphic conformal field theory structure on* $W = \coprod_{i=1}^{m} W^{a_i}$.

Let $\mathfrak{H}_1$ be the Banach space of all analytic functions on the closed unit disk $\{z \in \mathbb{C} \mid |z| \le 1\}$ and $K_{\mathfrak{H}_1}(n)$ for any $n \in \mathbb{N}$ the subset of $K(n)$ consisting of conformal equivalence classes containing spheres with $1 + n$ tubes such that the inverses of the local coordinates at punctures can be extended to the closed unit disk and the images of the closed unit disks under these inverses are disjoint. Let $K_{\mathfrak{H}_1} = \{K_{\mathfrak{H}_1}(n)\}_{n \in \mathbb{N}}$. Then the sewing operation is always defined in $K_{\mathfrak{H}_1}$. Also, the identity of K is in $K_{\mathfrak{H}_1}$ and the symmetric groups act on $K_{\mathfrak{H}_1}$. So $K_{\mathfrak{H}_1}$ is a suboperad of K (not just a partial suboperad of K). For any $n \in \mathbb{N}$ and $c \in \mathbb{C}$, let $\tilde{K}_{\mathfrak{H}_1}^c(n)$ be the restriction of the line bundle $\tilde{K}^c(n)$ to $K_{\mathfrak{H}_1}(n)$. Then $\tilde{K}_{\mathfrak{H}_1}^c = \{\tilde{K}_{\mathfrak{H}_1}^c(n)\}_{n \in \mathbb{N}}$ is a suboperad (not partial) of $\tilde{K}^c$.

Let $\overline{\tilde{K}_{\mathfrak{H}_1}^c}$ be the complex conjugate operad of $\tilde{K}_{\mathfrak{H}_1}^c$ defined in the obvious way and $\tilde{K}_{\mathfrak{H}_1}^c \otimes \overline{\tilde{K}_{\mathfrak{H}_1}^c}$ the tensor product operad of $\tilde{K}_{\mathfrak{H}_1}^c$ and $\overline{\tilde{K}_{\mathfrak{H}_1}^c}$. In general, algebras over the operad $\tilde{K}_{\mathfrak{H}_1}^c \otimes \overline{\tilde{K}_{\mathfrak{H}_1}^c}$ might not have any topological structure. But since $\tilde{K}_{\mathfrak{H}_1}^c \otimes \overline{\tilde{K}_{\mathfrak{H}_1}^c}$ consists of infinite-dimensional Banach manifolds, we are interested in algebras over it with topological structures. Consider the abelian category of Hilbert spaces over $\mathbb{C}$. Then the tensor product operation for Hilbert spaces gives a tensor category structure to this abelian category. Thus, for any Hilbert space H over $\mathbb{C}$, we have the endomorphism operad of H. This endomorphism operad has a topological structure. A *Hilbert algebra* over $\tilde{K}_{\mathfrak{H}_1}^c \otimes \overline{\tilde{K}_{\mathfrak{H}_1}^c}$ is a Hilbert space H over $\mathbb{C}$ and a continuous morphism of topological operads from $\tilde{K}_{\mathfrak{H}_1}^c \otimes \overline{\tilde{K}_{\mathfrak{H}_1}^c}$ to the endomorphism operad of H.

DEFINITION 5.4. A *genus-zero conformal field theory of central charge* c is a Hilbert algebra over $\tilde{K}_{\mathfrak{H}_1}^c \otimes \overline{\tilde{K}_{\mathfrak{H}_1}^c}$ such that the morphism from $\tilde{K}_{\mathfrak{H}_1}^c \otimes \overline{\tilde{K}_{\mathfrak{H}_1}^c}$ to the endomorphism operad of H is linear on the fibers of $\tilde{K}_{\mathfrak{H}_1}^c(n) \otimes \overline{\tilde{K}_{\mathfrak{H}_1}^c}(n)$, $n \in \mathbb{N}$.

Let $\mathcal{M}$ be a genus-zero modular functor and $\mathcal{M}_{\mathfrak{H}_1}(n)$, $n \in \mathbb{N}$, the restrictions of $\mathcal{M}(n)$ to $K_{\mathfrak{H}_1}(n)$. Then $\mathcal{M}_{\mathfrak{H}_1} = \{\mathcal{M}_{\mathfrak{H}_1}(n)\}_{n \in \mathbb{N}}$ is a suboperad (not partial) of $\mathcal{M}$. We also have the notion of *Hilbert algebra* over $\mathcal{M}_{\mathfrak{H}_1}$ defined in the obvious way.

DEFINITION 5.5. Let $\mathcal{M}$ be a unitary genus-zero modular functor. A genus-zero weakly holomorphic conformal field theory W over $\mathcal{M}$ is *unitary* if there is a positive-definite Hermitian form on W such that the morphism of partial pseudo-operads $\mathcal{M}$ to $\mathcal{H}_W$ induces a Hilbert algebra structure over $\mathcal{M}_{\mathfrak{H}_1}$ on the completion H_W^h of W (here h means holomorphic).

Note that if a genus-zero weakly holomorphic conformal field theory W constructed from irreducible modules for a vertex operator algebra V is unitary, then in particular, for any irreducible V-modules W_1, W_2, W_3, an intertwining operator of type $\binom{W_3}{W_1 W_2}$ evaluated at any complex number z satisfying $0 < |z| < 1$ must map $W_1 \otimes W_2$ to $H_{W_3}^h$, where $H_{W_3}^h$ is the closure of W_3 in H_W^h. If the vertex operator algebra V satisfies the conditions to use the tensor product theory, then using these conditions and the contragredient intertwining operators, one can show that such an intertwining operator evaluated at any nonzero complex number indeed maps $W_1 \otimes W_2$ to $H^h_{W_3}$. We have:

PROPOSITION 5.6. *The genus-zero weakly holomorphic conformal field theories constructed from the minimal Virasoro vertex operator algebras and from the vertex operator algebras associated to WZNW models are unitary.*

Let $\mathcal{M}$ be a unitary genus-zero modular functor. Then for any $n \in \mathbb{N}$ and $Q \in K(n)$, the corresponding positive-definite Hermitian form on the fiber of $\mathcal{M}(n)$ at Q identifies the dual space of this fiber with the fiber of $\overline{\mathcal{M}}(n)$ at Q and thus also identifies the dual space of the fiber of $\overline{\mathcal{M}}(n)$ at Q with the fiber of $\mathcal{M}(n)$ at Q. Thus the adjoint of the morphism from $\mathcal{M} \otimes \overline{\mathcal{M}}$ to $\tilde{K}^c \otimes \overline{\tilde{K}^c}$ gives a morphism from $(\tilde{K}^c \otimes \overline{\tilde{K}^c})^{-1}$ to $\mathcal{M} \otimes \overline{\mathcal{M}}$, where $(\tilde{K}^c \otimes \overline{\tilde{K}^c})^{-1} = \{(\tilde{K}^c \otimes \overline{\tilde{K}^c})^{-1}(n)\}_{n \in \mathbb{N}}$, and for any $n \in \mathbb{N}$, $(\tilde{K}^c \otimes \overline{\tilde{K}^c})^{-1}(n)$ is the line bundle whose fibers are the duals of the fibers of $\tilde{K}^c(n) \otimes \overline{\tilde{K}^c}(n)$. It is clear that $(\tilde{K}^c \otimes \overline{\tilde{K}^c})^{-1}$ is canonically isomorphic to $\tilde{K}^c \otimes \overline{\tilde{K}^c}$. Thus we obtain a morphism from $\tilde{K}^c \otimes \overline{\tilde{K}^c}$ to $\mathcal{M} \otimes \overline{\mathcal{M}}$.

Let W be a unitary genus-zero weakly holomorphic conformal field theory over a unitary genus-zero modular functor $\mathcal{M}$. Let $\overline{H}^h_W$ be the complex conjugate of H^h_W and $\mathcal{M}_{\mathfrak{H}_1}$ the complex conjugate operad of $\mathcal{M}_{\mathfrak{H}_1}$. Then $\overline{H}^h_W$ has a structure of Hilbert algebra over $\mathcal{M}_{\mathfrak{H}_1}$. Let $H_W = H^h_W \otimes \overline{H}^h_W$ where $\otimes$ is the Hilbert space tensor product. Then the Hilbert algebra structure over $\mathcal{M}_{\mathfrak{H}_1}$ on H^h_W and the Hilbert algebra structure over $\overline{\mathcal{M}}_{\mathfrak{H}_1}$ on $\overline{H}^h_W$ induce a Hilbert algebra structure over $\mathcal{M}_{\mathfrak{H}_1} \otimes \overline{\mathcal{M}}_{\mathfrak{H}_1}$ on H_W. Since $\mathcal{M}$ is unitary, we have a morphism from $\tilde{K}^c \otimes \overline{\tilde{K}^c}$ to $\mathcal{M} \otimes \overline{\mathcal{M}}$ by the discussion above. When restricted to $\tilde{K}^c_{\mathfrak{H}_1} \otimes \overline{\tilde{K}^c_{\mathfrak{H}_1}}$, it gives a morphism from $\tilde{K}^c_{\mathfrak{H}_1} \otimes \overline{\tilde{K}^c_{\mathfrak{H}_1}}$ to $\mathcal{M}_{\mathfrak{H}_1} \otimes \overline{\mathcal{M}}_{\mathfrak{H}_1}$. Combining this morphism and the Hilbert algebra structure over $\mathcal{M}_{\mathfrak{H}_1} \otimes \overline{\mathcal{M}}_{\mathfrak{H}_1}$ on H_W, we obtain (cf. [S]):

PROPOSITION 5.7. *Let W be a unitary genus-zero weakly holomorphic conformal field theory over a unitary genus-zero modular functor $\mathcal{M}$. Then H_W is a genus-zero conformal field theory.*

In particular, we have:

COROLLARY 5.8. *Let W be the unitary genus-zero weakly holomorphic conformal field theory constructed from a minimal Virasoro vertex operator algebra or from a vertex operator algebras associated to WZNW model. Then H_W is a genus-zero conformal field theory.*

REFERENCES

[BPZ] A. A. Belavin, A. M. Polyakov and A. B. Zamolodchikov, Infinite conformal symmetries in two-dimensional quantum field theory, *Nucl. Phys.* **B241** (1984), 333–380.

[B] R. E. Borcherds, Vertex algebras, Kac-Moody algebras, and the Monster, *Proc. Natl. Acad. Sci. USA* **83** (1986), 3068–3071.

[DMZ]   C. Dong, G. Mason and Y. Zhu, Discrete series of the Virasoro algebra and the moonshine module, in: *Algebraic Groups and Their Generalizations: Quantum and Infinite-Dimensional Methods*, ed. William J. Haboush and Brian J. Parshall, Proc. Symp. Pure. Math., American Math. Soc., Providence, 1994, Vol. 56, Part 2, 295–316.

[DL]   C. Dong and J. Lepowsky, *Generalized Vertex Algebras and Relative Vertex Operators*, Progress in Math., Vol. 112, Birkhäuser, Boston, 1993.

[F]   I. B. Frenkel, talk presented at the Institute for Advanced Study, 1988; and private communications.

[FHL]   I. B. Frenkel, Y.-Z. Huang and J. Lepowsky, On axiomatic approaches to vertex operator algebras and modules, preprint, 1989; *Memoirs Amer. Math. Soc.* **104**, 1993.

[FLM]   I. B. Frenkel, J. Lepowsky, and A. Meurman, *Vertex operator algebras and the Monster*, Academic Press, New York, 1988.

[FZ]   I. B. Frenkel and Y. Zhu, Vertex operator algebras associated to representations of affine and Virasoro algebras, *Duke Math. J.* **66** (1992), 123–168.

[H1]   Y.-Z. Huang, *On the geometric interpretation of vertex operator algebras*, Ph.D. thesis, Rutgers University, 1990.

[H2]   Y.-Z. Huang, Geometric interpretation of vertex operator algebras, *Proc. Natl. Acad. Sci. USA* **88** (1991), 9964–9968.

[H3]   Y.-Z. Huang, A theory of tensor products for module categories for a vertex operator algebra, IV, *J. Pure Appl. Alg.* **100** (1995), 173-216.

[H4]   Y.-Z. Huang, *Two-dimensional conformal geometry and vertex operator algebras*, Birkhäuser, to appear.

[H5]   Y.-Z. Huang, Virasoro vertex operator algebras, the (nonmeromorphic) operator product expansion and the tensor product theory, *J. Alg.* **182** (1996), 201–234.

[HL1]   Y.-Z. Huang and J. Lepowsky, Vertex operator algebras and operads, *The Gelfand Mathematical Seminar, 1990–1992*, ed. L. Corwin, I. Gelfand and J. Lepowsky, Birkhäuser, Boston, 1993, 145–161.

[HL2]   Y.-Z. Huang and J. Lepowsky, Operadic formulation of the notion of vertex operator algebra, in: *Mathematical Aspects of Conformal and Topological Field Theories and Quantum Groups, Proc. Joint Summer Research Conference, Mount Holyoke, 1992*, ed. P. Sally, M. Flato, J. Lepowsky, N. Reshetikhin and G. Zuckerman, Contemporary Math., Vol. 175, Amer. Math. Soc., Providence, 1994, 131-148.

[HL3]   Y.-Z. Huang and J. Lepowsky, A theory of tensor products for module categories for a vertex operator algebra, I, *Selecta Mathematica, New Series* **1** (1995), 699–756.

[HL4]   Y.-Z. Huang and J. Lepowsky, A theory of tensor products for module categories for a vertex operator algebra, II, *Selecta Mathematica, New Series* **1** (1995), 757–786.

[HL5]   Y.-Z. Huang and J. Lepowsky, Tensor products of modules for a vertex operator algebras and vertex tensor categories, in: *Lie Theory and Geometry, in honor of Bertram Kostant,* ed. R. Brylinski, J.-L. Brylinski, V. Guillemin, V. Kac, Birkhäuser, Boston, 1994, 349–383.

[HL6]   Y.-Z. Huang and J. Lepowsky, A theory of tensor products for module categories for a vertex operator algebra, III, *J. Pure Appl. Alg.* **100** (1995), 141-171.

[HL7]   Y.-Z. Huang and J. Lepowsky, On the D-module and formal-variable approaches to vertex algebras, in: *Topics in Geometry: In Memory of Joseph D'Atri*, ed. S. Gindikin, Progress in Nonlinear Differential Equations, Vol. 20, Birkhäuser, Boston, 1996. 175–202.

[HL8]   Y.-Z. Huang and J. Lepowsky, Affine Lie algebras and vertex tensor categories, to appear.

[J]   V. F. R. Jones, Hecke algebra representations of braid groups and link polynomials, *Ann. Math.* **126** (1987), 335–388.

[KZ]   V. G. Knizhnik and A. B. Zamolodchikov, Current algebra and Wess-Zumino models in two dimensions, *Nucl. Phys.* **B247** (1984), 83–103.

[K]   G. Köthe, *Topological vector spaces I*, Die Grundlehren der mathematischen Wissenschaften, Vol. 159, Springer-Verlag, New York, 1969.

[M]   J. P. May, Definitions: operads, algebras and modules, in this volumn.

[MS]   G. Moore and N. Seiberg, Classical and quantum conformal field theory, *Comm. in Math. Phys.* **123** (1989), 177–254.

[S]   G. B. Segal, The definition of conformal field theory, preprint, 1988.

[T]   H. Tsukada, *String path integral realization of vertex operator algebras*, PhD thesis, Rutgers University, 1988; *Memoirs Amer. Math. Soc.* **91**, 1991.

[TK]   A. Tsuchiya and Y. Kanie, Vertex operators in conformal field theory on $\mathbb{P}^1$ and monodromy representations of braid group, in: *Conformal Field Theory and Solvable Lattice Models, Advanced Studies in Pure Math.*, Vol. 16, Kinokuniya Company Ltd., Tokyo, 1988, 297–372.

[Wa]   W. Wang, Rationality of Virasoro vertex operator algebras, *International Mathematics Research Notices* (in *Duke Math. J.*) **7** (1993), 197–211.

[Wi]   E. Witten, Non-abelian bosonization in two dimensions, *Comm. Math. Phys.* **92** (1984), 455–472.

DEPARTMENT OF MATHEMATICS, RUTGERS UNIVERSITY, NEW BRUNSWICK, NJ 08903
E-mail address: yzhuang@math.rutgers.edu

Contemporary Mathematics
Volume **202**, 1997

MODULAR FUNCTOR AND REPRESENTATION
THEORY OF $\widehat{sl_2}$ AT A RATIONAL LEVEL

BORIS FEIGIN AND FEODOR MALIKOV

ABSTRACT. We define a new modular functor based on Kac-Waki-moto admissible representations and the corresponding $\mathcal{D}-$module on the moduli space of rank 2 vector bundles with parabolic structure. a new fusion functor arises which is related to the representation theory of the pair "$osp(1|2), sl_2$" in the same way as the fusion functor for the virasoro algebra is related to the representation theory of the pair "sl_2, sl_2".

1. INTRODUCTION

In this paper we define a new modular functor based on Kac-Waki-moto admissible representations over $\widehat{sl_2}$. The modular functor introduced by Segal [42] assigns a finite-dimensional vector space to the data consisting of a punctured curve, a rank 2 vector bundle and a collection of integral dominant highest weights attached to the punctures. Our modular functor does the same for Segal's data (with integral dominant highest weights replaced with admissible highest weights) extended by the lines in the fibers over the punctures. As the data " surface, vector bundle, punctures, lines in fibers over punctures" evolve, so does the corresponding finite dimensional vector space. This leads to a new $\mathcal{D}-$module on the moduli space of rank 2 vector bundles with parabolic structure (fixed lines in certain fibers). The main feature of this $\mathcal{D}-$module, as opposed to the standard one (see Tsuchiya-Ueno-Yamada [44], or Beilinson-Feigin-Mazur [4]), or Moore-Seiberg [37]), is that it is singular over a certain set of exceptional vector bundles. The latter is closely related to Hitchin's global nilpotent cone.

We also prove that our $\mathcal{D}-$module has (in a suitable sense) regular singularities at infinity and that the dimension of the generic fiber can be calculated by the usual combinatorial algorithm: by pinching the surface the problem is reduced to the case of a sphere with $\geq$ 3 punctures and further to a collection of spheres with 3 punctures. The dimension of the space attached to the datum "3 modules sitting at 3 points on a sphere" is calculated explicitly. It is a pure linear algebra calculation of the dimension of the space of coinvariants of

1991 *Mathematics Subject Classification*. Primary 22-XX, 81-XX.

a certain infnite dimensional algebra in a certain infinite dimensional representation. As the result is amusing we will record it here.

First of all, and this is important, in the genus zero case, one can work with modules at a generic level, as opposed to admissible representations which only exist when the level is rational. This is in complete analogy with the usual WZW model, where the famous theory of Knizhnik-Zamolodchikov equations arises from a collection of the so-called Weyl modules sitting on a sphere (the terminology is borrowed from [30]). The family of Weyl modules is good for the purpose of studying integrable representations because each integrable representation is a quotient of some Weyl module. This is no longer the case as far as admissible representations are concerned. A family of modules suitable for our needs is that of what we call *generalized Weyl modules*; the latter is defined to be a Verma module quotiented out by a singular vector.

Generalized Weyl modules are naturally parametrized by the symbols (V_r^ϵ, V_s) $r, s \geq 0$, $\epsilon \in \mathbf{Z}/2\mathbf{Z}$. Here V_r is to be thought of as the $(r+1)$−dimensional irreducible sl_2−module; the meaning of V_r^ϵ will be explained soon. It is appropriate to keep in mind that the conventional Weyl module is defined to be the module induced from V_r. Therefore usually Weyl modules are labelled by sl_2−modules. In our situation Weyl modules are those related to symbols (V_0^0, V_s).

According to Verlinde, the dimensions of the spaces associated to 3 modules on a sphere are structure constants of the Verlinde algebra. The result of calculation of the Verlinde algebra in our situation is as follows:

$$(1) \quad (V_{r_1}^\alpha, V_{s_1}) \circ (V_{r_2}^\beta, V_{s_2})$$
$$= (V_{r_1+r_2}^{\alpha+\beta}, V_{s_1} \otimes V_{s_2}) + (V_{r_1+r_2-1}^{\alpha+\beta+1}, V_{s_1} \otimes V_{s_2})$$
$$+ (V_{r_1+r_2-2}^{\alpha+\beta}, V_{s_1} \otimes V_{s_2}) + \cdots + (V_{|r_1-r_2|}^{\alpha+\beta}, V_{s_1} \otimes V_{s_2}).$$

Recall that the usual Verlinde algebra built on Weyl modules is as follows:

$$V_{s_1} \circ V_{s_2} = V_{s_1} \otimes V_{s_2},$$

i.e. it is the Grothendieck ring of the category of finite dimensional representations of sl_2. Observe that our formula agrees with the latter on Weyl modules.

The first component of the right hand side of our formula is equally easy to interpret. It is known that the symbols V_r^ϵ naturally parametrize finite dimensional representations of the simplest rank 1 superalgebra $osp(1|2)$. The category of finite dimensional $osp(1|2)-$modules is a tensor category and (1) reads as follows: the Verlinde algebra is isomorphic to the product of the Grothendieck rings of the categories of finite dimensional representations of $osp(1|2)$ and sl_2.

It is known in principle what to do when passing from modules to their quotients, in our case from generalized Weyl modules at a generic level to admissible representations at a rational level: one has to replace Lie algebras with quantized universal enveloping algebras at roots of unity and consider the Grothendieck rings of the corresponding semisimple "quotient categories". Examples: the Verlinde algebra built on integrable $\widehat{sl_2}-$modules has to do with sl_2 in this way, and the Verlinde algebra built on minimal representations of Virasoro algebra in this way has to do with 2 copies of sl_2. The calculation of coinvariants shows that one more example can be added to the list: the Verlinde algebra built on admissible representations is related to the pair $(osp(1|2), sl_2)$ in exactly the same way as the $Vir-$Verlinde algebra is related to the pair of sl_2's.

The interest in admissible representation originates in the fact that the characters of admissible representations representations at a fixed level give a representation of the modular group. However, the realization of this fact immediately gave rise to two puzzles:

(i) Given a representation of the modular group, the Verlinde formula produces the structure constants of the Verlinde algebra; in the case of admissible representations some of the structure constants are negative. This does not make much sense as they are supposed to count dimensions.

(ii) Quantum Drinfeld-Sokolov reduction provides a functor from the category of $\widehat{sl_2}$ - modules to the category of $Vir-$modules, which sends admissible representations to minimal representations. It should give an epimorphism (or some weakened version of it) of a suitably defined Verlinde algebra for $\widehat{sl_2}$ on the well-known Verlinde algebra for Vir.

We are able to give an answer to (ii), and a partial answer to (i).

As far as (ii) is concerned, let us for simplicity step aside and consider $Vir-$modules at a generic (not necessarily rational) level. Then there is an analogue of a generalized Weyl module – Verma module quotiented out by a singular vector – and these are naturally parametrized by the symbols (V_r, V_s). The desired epimorphism is given by:

$$(V_r^\epsilon, V_s) \mapsto (V_r, V_s) + (V_{r-1}, V_s).$$

This map is naturally related to the Drinfeld-Sokolov reduction in the following way. As we have fixed the category of representations, we have triangular decomposition of sl_2; in particular we have 2 opposite nilpotent subalgebras, $\mathbf{C}e$, $\mathbf{C}f$. Therefore there are in fact 2 Drinfeld-Sokolov functors, ϕ_e, ϕ_f. It happens that the map above is induced by the direct sum $\phi_e \oplus \phi_f$. Further, this map is even easier to interpret from the point of view of finite dimensional representation theory. Indeed, $osp(1|2)$ contains $sl(2)$ as the even part. Therefore there arises the forgetful morphism of the categories of finite dimensional representations $Rep(osp(1|2)) \to Rep(sl(2))$. It is easy to see that the forgetful morphism is exactly our map $V_r^\epsilon \to V_r \oplus V_{r-1}$.

As for (i), the situation is as follows. The structure constants naturally arrange in a tensor $\{c_{ij}^r\}$, the indices running through a set of representations in question. Let us compare the set $\{c_{ij}^r\}$ of structure constants of our algebra and the set $\{b_{ij}^r\}$ of structure constants of the algebra calculated by the Verlinde formula.

If our $c_{ij}^r = 0$, then $b_{ij}^r = 0$. If $c_{ij}^r \neq 0$, then b_{ij}^r is "almost certainly" zero, however in some exceptional cases it is non-zero. The latter cases in our situation are interpreted in the following way. Recall that we have not only 3 modules, i, j, r, but also 3 Borel subalgebras, $\mathbf{b}_i, \mathbf{b}_j, \mathbf{b}_r$, which vary. Now as $c_{ij}^r \neq 0$, the fiber of our $\mathcal{D}$−module is $\neq 0$ (in fact it is 1-dimensional), if the 3 Borel subalgebras are pairwise different. If however 2 of them meet, the fiber usually vanishes, but sometimes survives. It survives if and only if $b_{ij}^r \neq 0$. If non-zero, b_{ij}^r can be ± 1. There is no doubt that b_{ij}^r is a result of some cohomological calculation related to the $\mathcal{D}$−module. Unfortunately we cannot make this more precise at the moment.

Just as in the usual case n Weyl modules on a sphere produce a trivial vector bundle with the flat (Knizhnik-Zamolodchikov)connection on a space of dimension n, in our case we get a bundle with a flat connection on a space of dimension $2n$. The extra coordinates come from the flag manifold – recall that we are dealing with moduli of vector bundles with parabolic structure. Horizontal sections of this connection satisfy a system of differential equations; we get twice as many equations as there are KZ equations: half of them are indeed KZ equations and the other half comes from singular vectors in Verma modules over $\widehat{sl}_2$. The latter is only natural – it is exactly one of the lessons of the pioneering work [6]. This allows us to put the integral formulas for solutions of the Knizhnik-Zamolodchikov equations, which we wrote in [20], in a proper context: they give horizontal sections of this new connection. We conjecture that our methods in fact provide all horizontal sections.

The relation of our formulas to those in [43] is that the latter are necessarily polynomials as functions on the flag manifold while ours are not.

An important representation theoretic fact behind all the mentioned results is a theorem on the singular support of admissible representations. It claims that a representation of $\widehat{sl}_2$ is admissible if and only if its singular support is contained in the space of 1-forms with values in the nilpotent cone of sl_2. The result was discovered at least conjecturally and at least for the vacuum representation a few years ago by E.Frenkel and Feigin. The proof we propose here uses, in particular, some results of our work [19]. As an application we get a construction of an infinite collection of elements of the annihilating ideal of admississible representations at a given level.

We wish to acknowledge that there has been a number of works approaching WZW model for admissible representation from different points of view, see for example [1, 17, 24, 38, 40]. It would be interesting to relate our integral formulas with those in [38] and the new Hopf algebra of [40] to the above mentioned "$osp(1|2) \times sl_2$" at roots of unity. To the best of our knowledge, the Verlinde algebras proposed in these works do not solve (ii) above – those algebras are rather trivial when compared to the $Vir-$analogue. Our starting point, see [19], was the work [1], where the Verlinde algebra for admissible representations was first calculated (in a form equivalent but much less illuminating than the one described above), using a language which left completely open the problem of existence of a $\mathcal{D}-$module, such that the dimension of the fiber is calculated through this algebra.

Acknowledgments. Parts of this work were reported at the AMS meeting in Hartford, March, 1995, and at Service de Physique Theorique at Saclay, in November, 1994. We are grateful to J.-B. Zuber for invitation and warm hospitality. A considerable part of this work was done over the 2 years one of us spent at Yale. The inspiring and friendly atmosphere at the Department of Mathematics contributed a lot – and so did the discussions with Igor Frenkel, Ian Grojnowski and Gregg Zuckerman. We are grateful to Itzhak Bars for bringing to our attention the paper [38], to Sanjaye Ramgoolam for sending his work, to David Kazhdan for the interesting conversation at Harvard, to Susan Montgomery and Horia Pop who helped us to track the work [32], and to Edward Frenkel for consultations concerning the Drinfeld-Sokolov reduction. Work of Malikov was partially supported by a grant of NSF.

For a variety of reasons the work on this paper has extended over a few years. The final agony took place in June, 1996 while Malikov was participating in the school "Representation theory and mathematical physics" at the E.Schrodinger Institute in Vienna. We are grateful to I.Penkov and J.Wolf for invitation and to the Institue for support. It was a pleasure to discuss our results with our colleagues at IHES in July 1996.

2. Notations and known results

2.1.   Some notations from commutative algebra are as follows:

$\mathbf{C}[t]$ is a polynomial ring, $\mathbf{C}[[t]]$ is its completion by positive powers of t; $\mathbf{C}[t, t^{-1}]$ is the ring of Laurent polynomials and $\mathbf{C}((t))$ is its completion by positive powers of t.

By functions on the formal (punctured) neighborhood of a nonsingular point on a curve we will mean a ring isomorphic to $\mathbf{C}[[t]]$ ($\mathbf{C}((t))$ resp.); to specify such an isomorphism means to pick a local coordinate t. The analogous meaning will be given to the phrase " sections of a vector bundle on the formal (punctured) neighborhood of a non-singular point on a curve".

2.2.   Set $\mathrm{g} = sl_2$, $\hat{\mathrm{g}} = \widehat{sl_2} = sl_2 \otimes \mathbf{C}[z, z^{-1}] \oplus \mathbf{C}c$. Choose a basis e, h, f of g satisfying the standard relations $[h, e] = 2e$, $[h, f] = -2f$, $[e, f] = h$. We say define

$\mathrm{g}_{\geq} = \mathbf{C}e \oplus \mathbf{C}h$ and $\hat{\mathrm{g}}_{\geq} = \mathrm{g} \otimes z\mathbf{C}[[z]] \oplus \mathrm{b} \oplus \mathbf{C}c$ to be the standard Borel subalgebras of g and ĝ resp;

$\mathrm{g}_{>} = \mathbf{C}e$ and $\hat{\mathrm{g}}_{>} = \mathrm{g} \otimes z\mathbf{C}[[z]] \oplus \mathrm{g}_{>}$ to be the standard "maximal nilpotent subalgebras" of g and ĝ resp.;

$\mathbf{C}h$ and $\mathbf{C}h \oplus \mathbf{C}c$ to be the standard Cartan subalgebras of g and ĝ resp.

The Verma module $M_{\lambda, k}$ is the module induced from the character of $\mathrm{g} \otimes z\mathbf{C}[[z]] \oplus \mathrm{b} \oplus \mathbf{C}c$ annihilating $\mathrm{g} \otimes z\mathbf{C}[z] \oplus \mathbf{C}e$ and sending h and c to λ and k resp. k is often referred to as the level. The generator of $M_{\lambda, k}$ is usually denoted by $v_{\lambda, k}$. A quotient of a Verma module is called a highest weight module.

The algebra ĝ is $\mathbf{Z}_{+}^2$–graded by assigning $f \otimes z^n \mapsto (1, -n)$, $e \otimes z^n \mapsto (-1, -n)$ and so is a Verma module (as well as its quotients): $M_{\lambda, k} = \oplus_{i, j} M_{\lambda, k}^{i, j}$.

There is a canonical antiinvolution $\omega : \hat{\mathrm{g}} \to \hat{\mathrm{g}}$ interchanging $\hat{\mathrm{g}}_{>}$ and $\hat{\mathrm{g}}_{<}$ and constant on the Cartan subalgebra. For any highest weight module V denote by V^c and call *contragredient* the module equal to the restricted dual V^* as a vector space with the following action of $\hat{\mathfrak{g}}$:

$$< gx, y >=< x, \omega(g)y >, \; g \in \hat{\mathfrak{g}}, x \in V^*, y \in V.$$

If a highest weight module V is irreducible then it is isomorphic to V^c. A morphism of highest weight modules $V_1 \to V_2$ naturally induces a morphism of the corresponding contragredient modules: $V_2^c \to V_1^c$.

A morphism of Verma modules $M_{\lambda,k} \to M_{\mu,k}$ is determined by the image of $v_{\lambda,k}$. The image can be written as $Sv_{\mu,k}$ for a uniquely determined element S of the universal enveloping algebra of $\mathfrak{g} \otimes z^{-1}\mathbf{C}[z^{-1}] \oplus \mathbf{C}f$. If non-zero, the vector $Sv_{\mu,k}$, or even S for that matter, is called *singular*. The singular vector can be equivalently defined as an eigenvector of the Cartan subalgebra of $\hat{\mathfrak{g}}$ annihilated by $\hat{\mathfrak{g}}_>$. In this form the definition applies to an arbitrary $\hat{\mathfrak{g}}$−module.

2.3. **Singular vector formula.** It follows from the Kac-Kazhdan determinant formula that a singular vector generically appears in the homogeneous components of degree either $n(-1, m)$, $m > 0$, $n > 0$ or $n(1, m)$, $m \geq 0, n > 0$. Denote the corresponding singular vectors by $S^1_{n,m}$ and $S^0_{n,m}$ resp.

Singular vectors S^i_{nm} were found in [33] in an unconventional form containing non-integral powers of elements of $\hat{\mathfrak{g}}$ (see also [3] for another approach):

$$(2) \quad S^0_{nm} = (e \otimes z^{-1})^{n+mt} f^{n+(m-1)t}(e \otimes z^{-1})^{n+(m-2)t} \cdots (e \otimes z^{-1})^{n-mt},$$

$$(3) \qquad S^1_{nm} = f^{n+mt}(e \otimes z^{-1})^{n+(m-1)t} f^{n+(m-2)t} \cdots f^{n-mt},$$

where $t = k + 2$.

This form is not always convenient to calculate a singular vector. It is, however, a useful tool to derive properties of a singular vector. For example, denoting by $\pi : \hat{\mathfrak{g}} \to \mathfrak{g}$, $g \otimes z^n \mapsto \mathfrak{g}$ the evaluation map, one uses (2), (3) to derive that (see [23], also [34] for the proof in a more general quantum case):

$$(4) \qquad \pi S^0_{nm} = (\prod_{i=1}^{m} \prod_{j=1}^{n} P(-it - j))e^n$$

$$(5) \qquad \pi S^1_{nm} = (\prod_{i=1}^{m} \prod_{j=0}^{n-1} P(it + j))f^n,$$

where $P(t) = ef - (t + 1)h - t(t + 1)$.

2.4. **Generalized Weyl modules and admissible representations.**
The structure of Verma modules over $\hat{\mathfrak{g}}$ is known in full detail ([35]).
Outside the critical level ($k = -2$) a Verma module is generically irreducible. $M_{\lambda,k}$ happens to be reducible if and only if it contains a
singular vector. If $M_{\lambda,k}$ is reducible then the following 2 cases arise:

(i) k is generic (not rational) and $M_{\lambda,k}$ contains only one singular
vector;

(ii) $k + 2 = p/q > 0$ is a ratio of 2 positive integers and $M_{\lambda,k}$ contains
infinitely many singular vectors.

It can of course happen that $k + 2 = p/q < 0$. We will not be
interested in this case and confine ourselves to mentioning that here
the situation is in a sense dual to (ii).

2.4.1. *Case (i).* $M_{\lambda,k}$ contains a unique proper submodule M generated
by the singular vector. M is, in fact, a Verma module.

Definition. The irreducible quotient $V_{\lambda,k}$ is called *a generalized Weyl
module.*   □

There arises the exact sequence

(6) $$0 \to M \to M_{\lambda,k} \to V_{\lambda,k} \to 0.$$

A simple property of the Kac-Kazhdan equations [26] is that, given
(6), the module M is irreducible and does not project on any generalized Weyl module. Note that if the composition series of a $\hat{\mathfrak{g}}$−module
only consists of generalized Weyl modules then this module breaks up
into a direct sum of its components. (This can be proved by methods
of Deodhar-Gabber-Kac [10].)

It is an exercise on the Kac-Kazhdan equations to derive that the
highest weight (λ, k) of a generalized Weyl module $V_{\lambda.k}$ belongs to either
the line

(7) $$\lambda = -it + j - 1, \; k = t - 2,$$

for some $i \geq 0, j \geq 1$, or to the line

(8) $$\lambda = it - j - 1, \; k = t - 2,$$

for some $i, j \geq 1$; in both cases t is regarded as a parameter. Formula
(7) corresponds to the case when $V_{\lambda,k}$ is obtained from $M_{\lambda,k}$ by quotienting out the singular vector $S_{i,j}^0$; analogously, (8) corresponds to the
case when $V_{\lambda,k}$ is obtained from $M_{\lambda,k}$ by quotienting out the singular
vector $S_{i,j}^1$.

We see that for a fixed level k generalized Weyl modules are parametrized by the triples consisting of a pair of nonnegative numbers, i, j in
the formulas above, and an element taking one of the 2 values needed

to distinguish between (7) and (8). To be more precise, denote by V_i the $(i+1)$−dimensional irreducible representation of g.

Notation. Assign to $V_{\lambda,k}$ either the symbol (V_i^0, V_{j-1}), $i \geq 0, j \geq 1$, if (λ, k) satisfies (7), or the symbol (V_{i-1}^1, V_{j-1}), $i, j \geq 1$, if (λ, k) satisfies (8). $\quad\square$

This gives us a one-to-one correspondence between the set of generalized Weyl modules at a fixed generic level and the set of symbols (V_i^ϵ, V_j), where ϵ is understood as an element of $\mathbf{Z}/2\mathbf{Z}$.

Observe that the conventional Weyl module of level k is defined to be the induced representation

$$\mathrm{Ind}^{\hat{g}}_{g[[z]]\oplus \mathbf{C}c} V_n,$$

where $g[[z]]$ operates on V_n via the evaluation map $g[[z]] \to g$ and $c \mapsto k$. From our point of view the Weyl module is a quotient of the Verma module $M_{n,k}$ by the submodule generated by the singular vector $f^{n+1}v_{\lambda,k}$. In other words, Weyl modules are associated to the symbols (V_0^0, V_n). This partially explains the appearance of g−modules in our notations.

2.4.2. *Case (ii)*. A Verma module contains infinitely many singular vectors and is embedded in finitely many other Verma modules. Among all singular vectors in $M_{\lambda,k}$ there are 2 independent ones and these generate the maximal proper submodule. Although formally all such Verma modules look alike a special role is played by those which can only embed (non-trivially) in themselves. The highest weights of such modules were called by Kac and Wakimoto *admissible* ([29]) and are described as follows.

Let $k + 2 = p/q$, where p, q are relatively prime positive integers. The set of admissible highest weights at level $k = p/q - 2$ is given by

$$\Lambda_k = \{\lambda(m, n) = m\frac{p}{q} - n - 1 : 0 < m \leq q, 0 \leq n \leq p - 1\}.$$

What is said above about the structure of Verma modules implies that any Verma module appears in an exact sequence of the form

$$(9) \quad 0 \leftarrow L_{\lambda_0,k} \leftarrow M_{\lambda_0,k} \xleftarrow{d_0} M_{\lambda_1,k} \oplus M_{\mu_1,k} \xleftarrow{d_1} M_{\lambda_2,k} \oplus M_{\mu_2,k} \xleftarrow{d_2} \cdots ,$$

where λ_0 is an admissible weight at level k and $L_{\lambda_0,k}$ is the corresponding irreducible module. $L_{\lambda_0,k}$ is also called *admissible*. The exact sequence (9) is called the Bernstein - Gel'fand - Gel'fand (BGG) resolution.

Again cohomological arguments show (see e.g. [29]) that if the composition series of a $\hat{g}$–module consists only of admissible representations then the module is completely reducible.

The parametrization of the set of admissible representations we are going to use is as follows. Two different generalized Weyl modules project onto one and the same admissible representation: formula (9) implies that the the two modules projecting onto $L_{\lambda_0,k}$ are $M_{\lambda_0,k}/M_{\lambda_1,k}$ and $M_{\lambda_0,k}/M_{\mu_1,k}$. Therefore two different triples (V_m^ϵ, V_n) are related to the same admissible represenation. Introduce the equivalence relation $\approx$ by $(V_m^\epsilon, V_n) \approx (V_{q-1-m}^{\epsilon+1}, V_{p-2-n})$, $0 \leq m \leq q-1$, $0 \leq n \leq p-2$. Denote by $(V_m^\epsilon, V_n)^\sim$ the equivalence class of (V_m^ϵ, V_n).

It easy to check that admissible representations are parametrized by the equivalence classes of the triples:

$$(10) \qquad \{\text{ admissible representations }\} \iff \{(V_m^\epsilon, V_n)^\sim\}.$$

2.5. A considerable part of the above carries over to the arbitrary Kac-Moody algebra case. Here, for example, is the definition of an admissible representation. Drop the condition that $g = sl_2$, let $M_{\lambda,k}$ be a Verma module over $\hat{g}$ and $L_{\lambda,k}$ be its irreducible quotient. Call (λ, k) admissible if $M_{\lambda,k}$ satisfies the following projectivity condition: ifthe composition series of a $\hat{g}$–module W contains $L_{\lambda,k}$ then $M_{\lambda,k}$ maps non-trivially into W.

Unfortunately we do not have a reasonable definition of a generalized Weyl module in the higher rank case. (Actually we have but cannot prove it.) This is one of the reasons for which we have to confine ourselves mostly to the sl_2–case.

2.6. **Loop modules.** We will also be using $\hat{g}$–modules different from Verma modules or the corresponding irreducible ones.

Denote by $\mathcal{F}_{\alpha\beta}$ the g-module with the basis F_i, $i \in \mathbf{Z}$, and the action given by

$$eF_i = -(\alpha + i - \beta)F_{i+1}, \ hF_i = (2\alpha + 2i - \beta)F_i, \ fF_i = (-\alpha - i)F_{i-1}.$$

The space $\mathcal{F}_{\alpha\beta}^{\mathbf{C}^*} = \mathcal{F}_{\alpha\beta} \otimes \mathbf{C}[z, z^{-1}]$ is endowed with the natural $\hat{g}$–module structure. The elements $F_{ij} = F_i \otimes z^j$, $i, j \in \mathbf{Z}$ serve as a natural basis of it.

Recall (see 2.3) that S_{nm}^1, S_{nm}^0 stand for a singular vector of degree $n(-1, m)$ or $n(1, m)$ resp. in a Verma module. The following formulas are proved by using (4,5):

(11) $S^1_{nm} F_{n,nm}$

$$= \{\prod_{i=1}^{m}\prod_{j=1}^{n}(-it-j-\alpha+\beta)(-it-j-\alpha)\}\{\prod_{s=1}^{n}(\alpha+s)\}F_{00}$$

(12) $S^0_{nm} F_{-n,m}$

$$= \{\prod_{i=1}^{m}\prod_{j=1}^{n}(it+j-\alpha+\beta)(-it-j-\alpha)\}\{\prod_{s=1}^{n}(\alpha-\beta-s)F_{00},$$

where $t = k+2$.

3. Construction of the modular functor.

Although most of our results have to do with sl_2, up to some point it is no extra effort to work in greater generality. So until sect.4, g will stand for sl_n unless otherwise stated.

3.1. Algebra $\hat{g}^A$ and categories of $\hat{g}^A$-modules.

3.1.1. Let $\mathcal{C}$ be a smooth compact algebraic curve and $\rho : \mathcal{E} \to \mathcal{C}$ be a rank n vector bundle with a flat connection. The connection associates to a section s of any bundle $\mathcal{A}$ associated with $\mathcal{E}$ the section dc of $\Omega \otimes \mathcal{A}$, where Ω is the sheaf of differential forms over $\mathcal{C}$. A typical example of $\mathcal{A}$ is the bundle $\operatorname{End} \mathcal{E}$ of fiberwise endomorphisms of $\mathcal{E}$. The sheaf of sections of $\operatorname{End} \mathcal{E}$ is naturally a sheaf of Lie algebras over $\mathcal{C}$.

For a point $P \in \mathcal{C}$ let g^P be the algebra of sections of $\operatorname{End}\mathcal{E}$ over the formal neighborhood of P. For a finite subset $\bar{A} = \{P_1, P_2, \ldots, P_m\} \subset \mathcal{C}$ set $g^{\bar{A}} = \oplus_{i=1}^{m} g^{P_i}$. Define $\hat{g}^{\bar{A}}$ to be the central extension of $g^{\bar{A}}$ by the cocycle

$$< x, y > = \sum_{i=1}^{m} \operatorname{Res}_{P_i} \operatorname{Tr} dx \cdot y.$$

In particular, we obtain the splitting

(13) $$\hat{g}^{\bar{A}} = g^{\bar{A}} \oplus \mathbf{C} \cdot c.$$

Consider a finite set $A = \{(P_1, b_1), \ldots, (P_m, b_m)\}$ where $P_i \in \mathcal{C}$ are pairwise different and b_i is a Borel subalgebra of the algebra of traceless linear transformations of the fiber $\rho^{-1}P_i$ $(1 \leq i \leq m)$. Let $\bar{A}$ be the projection of A on $\mathcal{C}$. Set $\hat{g}^A = \hat{g}^{\bar{A}}$.

3.1.2. Given A as above, set $n_i = [b_i, b_i]$. Denote by $\hat{g}_>^A$ the subalgebra consisting of sections $x(\cdot)$ such that $x(P_i) \in n_i$, $1 \leq i \leq m$, and by $\hat{g}_\geq^A$ the subalgebra spanned by the space of sections $x(.)$ such that $x(P_i) \in b_i$, $1 \leq i \leq m$, and the central element c. These are analogues of the maximal "nilpotent" and maximal "solvable" subalgebras for $\hat{g}^A$, c.f.2.2.

Denote by $\mathcal{O}_k^A$, $k \in \mathbf{C}$, the category of finitely generated $\hat{g}^A$-modules satisfying the conditions:

(i) c acts as multiplication by k;

(ii) the action of the subalgebra $\hat{g}_>^A$ is locally finite.

In much the same way as in 2.2 one defines Verma and generalized Weyl modules over $\hat{g}^A$:

Definition.

(i) We will say that (λ, k) is a highest weight of $\hat{g}^A$ if λ is a functional on $\oplus_i b_i / n_i$ and k is a number.

(ii) A highest weight (λ, k) naturally determines a character of $\hat{g}_\geq^A$ sending c to k and annihilating $\hat{g}_>^A$. Denote by $\mathbf{C}_{\lambda,k}$ the corresponding 1-dimensional representation.

(iii) Define the Verma module $M_{\lambda,k}^A$ to be the induced representation

$$\operatorname{Ind}_{\hat{g}_\geq^A}^{\hat{g}^A} \mathbf{C}_\lambda. \quad \square$$

There is an isomorphism

$$M_{\lambda,k}^A \approx \otimes_{i=1}^m M_{\lambda_i,k}^{P_i, b_i}.$$

Suppose now that each $M_{\lambda_i,k}^{P_i, b_i}$ has at least one singular vector. If $k \in \mathbf{C} \setminus \mathbf{Q}$ then this singular vector is unique for each i. Quotienting out all of them one obtains the *generalized Weyl module* $V_{\lambda,k}^A$. As above there is an isomorphism

$$V_{\lambda,k}^A \approx \otimes_{i=1}^m V_{\lambda_i,k}^{P_i, b_i}.$$

If k is not a rational number then any generalized Weyl module is irreducible. Denote by $\tilde{\mathcal{O}}_k$ the full subcategory of $\mathcal{O}_k$ consisting of all $\hat{g}^A$−modules whose composition series consist of generalized Weyl modules. Again if k is not a rational number then $\tilde{\mathcal{O}}_k$ is semisimple.

If k is rational then there arises the admissible representation $L_{\lambda,k}^A$ if (λ, k) is admissible. If the composition series of a module V^A consists only of admissible representations, then V^A is completely reducible.

3.1.3.   Let A be as in 3.1.1. Let $g(\mathcal{C}, A)$ be the Lie algebra of meromorphic sections of $\operatorname{End}\mathcal{E}$ holomorphic outside $\bar{A}$. The maps of restriction to formal neighborhoods give rise to the Lie algebra morphism

$$(14) \qquad\qquad g(\mathcal{C}, A) \to g^A$$

The splitting (13) provides us with the section $s_A : g^A \to \hat{g}^A$. Composition of (14) with s_A gives the linear morphism

$$(15) \qquad\qquad g(\mathcal{C}, A) \to \hat{g}^A.$$

The residue theorem implies that (15) is a Lie algebra morphism (even though s_A is not!).

By (15), the standard pullback makes each object of $M^A \in \mathcal{O}_k^A$ into a $g(\mathcal{C}, A)$-module. Hence there arises the space of coinvariants

$$(M^A)_{g(\mathcal{C}, A)} = M/g(\mathcal{C}, A)M.$$

3.2. **Localization of $\hat{g}^A$-modules.**

3.2.1.   Let V^A be a $\hat{g}^A-$module and suppose for simplicity that A consists of 1 element (P, b). Consider a family of data $\{P, \mathcal{E} \to \mathcal{C}\}$ – let us not care about Borel subalgebras for the moment. One expects that the corresponding family of vector spaces arranges then into a locally trivial vector bundle. An obstacle to obtaining this is that we have defined V^P up to an isomorphism but have not specified any such isomorphism. For example, an attempt to choose a basis in V^P requires one to choose (in particular) a local coordinate z at P, such that $z(P) = 0$. Different choices of z are essentially different as the group $\operatorname{Diff}(P)$ of diffeomorphisms of the formal neighborhood of P does not in general act on V^P. However the subgroup $\operatorname{Diff}(P)_1 \subset \operatorname{Diff}(P)$ of diffeomorphisms preserving the 1-jet of the parameter does act on V^P. We see that V^P, in fact, depends on the 1-jet of parameter at P.

To take care of Borel subalgebras, let us recall that to an $n-$dimensional vector space W one associates the flag manifold $F(W) = GL(n, \mathbf{C})/B$ and *the base affine space* $Base(W) = GL(n, \mathbf{C})/N$, where B is a Borel subgroup and N the unipotent subgroup of B. The natural map $Base(W) \to F(W)$ is a principal $(\mathbf{C}^*)^{\times n}$-bundle.

Similar arguments applied to b show that

the module $V^A = V^{P,\mathrm{b}}$ depends on the quadruple (P, b, j, x) such that j is the 1-jet of a parameter at P and $x \in Base(\mathbf{C}^n)$ belongs to the preimage of b.

One concludes that we do get a locally trivial vector bundle after pull-back to the space of pairs "1-jet of parameter at P, element of the maximal torus of the Borel group related to b". Let us now be more precise.

3.2.2. Let $\bar{\pi} : \mathcal{C}_S \to S$ be a family of smooth projective curves and $\rho_S : \mathcal{E}_S \to \mathcal{C}_S$ be a rank n vector bundle. There arise 2 more bundles:

(i) the bundle $Base(\rho_S) : Base(\mathcal{E}_S) \to \mathcal{C}_S$ with the fiber over any $x \in \mathcal{C}_S$ equal to the base affine space of the vector space $\rho_S^{-1}x$;

(ii) the $\mathbf{C}^*$−bundle $J^{(1)}(\mathcal{C}_S) \to \mathcal{C}_S$ of 1-jets of coordinates along the fibers of $\bar{\pi}$.

Consider the fiber product $Base(\mathcal{E}_S) \times_{\mathcal{C}_S} J^{(1)}(\mathcal{C}_S)$ and the natural map

$$\pi : Base(\mathcal{E}_S) \times_{\mathcal{C}_S} J^{(1)}(\mathcal{C}_S) \to S.$$

Pick a non-empty finite set A_S of sections of π satisfying the condition:

for any $s \in S$ the natural projection of the set $A_S(s) = \{a(s),\ a \in A_S\}$ on $\bar{\pi}^{-1}(s)$ is an injection.

Pick an arbitrary curve, say $\mathcal{C}_{s_0}$, from our family. Consider a highest weight module M^A over $\hat{g}^A$, where we write A instead of the lengthy $A_S(s_0)$; what follows is obviously independent of the choice of s_0.

By 3.1.2 and 3.2.1, we get a $\hat{g}^{A_S(s)}$−module $M^{A_S(s)}$ for any $s \in S$ and the collection $\{M^{A_S(s)},\ s \in S\}$ arranges into a locally trivial vector bundle. With each $s \in S$ we can further associate a vector space, that is, the space of coinvariants

$$\left(M^{A_S(s)}\right)_{\mathrm{g}(\pi^{-1}S, A_S(s))},$$

see 3.1.3.

Theorem 3.2.1. *Suppose the collection $\psi = (M^A, \pi, A_S)$, satisfying the conditions imposed above, is given. Then there is a twisted $\mathcal{D}$−module (that is, a sheaf of modules over a certain algebra of twisted differential operators) on S such that its fiber over $s \in S$ is $\left(M^{A_S(s)}\right)_{\mathrm{g}(\pi^{-1}S, A_S(s))}$.*

This theorem is an immediate consequence of [4] and [5, 8] . Briefly, the construction is as follows. Take a vector field ξ on $U \subset S$. It lifts to a meromorphic vector field on $\mathcal{C}_S - A_S(S)$ over U, and further to a meromorphic vector field on $\pi^{-1}(U) \subset Base(\mathcal{E}_S) \times_{\mathcal{C}_S} J^{(1)}(\mathcal{C}_S)$; denote this vector field by ξ^*. Trivializing the infinitesimal neighborhood of $A_S(U) \subset \mathcal{C}_S$ by choosing, locally with respect to $U \subset S$, coordinates in the fibers, one gets vertical components $\{\xi^*_{vert;i}\}$, such that $\xi^*_{vert;i}$ is the vertical component in the formal neighborhood of the i−th section. Projecting $\xi^*_{vert;i}$ on $Base(\mathcal{E}_S)$ one gets some element of $U(\mathrm{g})$, say u_i; projecting $\xi^*_{vert;i}$ on $J^{(1)}(\mathcal{C}_S)$ one gets some vector field, say v_i. Both

u_i, v_i act on our $\hat{g}^A$-module M^A: u_i naturally, v_i by means of the Sugawara construction. Going over the definitions one gets that this defines a twisted $\mathcal{D}$-module with the fiber as in the theorem.   □

Denote the constructed $\mathcal{D}$-module by $\Delta_\psi(M^A)$.

In the case when M^A is an admissible representation the following result is valid.

Theorem 3.2.2. *If $n = 2$ and M^A is an admissible $\hat{g}^A$-module then $\Delta_\psi(M^A)$ is* holonomic *for almost any vector bundle $\mathcal{E}_S$ (i.e. as a sheaf $\Delta_\psi(M^A)$ is isomorphic to the sheaf of sections of a certain finite rank vector bundle over some open set in S).*

Proof.

To prove this theorem essentially means to show that the spaces $(M^{A_S(s)})_{g(\pi^{-1}S, A_S(s))}$, $s \in S$, are all finite dimensional. This will be done in 4.3.3, Proposition 4.3.2 in the higher genus case and in 4.4.2, Proposition 4.4.2 for $\mathbf{CP^1}$. . We will also give there a precise meaning to the phrase "almost any vector bundle" in Theorem 3.2.2.   □

Results of 4.7 will show that the standard combinatorial algorithm can be used to calculate the dimension of the fiber of our $\mathcal{D}$-module using the dimensions of the spaces of coinvariants on a sphere with 3 punctures. The latter dimensions will be calculated in 4.5.4.

4. THE SPACES OF COINVARIANTS

In this section we will be concerned with the space of coinvariants $(M^A)_{g(\mathcal{C}, A)}$ (or spaces closely related to it) in the case when M^A is either a generalized Weyl module or an admissible representation. The standard tool to get finiteness results about coinvariants is the notion of *singular support*.

4.1. Singular support and coinvariants.
Let a be a Lie algebra. The universal enveloping algebra Ua is filtered in the standard way so that the associated graded algebra is Sa. Let V be an a-module carrying a filtration compatible with the filtration of Ua. Then the graded space, GrV, becomes naturally an Sa-module.

Definition The singular support, SSV, of V is the zero set of the vanishing ideal of the Sa-module $Gr\,V$.   □

Obviously, SSV is a conical subset of a*.

For a subalgebra n $\subset$ a, call V an (a, n)-module if it is an a-module and n acts on V locally nilpotently. Typical example: any module from the $\mathcal{O}$-category is a $(\hat{g}, \hat{g}_>)$-module.

Lemma 4.1.1. (see [4]) Let a be a Lie algebra and $p \subset a$ be a sub-algebra. Denote by $p^\perp$ the annihilator of p in a^*. Let V be a finitely generated (a, n)-module. If $SSM \cap p^\perp = \{0\}$ and $\dim a/n \oplus p < \infty$ then $\dim M_p < \infty$.

Recall that from now on $g = sl_2$ unless otherwise stated.

4.2. **Singular support of $\hat{g}^A$–modules.** Observe that there is an involution σ of $\hat{g}$ sending f to $e \otimes z^{-1}$ and $e \otimes z^{-1}$ to f, see 2.2 for notations. There arises the involution, also denoted by σ, acting on the algebras $\hat{g}^A$ and their duals. This involution is not canonical but we do not have to care, as our considerations here are purely local.

Denote by Ω^A the space of g-valued differential forms on the formal neighborhoods of the points from A. There is a natural embedding $\Omega^A \hookrightarrow (g^A)^*$ ("take the traces and then sum up all the residues!")

We will make use of 2 subspaces of Ω^A: Ω_{reg}^A is all regular forms and Ω_{nilp}^A is all forms with values in the nilpotent cone.

Theorem 4.2.1. (i) If M^A is a generalized Weyl module then $SSM^A = \Omega_{reg}^A \cup \sigma\Omega_{reg}^A$.

(ii) If M^A is an admissible representation then $SSM^A = \Omega_{nilp}^A \cup \sigma\Omega_{nilp}^A$.

Remark 4.2.2. (i) It is easy to see that although σ is not determined uniquely the spaces $\sigma\Omega_{reg}^A$, $\sigma\Omega_{nilp}^A$ are canonical. For example, $\sigma\Omega_{reg}^A$ is the space of forms such that:

they have at most order 1 poles at $\bar{A}$;

their residue at each $P_i \in \bar{A}$ belong to n_i;

at each $P_i \in \bar{A}$ their constant term belongs to b_i.

(ii) Statement (ii) of Theorem 4.2.1 is easy to invert using cohomological characterization of admissible representations obtained in [19]. We will not need this result in what follows.

Proof of Theorem 4.2.1.

First of all, as the statements are purely local, we will assume that $\#A = 1$ and forget about the curve.

(i) It follows from the definition of singular support that if V is an a-module on 1 generator v, then SSV is the zero set of all symbols of elements of $U(a)$ annihilating v. Fo example, if V is a Verma module over $\hat{g}$, then SSV consists of all $x \in \hat{g}^*$ satisfying $x(g) = 0$ for all $g \in \hat{g}_\geq$. This system is easy to solve and we get the following: SSV consists of all differential g–valued forms on the formal neighborhood of 0 having at most an order 1 pole at 0, the residue lying in $[b, b]$, where $b \subset g$ is the Borel from which our module is induced.

If, however, V is a generalized Weyl module, then one more equation is to be added, that is, the one coming from the unique singular vector S in the corresponding Verma module. The symbol of S is equal to $e_{-1}^m f^n$ (in natural notations). So we obtain one more equation and solving it get the desired result.

(ii) Here we have to consider 2 singular vectors and because of non-commutativity of the universal enveloping algebra the argument becomes more involved. Consider first what is known as the *vacuum module*. Such a module arises when the level $k + 2 = p/q$, and in notations of sect.2.4.2, (10) it is (V_{q-1}^1, V_{p-2}) , or, in more standard notation from representation theory (see the same section, a few lines before), $L_{0,k}$. An essential property of this module is that its generator is annihilated by g and so, by (i), $SSL_{0,k}$ consists of forms regular at 0. One more equation arises, the one coming from another singular vector in the corresponding Weyl module $V_{0,k}$. This vector can be calculated using (5), its weight is $(p-1)(-1, q)$.

Start with the case $p = 2$. The mentioned property of the vacuum module implies that under the action of g, this singular vector generates a g–submodule isomorphic to g. On the other hand, this singular vector is naturally identified with an element of $U(\oplus_{i>0} \text{g} \otimes z^{-i})$. (As we are dealing with a Weyl module, we can disregard elements of g.)

Lemma 4.2.3. *Symbol of the singular vector belongs to* $S^q(\text{g} \otimes z^{-1})$.

We postpone the proof of this lemma for a moment and use it to derive the statement of the theorem in this case. It follows that the symbols of the elements of the g–submodule generated by the singular vector determine an embedding of g in $S^q(\text{g} \otimes z^{-1})$. By the well-known theorem of B.Kostant there is only one such embedding. It is given by

$$\text{g} \ni g \mapsto C^{(q-1)/2} g,$$

where C is the quadratic Casimir element. (It is easy to see that for the vacuum representation under consideration q is necessarily odd.) Therefore we get dim g $= 3$ more equations and immediately see that the set of common zeros is given by 1: $C = 0$, where C is regarded as a quadratic function on g. As is well-known, the last equation determines the nilpotent cone of g. This proves the desired result at one point, 0; to complete the proof one notices that the submodule generated by the singular vector is closed under the differentiation d/dz (by the Sugawara construction).

If $p > 2$, then the arguments remain essentially unchanged. By looking at the singular vector formula (5) one realizes that the symbol of the singular vector simply gets raised to the $(p-1)$-st power, and instead of the embedding $g \to S^q(g \otimes z^{-1})$, the singular vector determines an embedding $V_{2(p-1)} \to S^{q(p-1)}(g \otimes z^{-1})$. So, instead of zeros of a collection of functions we get zeros of the $(p-1)$-st powers of the same collection of functions. The theorem for vacuum representations follows.

Proof of Lemma 4.2.3. This seemingly benign statement is not quite trivial. The equivalent reformulation is as follows: the degree (we set $deg\, g = 1$, $g \in \hat{g}$) of the symbol of the singular vector equals q. The easiest way to see that this is true is to apply this singular vector to an element of a certain module $\mathcal{F}_{\alpha,\beta}^{\mathbf{C}^*}$, and calculate the result using (12), see 2.6. Observe that in that formula the singular vector, S_{1q}^0, is understood as an element of $U(\hat{g}_<)$. To kill elements of g, we impose 1 linear condition on the parameters α, β, see again (12). One parameter survives, say α, and the result happens to be a polynomial of degree q. $\square$

This completes the proof of Theorem 4.2.1 for vacuum representations. To include other admissible representations we need 2 more ingredients: calculation of coinvariants for 3 admissible modules on the sphere, and a construction of annihilating ideals in admissible representations. The end of the proof is to be found in sect. 5.2.3.

4.3. **Finiteness of coinvariants – the higher genus case.**

4.3.1. *Hitchin's theorem.* First recall a well-known result of Hitchin [25]. To a vector bundle $\mathcal{E} \to \mathcal{C}$ associate the map

$$(16) \qquad H(\mathcal{E}): \; H^0(\mathcal{C}, \Omega \otimes \mathrm{End}\,\mathcal{E}) \to \oplus_{i=2}^n H^0(\mathcal{C}, \Omega^{\otimes i}),$$
$$X \mapsto \mathrm{Tr}\, X^i$$

Call a bundle $\mathcal{E}$ *exceptional* if $\ker H(\mathcal{E}) \neq 0$. Obviously $\ker H(\mathcal{E})$ is exactly the space of global differential forms with values in nilpotent endomorphisms of the vector bundle $\mathcal{E}$.

Theorem 4.3.1. *(Hitchin [25]) The zero set of the map (16) is a maximal Lagrangian submanifold in the cotangent bundle of the moduli space of vector bundles over $\mathcal{C}$. In particular, exceptional vector bundles form a positive codimension algebraic subset of the moduli space of vector bundles.*

For us, the importance of Theorem 4.3.1 is in that generically a vector bundle does not allow a non-trivial global differential form with coefficients in nilpotent endomorphisms of the bundle.

4.3.2. *Subtracting lines from rank 2 vector bundles.* An analogue of subtracting a point from a line bundle (or, rather, from its divisor) is the operation of subtracting a line from a rank 2 vector bundle.

To a rank 2 vector bundle $\mathcal{E} \to \mathcal{C}$ one can associate a module over the sheaf of regular functions – the sheaf of sections of $\mathcal{E}$; denote this sheaf by $Sect(\mathcal{E})$. This establishes a one-to-one correspondence between rank 2 vector bundles and rank 2 locally free modules over the sheaf of regular functions. Now fix a line, l, in the fiber of $\mathcal{E}$ over some point $P \in \mathcal{C}$. Denote by $S(l)$ the sheaf such that:

(i) $S(l)|_U = Sect(\mathcal{E})|_U$ if P does not belong to U;

(ii) $S(l)|_U, P \in U$, is the space of meromorphic sections of $\mathcal{E}$ over U regular outside P, having at most an order 1 pole at P and such that their residues at P belong to the fixed line l.

It is obvious that $S(l)$ is a rank 2 locally free module. Therefore it defines a rank 2 vector bundle. Denote this vector bundle by $\mathcal{E}(l)$. If a collection of lines $l_1, l_2, ..., l_m$ is subtracted, then denote the corresponding vector bundle by $\mathcal{E}(l_1 + \cdots + l_m)$.

Suppose we have a moduli space of rank 2 vector bundles with parabolic structure and fixed determinant. The elements of such a space are isomorphism classes of the data (vector bundle $\mathcal{E}$, fixed lines $l_1, ..., l_m$ in some fibers.) It is rather clear that the map $(\mathcal{E}, l_1, ..., l_m) \mapsto (\mathcal{E}(l_1 + \cdots + l_m), l_1, ..., l_m)$ is a homeomorphism of 2 moduli spaces with different determinants.

Definition. Call the data $(\mathcal{E}, l_1, ..., l_m)$ generic if $\mathcal{E}(l_{i_1} + \cdots + l_{i_s})$ is not exceptional for any subset $\{i_1, ..., i_s\} \subset \{1, 2, ..., m\}$. $\square$

It follows from Theorem 4.3.1 that the set of generic vector bundles is open and everywhere dense.

4.3.3. *Finiteness of coinvariants.* Suppose we are in the situation of 3.1.2: we have an admissible $\hat{g}^A$–module M^A on the curve $\mathcal{C}$ with a vector bundle $\mathcal{E} \to \mathcal{C}$. As A is a collection of Borel subalgebras $b_1, ..., b_m$ operating in fixed fibers, we have a parabolic structure – lines $l_1, ..., l_m$ in the corresponding fibers preserved by the b_i's. Call the data $(\mathcal{E}, A)$ generic if the data $(\mathcal{E}, l_1, ..., l_m)$ is generic in the sense of 4.3.2.

Recall that we are interested in the space of coinvariants $M^A_{g(\mathcal{C}, A)}$, where $g(\mathcal{C}, A)$ is an algebra of endomorphisms of the bundle $\mathcal{E}$ regular outside points from the corresponding $\bar{A}$, see 3.1.3 and 3.2.2, Theorem 3.2.1.

Proposition 4.3.2. *Let $(\mathcal{E}, A)$ be generic. Then*

$$dim M^A_{g(\mathcal{C}, A)} < \infty.$$

Proof. One extracts from the definitions that the annihilator $g(\mathcal{C}, A)^{\perp}$ of the algebra $g(\mathcal{C}, A)$ is the space $\Omega_{\mathcal{C},A}(\mathcal{E})$ of global meromorphic $End(\mathcal{E})$−valued differential forms regular outside $\bar{A} \subset \mathcal{C}$.

By Theorem 4.2.1(ii) we get that $SSM^A \cap g(\mathcal{C}, A)^{\perp} = \Omega_{nilp}(\mathcal{E}) \cup \sigma\Omega_{nilp}(\mathcal{E})$, where $\Omega_{nilp}(\mathcal{E})$ is the space global nilpotent transformations of $\mathcal{E}$, and σ is the twist introduced in 4.2.

The genericity condition means that $\Omega_{nilp}(\mathcal{E}) = 0$, see 4.3.1 and 4.3.2.

On the other hand it is easy to see that the operation of subtracting a line generates the twist σ on endomorphisms. (In fact one has to compose subtracting of a line with a reflection in the fiber, but this does not change the isomorphism class of the bundle.) Therefore the genericity condition also implies that $\sigma\Omega_{nilp}(\mathcal{E}) = 0$.

Hence we get that $SSM^A \cap g(\mathcal{C}, A)^{\perp} = 0$. And, as the space $\hat{g}^A_{>} + g(\mathcal{C}, A)$ is of finite codimension in $\hat{g}^A$, application of Lemma 4.1.1, see 4.1, completes the proof.   □

In order to study quadratic degenerations we will need the following stronger finiteness result. Along with the set $A = \{(P_1, b_1), ..., (P_m, b_m)\}$, consider the set $A_2 = \{(P_{m+1}, b_{m+1}), (P_{m+2}, b_{m+2})\}$ such that the points $P_1, ..., P_{m+2} \in \mathcal{C}$ are different. Denote by $g(\mathcal{C}, A, A_2)$ the sub-algebra of $g(\mathcal{C}, A)$ consisting of functions taking values in $n_i = [b_i, b_i]$ at the point P_i, $i = m+1, m+2$.

Proposition 4.3.3. *If $(\mathcal{E}, A \bigsqcup A_2)$ is generic and M^A is admissible, then*

$$dim\, (M^A)_{g(\mathcal{C},A,A_2)} < \infty.$$

Proof. We are again going to apply Lemma 4.1.1. Observe that $g(\mathcal{C}, A, A_2)^{\perp}$ consists of meromorphic forms on $\mathcal{C}$ with values in $End(\mathcal{E})$, regular outside $\{P_1, ..., P_{m+2}\} \subset \mathcal{C}$, having at most order 1 poles at P_{m+1}, P_{m+2}, their residues at the latter points lying in b_1 (b_2 resp.).

By Theorem 4.2.1(ii), $g(\mathcal{C}, A, A_2)^{\perp} \cap SSM^A$ consists of forms with values in nilpotent endomorphisms, satisfying the above listed global conditions. This implies, in particular, that actually the residues of our forms belong to n_{m+1}, n_{m+2} at P_{m+1}, P_{m+2} resp..

Given an element $\omega \in g(\mathcal{C}, A, A_2)^{\perp} \cap SSM^A$, subtract some lines from $\mathcal{E}$ so as to make ω be everywhere regular. The genericity condition then implies that $\omega = 0$, and application of Lemma 4.1.1 completes the proof.   □

4.4. Finiteness of coinvariants − the case of $\mathbf{CP^1}$.

4.4.1. *Generic vector bundles on* $\mathbf{CP^1}$. Let $O(n)$ be the degree n line bundle over $\mathbf{CP^1}$. It is known, e.g. [39], that any rank 2 vector bundle over $\mathbf{CP^1}$ is a direct sum $O(r) \oplus O(s)$ for some r, s.

As there are no moduli, it is hard to speak about generic vector bundles. Nevertheless we will call $O(r) \oplus O(s)$ *exceptional* if $|r - s| > 1$. Here is a justification.

Lemma 4.4.1. *Let* $\mathcal{E} = O(r) \oplus O(s)$ *and* $(\mathcal{E}, l_1, ..., l_m)$, $m \geq |r - s|$, *a vector bundle with parabolic structure. Then generically with respect to* $l_1, ..., l_m$ *the bundle* $\mathcal{E}(l_1 + \cdots + l_m)$ *is not exceptional:*

$$\mathcal{E}(l_1 + \cdots + l_m) = \begin{cases} O(p+1) \oplus O(p) & \text{if } r + s - m = 2p + 1 \\ O(p) \oplus O(p) & \text{if } r + s - m = 2p. \end{cases}$$

Lemma 4.4.1 seems to be common knowledge, although we failed to find a reference with its proof.

Proceed just like we did in 4.3.2: call $(\mathcal{E}, l_1, ..., l_m)$ generic if $\mathcal{E}(l_{i_1} + \cdots + l_{i_s})$ is not exceptional for any subset $\{i_1, ..., i_s\} \subset \{1, 2, ..., m\}$.

4.4.2. *Finiteness of coinvariants.* A specific feature of the genus zero case is that we do not necessarily have to consider admissible representations – generalized Weyl modules, see 2.4.1, will also do.

Let us again consider a vector bundle $\mathcal{E}$ over $\mathbf{CP^1}$ and a $\hat{\mathfrak{g}}^A$−module M^A. As in 4.3.3, A determines a parabolic structure on $\mathcal{E}$, say $(\mathcal{E}, l_1, ..., l_m)$. Call the data $(\mathcal{E}, A)$ generic if $(\mathcal{E}, l_1, ..., l_m)$ is also.

Proposition 4.4.2. *If* $(\mathcal{E}, A)$ *is generic and* M^A *is either admissible or generalized Weyl module, then*

$$dim \, (M^A)_{g(\mathbf{CP^1}, A)} < \infty.$$

Proof It is a simplified version of the proof of Proposition 4.3.2 in 4.3.3. The new features are as follows: to include generalized Weyl modules one uses Theorem 4.2.1(i) in addition to Theorem 4.2.1(ii); instead of Hitchin's theorem one uses the "observation" that $O(n)$ has no non-zero global sections if $n < 0$. □

Corollary 4.4.3. *If* M^A *is a generalized Weyl module then there is a holonomic twisted* $\mathcal{D}$-*module living in the space* $(J^{(1)}(\mathbf{CP^1}) \times J^{(1)}(\mathbf{CP^1}))^{\times m}$ *with the fiber* $M^A_{g(\mathbf{CP^1}, A)}$.

Proof. Repeating word for word proof of Theorem 3.2.2 one derives from Proposition 4.4.2 existence of a twisted $\mathcal{D}$−module on the space $(Base(\mathbf{C^2}) \times J^{(1)}(\mathbf{CP^1}))^{\times m}$. But for sl_2, the flag manifold is $\mathbf{CP^1}$ and the base affine space $Base(\mathbf{C^2})$ coincides with $J^{(1)}(\mathbf{CP^1}))$. □

As in 4.3.3, we want to prove a generalization of Proposition 4.4.2 in order to prepare grounds for studying quadratic degeneration.

Along with $A = \{(P_1, b_1), \ldots, (P_m, b_m)\}$ consider 2 sets $A_1 = \{(P_{m+1}, b_{m+1})\}$ and $A_2 = \{(P_{m+1}, b_{m+1}), (P_{m+2}, b_{m+2})\}$ such that $P_1, \ldots, P_{m+2}$ are different points in $\mathcal{C}$.

With A_1 and A_2 associate the following 2 subalgebras of g($\mathbf{CP^1}, A$): g($\mathbf{CP^1}, A, A_1$) consists of all functions taking values in $n_{m+1} = [b_{m+1}, b_{m+1}]$ at the point P_{m+1}; g($\mathbf{CP^1}, A, A_2$) consists of all functions taking values in $n_i = [b_i, b_i]$ at the point P_i, $i = m + 1$, $m + 2$.

Proposition 4.4.4. *Let* $(\mathcal{E}, A_2)$ *be generic. Then*
   (i) *If* M^A *is a generalized Weyl module over* $\hat{g}^A$, *then* $dim(M^A)_{g(\mathbf{CP^1}, A, A_1)} < \infty$;
   (ii) *If* M^A *is an admissible representation of* $\hat{g}^A$, *then* $dim(M^A)_{g(\mathbf{CP^1}, A, A_2)} < \infty$.

Proof of (ii) repeats almost word for word that of Proposition 4.3.3 in 4.3.3 with simplifications analogous to those indicated in the proof of Proposition 4.4.2.

As to (i), its proof is again an application of the same technique in a slightly different form: one has to take a form $\omega \in$ g($\mathbf{CP^1}, A, A_2$)$^\perp \cap SSM^A$ and to subtract lines from $\mathcal{E}$ so as to make ω into a form with either one pole (at P_{m+1}) or 2 poles (one of them is again at P_{m+1}) in such a way that the bundle obtained is $O(n) \oplus O(n)$. The 2 cases are of course distinguished by the parity of the difference between the degrees of the determinant of $\mathcal{E}$ and ω. In both cases it is easy to prove that $\omega = 0$ using the fact that any differential form with trivial coefficients has at least 2 poles.   □

4.4.3. *Holonomic $\mathcal{D}$-module on* $(\mathbf{C} \times \mathbf{C})^{\times m}$. We will now get rid of twisted differential operators in Corollary 4.4.3 under the assumption that the vector bundle $\mathcal{E} \to \mathbf{CP^1}$ is trivial. Consider the set $A' = A \bigsqcup (P_\infty, b_\infty)$. Attach to the point (P_∞, b_∞) the module (V_0^0, V_0) known as the vacuum representation, see 2.4.1 for notations. (P_∞, b_∞) can be redefined as the module induced from the trivial representation (see also 2.4.1) and therefore there is an isomorphism $M^A_{g(\mathbf{CP^1}, A)} \approx M^{A'}_{g(\mathbf{CP^1}, A')}$. Now consider the twisted D–module with fiber $M^{A'}_{g(\mathbf{CP^1}, A')}$ on the space $(J^{(1)}(\mathbf{CP^1}) \times J^{(1)}(\mathbf{CP^1}))^{\times m+1}$. Restrict it to the space $(J^{(1)}(\mathbf{CP^1}) \times J^{(1)}(\mathbf{CP^1}))^{\times m}$ by fixing the point (P_∞, b_∞). The result of this operation is that the bundles in question trivialize: $\mathbf{CP^1} - b_\infty = \mathbf{CP^1} - P_\infty = \mathbf{C}$ and $J^{(1)}(\mathbf{C}) = \mathbf{C}^* \times \mathbf{C}$. Using this trivialization one gets a $\mathcal{D}$–module over the space $(\mathbf{C} \times \mathbf{C})^m$. Observing that it is the appearance of the bundle $J^{(1)}(\mathbf{CP^1}) \to \mathbf{CP^1}$ which was responsible for the twisting of the $\mathcal{D}$−module, one argues that we get a usual holonomic $\mathcal{D}$-module on $(\mathbf{C} \times \mathbf{C})^m$ with fiber $M^A_{g(\mathbf{CP^1},A)}$. In particular, we get a bundle with flat connection over an open subset of $(\mathbf{C} \times \mathbf{C})^m$.

Notation. Denote the bundle with flat connection constructed in this way by $\Delta(M^A)$. $\quad\square$

We are unable to describe this open subset explicitly at present. However it follows from the requirement that $(\mathcal{E}, A)$ be generic in all our finiteness results that the diagonals should be thrown away, meaning that $P_i \neq P_j$ and $\mathbf{b}_i \neq \mathbf{b}_j$ for all $i \neq j$.

One may want to write down differential equations satisfied by horizontal sections of this bundle. We will show in 5.3 that horizontal sections satisfy a system of $2m$ differential equations of which m equations are Knizhnik-Zamolodchikov equations and the other m are obtained from singular vectors of the Verma module projecting onto M^A.

Everything said here holds true for an admissible representation. It is easy to see that the bundle associated with an admissible representation is a quotient of the just constructed bundle for the corresponding generalized Weyl module.

4.5. Calculation of the dimensions of coinvariants. Fusion algebra. Let $\mathcal{E} \to \mathbf{CP^1}$ be the rank 2 trivial vector bundle and M^A be a $\hat{g}^A$−module. Here we will calculate the dimension of the space $(M^A)_{g(\mathbf{CP^1},A)}$, $\sharp A = 3$, in the following 2 cases: (i) the level k is not rational and M^A is a generalized Weyl module; (ii) $k+2 = p/q$, p and q being positive integers, and M^A is an admissible representation. Without loss of generality we can:

fix a coordinate z on $\mathbf{CP^1}$; assume that $A = \{(0, \mathbf{b}_0), (1, \mathbf{b}_1), (\infty, \mathbf{b}_\infty)\}$, where $\mathbf{b}_0 = \mathbf{C}e \oplus \mathbf{C}h$, $\mathbf{b}_\infty = \mathbf{C}f \oplus \mathbf{C}h$ and $\mathbf{b}_1 = \mathbf{C}(e-h-f) \oplus \mathbf{C}(h+2f)$.

(In fact, for any $\mathbf{b}_0 \neq \mathbf{b}_\infty$ we can always choose a basis of g so that $\mathbf{b}_0, \mathbf{b}_\infty$ are as above. As to $\mathbf{b}_1$, there really is some freedom but it is easy to see that all the calculations below are independent of the choice. We have set $\mathbf{b}_1 = (\exp f)\mathbf{b}_0(\exp -f)$.)

4.5.1. *The generic level case.* So by 3.1.2 we are given three irreducible generalized Weyl modules $V^0_{\lambda_0,k}, V^1_{\lambda_1,k}, V^\infty_{\lambda_\infty,k}$. Recall, see 2.4.1, that generalized Weyl modules are parametrized by symbols (V^ϵ_m, V_n), where m, n are nonnegative integers, $\epsilon \in \mathbf{Z}/2\mathbf{Z}$ and V_m is an $m+1$-dimensional g−module. Therefore we can and will assume that we have

$$(V^{\epsilon_i}_{m_i}, V_{n_i}), \ i = 0, 1, \infty.$$

It is convenient to interpret the result of calculation of $\dim (\otimes_i(V_{m_i}^{\epsilon_i}, V_{n_i}))_{\mathrm{g}(\mathbf{CP^1},A)}$ in terms of the *fusion algebra*. The latter is defined as follows. Suppose that for any pair of generalized Weyl modules, say $(V_{r_i}^{\alpha_i}, V_{s_i})$, $i = 0, 1$, there is only finite number of $(V_{r_\infty}^{\alpha_\infty}, V_{s_\infty})$ such that

$$\dim (\otimes_{i=0,1,\infty}(V_{r_i}^{\alpha_i}, V_{s_i}))_{\mathrm{g}(\mathbf{CP^1},A)} \neq 0.$$

Now view the symbols (V_m^ϵ, V_n) as generators of a free abelian group. Then there naturally arises an algebra (over $\mathbf{Z}$) with the operation of multiplication $\circ$ defined by

$$(17) \quad (V_{r_0}^{\alpha_0}, V_{s_0}) \circ (V_{r_1}^{\alpha_1}, V_{s_1})$$
$$= \sum_{(r_\infty, s_\infty, \alpha_\infty)} \dim \{(\otimes_{i=0,1,\infty}(V_{m_i}^{\epsilon_i}, V_{n_i}))_{\mathrm{g}(\mathbf{CP^1},A)}\}(V_{r_\infty}^{\alpha_\infty}, V_{s_\infty}).$$

The algebra defined in this way is called the *fusion algebra*. Of course the structure constants of the fusion algebra determine the dimensions of the spaces of coinvariants.

One last piece of notation: in the following theorem we formally set $(X \oplus Y, Z) = (X, Z) + (Y, Z)$ and $(X, Y \oplus Z) = (X, Y) + (X, Z)$. Recall also that in the category of g−modules one has

$$V_r \otimes V_s \approx V_{r+s} \oplus V_{r+s-2} \oplus \cdots \oplus V_{|r-s|}.$$

Theorem 4.5.1. *(i) For any triple of generalized Weyl modules the space $(V_{m_i}^{\epsilon_i}, V_{n_i}))_{\mathrm{g}(\mathbf{CP^1},A)}$ is finite dimensional.*

(ii) The fusion algebra is well-defined, multiplication being given by the following formula

$$(18) \quad (V_{r_1}^\alpha, V_{s_1}) \circ (V_{r_2}^\beta, V_{s_2}) =$$
$$(V_{r_1+r_2}^{\alpha+\beta}, V_{s_1} \otimes V_{s_2}) + (V_{r_1+r_2-1}^{\alpha+\beta+1}, V_{s_1} \otimes V_{s_2}) + (V_{r_1+r_2-2}^{\alpha+\beta}, V_{s_1} \otimes V_{s_2})$$
$$+ \cdots + (V_{|r_1-r_2|}^{\alpha+\beta}, V_{s_1} \otimes V_{s_2}).$$

4.5.2. Proof of Theorem 4.5.1.

Throughout the proof A will stand for $\{(0, b_0), (1, b_1)\}$, A_1 for $\{(\infty, b_\infty)\}$. Along with the algebras $\mathrm{g}(\mathbf{CP^1}, A)$, $\mathrm{g}(\mathbf{CP^1}, A, A_1)$ (see 4.4) introduce the algebra $\bar{\mathrm{g}}(\mathbf{CP^1}, A, A_1) \subset \mathrm{g}(\mathbf{CP^1}, A)$ consisting of all functions taking values in b_∞ at the point ∞.

Of course $\mathrm{g}(\mathbf{CP^1}, A, A_1) \subset \bar{\mathrm{g}}(\mathbf{CP^1}, A, A_1)$ is an ideal and $\dim \bar{\mathrm{g}}(\mathbf{CP^1}, A, A_1)/\mathrm{g}(\mathbf{CP^1}, A, A_1) = 1$. Define $\bar{h}_\infty$ to be a basis element of $\bar{\mathrm{g}}(\mathbf{CP^1}, A, A_1)/\mathrm{g}(\mathbf{CP^1}, A, A_1)$. It is a standard (and simple) fact of the Lie algebra cohomology theory that $\bar{h}_\infty$ acts on $(M^A)_{\mathrm{g}(\mathbf{CP^1},A,A_1)}$.

Lemma 4.5.2. *Let M^A be a generalized Weyl module. The element $\bar{h}_\infty$ has a simple spectrum as an operator acting on $(M^A)_{\mathrm{g}(\mathbf{CP^1},A,A_1)}$. Further, if $M^A = (V^\alpha_{r_1}, V_{s_1}) \otimes (V^\beta_{r_2}, V_{s_2})$ then the set of eigenvalues of $\bar{h}_\infty$ is the set of the highest weights of the modules appearing in the right-hand side of Theorem 4.5.1(ii).*

The proof of this lemma is essentially the same as that of Theorem 4.4 in [19] and mainly consists of solving a system of 2 equations related to 2 singular vectors – one in $(V^\alpha_{r_1}, V_{s_1})$, the other in $(V^\beta_{r_2}, V_{s_2})$. We will discuss it in 4.5.3. The derivation of Theorem 4.5.1 from Lemma 4.5.2 is again very similar to that of Theorem 3.2 from Theorem 4.4 in *loc. cit* and uses Verma modules as follows.

Lemma 4.5.3. *(i) Let $(M^A)^\mu_{\mathrm{g}(\mathbf{CP^1},A,A_1)} \in (M^A)_{\mathrm{g}(\mathbf{CP^1},A,A_1)}$ be the eigenspace associated to the eigenvalue μ of $\bar{h}_\infty$. Then $(M^A)^\mu_{\mathrm{g}(\mathbf{CP^1},A,A_1)} \approx (\bar{M}^{\bar{A}} \otimes M^{\infty,\mathrm{b}_\infty}_{\mu,k})_{\mathrm{g}(\mathbf{CP^1},A\cup A_1)}$.*

(ii) Projection of a Verma module $M^{\infty,\mathrm{b}_\infty}_{\mu,k}$ onto a generalized Weyl module W induces an isomorphism of the coinvariants

$$(M^A \otimes M^{\infty,\mathrm{b}_\infty}_{\mu,k})_{\mathrm{g}(\mathbf{CP^1},A\cup A_1)} \approx (M^A \otimes W)_{\mathrm{g}(\mathbf{CP^1},A\cup A_1)}.$$

Proof of Lemma 4.5.3

(i) A Verma module sitting at a point is induced from the 1-dimensional representation of the algebra of functions on the formal disk whose value at the point belongs to the corresponding Borel subalgebra. Therefore (i) follows from Frobenius duality.

(ii) Consider the resolution of W by Verma modules (see 2.4.1, formula (6)):

$$0 \to M \to M^{\infty,\mathrm{b}_\infty}_{\mu,k} \to W \to 0$$

and tensor it with M^A. There arises the long exact sequence of homology groups of which we consider the following part:

$$(19) \quad (M^A \otimes M)_{\mathrm{g}(\mathbf{CP^1},A\cup A_1)} \to (M^A \otimes M^{\infty,\mathrm{b}_\infty}_{\mu,k})_{\mathrm{g}(\mathbf{CP^1},A\cup A_1)}$$
$$\to (M^A \otimes W)_{\mathrm{g}(\mathbf{CP^1},A\cup A_1)} \to 0.$$

Since $M^{\infty,\mathrm{b}_\infty}_{\mu,k}$ projects onto a Weyl module, the Verma module M does not, see 2.4.1. Lemma 4.5.2 now gives that $(M^A \otimes M)_{\mathrm{g}(\mathbf{CP^1},A\cup A_1)} = \{0\}$. $\square$

To complete the proof of Theorem 4.5.1 observe that Lemma 4.5.2 and Lemma 4.5.3 together is a reformulation of Theorem 4.5.1. $\square$

Corollary 4.5.4. *Let $\sharp A = 1$ and let A_1 and A_2 be as in 4.4. The following conditions are equivalent:*

(i) M^A is a direct sum of generalized Weyl modules;

(ii)$SSM^A = \Omega^A_{reg} \cup \sigma\Omega^A_{reg}$;

(iii) For any Verma module W^{A_1}, $\dim(M^A \otimes W^{A_1})_{g(\mathbf{CP^1}, A \cup A_1, A_2)} < \infty$.

4.5.3.   Here we sketch the proof of Lemma 4.5.2. First of all replace M^A with the corresponding Verma module $\bar{M}^A$. Then pass from the space $(\bar{M}^A)_{g(\mathbf{CP^1}, A, A_1)}$ to its dual, that is, to the space of $g(\mathbf{CP^1}, A, A_1)$–invariant functionals on $\bar{M}^A$. Choose $h \otimes (1 - z^{-1})$ to be a representative of $\bar{h}_\infty$. Let Ψ be the eigenvector of $h \otimes (1 - z^{-1})$. By definition Ψ is a linear functional on $M^{0, b_0}_{\lambda_0, k} \otimes M^{1, b_1}_{\lambda_1, k}$. It is an excersise in Frobenius duality to show that such a functional exists and unique.

Define F to be the following linear functional on $M^{0, b_0}_{\lambda_0, k}$: $F(w) = \Psi(w \otimes v_{\lambda_1})$, where, as usual, v_{λ_1} is the vacuum vector of $M^{1, b_1}_{\lambda_1, k}$. As $M^{0, b_0}_{\lambda_0, k}$ is $\mathbf{Z}_+ \times \mathbf{Z}_+$–graded (see 2.2), we denote by F_{ij} the restriction of F to the (i, j)–component. Direct calculations show that with respect to the natural action of $\hat{g}$ on $M^{0, b_0}_{\lambda_0, k}$:

$$(20) \qquad\qquad \oplus_{i, j \in \mathbf{Z}} \mathbf{C} F_{ij} \approx \mathcal{F}^{\mathbf{C}^*}_{\alpha\beta}$$

$$(21) \qquad\qquad \text{where } \alpha = \frac{\lambda_\infty - \lambda_1 - \lambda_0 - 2}{2}, \; \beta = \lambda_1$$

The functional F factors through the projection $M^{0, b_0}_{\lambda_0, k} \to V^{0, b_0}_{\lambda_0, k}$ if and only if it vanishes on the singular vector of $M^{0, b_0}_{\lambda_0, k}$. In other words, if this singular vector, say S, has degree (i, j) then the following equation holds

$$S F_{ij} = 0.$$

The latter equation can be written down and solved explicitly using formulas (11 or 12). Similar arguments go through for the module $M^{1, b_1}_{\lambda_1, k}$ giving another equation, say

$$S' F_{i'j'} = 0.$$

Simultaneous solutions to these 2 equations give the desired result. By the way, as (11), (12) show, each of the expressions $S F_{ij}$, $S' F_{i'j'}$ splits in a product of linear factors; therefore geometrically the solution is a collection of intersection points of 2 families of lines in the plane. $\square$

4.5.4. *The rational level case.* Suppose $k + 2 = p/q$, p and q being positive integers. Now instead of 3 generalized Weyl modules sitting at 3 points in $\mathbf{CP^1}$ we are given 3 admissible representations sitting at 3 points on $\mathbf{CP^1}$. Recall, see 2.4.2, that admissible representations are parametrized by symbols (V_m^ϵ, V_n), $0 \le m \le q - 1, 0 \le n \le p - 2$ modulo the relation $(V_m^\epsilon, V_n) = (V_{q-1-m}^{\epsilon+1}, V_{p-2-n})$. Denote by $(V_m^\epsilon, V_n)^\sim$ the equivalence class of (V_m^ϵ, V_n). We assume that $(V_m^\epsilon, V_n)^\sim$ satisfies the same bilinear condition (V_m^ϵ, V_n) as in Theorem 4.5.1 does.

The definition of the fusion (Verlinde) algebra in this case repeats word for word that in 4.5.1.

Recall finally that the Kazhdan-Lusztig fusion functor [30] gives

$$V_r \dot{\otimes}_k V_s = V_{|m-n|} \oplus V_{|m-n|+2} \cdots \oplus V_{\min\{2k-r-s,r+s\}}.$$

The following theorem was proved in [19] in an equivalent but much less illuminating form.

Theorem 4.5.5. *(i) For any triple of admissible representations the space $(V_{m_i}^{\epsilon_i}, V_{n_i}))_{\mathrm{g}(\mathbf{CP^1}, A)}$ is finite dimensional.*

(ii) The fusion algebra is well-defined, multiplication being given by the following formula:

$$(V_{r_1}^\alpha, V_{s_1})^\sim \circ (V_{r_2}^\beta, V_{s_2})^\sim \; =$$

$$(V_{|r_1-r_2|}^{\alpha+\beta}, V_{s_1} \dot{\otimes}_{p-2} V_{s_2})^\sim \quad + \quad (V_{|r_1-r_2|+1}^{\alpha+\beta}, V_{s_1} \dot{\otimes}_{p-2} V_{s_2})^\sim +$$

$$(V_{|r_1-r_2|+2}^{\alpha+\beta}, V_{s_1} \dot{\otimes}_{p-2} V_{s_2})^\sim \quad + \quad \cdots + (V_N^{\alpha+\beta}, V_{s_1} \dot{\otimes}_{p-2} V_{s_2})^\sim,$$

where $N = min\{2q - 2 - r - s, r + s\}$.

It is an easy exercise to derive this theorem from Theorem 4.5.1. For future purposes, however, we now sketch its original proof. Set $A = \{(\infty, b_\infty)\}$, $A_2 = \{(0, b_0), (1, b_1)\}$. In addition to the algebras $\mathrm{g}(\mathbf{CP^1}, A), \mathrm{g}(\mathbf{CP^1}, A, A_2)$ as in 4.4, we introduce an algebra $\bar{\mathrm{g}}(\mathbf{CP^1}, A, A_2) \subset \mathrm{g}(\mathbf{CP^1}, A)$. The latter consists of all functions whose values at the points 0 (1 resp.) belong to b_0 (b_1 resp.). Obviously $\mathrm{g}(\mathbf{CP^1}, A, A_2) \subset \bar{\mathrm{g}}(\mathbf{CP^1}, A, A_2)$ is an ideal and the quotient algebra $\bar{\mathrm{g}}(\mathbf{CP^1}, A, A_2)/\mathrm{g}(\mathbf{CP^1}, A, A_2)$ is commutative and 2-dimensional. This algebra naturally operates on the space $(M^A)_{\mathrm{g}(\mathbf{CP^1}, A, A_2)}$. Let $\bar{h}_0, \bar{h}_1$ be a basis of $\bar{\mathrm{g}}(\mathbf{CP^1}, A, A_2)/\mathrm{g}(\mathbf{CP^1}, A, A_2)$.

Lemma 4.5.6. *([19])*

(i) $dim(M^A)_{\mathrm{g}(\mathbf{CP^1}, A, A_2)} < \infty$ if and only if M^A is an admissible representation.

(ii) Let M^A be an admissible representation. The elements $\bar{h}_0, \bar{h}_1$ have simple spectra as operators acting on $(M^A)_{\mathrm{g}(\mathbf{CP^1}, A, A_2)}$. Their eigenvalues recover the structure constants of the fusion algebra.

"Inserting" Verma modules and using the BGG resolution one derives Theorem 4.5.5 from Lemma 4.5.6 in a way similar to that we used in 4.5.1.

Another important corollary of Lemma 4.5.6 is as follows.

Corollary 4.5.7. *Let* $\sharp A = 1$. *The following conditions are equivalent:*
(i) M^A *is a sum of admissible representations;*
(ii) $SSM^A = \Omega^A_{nilp} \cup \sigma\Omega^A_{nilp}$;
(iii) $\dim(M^A)_{g(\mathbf{CP^1},A,A_2)} < \infty$.

4.5.5. *Classical and quantum* $osp(1|2)$. *Fusion algebra as a Grothendieck ring.* **A.** $osp(1|2)$ is a rank 1 superalgebra – one of the superanalogues of sl_2. It can be defined as an algebra on 2 odd generators, x_+, x_-, one even generator, h, and relations

$$[x_+, x_-] = h, \quad [h, x_{\pm}] = \pm x_{\pm}.$$

The even part of this algebra is sl_2 and is generated by $x_{\pm}^2$; the odd part is V_1 as an sl_2- module, its basis is $\{x_+, x_-\}$.

From this it is easy to obtain the following classification of all simple finite dimensional $osp(1|2)-$modules. (It is even simpler to do this in a way modelling the sl_2-case – by starting with Verma modules and then quotienting out a singular vector; for details see [32]). Each $osp(1|2)-$module W is a sum of an even and odd part $W = {}^{even}W \oplus {}^{odd}W$; each component is an sl_2-module, i.e. a direct sum of V_n's. These are generalities. But in reality each irreducible $osp(1|2)-$module is of one of the 2 following types:

$$V_n^0 \text{ such that } {}^{even}V_n^0 = V_n, \quad {}^{odd}V_n^0 = V_{n-1};$$

$$V_n^1 \text{ such that } {}^{even}V_n^0 = V_{n-1}, \quad {}^{odd}V_n^0 = V_n.$$

The fact that the dimensions of the even and odd parts differ by 1 is a consequence of the fact that the odd part of the algebra is V_1.

We see that each irreducible $osp(1|2)-$module is odd-dimesional; further V_n^0 and V_n^1 are isomorphic as modules and obtained from each other by the change of parity. This is the category of finite dimensional representations of $osp(1|2)$; denote it by $Rep(osp(1|2))$. As in the sl_2-case, one proves that $Rep(osp(1|2))$ is semisimple.

The universal enveloping algebra $Uosp(1|2)$ is in fact a Hopf algebra, for example the comultiplication is given by the standard formula $g \mapsto g \otimes 1 + 1 \otimes g$, $g \in osp(1|2)$. This makes $Rep(osp(1|2))$ a tensor category: $Rep(osp(1|2)) \times Rep(osp(1|2)) \to Rep(osp(1|2))$, $A, B \mapsto A \otimes B$, where the $osp(1|2)-$module structure on $A \otimes B$ is determined through the comultiplication (and the rule of signs!). Decomposing the tensor product of 2 irreducible modules one gets the Grothendieck ring of $Rep(osp(1|2))$.

Lemma 4.5.8.

$$V_{r_1}^\alpha \otimes V_{r_2}^\beta = V_{r_1+r_2}^{\alpha+\beta} + V_{r_1+r_2-1}^{\alpha+\beta+1} + V_{r_1+r_2-2}^{\alpha+\beta} + \cdots + V_{|r_1-r_2|}^{\alpha+\beta}.$$

Proof. Direct calculations show that $V_{r_1}^\alpha \otimes V_{r_2}^\beta$ contains one and only one singular (annihilated by x_+) vector of each weight from $|r_1 - r_2|$ to $r_1 + r_2$ and that the submodules generated by these vectors are irreducible. Proof is completed by counting dimensions. $\square$

Theorems 4.5.1 and 4.5.5 provide us with 2 commutative algebras. Here we interpret these algebras as Grothendieck rings of certain categories. Start with the algebra of Theorem 4.5.1 and denote it by $\mathcal{A}^{gen}$. Obviously $\mathcal{A}^{gen} = \mathcal{A}_0 \otimes \mathcal{A}$, where $\mathcal{A}$ is the Grothendieck ring of the category of finite-dimensional representations of g (its multiplication law is defined by the formula preceding Theorem 4.5.1) and $\mathcal{A}_0$ is the algebra with basis V_i^α, $i \geq 0, \alpha \in \mathbf{Z}/2\mathbf{Z}$, multiplication being given by

$$(22) \qquad V_{r_1}^\alpha \circ V_{r_2}^\beta = V_{r_1+r_2}^{\alpha+\beta} + V_{r_1+r_2-1}^{\alpha+\beta+1} + V_{r_1+r_2-2}^{\alpha+\beta} + \cdots + V_{|r_1-r_2|}^{\alpha+\beta}.$$

Comparing (22) with Lemma 4.5.8 we get the following.

Proposition 4.5.9. $\mathcal{A}_0$ *is the Grothendieck ring of the category of finite-dimensional representations of the superalgebra* $osp(1|2)$.

The appearance of $osp(1|2)$ here, artificial though it may seem to be, has deep reasons behind it. To see this we will analyze the rational level case using quantized enveloping algebras.

Remark 4.5.10. *It follows from Lemma 4.5.8 that the forgetful functor* $Rep(osp(1|2)) \to Rep(sl_2)$, $V_m^\alpha \mapsto V_m \oplus V_{m-1}$ *induces an epimorphism of the Grothendieck rings.*

B. Both Usl_2 and $Uosp(1|2)$ admit quantization, $U_t sl_2$ and $U_t osp(1|2)$ resp.. Let us recall the relevant formulas. The Drinfeld-Jimbo (see [9, 28]) algebra $U_t sl_2$, $t \in \mathbf{C}$ is defined to be the associative algebra on the generators $E, F, K^{\pm 1}$ with relations

$$EF - FE = \frac{K - K^{-1}}{t - t^{-1}}, \ KEK^{-1} = t^2 E, \ KFK^{-1} = t^{-2}F.$$

$U_t osp(1|2)$ is similarly defined [32] as the associative algebra on the generators $X_+, X_-, K^{\pm 1}$ with relations

$$X_+X_- + X_-X_+ = \frac{K - K^{-1}}{t - t^{-1}}, \ KX_\pm K^{-1} = t^{\pm 1}X_\pm.$$

The representation theory of sl_2 and $osp(1|2)$ "deforms to" the representation theory of $U_t sl_2$ and $U_t osp(1|2)$ resp. We will continue denoting by V_m the $m+1$−dimensional module over $U_t sl_2$, and by V_m^0, V_m^1 the $2\ (2m+1)$− dimensional modules over $U_t osp(1|2)$. For generic t these modules are irreducible, and the categories of finite dimensional representations, $Rep(U_t sl_2)$ and $Rep(U_t osp(1|2))$, generated by these modules are semisimple.

The deformations $U_t sl_2$ and $U_t osp(1|2)$ are especially remarkable in that they afford simultaneous deformation of the Hopf algebra structure. We get 2 tensor categories $Rep(U_t sl_2)$ and $Rep(U_t osp(1|2))$. What has been said implies that the Grothendieck rings of $Rep(U_t sl_2)$ and $Rep(U_t osp(1|2))$ are isomorphic to the Grothendieck rings of the corresponding classical objects if t is generic.

If however t is a root of unity, things change dramatically. Suppose for simplicity that t is a primitive l-th root of unity, l being odd. Then

(23)   (i) V_m is irreducible if and only if $m < l$;

(24)   (ii) V_m^ϵ is irreducible if and only if $m < l$.

(Both statements are proved by direct computations.)

What is even more important is that the categories $Rep(U_t sl_2)$ and $Rep(U_t osp(1|2))$ are no longer semisimple. For example, the tensor product of 2 irreducible representations is not semisimple. Things, however, are still very much under control.

Lemma 4.5.11. *Let t be a primitive l-th root of unity, l being odd, $m, n < l$. Then*

(i) $V_m \otimes V_n = V_{|m-n|} \oplus V_{|m-n|+2} \cdots \oplus V_{min\{2(l-1)-m-n, m+n\}} \oplus W$,

where W is not semisimple.

(ii) $V_m^\alpha \otimes V_n^\beta = V_{|m-n|}^{\alpha+\beta} \oplus V_{|m-n|+1}^{\alpha+\beta+1} \cdots \oplus V_{min\{2(l-1)-m-n, m+n\}}^{\alpha+\beta} \oplus W$,

where W is not semisimple.

Sketch of Proof. (i) is well-known, see [41]. We will however review both cases as at our level of brevity there will no difference between them. First, direct calculations as in the proof of Lemma 4.5.8 show that, regardless of t, in each weight space there can always be only one singular vector. Now decomposition of Lemma 4.5.8, statements (23, 24) and this uniqueness result show that the submodules V_{l-1+i} and V_{l-1-i} (or V_{l-1+i}^ϵ and V_{l-1-i}^ϵ), $i \le m + n - l + 1$ are non-trivially tangled. Other V_j coming from generic t are still irreducible and appear as direct summands. □

Definition.

(i) Define $Rep(U_t sl_2)^{(l)}$ and $Rep(U_t osp(1|2))^{(l)}$ to be the subcategories of $Rep(U_t sl_2)$ and $Rep(U_t osp(1|2))$ resp. consisting of direct sums of irreducible modules V_m (or V_m^α resp.), $m < l$.

(ii) Define functors

$$Rep(U_t sl_2)^{(l)} \times Rep(U_t sl_2)^{(l)} \to Rep(U_t sl_2)^{(l)},\ A, B \mapsto A \dot\otimes B,$$

$$Rep(U_t osp(1|2))^{(l)} \times Rep(U_t osp(1|2))^{(l)} \to Rep(U_t osp(1|2))^{(l)},$$

$$A, B \mapsto A \dot\otimes B,$$

by taking the usual tensor product and then throwing away W on the right hand side of formulas in Lemma 4.5.11.   □

We get tensor categories $Rep(U_t sl_2)^{(l)}$ and $Rep(U_t osp(1|2))^{(l)}$.

C. It is easy now to interpret the fusion algebra at the rational level in terms of the Grothendieck rings of $Rep(U_t sl_2)^{(l)}$ and $Rep(U_t osp(1|2))^{(l)}$. In view of Lemma 4.5.11 and the Definition above, Theorem 4.5.5 reads as follows.

Proposition 4.5.12. *The fusion algebra at level $k + 2 = p/q$ is a quotient of the tensor product of the Grothendieck rings of the categories $Rep(U_{t_1} sl_2)^{(p-1)}$ and $Rep(U_{t_2} osp(1|2))^{(q)}$. Further, the fusion algebra always contains the Grothendieck ring of $Rep(U_{t_2} osp(1|2))^{(q)}$ via the classes of symbols V_m^α, V_0.*

4.5.6. *Kac-Moody vs. Virasoro.* The Virasoro algebra, Vir, is defined to be the vector space with basis $\{L_i, z,\ i \in \mathbf{Z}\}$ and bracket

$$[L_i, L_j] = (i - j)L_{i+j} + \delta_{i,-j}\frac{i^3 - i}{12}z.$$

The representation theory of the Virasoro algebra is to a great extent parallel to that of $\hat{g}$. We will confine ourselves to essentials, making reference to [15].

One defines the Verma module $M_{h,c}$, where (h, c) is a highest weight, i.e. eigenvalues of L_0, z resp. determined by the vacuum vector; c is sometimes referred to as level. A Verma module is reducible if and only if it contains a singular vector. $M_{h,c}$ generically has no singular vectors. By the Kac determinant formula, there is a family of hyperbolas labelled by pairs of positive integers m, n in the plane with coordinates (h, c) such that if $M_{h,c}$ contains a singular vector, then (h, c) belongs to one of these hyperbolas; generically along the hyperbolas the singular vector is unique. Denote the singular vector arising in $M_{h,c}$ if (h, c) belongs to the hyperbola with the label m, n by S_{mn}. There arises the Vir-analogue of the generalized Weyl module $M_{h,c}/ < S_{mn} = 0 >$. Attach to $M_{h,c}/ < S_{mn} = 0 >$ the symbol (V_{n-1}, V_{m-1}). Further, for c fixed there arises a one-to-one correspondence between the Vir-analogues of generalized Weyl modules and the symbols (V_{n-1}, V_{m-1}). This has all been in precise analogy with 2.4.1.

It has hardly been written anywhere, but is nevertheless known, that the $Vir-$ analogue of the fusion algebra from 4.5.1, i.e. at a generic level, is as follows:

$$(25) \qquad (V_{n_1}, V_{m_1}) \circ (V_{n_2}, V_{m_2}) = (V_{n_1} \otimes V_{n_2}, V_{m_1} \otimes V_{m_2}).$$

(The interested reader can prove this result using methods of [16]; our treatment of the $\hat{g}$-fusion algebra in 4.5.1 is also a direct analogue of these.)

There is a functor sending $\hat{g}-$modules to $Vir-$modules – quantum Drinfeld-Sokolov reduction. One of the prerequisites for it is a choice of a nilpotent subalgebra of sl_2. The two obvious possibilities are $\mathbf{C}e$ and $\mathbf{C}f$. Denote the corresponding functors by ϕ_e and ϕ_f. It can be extracted from [13] that both functors send generalized Weyl modules to generalized Weyl modules. In our terminology one gets

$$\phi_e : \begin{array}{ccc} (V_m^0, V_n) & \mapsto & (V_m, V_n) \\ (V_m^1, V_n) & \mapsto & (V_{m-1}, V_n), \end{array}$$

$$\phi_f : \begin{array}{ccc} (V_m^0, V_n) & \mapsto & (V_{m-1}, V_n) \\ (V_m^1, V_n) & \mapsto & (V_m, V_n), \end{array}$$

where the symbol V_{-1}, if it arises, is understood as zero.

The $Vir-$analogue of admissible representations is the celebrated minimal representations. The latter can be defined as quotients of generalized Weyl modules by repeating word for word the definition of admissible representations from 2.4.2. It is known that minimal representations arise only when

$$c = c_{pq} = 1 - \frac{6(p-q)^2}{pq},$$

where p, q are relatively prime positive integers. There are again 2 generalized Weyl modules projecting on a given minimal representation. Therefore minimal representations are labelled by equivalence classes of symbols (V_m, V_n). It can be shown that the equivalence relation is as follows: $(V_m, V_n) \approx (V_{q-2-m}, V_{p-2-n})$ for $c = c_{pq}$. From this and (25) one can easily calculate the fusion algebra. We will not write down the relevant formulas here and confine ourselves to mentioning that the algebra is related to the product of Grothendieck rings of 2 quantum $U_t(\mathrm{sl}_2)$ at appropriate roots of unity in much the same way as the fusion algebra for $\hat{g}$ is related to the product of Grothendieck rings of $U_t(osp(1|2))$ and $U_t(\mathrm{sl}_2)$. Recall also that the Vir-fusion algebra was calculated in [6]; mathematically acceptable exposition can be found in [16].

Another property of the Drinfeld-Sokolov reduction is that both ϕ_e and ϕ_f send admissible representations at level $k = 2 - p/q$ of $\hat{g}$ to minimal representations of Vir at the level c_{pq}, see [18] .

Proposition 4.5.13. *The functor $\phi_e \oplus \phi_f$ determines an epimorphism of the $\hat{g}-$fusion algebra onto the $Vir-$fusion algebra at both the generic and rational levels.*

Proof. The generic level case follows from Remark 4.5.10 and formula (25) above. In the rational level case, the statement follows from the fact that both $\hat{g}-$ and Vir-fusion algebras are obtained from their generic level counterparts by imposing the equivalence relations and that the 2 equivalence relations agree with each other. □

4.6. Fusion functor. This part is an announcement, proofs will appear elsewhere.

Suppose we have a trivial vector bundle $\mathcal{E} \to \mathbf{CP^1}$, $A = \{(P_1, b_1), (P_2, b_2)\}$, $B = \{(P_3, b_3)\}$, so that $(\mathcal{E}, A \bigsqcup B)$ is generic. There is a construction which to a $\hat{g}^A-$module associates a $\hat{g}^B-$module. This construction is a natural adaptation of the Kazhdan-Lusztig tensoring [30] to our needs.

Denote by $g(\mathbf{CP^1}, A, B)$ the subalgebra of $g(\mathbf{CP^1}, A)$ consisting of functions taking values in $n_3 = [b_3, b_3]$ – just like we did in 4.4.2. For a $\hat{g}^A-$module M^A, denote by M_N^A the subspace of $(M^A)^*$ annihilated by $g(\mathbf{CP^1}, A, B)^N$. Obviously $M_N^A \subset M_{N+1}^A$, $N \geq 1$. Set

$$F^{A \to B}(M^A) = \cup_{N \geq 1} M_N^A.$$

One can show that the vector space $F^{A \to B}(M^A)$ affords in a natural way a structure of an $\hat{g}^B-$module at the same level; this is easy to show in the spirit of [30, 4]. Using our methods one can show that

(i) if M^A is from the $\mathcal{O}$−category, or further a generalized Weyl module, or further an admissible representation, then $F^{A \to B}(M^A)$ is also as a $\hat{\mathfrak{g}}^B$−module;

(ii) the Grothendieck rings arising in this way coincide with those in Theorem 4.5.1 or Theorem 4.5.5 if the level is generic or rational resp .

This generalizes the statement for the integrable representations, see [11].

Problem. Describe the tensoring which arises in the spirit of Kazhdan‐Lusztig.

4.7. **Quadratic degeneration.**

4.7.1.   The setup here will be the following version of 3.2.2:

(i) Let $\bar{\pi} : \mathcal{C}_S \to S$ be a family of curves over a formal disk S, such that the fiber over the generic point of S ("outside the origin") is a smooth projective curve, and over the origin, O, the fiber is a curve $\mathcal{C}_O$ with exactly one quadratic singularity;

(ii) $\rho_S : \mathcal{E}_S \to \mathcal{C}_S$ is a rank 2 vector bundle.

As in 3.2.2, we complete these data to the localization data with logarithmic singularities, say $\tilde{\psi}$. In the standard way, Theorem 3.2.1 rewrites to give a $\mathcal{D}$−module over S with logarithmic singularities at O; call it $\Delta_{\tilde{\psi}}(M^A)$. This is because $Spec(S)$ is $\mathbf{C}[[t]]$ and the vector fields vanishing at $q = 0$ are exactly those which can be lifted to $\mathcal{C}_S$.

Along with the family $\bar{\pi} : \mathcal{C}_S \to S$ consider the family $\bar{\pi}^\vee : \mathcal{C}_S^\vee \to S$, obtained from $\bar{\pi} : \mathcal{C}_S \to S$ by replacing the singular fiber $\mathcal{C}_O$ with its normalization $\mathcal{C}_O^\vee$ (i.e. be pulling $\mathcal{C}_O$ apart at the self-intersection point). There is a projection $\mathcal{C}_O^\vee \to \mathcal{C}_O$ and the preimage of the self-intersection point $a \in \mathcal{C}_O$ consists of 2 points $a_0, a_\infty \in \mathcal{C}_O^\vee$.

It is obvious that the datum $\mathcal{E} \to \mathcal{C}_S$ is equivalent to the data "$\rho_S^\vee : \mathcal{E}_S^\vee \to \mathcal{C}_S^\vee$, equivalence $(\rho_S^\vee)^{-1}(a_0) \approx (\rho_S^\vee)^{-1}(a_\infty)$". The localization data with logarithmic singularities $\tilde{\psi}$ rewrites to give "normalized" localization data $\psi^\vee$.

In addition fix 2 different lines l_0, l_∞ in the fiber of $\mathcal{E}_S$ over the point $a \in \mathcal{C}_O$. This determines 2 Borel subalgebras, $\mathrm{b}_0, \mathrm{b}_\infty$ operating in the fiber over a.

After normalization these additional data determine the line l_0 and the Borel subalgebra b_0 operating in the fiber of $\mathcal{E}_S^\vee$ over a_0, as well as the line l_∞ and the Borel subalgebra b_∞ operating in the fiber over a_∞. We also get a distinguished Cartan subalgebra $\mathrm{h} = \mathrm{b}_0 \cap \mathrm{b}_\infty$. Set $A^\vee = A \bigsqcup \{(a_0, \mathrm{b}_0), (a_\infty, \mathrm{b}_\infty)\}$.

Now with a $\hat{g}^A$–module M^A at the level k and an admissible weight $\lambda \in h^*$ we associate the $\hat{g}^{A^\vee}$–module $M^A \otimes L_{\lambda,k}^{P_0,b_0} \otimes L_{\lambda,k}^{P_\infty,b_\infty}$. We get a $\mathcal{D}$–module for the "normalized"localization data:

$$\oplus_\lambda \Delta_{\psi^\vee}(M^A \otimes L_{\lambda,k}^{P_0,b_0} \otimes L_{\lambda,k}^{P_\infty,b_\infty}).$$

Proposition 4.7.1. *Generically with respect to l_0, l_∞, if $\Delta_{\tilde{\psi}}(M^A)$ is smooth then $\oplus_\lambda \Delta_{\psi^\vee}(M^A \otimes L_{\lambda,k}^{P_0,b_0} \otimes L_{\lambda,k}^{P_\infty,b_\infty})$ is also and there is an isomorphism of $\mathcal{D}$–modules*

$$\Delta_{\tilde{\psi}}(M^A) \approx \oplus_\lambda \Delta_{\psi^\vee}(M^A \otimes L_{\lambda,k}^{P_0,b_0} \otimes L_{\lambda,k}^{P_\infty,b_\infty}).$$

4.7.2. *Proof.* (i) Begin with the genus zero case.Observe that the algebra of regular functions on the neighborhood of the point a is $\mathbf{C}[t_0, t_\infty][[t]]/ < t_0 t_\infty = t >$ where t is a coordinate on S; $\mathcal{C}_O^\vee$ in this case is just a union of 2 spheres. Therefore the set A splits in two: A' and A'', each of which has to do with one of the spheres.

Hence the algebra $g(\bar{\pi}^{-1}(s), A)$ can be degenerated into the following one as s "approaches" O:

$$(g(\mathbf{CP^1}, A', (P_0, b_0)) + h) \oplus_h (h + g(\mathbf{CP^1}, A'', (P_\infty, b_\infty))).$$

The meaning of the last expression is as follows: recall, see 4.4, that $g(\mathbf{CP^1}, A', (P_0, b_0))$ consists of functions regular outside $\bar{A}$ and sending P_0 to n_0; $g(\mathbf{CP^1}, A'', (P_\infty, b_\infty))$ is defined similarly with P_0, n_0 replaced with P_∞, n_∞; further the algebra $g(\mathbf{CP^1}, A', (P_0, b_0)) + h$ is the algebra of functions sending P_0 to b_0, the same is true for $h + g(\mathbf{CP^1}, A'', (P_\infty, b_\infty))$; finally "$\oplus_h$" means direct product over h.

Therefore the coinvariants degenerate into the space

$$((M^{A'})_{g(\mathbf{CP^1}, A', (P_0, b_0))} \otimes (M^{A''})_{g(\mathbf{CP^1}, A'', (P_\infty, b_\infty))})_h,$$

where h acts by means of the diagonal embedding; this makes sense as the fibers are identified.

By Proposition 4.4.4, the space

$$(M^{A'})_{g(\mathbf{CP^1}, A', (P_0, b_0))} \otimes (M^{A''})_{g(\mathbf{CP^1}, A'', (P_\infty, b_\infty))}$$

is finite dimensional. It is easy to extract from Lemma 4.5.2 that as an h-module this space is semisimple and therefore is isomorphic to

$$\oplus_\lambda (M^A \otimes M_{\lambda,k}^{P_0,b_0} \otimes M_{\lambda,k}^{P_\infty,b_\infty})_{g(\mathcal{C}_O^\vee, A^\vee)}.$$

By Lemma 4.5.6, in the last formula λ can be chosen to be admissible and the Verma modules can be replaced with the corresponding admissible representations.

This proves that $\oplus_\lambda \Delta_{\psi^\vee}(M^A \otimes L_{\lambda,k}^{P_0,b_0} \otimes L_{\lambda,k}^{P_\infty,b_\infty})$ is smooth and gives a morphism

$$\Delta_{\tilde\psi}(M^A) \to \oplus_\lambda \Delta_{\psi^\vee}(M^A \otimes L_{\lambda,k}^{P_0,b_0} \otimes L_{\lambda,k}^{P_\infty,b_\infty}).$$

That this is an isomorphism can be shown in the standard way by constructing the inverse map using the formal character of $L_{\lambda,k}$, see [4].

(ii) The higher genus case is not much different. For example, pinching makes a torus into a sphere. Therefore in this case the proof is literally the same. It also proves an analogue of Lemma 4.5.6 for a torus. This provides a basis for induction.

In genus ≥ 2 at an appropriate place instead of Proposition 4.4.4 one has to make reference to Proposition 4.3.3 and then use induction. $\square$

4.7.3. *Remarks.* (i) The meaning of Proposition 4.7.1 is that the dimension of the generic fiber of the $D-$module $\Delta_\psi(M^A)$ can be calculated by the usual combinatorial algorithm: by pinching the surface and further inserting all possible representations the problem is reduced to the case of a sphere with three punctures and in the latter case explicit numerical results are available.

(ii) In the genus 0 case the analogue of Proposition 4.7.1 for generalized Weyl modules is valid. To see this it is enough to examine part (i) of the proof and to convince oneself that the only requirement on M^A used there was that M^A be a generalized Weyl module; in fact at an appropriate place instead of Lemma 4.5.6 one has to use Lemma 4.5.3.

(iii) Quadratic degeneration for generalized Weyl modules on the sphere allows to write horizontal sections of the corresponding bundle as a product of vertex operators. This will be explained in sect.5.

5. Screening operators and correlation functions

In this section we will study in detail the situation described in 4.4.3: we have the trivial rank 2 bundle $\mathcal{E} \to \mathbf{CP}^1$, a generalized Weyl module M^A, and a holonomic $\mathcal{D}-$module $\Delta(M^A)$ on the space $\mathbf{C}^m \times \mathbf{C}^m$ with fiber $(M^A)_{g(\mathbf{CP}^1,A)}$. For the reasons which will become clear later we replace this bundle with the dual one, its fiber being $((M^A)^*)^{g(\mathbf{CP}^1,A)}$. Denote the corresponding $D-$module by $\Delta(M^A)^*$. Using our results on quadratic degeneration we rewrite horizontal sections of the corresponding bundle with flat connection as matrix elements of vertex operators, which serves a two-fold purpose: we find that the differential equations satisfied by horizontal sections are provided by the singular vectors of the corresponding Verma module and we write down integral representations for solutions to these differential equations.

5.1. Vertex operators and corelation functions.
An alternative to the language of coinvariants in the genus zero case is the language of *vertex operators*.

Definition. A vertex operator is a $\hat{g}$–morphism

$$(26) \qquad Y : \mathcal{F}_{\alpha\beta}^{\mathbf{C}^*} \otimes V_1 \to V_2,$$

where $\mathcal{F}_{\alpha\beta}^{\mathbf{C}^*}$ is a loop module (see 2.6) and $V_1, V_2 \in \mathcal{O}_k$ are highest weight modules. $\qquad \square$

In other words, a vertex operator is an embedding $\mathcal{F}_{\alpha\beta}^{\mathbf{C}^*} \hookrightarrow \text{Hom}_{\mathbf{C}}(V_1 \to V_2)$. The space $\mathcal{F}_{\alpha\beta}^{\mathbf{C}^*}$ has the basis $\{F_{ij} = F_i \otimes z^j, \ i, j \in \mathbf{Z}\}$, where $\{F_i, \ i \in \mathbf{Z}\}$ is a basis in $\mathcal{F}_{\alpha\beta}$, see 2.6. Given a vertex operator Y, consider the generating function

$$Y(x, z) = x^{\Delta_1} z^{\Delta_2} \sum_{i,j=-\infty}^{\infty} F_{ij} x^{-i} z^{-j},$$

the "monodromy coefficients" Δ_1, Δ_2 are defined by:

$$\Delta_1 = \frac{-\lambda_2 + \lambda_1 + \beta}{2}, \ \Delta_2 = \frac{-C(\lambda_2) + C(\lambda_1) + C(\beta)}{2},$$

where λ_i is the highest weight of V_i and $C(\lambda) = \lambda(\lambda + 2)/2$. ($\Delta_1, \Delta_2$ will later appear as genuine monodromy coefficients of a certain flat connection.)

The formal series $Y(x, z)$ is, of course, an element of $\text{Hom}_{\mathbf{C}}(V_1, V_2 \otimes x^{\Delta_1} z^{\Delta_2} \mathbf{C}[[x^{\pm 1}, z^{\pm 1}]])$. Further, for any $g \in \mathfrak{g}$ the commutator $[g \otimes z^n, Y(x, z)]$ is also a well-defined element of $\text{Hom}_{\mathbf{C}}(V_1, V_2 \otimes x^{\Delta_1} z^{\Delta_2} \mathbf{C}[[x^{\pm 1}, z^{\pm 1}]])$. For the standard basis of $\mathfrak{g}$, see 2.2, one derives from the definition of a vertex operator that

$$(27) \qquad [e \otimes z^n, Y(x, z)] = z^n (-x^2 \frac{\partial}{\partial x} + \beta x) Y(x, z),$$

$$(28) \qquad [f \otimes z^n, Y(x, z)] = z^n \frac{\partial}{\partial x} Y(x, z),$$

$$(29) \qquad [h \otimes z^n, Y(x, z)] = z^n (2x^2 \frac{\partial}{\partial x} - \beta x) Y(x, z).$$

We conclude that for any $g \in \mathfrak{g}$ there is a differential operator $D_g(x)$ in x such that

$$(30) \qquad\qquad [g \otimes z^n, Y(x, z)] = z^n D_g(x) Y(x, z),$$

Suppose now we are given a collection of vertex operators

$$Y_i : \mathcal{F}_{\lambda_i \mu_i}^{\mathbf{C}^*} \otimes V_{i-1/2} \to V_{i+1/2}, \; 1 \le i \le m.$$

The product of the corresponding generating functions

$$Y_m(x_m, z_m) \cdots Y_2(x_2, z_2) Y_1(x_1, z_1)$$

is a well-defined element of

$$Hom_{\mathbf{C}}(V_{1/2}, V_{m+1/2} \otimes \prod_i x_i^{\Delta_{i,1}} z_i^{\Delta_{i,2}} \mathbf{C}[[x_1^{\pm 1}, \ldots x_m^{\pm 1}, z_1^{\pm 1}, \ldots z_m^{\pm 1}]]).$$

The matrix element

$$< v^*, Y_m(x_m, z_m \cdots Y_2(x_2, z_2) Y_1(x_1, z_1) v >, \; v \in V_{1/2}, v^* \in V_{m+1/2}^*$$

is, therefore, a formal Laurent series in x_i, z_i, $1 \le i \le m$.

Definition Suppose $Y_i(x_i, z_i)$, $1 \le i \le m$ are as above. Then the matrix element

$$(31)$$
$$\Psi(x_1, \ldots, x_m, z_1, \ldots z_m) = < v^*, Y_m(x_m, z_m) \cdots Y_2(x_2, z_2) Y_1(x_1, z_1) v >$$

is called *a correlation function* if $V_{1/2}, \ldots, V_{m+1/2}$ are irreducible generalized Weyl modules, $V_{1/2}$ is the vacuum module, v is the highest weight vector of $V_{1/2}$ and v^* is the dual to the highest weight vector of $V_{m+1/2}$. (The latter condition is meaningful in view of the weight space decomposition of a highest weight module.) $\square$

A correlation function has been understood as a formal power series. We will show that, in fact, it is a holomorphic function satisfying a certain holonomic system of partial differential equations. In order to do that we will interpret vertex opeartors as horizontal sections of a line bundle with a flat connection provided by three modules on $\mathbf{CP}^1 \times \mathbf{CP}^1$.

5.2. From coinvariants to vertex operator algebra.

5.2.1. We return to the setup of 4.5.1. In the cartesian product $\mathbf{CP^1} \times \mathbf{CP^1}$ fix a coordinate system (x, z). Attach to the point x in the first factor the Borel subalgebra $\mathfrak{b}_x$ spanned by the vectors $e_x = e - xh - x^2 f$, $h_x = h + 2xf$. This means, in particular, that $\mathfrak{b}_0$ is the standard Borel subalgebra $\mathbf{C}e \oplus \mathbf{C}h$ (see 2.2) and $\mathfrak{b}_\infty$ is the opposite one. Set $A = \{(0, 0), (x, z), (\infty, \infty)\}$. Let $V^A = V_0^{\mathfrak{b}_0, 0} \otimes V_1^{\mathfrak{b}_x, z} \otimes V_\infty^{\mathfrak{b}_\infty, \infty}$ be a generalized Weyl module over $\hat{g}^A$. Consider the space of invariants $((V^A)^*)^{\mathfrak{g}(\mathbf{CP^1}, A)}$. By Theorem 4.5.1 this space is either 0- or 1-dimensional. Suppose the latter possibility is the case. Then by Theorem 3.2.2 we get a line bundle with flat connection over $\mathbf{C}^* \times \mathbf{C}^*$ whose fiber over the point $(x, z) \in \mathbf{C}^* \times \mathbf{C}^*$ is $((V^A)^*)^{\mathfrak{g}(\mathbf{CP^1}, A)}$. There arises an embedding

$$V_1^{\mathfrak{b}_x, z} \hookrightarrow \mathrm{Hom}_{\mathbf{C}}(V_0^{\mathfrak{b}_0, 0} \otimes V_\infty^{\mathfrak{b}_\infty, \infty}, \mathbf{C}).$$

The dual space $(V_\infty^{\mathfrak{b}_\infty, \infty})^*$ as a $\hat{g}$–module is isomorphic to the contragredient module $(V_\infty^{\mathfrak{b}_\infty, \infty})^c$, see 2.2. As the level is generic, the latter module is irreducible and is, therefore, isomorphic to a certain generalized Weyl module $V_\infty^{\mathfrak{b}_0, 0}$. Hence we get an embedding

$$V_1^{\mathfrak{b}_x, z} \hookrightarrow \mathrm{Hom}_{\mathbf{C}}(V_0^{\mathfrak{b}_0, 0}, V_\infty^{\mathfrak{b}_0, 0} \otimes x^{\Delta_1} z^{\Delta_2} \mathbf{C}[[x^{\pm 1}, z^{\pm 1}]]),$$

where Δ_1, Δ_2 are monodromy coefficients of the flat connection. We conclude that any $w \in V_1^{\mathfrak{b}_x, z}$ can be looked upon as a c ertain generating function

$$w(x, z) = x^{\Delta_1 - n} z^{\Delta_2 - l} \sum_{i, j \in \mathbf{Z}} w_{ij} x^{-i} z^{-j}$$

of a family of operators $\{w_{ij} \subset \mathrm{Hom}_{\mathbf{C}}(V_0^{\mathfrak{b}_0, 0}, V_\infty^{\mathfrak{b}_0, 0})\}$, where (n, l) is the bidegree of w as an element of $V_1^{\mathfrak{b}_x, z}$.

Lemma 5.2.1. *Suppose $v_1 \in V_1^{\mathfrak{b}_x, z}$ is the highest weight vector. Then*

(i) $v_1(x, z)$ is a generating function of a certain vertex operator as in 5.1;

(ii) any vertex operator is obtained in this way.

Proof is a direct and simple calculation using definitions, see also 4.5.3 formula (20). $\square$

Let us now relate correlation functions to horizontal sections of the bundle built on the generalized Weyl module M^A, $\Delta(M^A)^*$, see beginning of sect.5 for notations. Suppose that M^A is the tensor product of "individual" generalized Weyl modules

$$\otimes_{i=1}^m V_i^{z_i, \mathfrak{b}_{x_i}}.$$

Consider all possible correlation functions
$$< v^*, v_m(x_m, z_m) \cdots v_1(x_1, z_1) v >,$$
where $v_i(x_i, z_i)$ is a generating function of a vertex operator related to
the highest weight vector $v_i \in V_i^{z_i, b_{x_i}}$.

Corollary 5.2.2. *Let M^A be as above. Over a suitable open con-tractible subset U of $\mathbf{C}^m \times \mathbf{C}^m$, there is an isomorphism between the space of horizontal sections of the bundle $\Delta(M^A)^*$ and the space of correlation functions*
$$< v^*, v_m(x_m, z_m) \cdots v_1(x_1, z_1) v > .$$

Proof. Intertwining properties of vertex operators imply that a correlation function is a horizontal section of $\Delta(M^A)^*$ in a formal sense. This give a map in one direction. A map in the opposite direction in provided by quadratic degeneration, see Proposition 4.7.1.   □

5.2.2. By Lemma 5.2.1 coinvariants recover vertex operators. In fact they give us much more: the collection of generating functions $w(x, z)$, $w \in V_1^{b_x, z}$, affords a kind of *vertex operator algebra* structure. We will not discuss the latter in detail (see [22]) and only explain how one can get explicit formulas for $w(x, z)$, $w \in V_1^{b_x, z}$ in terms of the vertex operator $v_1(x, z)$ related to the highest weight vector v_1.

For any $g \in \mathfrak{g}$ set $g(i) = g \otimes z^i \in \hat{\mathfrak{g}}$. Define the *current* $g(z)$ to be
$g(z) = \sum_{i \in \mathbf{Z}} g(i) z^{-1-i} \in \hat{\mathfrak{g}} \otimes \mathbf{C}[[z^{\pm 1}]]$. Define $g(z)^{(l)}$ to be the l-th
(formal) derivative of $g(z)$ with respect to z. For any $g(z)^{(l)}$ set

$$(g(z)^{(l)})_+ = (\frac{d}{dz})^l \sum_{i=0}^{\infty} g_{-i-1} z^i, \quad (g(z)^{(l)}_- = g(z)^{(l)} - (g(z)^{(l)})_+.$$

Observe that for any $w(x, z) \in \mathrm{Hom}_{\mathbf{C}}(V_0^{bo,0}, V_\infty^{bo,0} \otimes x^{\Delta_1} z^{\Delta_2} \mathbf{C}[[x^{\pm 1}, z^{\pm 1}]])$
and any $g \in \mathfrak{g}$, the products $(g(z)^{(l)})_- w(x, z), w(x, z)(g(z)^{(l)})_+$ are also well-defined elements of $\mathrm{Hom}_{\mathbf{C}}(V_0^{bo,0}, V_\infty^{bo,0} \otimes x^{\Delta_1} z^{\Delta_2} \mathbf{C}[[x^{\pm 1}, z^{\pm 1}]])$.

Define for any $g \in \mathfrak{g}$,
$$w(x, z) \in \mathrm{Hom}_{\mathbf{C}}(V_0^{bo,0}, V_\infty^{bo,0} \otimes x^{\Delta_1} z^{\Delta_2} \mathbf{C}[[x^{\pm 1}, z^{\pm 1}]])$$

(32) $: g(z)^{(k)} w(x, z) := (g(z)^{(k)})_- w(x, z) + w(x, z)(g(z)^{(k)})_+.$

Lemma 5.2.3. *Let $g \in \mathfrak{g}$, $w \in V_1^{b_x, z}$, $w(x, z)$ the corresponding ele-ment of $\mathrm{Hom}_{\mathbf{C}}(V_0^{bo,0}, V_\infty^{bo,0} \otimes x^{\Delta_1} z^{\Delta_2} \mathbf{C}[[x^{\pm 1}, z^{\pm 1}]])$. Then*
(i) $(g \cdot w)(x, z) = [g, w(x, z)]$;
(ii) $(g(-l) \cdot w)(x, z) = (1/(l-1)!) : g(z)^{(l-1)} w(x, z) :, \ l > 0;$

Proof is a direct calculation of matrix elements of the operator $(g(-l) \cdot w)(x, z)$ based on the definition of the space of coinvariants. □

5.2.3. *Application: annihilating ideals and singular support of admissible representations.* There is a well-known theorem of M.Duflo (see e.g. [7]), which establishes essentially a one-to-one correspondence between 2-sided primitive ideals of $U(\mathrm{g})$ (one actually has to quotient out a maximal ideal of the center) and the annihilating ideals of irreducible highest weight g-modules. There are various reasons for which this correspondence does not survive for $\hat{\mathrm{g}}$−modules. Ideas from conformal field theory provided a substitute – a remarkable construction of elements in a certain completion of $\tilde{U}(\hat{\mathrm{g}})$ of $U(\hat{\mathrm{g}})$ which annihilate some classes of irreducible highest weight modules. This construction is well-known for integrable highest weight modules. Our results allow us to extend it to admissible representations over $\widehat{sl_2}$. As an application, we will finish the proof of the theorem on singular support in (almost) one line.

Let Vac be the vacuum representation at level k: $Vac = Ind^{\hat{\mathrm{g}}}_{\hat{\mathrm{g}}_\geq \oplus \mathbf{C}c} \mathbf{C}.$ Consider an embedding

(33) $$Vac \to Hom(V, V \otimes \mathbf{C}[[z, z^{-1}]]),$$

where V is a generalized Weyl module. Observe that since Vac is induced from the trivial g-module, there is no dependence on a Borel subalgebra, and no x in the notation.

Suppose now k becomes rational. Then Vac, V become reducible. The map (33) comes from the invariant functional, and this functional pushes forward on the corresponding irreducible modules if and only if it vanishes on the singular vectors. When this is possible is the question addressed in sect.4.5.4, Theorem 4.5.5. The coinvariant survives if and only if V projects onto an admissible representation (Vac always does so). The tiny part of Theorem 4.5.5 we have used simply means that Vac is the identity of the fusion algebra.

On the other hand, if S is the unique singular vector appearing in Vac, when k becomes rational, then the "field" $S(z)$ related to S under (33) is zero. Therefore all its Fourier components act as 0 on the quotient of V. Of course for the same reason all elements of the submodule of Vac generated wby S will analogously give rise to a huge collection of operators annihiliting admissible representations. Explicit formulas for these elements can in principle be written down using Lemma 5.2.3.

Example If k is a positive integer, then $S = (e \otimes z^{-1})^{k+1}$. Formulas of the previous section show that in this case $S(z) = e(z)^{k+1}$. The same is easy to do for an arbitrary affine Lie algebra to get $e_\theta(z)^{k+1}$, where e_θ is the highest root vector. This is the formula mentioned in the beginning of this section.

Back to Proof of Theorem 4.2.1 Consider elements $[g, S]$, $g \in \mathfrak{g}$. Their symbols were calculated in the beginning of proof of Theorem 4.2.1, and the set of their common zeros was shown to lie in the nilpotent cone.

Consider now $\lim_{z \to 0}[g, S](z)v$, $g \in \mathfrak{g}$, and identify it with an element of $U(\hat{\mathfrak{g}}_<)$ (divide out v!) , where v is the highest weight vector of V. From formulas provided by Lemma 5.2.3 it follows that the symbol of this element is essentially the same as that of $[g, S]$. (There will actually be one more term which is easy to kill using another singular vector.) Therefore common zeros of elements $\lim_{z \to 0}[g, S](z)v$, $g \in \mathfrak{g}$, also lie in the nilpotent cone.   □

5.3. Differential equations satisfied by correlation functions.

We return to the setup of 5.1 and consider a correlation function

$$\Psi(x_1, \dots, x_m, z_1, \dots z_m) = < v^*, Y_m(x_m, z_m) \cdots Y_2(x_2, z_2)Y_1(x_1, z_1)v >,$$

coming from the product of vertex operators

$$Y_i : \mathcal{F}^{\mathbf{C}^*}_{\lambda_i \mu_i} \otimes V_{i-1/2} \to V_{i+1/2}, \ 1 \le i \le m.$$

Using Lemma 5.2.1 we assume that there are generalized Weyl modules V_i, $1 \le i \le m$, with highest weight vectors v_i, $1 \le i \le m$, such that $Y_i(x, z) = v_i(x, z)$. An advantage of this point of view is that for any collection of elements $w_i \in V_i$, $1 \le i \le m$, we can consider the matrix element

$$< v^*, w_m(x_m, z_m) \cdots w_2(x_2, z_2)w_1(x_1, z_1)v > .$$

Lemma 5.3.1. *For any* $w_i \in V_i$, $1 \le i \le m$

$$< v^*, w_m(x_m, z_m) \cdots w_2(x_2, z_2)w_1(x_1, z_1)v >$$
$$= D \cdot \Psi(x_1, \dots, x_m, z_1, \dots, z_m),$$

where D is a differential operator in the x's with coefficients being rational functions in the z's.

Proof. Start with the function

$$(34) \quad < v^*, v_m(x_m, z_m) \cdots v_{i+1}(x_{i+1}, z_{i+1})(g(-l)v_i)(x_i, z_i)$$
$$v_{i-1}(x_{i-1}, z_{i-1}) \cdots v_1(x_1, z_1)v >, \ l > 0.$$

By Lemma 5.2.3 (ii) this rewrites as

$$(35) \quad < v^*, v_m(x_m, z_m) \cdots v_{i+1}(x_{i+1}, z_{i+1})(g(z)_-^{(l-1)} v_i(x_i, z_i)$$
$$- v_i(x_i, z_i)g(z)_+^{(l-1)})v_{i-1}(x_{i-1}, z_{i-1}) \cdots v_1(x_1, z_1)v >, \; l > 0.$$

Then commute all g_i, $i < 0$, through to the right and all g_i, $i \geq 0$, to the left in a standard way, c.f.[21] and use commutation relations (27),(28),(29). The case $l = 0$ is treated in a similar and simpler way using Lemma 5.2.3 (i). Further argue by induction again using Lemma 5.2.3. □

By definition each V_i is a quotient of a Verma module and therefore there are elements, singular vectors in the corresponding Verma module (see 2.2) $S_i \in U(\hat{g})$, such that $S_i v_i = 0$, $1 \leq i \leq m$. On the other hand, by Lemma 5.3.1 there are differential operators D_i, $1 \leq i \leq m$ such that

$$D_i < v^*, v_m(x_m, z_m) \cdots v_2(x_2, z_2)v_1(x_1, z_1)v >=$$
$$< v^*, v_m(x_m, z_m) \cdots v_{i+1}(x_{i+1}, z_{i+1})(S_i v_i)(x_i, z_i)$$
$$v_{i-1}(x_{i-1}, z_{i-1}) \cdots v_1(x_1, z_1)v >, \; 1 \leq i \leq m.$$

We arrive at the following result.

Lemma 5.3.2. *The correlation function*

$$\Psi(x_1, \ldots, x_m, z_1, \ldots, z_m) =< v^*, v_m(x_m, z_m) \cdots v_2(x_2, z_2)v_1(x_1, z_1)v >$$

satisfies the system of equations

$$(36) \quad D_i \Psi(x_1, \ldots, x_m, z_1, \ldots, z_m) = 0, \; 1 \leq i \leq m.$$

Observe that, although there are in general no explicit formulas for D_i, the fact that $[D_i, D_j] = 0$ is an obvious consequence of the definition.

We have obtained m equations satisfied by our function of $2m$ variables. The rest is, of course, the Knizhnik-Zamolodchikov equations. Let us write them down explicitly. Recall that we can look upon $\Psi(x_1, \ldots, x_m, z_1, \ldots z_m)$ as a function of $z_1, \ldots z_m$ with coefficients in a completed tensor product of m g−modules. (The variables $x_1, \ldots, x_m$ are responsible for that, see (27,28,29).) For any $A = \sum_s a_i \otimes b_s \in g \otimes g$ denote by $A_{ij}, 1 \leq i, j \leq m$, the operator acting on the m−fold tensor product of g-modules by the formula

$$A_{ij} \cdot w_1 \otimes \cdots w_m = \sum_s w_1 \otimes a_s w_i \otimes \cdots b_s w_j \otimes \cdots w_m.$$

The formula (30) implies that A_{ij} is a differential operator in x_i, x_j. Set $\Omega = ef + fe + h^2/2$.

Lemma 5.3.3. *([31])*

The correlation function $\Psi = \Psi(x_1, \ldots, x_m, z_1, \ldots z_m)$ satisfies the system of Knizhnik-Zamolodchikov equations

$$(37) \qquad (k+2)\frac{\partial}{\partial z_i}\Psi = \sum_{j \neq i} \frac{\Omega_{ij}}{z_i - z_j}\Psi, \ 1 \leq i \leq m.$$

There is no need to prove this lemma here as one can repeat word for word the known proofs. However we point out that if one considers a highest weight module V as a module over the semi-direct product of $\hat{g}$ and the Virasoro algebra Vir then V is annihilated by the element $d/dz - L_{-1}$, where L_{-1} is one of the Sugawara elements. One then shows that the singular vectors $(d/dz - L_{-1})v_i$, where v is a highest weight vector of V_i, give rise to the equations (37) in exactly the same way the singular vectors S_i gave rise to the equations (36). An immediate consequence of this proof is that the system of equations (37), (36) is consistent.

5.4. Screening operators and integral representations of correlation functions.

5.4.1. Suppose a function $\Psi_{old} = \Psi(x_1, \ldots, x_m, z_1, \ldots z_m)$ is the matrix element of the product of vertex operators

$$\Psi_{old} = < v^*, \pi \circ Y(x_m, z_m) \cdots Y_1(x_1, z_1)v_0 >,$$

$$Y_i : \mathcal{F}^{\mathbf{C}^*}_{\lambda_i \mu_i} \otimes V_{i-1/2} \to V_{i+1/2}, \ 1 \leq i \leq m,$$

satisfying the same conditions as the expression in (31), see 5.1, except that instead of assuming that $V_{m+1/2}$ is a generalized Weyl module we assume that $V_{m+1/2}$ is a contragredient Verma module, see 2.2. (Why "old" will become clear in a moment.) It is easy to see that $\Psi_{old} = \Psi(x_1, \ldots, x_m, z_1, \ldots, z_m)$ satisfies the same system of equations (36),(37). Suppose in addition that there is a projection $\pi : V_{m+1/2} \to W$ onto another contragredient Verma module W. Denoting by w^* the element dual to the highest weight vector $w \in W$ one can consider the matrix element

$$\Psi_{new} = < w^*, \pi \circ Y(x_m, z_m) \cdots Y_1(x_1, z_1)v_0 > .$$

We again observe that Ψ_{old} is a solution to the same system (36,37). This new solution can be calculated as follows.

There arises the dual map $\pi^* : W^* \to V^*_{m+1/2}$ and by definition there is an element S of $U(\hat{g}_>)$ such that $\pi^*(w^*) = Sv^*$. We now take the definition of Ψ_{new}, replace in it $\pi^*(w^*)$ with Sv^* and get

$$(38) \qquad \Psi_{new} = < S \cdot v^*, Y_m(x_m, z_m) \cdots Y_1(x_1, z_1) v_0 > .$$

Then we commute S through to the right. The intertwining properties of vertex operators tell us that

$$(39) \qquad \Psi_{new} = S^t \cdot \Psi_{old},$$

where ^{t} signifies the canonical anti-involution on a Lie algebra $(g_1 g_2 \cdots g_n \to g_n g_{n-1} \cdots g_1)$ and the action is determined by the following condition: if $g \in \mathrm{g}$ then

$$(g \otimes z^n) \cdot \Psi_{old} = \sum_{i=1}^m D_g(x_i) z_i^n \Psi_{old},$$

see (30).

We intend to use (39) in the case when π and therefore S do not exist!

5.4.2. *Screening operators.* Let $V_{\lambda_\infty, k}$ be a highest weight module and $v \in V_{\lambda_\infty, k}$ a highest weight vector. If the obvious integrality conditions are satisfied then the vectors $f^{\lambda_\infty + 1} v$, $(e \otimes z^{-1})^{k - \lambda_\infty + 1} v$ are singular and give rise to embeddings of the type $W \hookrightarrow V_{\lambda_\infty, k}$. Now take 3 highest weight modules $V_{\lambda_i, k}$, $i = 0, 1, \infty$, attach them to 3 points in $\mathbf{CP^1}$ and consider the space of coinvariants

$$(\otimes_{i=0,1,\infty} V_{\lambda_i, k}^{b_i, i})_{\mathrm{g}(\mathbf{CP^1}, \{0,1,\infty\})}.$$

Of course an embedding $W \hookrightarrow V_{\lambda_\infty, k}$ gives rise to a map

$$(W^{b_\infty, \infty} \otimes_{i=0,1} V_{\lambda_i, k}^{b_i, i})_{\mathrm{g}(\mathbf{CP^1}, \{0,1,\infty\})} \hookrightarrow (\otimes_{i=0,1,\infty} V_{\lambda_i, k}^{b_i, i})_{\mathrm{g}(\mathbf{CP^1}, \{0,1,\infty\})}.$$

It is remarkable that even if the embedding $W \hookrightarrow V_{\lambda_\infty, k}$ does not exist the last map still does. In the language of vertex operators this phenomenon was explained in great detail in [20].

Therefore with each of the formal singular vectors – $f^{\lambda_\infty + 1} v$ or $(e \otimes z^{-1})^{k - \lambda_\infty + 1} v$ – we have associated an operator acting on coinvariants. Call these operators *screenings* and denote them by R_1 and R_0 respectively.

Let us calculate the action of the screenings explicitly. By definition $R_j (\otimes_{i=0,1,\infty} V_{\lambda_i, k}^{b_i, i})_{\mathrm{g}(\mathbf{CP^1}, \{0,1,\infty\})}$ only depends on $V_{\lambda_\infty, k}$ so we will be simply writing $R_j(V_{\lambda_\infty, k})$. Now formulas for the related singular vectors

$(f^{\lambda_\infty+1}v, \ (e \otimes z^{-1})^{k-\lambda_\infty+1}v)$ and a very simple calculation using the formulas (7,8) give the following result:

$$(40) \qquad\qquad R_1((V_m^0, V_n)) = (V_{m-1}^1, V_n)$$

$$(41) \qquad\qquad R_1((V_m^1, V_n)) = (V_{m+1}^0, V_n)$$

$$(42) \qquad\qquad R_0((V_m^0, V_n)) = (V_{m+1}^1, V_n)$$

$$(43) \qquad\qquad R_0((V_m^1, V_n)) = (V_{m-1}^0, V_n)$$

Suppose we are given 2 generalized Weyl modules and a vertex operator acting between them. Suppose in addition that this vertex operator is related to a highest weight in the third generalized Weyl module, say (V_m^ϵ, V_n). Theorem 4.5.1 tells us that given such a vertex operator our screenings give us all the others of the type (V_i^α, V_n) – we cannot only change the value of n. But then there is the standard screening operator – S – which takes care of n, see e.g. [14]. So these three – R_1, R_2, S – provide us with all vertex operators. This has an important application to the calculation of correlation functions.

Start with a simple correlation function given by the product of vertex operators, each of which is characterized by the condition $m = 0$. Then applying S an appropriate number of times one gets all vertex operators and, hence, all correlation functions in the spirit of Varchenko-Schechtman, see[2].

Now take a Varchenko-Schechtman correlation function Ψ_{old}. It comes from a product of vertex operators:

$$\Psi_{old} = < v^*, Y(x_m, z_m) \cdots Y_1(x_1, z_1) v_0 >,$$

$$Y_i : \ \mathcal{F}_{\lambda_i \mu_i}^{\mathbf{C}^*} \otimes V_{i-1/2} \to V_{i+1/2}, \ 1 \leq i \leq m.$$

Let W_i, $0 \leq i \leq m$, be words on 2 letters R_1 and R_2. Replacing $V_{i+1/2}$ with $W_i(V_{i+1/2})$ we get a new correlation function Ψ_{new}. Doing this with all Ψ_{old} and sufficiently many W_i, $0 \leq i \leq m$, we get all solutions to (36,37). In principle all these solutions can be written down explicitly. It is especially simple to do so in the case when we keep $V_{i+1/2}$, $0 \leq i \leq m - 1$, and only change $V_{m+1/2}$.

So assume that Ψ_{old} is as above and replace $V_{m+1/2}$ with $R_j(V_{m+1/2})$, $j = 0, 1$. Then by (38) one is to expect that

$$\Psi_{new} = < X^\alpha \cdot v^*, \circ Y(x_m, z_m) \cdots Y_1(x_1, z_1) v_0 >,$$

where X is either e or $f \otimes z$ if $j = 1$ or 0 resp., and α is either $\lambda + 2$ or $k - \lambda + 2$ resp., where λ is the highest weight of $V_{m+1/2}$.

Of course if α is not a nonnegative integer then the last formula does not make much sense. Nevertheless, using it and (39) as a motivation we arrive at

$$\Psi_{new} = X^\alpha \Psi_{old}.$$

Now the right-hand side of the last equality does make sense: X is a first order differential operator, see 5.4.1, therefore we can set in a rather straightforward manner

$$X^\alpha \Psi_{old} = \int t^{-\alpha-1}\{\exp(-Xt)\Psi_{old}\}dt$$

and get a nice integral operator; for details see [20].

This procedure can be easily iterated to provide the functions

(44)
$$\int \prod_{i=1}^{n} t_i^{-\alpha_i-1}\{\exp(-X_1 t_1)\exp(-X_2 t_2)\cdots\exp(-X_n t_n)\Psi_{old}\}\prod_{i=1}^{n} dt_i,$$

where $X_1, X_2, \ldots$ is either $e, f \otimes z, e, \ldots$ or $f \otimes z, e, f \otimes z, \ldots$.

Lemma 5.4.1. *The functions (44) are solutions to (36,37).*

Proof is the same as the proof of the analogous statement in [19]. In fact it is an easy exercise to make the heuristic arguments which have led us to the formula (44) into a precise proof. $\square$

Integrating functions (44) with respect to the x's (or doing something similar but more esoteric) one is supposed to get the Dotsenko-Fateev correlation functions for the Virasoro algebra. It would be interesting to do this explicitly and to compare the result with the calculations in [24].

Conjecture 5.4.2. *(i) Formulas (44) provide all solutions to the system (36),(37).*

(ii) If the level k is rational, then there arises a subbundle of the bundle in question, the one with fiber $((L^A)^*)^{g(\mathbf{CP^1}, A)}$, where L^A is the corresponding admissible representation. We conjecture that in this case formulas (44) actually give horizontal sections of the latter bundle.

REFERENCES

[1] Awata H.,Yamada Y., *Fusion rules for the Fractional Level $\widehat{sl}(2)$ Algebra*, Modern Phys.Lett. A 7 (1992) 1185-1195

[2] Awata H., Tsuchiya A., Yamada Y., Nucl.Phys., **B365**(1991), 680-698

[3] Bauer M., Sochen N. Singular vectors by fusions in $A_1^{(1)}$ Phys.Lett. B 275(1992) 82-86

[4] Beilinson A., Feigin B., Mazur B. *Introduction to algebraic field theory on curves*, preprint

[5] Beilinson A., Bernstein J. *Localisation de g−modules*, C.R.Acad.Sci.Paris, **292** (1981), 15-18

[6] Belavin A.A., Polyakov A.M., Zamolodchikov A.B., Nuclear Physics **B241** (1984) 333-380

[7] Bernstein J.N., Gelfand S.I. Compositio Mathematica **41**, 2, (1980) 245-285

[8] Brylinski J-L., Kashiwara M. *Kazhdan-Lusztig conjecture and holonomic systems*, Invent.Math., **64** (1981), 387-410

[9] Drinfeld V.G. Proc.Int.Congr.Math. Berkeley **1** 1986

[10] Deodhar V.V., Gabber O., Kac V.G., Adv.in Math **45** (1982) 92-116

[11] Finkelberg M. Fusion categories, Ph.D. thesis, Harvard university,1993

[12] Feigin B., Fuchs D. Journal of Geom. and Phys. **5** (1988) 209-235

[13] Feigin B., Frenkel E. Phys.Lett.B **246** (1990) 75

[14] Feigin B., Frenkel E., in Physics and Mathematics of Strings, V.G.Knizhnik Memorial Volume, eds. L.Brink, et.al, 271-316, World Scientific, Singapore 1990

[15] Feigin B.L., Fuchs D.B. in Representations of Lie groups and related topics, eds. A.M.Vershik, D.P.Zhelobenko, 465-554, Gordon and Breach, New York 1990

[16] Feigin B.L., Fuchs D.B. J.Geom.Phys. **5**(1988) n.2

[17] Felder G. Nucl.Phys. **317** (1989)

[18] Frenkel E., Kac V., Wakimoto M. Comm.Math.Phys. **147**(1992) 295-328

[19] Feigin B., Malikov F. Letters in Math.Phys. **31** 1994, 315-325

[20] Feigin B., Malikov F., Advances in Sov.Math. **17**(1993) 15-63

[21] Frenkel I.B., Reshetikhin N.Yu Comm.Math.Phys. **146**(1992) 1-60

[22] Frenkel I., Lepowsky J., Meurman A. *Vertex Operator Algebras and the Monster*, Academic Press, Inc 1988

[23] Fuchs D.B. Funkc. Anal. i Ego Pril. **23** (1989) 2, 81-83

[24] Furlan P., Ganchev A.Ch., Paunov R., Petkova V.B., Phys.Letters **267** (1991) 63; Nucl.Phys. **B394** (1993) 665-706d 2

[25] Hitchin N. Duke Math.J., **54** (1987), n.1

[26] Kac V.G., Kazhdan D.A., Adv.in Math. **34** (1979) 97-108

[27] Iohara K., Malikov F., Modern Phys.Lett.A **8**(1993), No.38 3613-3624

[28] Jimbo M. Lett.Math.Phys. **10** 63-69 (1985)

[29] Kac V.G., Wakimoto M. Proc. Nat'l Acad. Sci. USA **1988** 4956

[30] Kazhdan D., Lusztig G. Duke Math.J. **62** 21-29

[31] Knizhnik V.G., Zamolodchikov A.B. Nucl.Phys. **B 247** (1984) 83-103

[32] Kulish P.P., Reshetikhin N.Yu. Lett.Math.Phys. **18** 143-149 (1989)

[33] Malikov F.G., Feigin B.L., Fuchs D.B., Funkc.Anal.i ego Pril. **20**(1988) 2, 25-37

[34] Malikov F., Infinite Analysis - Proceedings of the RIMS Research Project 1991 Part B, 623 - 645, World Scientific Co. Pte. Ltd.

[35] Malikov F. Leningrad Math.Journal **2** (1991) 269-286

[36] Mathieu P., Walton M. Prog. Theor. Phys. Suppl. **102** (1990) 229

[37] Moore G., Seiberg N. Comm.Math.Phys. **123** (1989) 177-254

[38] Petersen J.L., Rasmussen J., Yu M. Conformal blocks for admissible representations in $sl(2)$ current algebra, Nucl.Phys. B **457** (1995) 309-342

[39] Pressley A., Segal G. Loop groups, Oxford Mathematical Monographs, 1988

[40] Ramgoolam S. New modular Hopf algebras related to rational level $\widehat{sl}_2$ YCTP-P2-93, hep-th/9301121

[41] Reshetikhin N.Yu., Turaev V.G. Inv.Math. **103** 547-597 (1991)

[42] Segal G., in Proceedings of XI International Congress on Mathematical Physics, 17 - 27 July, 1988 Swansea, Adam Hilger Bristol and New York, 1988, 22-37

[43] Schechtman V., Varchenko A., Invent.Math. **106**(1991), 139-194

[44] Tsuchiya A., Ueno K., Yamada Y., Adv. Studies in Pure Math. **19** (1989) 459-566

[45] Verlinde E. Nucl.Phys. **B300** (1988) 360

LANDAU INSTITUTE FOR THEORETICAL PHYSICS

DEPARTMENT OF MATHEMATICS, UNIVERSITY OF SOUTHERN CALIFORNIA

Contemporary Mathematics
Volume **202**, 1997

Quantum generalized cohomology

Jack Morava

ABSTRACT. We construct a ring structure on complex cobordism tensored with
$\mathbb{Q}$, which is related to the usual ring structure as quantum cohomology is
related to ordinary cohomology. The resulting object defines a generalized
two-dimensional topological field theory taking values in a category of spectra.

Introduction

The conclusion of this paper is that the theory of two-dimensional topological gravity has a remarkably straightforward homotopy-theoretic interpretation in
terms of a generalized cohomology theory, completely analogous to the more familiar interpretation of quantum ordinary cohomology as a topological field theory.
Two-dimensional gravity originated in attempts to integrate over the space of metrics on a Riemann surface; it was reformulated by Witten in terms of an algebra
of generalized Miller-Morita-Mumford characteristic classes for surface bundles. In
the interpretation proposed here, this algebra is the coefficient ring of a generalized
cohomology theory, and topological gravity becomes a topological-field-theory-like
functor, which assigns invariants to families of algebraic curves just as a classical
topological field theory assigns invariants to individual curves; in this it resem-
bles algebraic K-theory, which assigns homotopy-theoretic invariants to families of
modules over a ring.

Here is an outline of the argument. After a preliminary section which collects
some background information, we define a generalized topological field theory in §2
in terms of (homotopy classes of) maps

$$\tau_{g,n} : \overline{M}_{g,n}^+ \to \mathbf{M}^{\wedge n}$$

from a compactified moduli space of curves marked with n points, to n-fold powers
of a module-spectrum $\mathbf{M}$. These maps preserve a monoidal structure (defined geometrically in the domain by glueing curves together at marked points, but defined
algebraically in the range); in the language of [8 §3,11 §1.7], $\tau_{*,*}$ is a representation of a certain cyclic operad. The existence of such a representation entails the
existence of a (quantum) multiplication on the module-spectrum $\mathbf{M}$ (cf. §2.4); in

---

1991 *Mathematics Subject Classification.* Primary 14H10, Secondary 55N35, 81R10.
Author was supported in part by NSF Grant #9504234.

familiar cases this is the multiplicative structure defined by the WDVV equation, and when $n = 0$ we recover Witten's tau-function for the moduli space of curves.

That topological gravity and quantum cohomology are closely related is clear from [29], but I suspect that the simplicity of the underlying geometry is not widely understood. The main technical lemma (§2.1) is a kind of splitting theorem (cf. [15,25]) for generalized Gromov-Witten classes; it is quite natural, in light of Kontsevich's ideas about stacks of stable maps. These moduli objects have not yet been shown to be smooth for curves of all genus for any variety more complicated than a point [15, §1], so our main example is still conjectural, but in fact smoothness is much more than we need; the constructions of this paper require only that the generalized Gromov-Witten maps of §1.3 be local complete intersection morphisms. This is a convenient working hypothesis, which can be weakened further by elaborating the cohomological formalism; but this paper concerned with the consequences of this assumption, not with proving it. For curves of genus zero the moduli stacks are known to be smooth, if the target manifolds are convex in a suitable sense [2]; this leads to a simple proof (§2.2) of the associativity of the quantum multiplication, and when the defining variety is a point, we can calculate the corresponding coupling constant (§2.3).

A short final section discusses some related issues. In particular, there is reason to think that the Virasoro algebra is a ring of 'quantum generalized cohomology operations' for the main example. To state this more precisely requires a short digression about representations of the group of antiperiodic loops on the circle, which is included as an appendix.

It is a pleasure to thank Ralph Cohen, Yuri Manin, and Alexander Voronov for helpful conversations about the content of this paper; that the paper exists at all, however, is the consequence of helpful conversations with Graeme Segal and Edward Witten.

1. Notation and conventions

1.1. Let V be a simply-connected projective smooth complex algebraic variety of real dimension $2d$, with first Chern class $c_1(V)$, and let H denote its second integral homology group $H_2(V, \mathbb{Z})$. We will use a rational version

$$\Lambda = \mathbb{Q}[H \times \mathbb{Z}]$$

of the Novikov ring [18 §1.8] of V: its elements are Laurent polynomials

$$\sum_{k \in \mathbb{Z}, \alpha \in H} c_{\alpha,k} \alpha \otimes v^k$$

with coefficients $c_{\alpha,k} \in \mathbb{Q}$. This ring has a useful grading, in which v has (cohomological) degree two, and α has degree $2\langle c_1(V), \alpha \rangle$. We will also use the notation

$$v_{(k)} = \frac{v^k}{k!}$$

for the k-th divided power of v. If $u : \Sigma \to V$ is a map from a connected oriented surface to V, then the degree of u is the class $u_*[\Sigma] \in H$, where $[\Sigma] \in H_2(\Sigma, \mathbb{Z})$ is the fundamental class of the surface.

1.2. $MU_*(X)$ will denote the complex bordism of a CW-space X, and $\mathbf{MU}_\Lambda$ is the spectrum representing the homology theory defined on base-pointed finite complexes by

$$X \mapsto MU_*(X)\hat{\otimes}\Lambda = [S^*, X \wedge \mathbf{MU}_\Lambda];$$

the tensor product has been completed, so $MU^*(pt)\hat{\otimes}\Lambda$ is the graded ring of formal Laurent series in v, with coefficients from the graded group ring $MU^*(pt)[H]$. It will be convenient to write $\mathbf{MU}_\Lambda(V)$ for the function spectrum $F(V^+, \mathbf{MU}_\Lambda)$; the superscript $+$ indicates the addition of a disjoint basepoint. The fiber product of spaces (or schemes) X and Y over Z will be denoted $X \times_Z Y$, and the product of $MU^*_\Lambda(X)$ and $MU^*_\Lambda(Y)$ over $MU^*_\Lambda(Z)$ will be denoted $MU^*_\Lambda(X) \otimes_Z MU^*_\Lambda(Y)$. The spectrum $\mathbf{MU}_\Lambda(V)$ has an $\mathbf{MU}_\Lambda$-algebra structure, and

$$\mathbf{MU}_\Lambda(V^n) = \mathbf{MU}_\Lambda(V) \wedge_{\mathbf{MU}_\Lambda} \cdots \wedge_{\mathbf{MU}_\Lambda} \mathbf{MU}_\Lambda(V)$$

is its n-fold Robinson smash power [24] over $\mathbf{MU}_\Lambda$. There is a map

$$Tr_V : \mathbf{MU}_\Lambda(V) \to \mathbf{MU}_\Lambda$$

of $\mathbf{MU}_\Lambda$-module spectra, which represents the transfer map

$$MU^*_\Lambda(X \wedge V^+) \to MU^{*-2d}_\Lambda(X)$$

defined by the (complex oriented) projection $V \to pt$, followed by multiplication with v^d (to shift dimensions).

According to Quillen, a proper complex-oriented map $\Phi : P \to M$ between smooth manifolds defines an element $[\Phi]$ of $MU^k(M)$, where k is the codimension of Φ; more generally, a suitably oriented map between geometric cycles, with enough of a normal bundle to possess rational Chern classes, will define an element of $MU^k_{\mathbb{Q}}(M)$. Contravariant maps in cobordism are defined by fiber products, while covariant map are defined by the obvious compositions. Finally, the bilinear form

$$b_V : \mathbf{MU}_\Lambda(V) \wedge \mathbf{MU}_\Lambda(V) \to \mathbf{MU}_\Lambda(V) \to \mathbf{MU}_\Lambda$$

is the composition of the trace with the multiplication map of $\mathbf{MU}_\Lambda(V)$.

The graded ring $MU^*_\Lambda(V)$ is a technical replacement for $MU^*_{\mathbb{Q}[v,v^{-1}]}(V \times H)$, which is in some ways more natural; but the latter ring does not help with the usual convergence problems, which (in the present framework) are consequences of the failure of the map $H \times H \to pt$ to be proper.

1.3. A (marked) algebraic curve is stable if its group of automorphisms is finite; $\overline{M}_{g,n}$ will denote the Deligne-Mumford-Knudsen space of such curves of arithmetic genus g, marked with n ordered smooth points. These spaces are compact orbifolds, of complex dimension $3(g-1)+n \geq 0$; cases of low genus are thus sometimes exceptional. It is useful to understand n to be a finite ordered set (or ordinal number), so that permutations of n can act on $\overline{M}_{g,n}$. More generally, $\overline{M}_{g,n}(V, \alpha)$ will denote the stack [2 §3, 19] of stable maps of degree α from a curve of genus g marked with n ordered smooth points, to V; there is a morphism from $\overline{M}_{g,n}(V, \alpha)$ to $\overline{M}_{g,n}$ which assigns to a map (the stabilization of) its domain, and there is a morphism to V^n which evaluates a map at the marked points. The product of these is a perfect (finite Tor-dimension [7 II §1.2]) proper morphism

$$\Phi^V_{g,n,\alpha} : \overline{M}_{g,n}(V, \alpha) \to \overline{M}_{g,n} \times V^n$$

of stacks. At a point $u : \Sigma \to V$ of $\overline{M}_{g,n}(V, \alpha)$ defined by a smooth Σ, the relative tangent space to $\overline{M}_{g,n}(V, \alpha)$ over $\overline{M}_{g,n}$ defines a K-theory class

$$[H^0(\Sigma, u^*T_V)] - [H^1(\Sigma, u^*T_V)]$$

of complex dimension $d(1 - g) + \langle c_1(V), \alpha \rangle$, where d is the complex dimension of V. It seems likely that under reasonable hypotheses this map will be a local complete intersection morphism of stacks, and in particular that the K-theory class of its cotangent complex at a singular point (Σ, u) will equal the holomorphic Euler class of the pullback of u^*T_V to the normalization of Σ. [In fact, Kontsevich [13 §1.4] has already sketched something very close to a local complete intersection structure for this morphism.]. Similarly, let $\epsilon_{g,n,\alpha}^V(k)$ denote the proper complex-oriented map from $\overline{M}_{g,n+k}(V, \alpha)$ to $\overline{M}_{g,n}(V, \alpha)$ which forgets the final k marked points, and define

$$\Phi_{g,n,\alpha}^V(k) = \Phi_{g,n,\alpha}^V \circ \epsilon_{g,n,\alpha}^V(k) : \overline{M}_{g,n+k}(V, \alpha) \to \overline{M}_{g,n} \times V^n.$$

There are also generalizations

$$\mu_s^V : \overline{M}_{g,r+s}(V, \alpha) \times_{V^s} \overline{M}_{h,s+t}(V, \beta) \to \overline{M}_{g+h+s-1,r+t}(V, \alpha + \beta),$$

of Knudsen's glueing morphisms, cf. [12]. All these maps represent natural transformations between moduli functors, so their normal bundles are reasonably accessible. In diagrams below, complicated subscripts and superscripts will be supressed when they are redundant in context.

2. Generalized topological field theories

2.1.   We will be interested in generalized Gromov-Witten invariants defined by the cobordism classes of these morphisms. **I will assume that these maps are local complete intersection morphisms**; such maps between (possibly singular) varieties have most of the topological transversality properties of maps between smooth manifolds. In particular, they have well-behaved Gysin homomorphisms and normal bundles [1 IV §4] and they thus define elements in complex cobordism tensored with the rationals. [A variant approach is discussed below in §2.5.] Our generalized Gromov-Witten invariants are the classes

$$\phi_{g,n}^V(k) = \sum_{\alpha \in H} [\Phi_{g,n,\alpha}^V(k)] \alpha \otimes v_{(k)} \in MU_\Lambda^{2d(n+g-1)}(\overline{M}_{g,n} \times V^n);$$

permutations of k define cobordant elements. Summing these classes over k defines

$$\tau_{g,n}^V = v^{-d(n+g-1)} \sum_{k \geq 0} \phi_{g,n}^V(k) \in MU_\Lambda^0(\overline{M}_{g,n} \times V^n);$$

the convergence problems mentioned in the preceding section do not appear when the function $\alpha \mapsto [\Phi_{g,n,\alpha}^V(k)]$ is supported in a proper cone in H. The tau-function $\tau_{g,n}^V$ can be interpreted geometrically as the cobordism class of the 'grand canonical ensemble' of maps from a curve of genus g marked with n ordered smooth points, together with an indeterminate number of further distinct smooth (unordered) points, to V, but we will be more concerned with the homotopy class

$$\tau_{g,n}^V : \overline{M}_{g,n}^+ \to \mathbf{MU}_\Lambda(V^n)$$

it defines.

PROPOSITION 2.1. *The diagram*

$$\overline{M}^{+}_{g,r+s} \wedge \overline{M}^{+}_{h,s+t} \xrightarrow{\mu_s} \overline{M}^{+}_{g+h+s-1,r+t}$$

$$\downarrow{\tau \wedge \tau} \qquad\qquad\qquad \downarrow{\tau}$$

$$\mathbf{MU}_\Lambda(V^{r+s}) \wedge \mathbf{MU}_\Lambda(V^{s+t}) \xrightarrow{b^s_V} \mathbf{MU}_\Lambda(V^{r+t})$$

commutes up to homotopy; alternately,

$$\mu^*_s(\tau^V_{g+h+s-1,r+t}) = v^{sd} Tr^s_V(\tau^V_{g,r+s} \otimes_{V^s} \tau^V_{h,s+t}).$$

SKETCH PROOF, under the standing hypothesis above : The general case reduces by induction to the case $s = 1$, which can be stated as a coproduct formula

$$\mu^* \phi^V_{g+h,r+t}(k) = v^d \sum_{i+j=k} Tr_V(\phi^V_{g,r+1}(i) \otimes_V \phi^V_{h,1+t}(j)).$$

This can be reformulated as the assertion that the two diagrams

$$\bigsqcup_{i+j=k} \overline{M}_{g,i+r+1}(V,\alpha) \times_V \overline{M}_{h,1+t+j}(V,\beta) \xrightarrow{\mu_V} \overline{M}_{g+h,r+t+k}(V,\alpha+\beta)$$

$$\downarrow{\bigsqcup \epsilon(i) \times \epsilon(j)} \qquad\qquad\qquad\qquad \downarrow{\epsilon(k)}$$

$$\overline{M}_{g,r+1}(V,\alpha) \times_V \overline{M}_{h,1+t}(V,\beta) \xrightarrow{\mu} \overline{M}_{g+h,r+t}(V,\alpha+\beta)$$

and

$$\bigsqcup_{\alpha+\beta=\gamma} \overline{M}_{g,r+1}(V,\alpha) \times_V \overline{M}_{h,1+t}(V,\beta) \xrightarrow{\mu_V} \overline{M}_{g+h,r+t}(V,\gamma)$$

$$\downarrow{\bigsqcup \Phi_\alpha \times \Phi_\beta} \qquad\qquad\qquad\qquad \downarrow{\Phi_\gamma}$$

$$\overline{M}_{g,r+1} \times V^{r+1+t} \times \overline{M}_{h,1+t} \xrightarrow{\mu \times tr_V} \overline{M}_{g+h,r+t} \times V^{r+t}$$

are fiber products; the claim follows by stacking the first diagram on top of the second. The bottom diagram describes the stable maps of decomposable curves in terms of the restrictions to their components. In the top diagram, the union is to be taken over partitions of the set with k elements into subsets of cardinality i and j; on the level of functors, this diagram asserts that the ways of sprinkling points on a curve decomposed into two components correspond to the ways of sprinkling points on the components separately.

The last assertion is not entirely straightforward, because forgetting marked points may destabilize a genus zero component of a stable marked curve; the morphism $\Phi^V(k)$ will blow such components down to points. The union in the upper left corner of the diagram is thus not necessarily disjoint: the fiber product is obtained from the disjoint union by identification along certain divisors, cf. [21 §3]. The point, however, is that the cobordism classes are defined by maps rather than by subobjects; the fiber product class is equivalent to the sum of the classes defining the disjoint union. □

2.2. Proposition 2.1 states that the triple $(\tau^V_{*,*}, \mathbf{MU}_\Lambda(V), b_V)$ defines a topological field theory which takes values in the category of $\mathbf{MU}_\Lambda$-module spectra, where the usual monoidal structure defined by tensor product of modules over a ring is replaced by the smash product of module-spectra over a ring-spectrum. The domain of this generalized topological field theory is the monoidal category *(Stable Curves)* with finite ordered sets as objects; morphisms are finite unions of marked curves. [This category, however, does not possess identity maps for its objects.]

Both the domain and range of the generalized topological field theory are topological categories, and $\tau_{*,*}^V$ defines a homotopy class of maps from the space of morphisms of the domain category to the space of morphisms of the range. These homotopy classes preserve the composition of morphisms and thus define a functor. In this generality, we need a version of Proposition 2.1 for Knudsen glueing of two points on a connected curve, but the changes required for this are minor.

The construction involves the bilinear map b_V, but it has not otherwise used the multiplicative structure on $\mathbf{MU}_\Lambda(V)$. In fact the morphism

$$\tau_{0,3}^V : S^0 = \overline{M}_{0,3}^+ \to \mathbf{MU}_\Lambda(V^3)$$

defines a composition

$$*_V : \mathbf{MU}_\Lambda(V) \wedge S^0 \wedge \mathbf{MU}_\Lambda \xrightarrow{id \wedge \tau_{0,3} \wedge id} \mathbf{MU}_\Lambda(V^5) \xrightarrow{b_V^{\otimes 2} \wedge id} \mathbf{MU}_\Lambda(V),$$

and Proposition 2.1 has the following

COROLLARY 2.2. *The pair* $(\mathbf{MU}_\Lambda(V), *_V)$ *is a homotopy commutative and homotopy associative ring-spectrum.*

SKETCH PROOF. Smashing the morphism

$$A(2,2) := (id \wedge b_V \wedge id)(\tau_{0,2+1} \wedge \tau_{0,1+2}) : S^0 = \overline{M}_{0,2+1}^+ \wedge \overline{M}_{0,1+2}^+ \to \mathbf{MU}_\Lambda(V^4)$$

with the identity map of $\mathbf{MU}_\Lambda(V^3)$ defines a map from $\mathbf{MU}_\Lambda(V^3)$ to $\mathbf{MU}_\Lambda(V^7)$; arranging the seven copies of V into pairs and applying the trace map b_Λ three times defines a collection of maps from $\mathbf{MU}_\Lambda(V^3)$ to $\mathbf{MU}_\Lambda(V)$ indexed by the possible groupings of the factors. By our conventions $2 + 1$ and $1 + 2$ are isomorphic but not equal, so the notation for this associator class emphasizes that it depends on four points partitioned into two subsets, each containing two items. Ignoring obvious involutions, there are three different partitions, corresponding to the maps π_0, π_1, π_∞ from the 0-manifold $\overline{M}_{0,2+1} \times \overline{M}_{0,1+2}$ to $\overline{M}_{0,4}$ which send it to a degenerate curve of genus zero with two irreducible components, each carrying two marked points (aside from the node); the cross-ratio identifies these configurations with the standard points 0,1, and ∞ on the projective line. To verify associativity it suffices to show that the homotopy class $A(2,2)$ is independent of the way the four points are partitioned into pairs; but by the proposition, $A(2,2) = \tau_{0,4} \circ \pi_i$ factors through $\overline{M}_{0,4}$, where the three maps π_i become homotopic. $\square$

2.3. The morphism

$$\tau_{0,3} = v^{-2d} \sum_{k \geq 0} [\overline{M}_{0,k+3}(V) \to V^3] v_{(k)}$$

defining this quantum multiplication is essentially the Gromov-Witten potential [15]. Because the moduli spaces $\overline{M}_{g,n}$ are not defined when $3(g-1) + n$ is negative, however, it is not clear that the resulting multiplicative structure on $\mathbf{MU}_\Lambda(V)$ possesses a unit. The class

$$q_V := 1 *_V 1 = v^{-2d}(b_V \otimes id)(\tau_{0,3}^V)$$

is the coupling constant for the topological field theory defined by $MU_\Lambda^*(V)$; this theory assigns to a connected surface of genus g with one boundary component, the $2g$th power of 1 with respect to the product $*_V$.

COROLLARY 2.3. *When V is a point, the coupling constant of the resulting topological field theory is*

$$q = \sum_{k \geq 0} [\overline{M}_{0,k+3}] v_{(k)} \in MU_\Lambda^0(pt),$$

and the quantum product in $MU_\Lambda^*(pt)$ is $x * y = qxy$.

In this formula $[\overline{M}_{0,k+3}]$ is the cobordism class of the manifold of configurations of $k + 3$ points on a curve of genus zero; the sum is thus a cobordism analogue of Manin's Hodge-theoretic invariant φ [19 §0.3.1]. With this structure, $(MU_\Lambda^*(pt), *_{pt})$ is isomorphic to $MU_\Lambda^*(pt)$ with its usual multiplication, by a homomorphism which sends 1^{*2g} to q^{2g}. More generally, the operation $x \mapsto 1 *_V x$ is a module isomorphism, and it seems reasonable to hope that $(1*_V)^{-1}(1)$ will be a unit for $*_V$.

Knudsen glueing defines a pair-of-pants product

$$\mu^+ : \overline{M}_{g,1}^+ \wedge \overline{M}_{h,1}^+ \to \overline{M}_{g,1}^+ \wedge \overline{M}_{0,3}^+ \wedge \overline{M}_{h,1}^+ \to \overline{M}_{g+h,1}^+$$

and it follows from §2.1 and the arguments above that the diagram

$$
\begin{array}{ccc}
\overline{M}_{g,1}^+ \wedge \overline{M}_{h,1}^+ & \xrightarrow{\mu^+} & \overline{M}_{g+h,1}^+ \\
\downarrow{\tau \wedge \tau} & & \downarrow{\tau} \\
\mathbf{MU}_\Lambda \wedge \mathbf{MU}_\Lambda & \xrightarrow{*_V} & \mathbf{MU}_\Lambda
\end{array}
$$

is homotopy-commutative; in other words,

$$\tau_{*,1}^V : \overline{M}_{*,1} \to \mathbf{MU}_\Lambda$$

is a kind of homomorphism of monoids. This is probably the most intuitive way to think of the product in quantum cohomology, but from the present point of view it is a conclusion, not a definition.

2.4. A generalized topological field theory has an associated theory of topological gravity, which assigns invariants to proper flat families of stable curves. Such a family Z, say of topological type (g, n), is defined by its classifying map to $\overline{M}_{g,n}$; the pullback of $\tau_{g,n}^V$ along this morphism defines a class $\tau^V(Z) \in [Z^+, \mathbf{MU}_\Lambda(V^n)]$. If (for simplicity) we assume that V is a point, and write $[Z] \in H_*(Z)$ for the fundamental class of Z, then the image $\tau_*(Z)$ of $[Z]$ in $H_*(\mathbf{MU}, \Lambda)$ under the map induced on homology by $\tau(Z)$ is a kind of absolute invariant of the family, obtained by integrating $\tau(Z)$ over $[Z]$. In particular, the vacuum morphism

$$0 \to 0$$

is defined by the family of arbitrary finite unions of unmarked stable curves; the infinite symmetric product $SP^\infty(\bigsqcup_{g \geq 0} \overline{M}_{g,0})$ is a rational model for its parameter space. The resulting absolute invariant

$$\tau = \exp\left(\sum_{g \geq 0} \tau_*(\overline{M}_{g,0})\right) \in H_*(\mathbf{MU}, \mathbb{Q}[v, v^{-1}])$$

is Witten's tau-function for two-dimensional topological gravity [20]. The point is that the characteristic number homomorphism

$$MU^*(M) \to H^*(M, H_*(\mathbf{MU}))$$

sends the class $[\Phi : P \to M]$ to a sum of the form $\sum_I \Phi_* c_I(\nu) t^I$, where ν is the stable normal bundle of Φ, $c_I(\nu)$ is a certain polynomial (indexed by $I = i_1, \dots$) in its Chern classes, $t^I = t_1^{i_1} \dots$ is a product of elements in $H_*(\mathbf{MU}) = \mathbb{Z}[t_i | i \geq 1]$, and Φ_* is the covariant (Gysin) homomorphism induced by Φ. The stable normal bundle of $\Phi(k)$ is inverse to the tangent bundle along the fibers, which is the sum of the k line bundles defined by the tangent space to the universal curve at its k marked points, so its Chern polynomials can be expressed as polynomials in the Chern classes of these line bundles. Under the pushdown $\Phi(k)_*$, these become polynomials in the Mumford classes in the cohomology of $\overline{M}_g$.

2.5. Some aspects of Kontsevich-Witten theory suggest that Gromov-Witten invariants can be defined more naturally in K-theory than in ordinary cohomology. The algebraic K-theory of a reasonable stack, tensored with the rationals, agrees with the algebraic K-theory of its quotient space [3 §7], but (perhaps because of this) the K-theory of stacks seems to have received little attention otherwise. The following assumes that the standard direct image construction for perfect proper maps of schemes [27 §3.16.4] generalizes to stacks.

Let $\pi_{g,n} : C_{g,n} \to \overline{M}_{g,n}$ be the universal stable curve; because the range is smooth, this is a perfect proper morphism. Let

$$C_{g,n,\alpha}(V) := \overline{M}_{g,n,\alpha}(V) \times_{\overline{M}_{g,n}} C_{g,n} ;$$

from now on I will supress the subscripts. The projection $\bar{\pi} : C(V) \to \overline{M}(V)$ to the first factor, being the pullback of a perfect morphism, is again perfect. Let $U : C(V) \to V$ be the universal evaluation morphism; the vector bundle $U^* T_V$ defines an element of $K(C(V))$, and its hypothetical direct image $\bar{\pi}_* U^* T_V := \nu(\Phi^V) \in K(\overline{M}(V))$ is a reasonable candidate for the normal bundle of Φ^V.

Because Φ^V is itself proper and perfect, we can define generalized Gromov-Witten classes

$$\sum_I \Phi_*^V m_I(\nu(\Phi^V)) t^I \in K^*(\overline{M}_{g,n} \times V^n) \otimes_K K_* \mathbf{MU},$$

where I is a multiindex as above, $t^I = \prod_k t_k^{i_k}$ is a basis for $K_* \mathbf{MU}$, and m_I denotes the K-theory characteristic class associated to the monomial symmetric function by the correspondence which assigns gamma operations [3 V §3] to the elementary symmetric functions. By the Hattori-Stong theorem, such a sum can be identified with a class in the localization $MU^*(\overline{M}_{g,n} \times V^n)[\mathbb{C}P(1)^{-1}]$ of complex cobordism.

If Φ^V is a local complete intersection morphism, this approach to defining Gromov-Witten invariants agrees with the definition in §2.1. In any case some such hypothesis seems to be needed to make the arguments of Prop. 2.1 work.

3. Some questions

3.1. Kontsevich and Witten [13,17,29] show that the tau-function for the vacuum state of two-dimensional topological gravity is a lowest weight vector for a certain representation of the Virasoro algebra. This Lie algebra bears a striking resemblance to the Lie algebra defining the Landweber-Novikov algebra of operations in complex cobordism, but the relation between these two structures is not well-understood. I have included as an appendix a construction for the Kontsevich-Witten representation, starting from a representation of a certain loop group of antiperiodic functions on the circle, following [6]. One point of the appendix is that the representation theory of this loop group is essentially trivial.

On the other hand, the usual complex cobordism functor takes values in the monoidal category of $\mathbb{Z}/2\mathbb{Z}$-graded **G**-equivariant sheaves over the moduli scheme Spec $MU^*(pt)$ of formal group laws, with the Landweber-Novikov group **G** of formal coordinate transformations acting by change of coordinate; but this category is equivalent, after tensoring with $\mathbb{Q}[v, v^{-1}]$, to the category of $\mathbb{Z}/2\mathbb{Z}$-graded vector spaces. It therefore seems not completely unreasonable to conjecture that the group of antiperiodic loops is a kind of motivic group for the generalized quantum cohomology theory defined by $\mathbf{MU_\Lambda}$.

As for products in V, the functor $\tau_{*,*}^V$ seems to behave very naturally (cf. [14]); in particular, it is reasonable to expect that

$$\tau_{g,n}^{V_0 \times V_1} = \tau_{g,n}^{V_0} \otimes_{MU_\Lambda^*} \tau_{g,n}^{V_1}.$$

3.2. The work in this paper was originally motivated by a desire to understand topological gravity and quantum cohomology from the point of view of Floer homotopy theory [4,5], but such questions have been supressed here. It may be helpful, however, to observe that the circle group $\mathbb{T}$ acts on the universal cover $\widetilde{LV}$ of the free loopspace of V, with $V \times H$ as fixed point set, so we can think of $MU_\Lambda^*(V)$ as its $t_\mathbb{T}MU_\mathbb{Q}^*$-cohomology [10 §15]. The Tate cohomology $t_\mathbb{T}MU_\mathbb{Q}^*(\widetilde{LV})$ is a rough approximation to the Floer **MU**-homotopy type of $\widetilde{LV}$, and we might hope to understand the relation between these invariants as a localization theorem for Tate cohomology.

More specifically, given a compact pointed Riemann surface (Σ, x), let

$$(D, 0) \to (\Sigma, x)$$

be a holomorphically embedded closed disk; the boundary ∂D separates the surface into components $\bar{\Sigma}_0$ and $\bar{\Sigma}_\infty$, with x the point at infinity, as in [23 §8.11]. Let $Hol(\Sigma, D; V)$ denote the space of continuous maps from Σ to V which are holomorphic on Σ_0 and Σ_∞. This is a manifold, with tangent space

$$H^0(\bar{\Sigma}_0, u_0^* T_V) \oplus H^0(\bar{\Sigma}_\infty, u_\infty^* T_V)$$

at $u \in Hol(\Sigma, D; V)$; here the sections of the pullback bundles are to be holomorphic on the interior and smooth on the boundary. Restriction to the boundary defines a map $u \mapsto \partial u$ to the free loopspace of V, but the homotopy class of u_∞ defines a canonical contraction of ∂u, so this restriction map factors naturally through a lift to the universal cover $\widetilde{LV}$ of LV.

This map is Fredholm, with index equal to the holomorphic Euler characteristic of $u^* T_V$. [More precisely: since u will usually not be holomorphic, $u^* T_V$ can't be expected to be holomorphic either; but $u^* T_V$ restricts to a holomorphically trivial bundle on an annulus containing ∂D, so $u_0^* T_V$ extends to a holomorphic bundle $\tilde{u}^* T_V$ on Σ. Then $\chi(\tilde{u}^* T_V)$ is the index at u.] Moreover, away from maps which collapse ∂D to a point, this map appears to have a good chance to be proper.

We can elaborate this construction, by considering the space $Hol(\Sigma_{\hat{x}}, V)$ of holomorphic disks in Σ centered at x, together with a map to V, continuous and holomorphic away from ∂D as above; since we're enlarging things, we may as well include trivial disks too. This thickening has the same homotopy type as the preceding space, but now $\mathbb{T}$ acts by rotating loops. More generally, we can allow the moduli of Σ to vary as well, thus defining a space of maps over a thickening of the moduli space $\overline{M}$. Restriction maps this space to a similar thickening of $\overline{M} \times \widetilde{LV}$, defining a candidate for a proper $\mathbb{T}$-equivariant Fredholm map, and thus an element of $MU_{\mathbb{T}}^{-2\chi}(\overline{M} \times \widetilde{LV})$, which restricts to the classical Gromov-Witten invariant at the fixed point set of $\mathbb{T}$.

Appendix: $MU^*(pt)$ as a Virasoro-Landweber-Novikov bimodule

The Virasoro algebra is the Lie algebra of a central extension of the group $\mathbf{D}$ of diffeomorphisms of a circle; it is true generally [23 §13.4] that $\mathbf{D}$ acts projectively on a positive energy representation of a loop group, and in this appendix I will sketch the construction of an action of the double cover $\mathbf{D}(2)$ of $\mathbf{D}$ on the basic representation of the twisted loop group

$$L\mathbb{T}\mathrm{wist} = \{f \in L\mathbb{T} | f = \iota(f)\}$$

of functions from the circle $\mathbb{R}/\mathbb{Z}$ to $\mathbb{T} = \{z \in \mathbb{C} \, | \, |z| = 1\}$ which are invariant under the involution $\iota(f)(x) = f(x + \frac{1}{2})^{-1}$.

Because the loop functor preserves fibrations, the exact sequence of the exponential function $\mathbf{e}(x) = e^{2\pi i x}$ yields an exact sequence

$$0 \to \mathbb{Z} \to L\mathbb{R} \to L\mathbb{T}_0 \to 0$$

of abelian groups with involution, the group on the right being the identity component of the group of untwisted loops. The associated exact sequence

$$0 \to L\mathbb{R}^{\mathbb{Z}/2\mathbb{Z}} \to L\mathbb{T}\mathrm{wist} \to H^1(\mathbb{Z}/2\mathbb{Z}, \mathbb{Z}) \to 0$$

of cohomology groups presents the antiperiodic loops as a canonically split extension of the group $\mathbb{Z}/2\mathbb{Z}$ of constant loops with value plus or minus one, by a vector space of antiperiodic functions.

Now $L\mathbb{T}_0$ contains a subgroup $\mathbb{T}$ of constant loops, and $\mathbf{D}$ contains the subgroup R of rotations, so the lift of a positive-energy projective unitary representation of $L\mathbb{T}$ to an honest unitary representation of an extension $\widetilde{L\mathbb{T}}$ of $L\mathbb{T}$ by a circle group C restricts to a representation of a semidirect product $R : E$, where E is an extension of $\mathbb{Z} \times \mathbb{T}$ by C which splits over the identity component. The irreducible positive-energy projective representations of $L\mathbb{T}$ are classified by their restriction to representations of $R \times C \times \mathbb{T}$ [23 §9.3]; an irreducible representation of $\mathbb{T}$ is classified by its weight, and the corresponding integer defined by C is the level. However, the identity component of $\widetilde{L\mathbb{T}}\mathrm{wist}$ has a trivial subgroup of constant loops: its representation theory is effectively weightless.

There is, however, an interesting basic representation of $\widetilde{L\mathbb{T}}\mathrm{wist}$; one construction, modelled on [23 §2], begins with the skew bilinear form defined on $L\mathbb{R}^{\mathbb{Z}/2\mathbb{Z}}$ by

$$B(f_0, f_1) = \frac{2}{\pi} \int_0^1 f_0(x + \tfrac{1}{2}) f_1(x) dx.$$

The group $\mathbf{D}(2)$ of smooth orientation-preserving maps g of $\mathbb{R}$ to itself satisfying $g(x + \frac{1}{2}) = g(x) + \frac{1}{2}$ acts on this symplectic space, by

$$g, f \mapsto g'^{\frac{1}{2}} f \circ g.$$

The complexified space of antiperiodic functions admits the decomposition

$$L\mathbb{R}^{\mathbb{Z}/2\mathbb{Z}} \otimes \mathbb{C} = A_+ \oplus A_-,$$

A_+ being the subspace of functions on the circle which extend inside the unit disk. There is a standard [23 §9.5] unitary representation of $\widetilde{L\mathbb{T}}$wist on the symmetric algebra $S(A_+)$ associated to this polarization; the basis

$$a_n = -\pi i^{-\frac{1}{2}}(-\tfrac{1}{2})^{n+\frac{1}{2}}\Gamma(n+\tfrac{1}{2})^{-1}\mathbf{e}((n+\tfrac{1}{2})x)$$

for the complexification satisfies

$$B(a_n, a_m) = (2m+1)\delta_{n+m+1,0}.$$

The polarization is defined by a nonstandard complex struction in which conjugation acts by $\bar{a}_n = -ia_{-n-1}$, making $iB(\bar{a}, a)$ a positive-definite Hermitian form on A_+. [This complex structure differs from the standard one by a transformation which is diagonal in the basis a_n; this operator is real but unbounded.] The action of $\mathbf{D}(2)$ on $L\mathbb{R}^{\mathbb{Z}/2\mathbb{Z}}$ makes it reasonable to interpret antiperiodic functions as sections of a bundle of half-densities on the circle; the complexification of this bundle admits the nonvanishing flat section

$$(2\pi)^{\frac{1}{2}}\mathbf{e}(x+\tfrac{1}{8})(dx)^{\frac{1}{2}} = (dZ)^{\frac{1}{2}},$$

where $Z = \mathbf{e}(x)$. It follows from Euler's duplication formula that

$$a_n(dx)^{\frac{1}{2}} = (2n+1)!!Z^{-n-1}(dZ)^{\frac{1}{2}}$$

when n is nonnegative.

The Lie algebra of $\mathbf{D}(2)$ now acts on $S(A_+)$ with generators (cf. [14 §1.2, 29 §2])

$$L_k = \tfrac{1}{4}\sum_{n\in\mathbb{Z}} a_{k-n-1}a_n \qquad \text{if} \quad k \neq 0,$$

$$= \tfrac{1}{2}\sum_{n\geq 0} a_{-n-1}a_n + \tfrac{1}{16} \qquad \text{if} \quad k = 0.$$

Convenient polynomial generators t_n for the algebra $S(A_+)$, regarded as a ring of holomorphic functions on A_-, can be defined by expanding an element f of A_- as

$$\sum_{n\geq 0} t_n(f)Z^{-n-1}(dZ)^{\frac{1}{2}};$$

similar generators T_n, constructed by writing this element as

$$\sum_{n\geq 0} T_n(f)a_{-n-1},$$

satisfy the equation

$$t_n = (2n+1)!!T_n.$$

The Virasoro generators (which are not derivations) act on these elements so that

$$L_k T_n = (n-k+\tfrac{1}{2})T_{n-k} \quad \text{if} \quad n \geq k$$

$$= 0 \quad \text{otherwise}.$$

On the other hand the group $\mathbf{G}$ of invertible formal power series (under composition) in Z^{-1} acts on A_-, interpreted as a free module over the ring of power series in Z^{-1}; the Lie algebra of this group is spanned by vector fields

$$v_k = Z^{-k+1}d/dZ \quad \text{with} \quad k \geq 0,$$

which act on $S(A_+)$ (as derivations) such that

$$v_k t_n = (n - k + 1)t_{n-k} \quad \text{when} \quad n \geq k$$

$$= 0 \quad \text{otherwise.}$$

We can thus identify $MU^*_{\mathbb{C}}(pt)$ with $S(A_+)$ as a comodule over the Landweber-Novikov algebra, in a way which makes it a Virasoro representation as well. [The complex coefficients are only a technical convenience.] The resulting bimodule defines a kind of Morita equivalence of the category of cobordism comodules to the category of representations of $\widetilde{L\mathbb{T}}$wist, given the (weak) monoidal structure defined by the fusion product [27 §7] of representations.

References

1. P. Baum, W. Fulton, R. Mac Pherson, Riemann-Roch for singular varieties, Publ. Math. IHES 45 (1975) 101-145
2. K. Behrend, Yu. Manin, Stacks of stable maps and Gromov-Witten invariants, MPIM-Bonn preprint (1995)
3. P. Berthelot et al, Théorie des intersections et theoreme de Riemann-Roch [SGA 6] Lecture Notes in Mathematics 225, Springer (1971)
4. M. Betz, J. Rade, Products and relations in symplectic Floer homology, Stanford preprint (1995)
5. R.L. Cohen, J.D.S. Jones, G.B. Segal, Floer's infinite dimensional Morse theory and homotopy theory, Floer Memorial Volume, Birkhäuser, Progress in Mathematics 133 (1995) 297-326
6. R. Dijkgraaf, E. Verlinde, H. Verlinde, Loop equations and Virasoro constraints in nonperturbative two-dimensional topological gravity, Nucl. Phys. B348 (1991) 435-456
7. W. Fulton, R. Mac Pherson, Categorical framework for the study of singular spaces, Mem. AMS no. 243 (1981)
8. E. Getzler, M. Kapranov, Modular operads, MPIM-Bonn preprint, dg-ga/9408003
9. H. Gillet, Intersection theory on algebraic stacks and Q-varieties, J. Pure and Applied Algebra 34 (1984) 193-240
10. J. Greenlees, J.P. May, Generalized Tate cohomology, Mem. AMS no. 543 (1995)
11. T. Kimura, J. Stasheff, A.A. Voronov, The operad structures of moduli spaces and string theory, Comm. Math. Phys. 171 (1995) 1-25
12. F.F. Knudsen, The projectivity of the moduli space of stable curves II, Math. Scand. 52 (1983) 161-199
13. M. Kontsevich, Enumeration of rational curves via torus actions, in The Moduli Space of Curves, Birkhäuser, Progress in Mathematics 129 (1995) 335-368
14. M. Kontsevich, Intersection theory on the moduli space of curves and the matrix Airy function, Comm. Math. Phys. 147 (1992) 1-23
15. M. Kontsevich, Yu. Manin, Gromov-Witten classes, quantum cohomology, and enumerative geometry, Comm. Math. Phys. 164 (1994) 525-562
16. M. Kontsevich, Yu. Manin, Quantum cohomology of a product, Inventiones Math. 124 (1996) 313-339
17. E. Looijenga, Intersection theory on Deligne-Mumford compactifications [after Witten and Kontsevich], Sem. Bourbaki no. 768 (1993)
18. D. McDuff, D. Salamon, J-holomorphic curves and quantum cohomology, AMS University Lecture Series no. 6 (1995)
19. Yu. Manin, Generating functions in algebraic geometry and sums over trees, in The Moduli Space of Curves, Birkhäuser, Progress in Mathematics 129 (1995) 401-418
20. J. Morava, Topological gravity and algebraic topology, Proceedings of the Adams Symposium, vol. II, LMS Lecture Notes 176 (1992)
21. J. Morava, Primitive Mumford classes, Contemporary Math. 15 (1993) 291-302
22. R. Pandharipande, Notes on Kontsevich's compactification of the space of maps, U. of Chicago preprint (1995)
23. A. Pressley, G. Segal, Loop groups, Oxford University Press (1986)
24. A. Robinson, Derived tensor products in stable homotopy theory, Topology 22 (1983) 1-18

25. Y. Ruan, G. Tian, A mathematical theory of quantum cohomology, Math. Research Letters 1 (1994) 269-278
26. G. Segal, Unitary representations of some infinite-dimensional groups, Comm. Math. Phys. 80 (1981) 301-342
27. G. Segal, Two-dimensional conformal field theories and modular functors, in IXth International Congress on Mathematical Physics, ed. B. Simon, A. Truman, I.M. Davies, Adam-Hilger (1988)
28. R.W. Thomason, T. Trobaugh, Higher algebraic K-theory of schemes and of derived categories, in The Grothendieck Festschrift, vol III, ed. P. Cartier et al, Progress in Mathematics 88 (1990) 247-436, Birkhäuser
29. E. Witten, Two-dimensional gravity and intersection theory on moduli space, Surveys in Differential Geometry 1 (1991) 243

DEPARTMENT OF MATHEMATICS, JOHNS HOPKINS UNIVERSITY, BALTIMORE, MARYLAND 21218
E-mail address: jack@math.jhu.edu

Contemporary Mathematics
Volume **202**, 1997

NON-COMMUTATIVE RECIPROCITY LAWS
ASSOCIATED TO FINITE GROUPS[1]

J.-L. Brylinski and D. A. McLaughlin[2]

1. Introduction

Let G be a finite group. Starting from a characteristic class $\alpha \in H^4(BG; \mathbb{Z})$, Dijkgraaf and Witten construct a $2 + 1$-dimensional topological quantum field theory which serves as a prototype for studying WZW theories associated to compact groups [**Di-Wi**]. In their theory, the role of the loop group is played by the category G_{S^1}, whose objects correspond to isomorphism classes of principal G-bundles on S^1 and whose morphisms are bundle automorphisms. There is even a "reciprocity law" for extensions of the category G_{S^1} (see Theorem 2.2). This point of view was developed in our papers [**B-M1**][**B-M3**]. It is the exact analogue of the Segal-Witten reciprocity law for loop groups ([**A-D-K**][**B-M2**][**S1**][**S2**]) which is important for studying moduli of semi-stable bundles and representations of loop groups [**B-M4**]. In the case of the loop group $L\mathbb{C}^*$, one recovers the classical reciprocity law of Weil and Tate for Riemann surfaces [**De1**][**We**]. There is also an arithmetic version of Segal-Witten reciprocity due to Deligne [**De2**].

However, all these non-commutative reciprocity laws are *one-dimensional* in the sense that they are formulated over a Riemann surface or a scheme of dimension one. Elsewhere [**B-M7**], we develop the higher dimensional theory starting with the philosophy that a characteristic class of degree $2n$ gives rise to a reciprocity law over a $n - 1$-dimensional base which is a non-commutative generalization of the reciprocity laws of higher class field theory [**B-M6**][**K**][**P2**]. We expect that this will play a key role in the "higher dimensional Langlands program" and any generalization of conformal field theory to dimensions > 1. The present paper is to be regarded as a prototype for multidimensional reciprocity in much the same way as the Dijkgraaf-Witten theory gives a toy model for understanding WZW theories; all of the analytic technicalities fall out and one is left with a clear view of many of the essential features. We will concentrate on the case where the base has complex dimension 2.

The paper is organized as follows:

---

[1]AMS classification: primary 14J60, 14E20, 57M25, secondary 14F10

[2]This research was partially supported by N.S.F. grants DMS-9203517, DMS-9310433, DMS 9504522 and DMS 9504237.

In section 2, we recall the one-dimensional reciprocity law associated to a characteristic class $\alpha \in H^4(BG; \mathbb{Z})$, for G finite (Theorem 2.2). We show how this can be used to define a "modular functor" in the sense of Segal [S1]. We give explicit formulae for the genus zero case, as it is of particular interest to these conference proceedings; a conformal field theory at the tree level is a prototypical example of an algebra over an operad (e.g., see [K-S-V]).

Section 3 is devoted to establishing the basic properties of the transgression to the double loop space $LLBG$. We define a notion of reciprocity for cohomology classes in $H^3(LLBG; \mathbb{C}^*)$ and prove a reciprocity law (Theorem 3.2) which is the two-dimensional analogue of the Segal-Witten reciprocity law for a finite group.

In section 4, we consider a two-dimensional complex analytic space X and a Galois covering $E \to X$ with group G, unramified over some Zariski open set Z. For $\alpha \in H^6(BG; \mathbb{Z})$ and a pair (p, C) consisting of a point p lying on an irreducible curve C, we define the *symbol* $< E, \alpha >_{(p,C)}$. It is important to note that the symbol is *not* a number, rather it appears as a certain cohomology class in $H^3(BAut(E); \mathbb{C}^*)$, where $Aut(E)$ denotes the gauge group of the bundle $E \to Z$. We then have the reciprocity laws:

Theorem 4.2. (1) Fix $p \in X$. Then

$$\prod_C < E, \alpha >_{(p,C)}$$

is cohomologically trivial in $H^3(BAut(E); \mathbb{C}^*)$, where the product is taken over all irreducible curves containing p.

(2) Fix an irreducible curve C, then

$$\prod_p < E, \alpha >_{(p,C)}$$

is cohomologically trivial in $H^3(BAut(E); \mathbb{C}^*)$, where the product is taken over all points p in C.

Analyzing the proof of Theorem 4.2, we obtain

Theorem 4.3. Let $E \to X$ be a Galois covering with group G, defined over a two-dimensional complex-analytic space X. Suppose that E is unramified over some Zariski open set Z. Then each cohomology class $\alpha \in H^6(BG; \mathbb{Z})$ defines a *canonical central extension* by $\mathbb{C}^*$ of the automorphism group of the covering $E \to Z$.

In section 5, we give an application of this circle of ideas to knots. Let K^ϵ be a tubular neighbourhood of a framed knot K in S^3. A *peripheral system* $P(K)$ for the knot is given by the knot group $\pi_1(S^3 - K)$ together with a choice (up to isotopy) of a meridean and longitude on the 2-torus ∂K^ϵ. Starting with a pair $(P(K), \alpha)$ consisting of a peripheral system $P(K)$ and a characteristic class $\alpha \in H^5(BG; \mathbb{Z})$, we construct a *canonical* vector space $V(P(K), \alpha)$ (Theorem 5.1). Our construction is reminiscent of Segal's modular functor approach to the conformal blocks, which we have outlined in section 2. This is interesting, since a fundamental theorem of Waldhausen [Wa] asserts that any two knots can be distinguished using peripheral systems. In Theorem 5.2, we give a formula for the dimension of $V(P(K), \alpha)$ in terms of the "fundamental class" of the peripheral system. Unfortunately, for dimension reasons these constructions have no counterpart for compact groups.

Finally, we have included an appendix on some constructions in group cohomology which are used frequently throughout the paper. This is done in an effort to make the presentation self-contained.

We thank H. Trotter for useful conversations.

2. Review of the one-dimensional theory

The material in this section is based on §4 of [**B-M3**]. Let G be a finite group. For any topological space X, we introduce the category G_X defined as follows; the objects of G_X are principal G-bundles on X. The morphisms are isomorphisms of G-bundles, i.e., gauge transformations. It is easy to see that the classifying space BG_X is homotopy equivalent to the mapping space $Map(X, BG)$. In the case $X = S^1$, this gives $BG_{S^1} \cong LBG$–the free loop space of BG. The category G_{S^1} plays the role of the "loop group" of a finite group. It admits the following non-canonical description; choose representatives $g_1, ..., g_n$ for the conjugacy classes of G and let Z_{g_i} denote the centralizer of g_i. Then the isomorphism classes of objects of G_{S^1} are in 1-1 correspondence with the conjugacy classes of G and the automorphisms of a given G-bundle corresponding to g_i are given by elements of Z_{g_i}. Therefore

$$BG_{S^1} \cong LBG \cong \coprod_i BZ_{g_i}. \tag{2.1}$$

Next, we define a *central extension* of G_X by $\mathbb{C}^*$ to be a category $\widetilde{G_X}$ mapping to G_X, such that the automorphism groups in $\widetilde{G_X}$ are central extensions of automorphism groups in G_X. To be more precise, for each object A in $\widetilde{G_X}$, there should be a group homomorphism $\mathbb{C}^* \to Aut_{\widetilde{G_X}}(A)$ which satisfies the following; for any two objects A, B, the left and right actions of $\mathbb{C}^*$ on $Hom_{\widetilde{G_X}}(A, B)$ should coincide. Moreover, we require that $Hom_{\widetilde{G_X}}(A, B)/\mathbb{C}^*$ be isomorphic to $Hom(F(A), F(B))$, where $F : \widetilde{G_X} \to G_X$ is the canonical functor. Then for each object A, the group $Aut_{\widetilde{G_X}}(A)$ is a central extension of $Aut_{G_X}(F(A))$ by $\mathbb{C}^*$.

Later in this paper, we will be concerned with certain "representations" of $\widetilde{G_X}$. These will be functors $R : \widetilde{G_X} \to \mathbf{Vec}$ (the category of finite-dimensional vector spaces whose morphisms are linear transformations) with the property that for any object A of $\widetilde{G_X}$ and any $\phi \in \mathbb{C}^*$ lying in the kernel of $Aut_{\widetilde{G_X}}(A) \to Aut_{G_X}(F(A))$, the endomorphism $R(\phi)$ of the vector space $R(A)$ is equal to $\phi \cdot Id$. There is an obvious notion of the direct sum of two such functors inherited from the category $\mathbf{Vec}$ and we say that such a functor is *irreducible* if it cannot be decomposed as a direct sum in a non-trivial way. It is clear that every such functor admits a finite decomposition into irreducibles. Since the group G is finite, there are only finitely many such irreducible functors.

In the case $X = S^1$, an extension $\widetilde{G_{S^1}}$ can be described (non-canonically) by specifying a central extension of each subgroup Z_{g_i} by $\mathbb{C}^*$. From (2.1), we see that central extensions of G_{S^1} are classified by the cohomology group $H^2(LBG; \mathbb{C}^*)$.

We say that a central extension $\widetilde{G_{S^1}}$ has the *reciprocity property* if the following is true for every Riemann surface Σ, whose boundary $\partial\Sigma$ is a disjoint union of parametrized circles; the extension $\widetilde{G_{\partial\Sigma}}$ of $G_{\partial\Sigma}$ induced by Baer multiplication of the extensions on each boundary component, splits canonically when pulled back to the category G_Σ.

Now recall Segal's construction of a modular functor starting from an extension $\widetilde{G_{S^1}}$ which has the reciprocity property [S1]: to each parametrized circle (positively oriented), associate the set of finite-dimensional, irreducible representations of $\widetilde{G_{S^1}}$. The direct sum of all these representations is by definition the Hilbert space $H(S^1)$ of the conformal field theory. Set $H(-S^1) = H(S^1)^*$, where the minus sign denotes the negative orientation. The Hilbert space associated to a disjoint union of circles is defined to be the tensor product of the Hilbert spaces associated to each circle. We then have a Hilbert space $H(S)$ associated to any compact oriented 1-manifold S without boundary. Given a Riemann surface Σ, whose boundary $\partial\Sigma$ is a disjoint union of parametrized circles, we consider the extension of $\widetilde{G_\Sigma}$ obtained by pulling back the extension $\widetilde{G_{\partial\Sigma}}$. The Hilbert space $H(\partial\Sigma)$ then becomes a representation of $\widetilde{G_\Sigma}$. By our reciprocity assumption, we may "rescale" and obtain a representation of the category G_Σ itself on the Hilbert space $H(\partial\Sigma)$. Let $V(\Sigma)$ be the invariant part of this representation. Segal's modular functor in this context is then given by the assignment $\Sigma \longmapsto V(\Sigma)$ and the vector space $V(\Sigma)$ is sometimes called the space of conformal blocks.

The key tool to construct extensions of the category G_{S^1} which have the reciprocity property, is the transgression to the free loop space

$$\tau : H^i(X) \to H^{i-1}(LX).$$

This is defined to be the composition $\int_{S^1} \circ\, ev^*$, where $ev : LX \times S^1 \to X$ is the usual evaluation map and $\int_{S^1}$ denotes "integration over the fiber" S^1 (slant product with the fundamental class of the circle). We can now state the reciprocity law which is implicit in [Di-Wi].

Reciprocity Theorem 2.2. [B-M1] [B-M3] Those extensions of G_{S^1} which lie in the image of the natural transgression

$$\tau : H^3(BG; \mathbb{C}^*) \to H^2(LBG; \mathbb{C}^*)$$

have the reciprocity property.

This Theorem was also known to Segal [S2]. It implies that every characteristic class $\alpha \in H^4(BG; \mathbb{Z})$ can be used to define a modular functor. Since G is finite, the cohomology of G is torsion and we have

$$H^4(BG; \mathbb{Z}) \cong H^3(BG; \mathbb{C}^*) \cong H^3_{group}(G; \mathbb{C}^*).$$

Therefore α can be represented by a degree three group cocycle with coefficients in $\mathbb{C}^*$. It follows from the isomorphism

$$H^2(LBG; \mathbb{C}^*) \cong \oplus_i H^2_{group}(Z_{g_i}; \mathbb{C}^*),$$

that the transgression τ in Theorem 2.2 can be defined on the level of group cohomology by the formula

$$(\tau\alpha)_{g_i}(h, k) := \alpha(g_i, h, k) \cdot \alpha(h, g_i, k)^{-1} \cdot \alpha(h, k, g_i), \tag{2.3}$$

where $h, k \in Z_{g_i}$. It is easy to verify directly that $(\tau\alpha)_{g_i}$ defines a degree 2 group cocycle on Z_{g_i}. Without loss of generality, we will use *normalized* group cochains [E-M] throughout this paper. In view of Theorem 2.2, $\tau\alpha$ plays the role of a "symbol" and it even satisfies a multiplicative formula, a special case of which appeared in [B-M3]:

Proposition 2.4. Fix elements $x_1, ..., x_m$ in G. Then, for $h, k \in Z_{x_1} \cap Z_{x_2} \cap .... \cap Z_{x_m}$, we have

$$\tau\alpha_{x_1}(h, k) \cdot .... \cdot \tau\alpha_{x_m}(h, k) = \tau\alpha_{x_1 \cdots x_m}(h, k) \cdot \delta\mu(h, k),$$

where $\mu(y)$ is equal to

$$\prod_{i=1}^{m-1} \alpha^{-1}(y, x_1 \cdot x_2 \cdots x_i, x_{i+1}) \cdot \alpha(x_1 \cdot x_2 \cdots x_i, y, x_{i+1}) \cdot \alpha^{-1}(x_1 \cdots x_i, x_{i+1}, y).$$

In other words, $\tau\alpha_{x_1} \cdot .... \cdot \tau\alpha_{x_m}$ is cohomologous to $\tau\alpha_{x_1 \cdots x_m}$ on $Z_{x_1} \cap Z_{x_2} \cap .... \cap Z_{x_m}$.

We have seen that each 3-group cocycle α defines an extension $\widetilde{G_{S^1}}$ satisfying the reciprocity property and therefore gives a modular functor $V(\Sigma)$. The remainder of this section is devoted to finding an explicit description of the vector space $V(\Sigma)$.

To begin, we fix a Riemann surface Σ, whose boundary $\partial\Sigma$ is a disjoint union of $p + q$ parametrized circles, the first p being negatively oriented, the last q being positively oriented. There is an obvious conjugation action of G on homomorphisms from $\pi_1(\Sigma)$ to G and we let $S = \{f_1, ..., f_r\}$ be a complete list of representatives for the orbits of this action. Letting Z_f be the stabilizer of f, we have

$$BG_\Sigma \cong \coprod_{f \in S} BZ_f. \tag{2.5}$$

Let $\widetilde{G_\Sigma}$ be the central extension obtained by pulling back the extension $\widetilde{G_{\partial\Sigma}}$ defined by transgressing α. From (2.5), we see that $\widetilde{G_\Sigma}$ can be described by specifying a central extension $\widetilde{Z_f}$ of Z_f, for each $f \in S$. Fix such an element $f \in S$. This determines group elements $g_1, ..., g_{p+q}$ in G labelling each boundary circle of Σ. They satisfy the relation $g_1 \cdot .... \cdot g_{p+q} = 1$. Associated to the i-th boundary circle, we have the extension $\widetilde{Z_{g_i}}$ corresponding to the cocycle $\tau\alpha_{g_i}$. It follows that the central extension $\widetilde{Z_f}$ we are after is given by the cocycle $\tau\alpha_{g_1} \cdots \tau\alpha_{g_{p+q}}$. By the reciprocity law, the extension $\widetilde{Z_f}$ splits and so this cocycle is equal to a canonical coboundary $\delta\mu_f$. Using Proposition 2.4, it is easy to derive an explicit formula for μ_f. In case the surface Σ has genus zero, $\mu_f(y)$ is equal to

$$\prod_{i=1}^{p+q-1} \alpha^{-1}(y, g_1 \cdots g_i, g_{i+1}) \cdot \alpha(g_1 \cdots g_i, y, g_{i+1}) \cdot \alpha^{-1}(g_1 \cdots g_i, g_{i+1}, y), \tag{2.6}$$

for $y \in Z_f$.

Keeping our element $f \in S$ fixed, we consider the set $\{R_{l_i}^{(i)}\}$ of finite dimensional, irreducible representations of $\widetilde{Z_{g_i}}$. Let $\{\rho_{l_i}^{(i)}\}$ be the corresponding set of characters. A typical "block" $R_{l_1 \ldots l_{p+q}}$ of the Hilbert space $H(\partial\Sigma)$ is a representation of $\widetilde{Z_{f|\partial\Sigma}}$ with character

$$\rho_{l_1 \ldots l_{p+q}} := \overline{\rho}_{l_1}^{(1)} \otimes ... \otimes \overline{\rho}_{l_p}^{(p)} \otimes \rho_{l_{p+1}}^{(p+1)} \otimes ... \otimes \rho_{l_{p+q}}^{(p+q)}.$$

By pulling back from the boundary, $R_{l_1 \dots l_{p+q}}$ becomes a representation of $\widetilde{Z}_f$ with corresponding cocycle $\delta\mu_f$. Rescaling this action via

$$\rho_{(l_1 \dots l_{p+q})} \longmapsto \rho_{(l_1 \dots l_{p+q})} \cdot \mu_f^{-1},$$

we obtain a genuine (i.e. non-projective) representation of Z_f. Let $V^f_{l_1 \dots l_{p+q}}$ be the invariant part and let $n^f_{l_1 \dots l_{p+q}}$ be the dimension of this vector space. We then have the following description of the modular functor.

Theorem 2.7. With the above notation,

$$V(\Sigma) = \oplus_{f \in S} \ \oplus_{l_1 \dots l_{p+q}} \ V^f_{l_1 \dots l_{p+q}}.$$

Using Schur orthogonality for characters, we also obtain a formula for the dimension of $V(\Sigma)$.

Theorem 2.8. With the above notation,

$$dim \ V(\Sigma) = \sum_f \sum_{l_1 \dots l_{p+q}} n^f_{l_1 \dots l_{p+q}},$$

where

$$n^f_{l_1 \dots l_{p+q}} = \frac{1}{|Z_f|} \sum_{x \in Z_f} \rho_{l_1 \dots l_{p+q}}(x) \cdot \mu_f^{-1}(x).$$

In the case where Σ has genus zero, μ_f is given by formula (2.6).

3. A two-dimensional transgression

Let $T := S^1 \times S^1$ denote the standard torus and consider the category G_T. From section 2, we know that BG_T is homotopy equivalent to the double loop space $LLBG := Map(T, BG)$. We can give the following non-canonical description of the "double loop group" G_T. Let the finite group G act on the set $\{(g, h) : [g, h] = 1\}$ of commuting pairs by simultaneous conjugation of both factors. Choose representatives (g_i, h_i), $1 \le i \le n$ for each orbit of this action and let $Z_{(g_i, h_i)}$ denote the stabilizer of the pair (g_i, h_i). Then the objects of G_T correspond to the orbits and the automorphisms of a bundle lying in the orbit of (g_i, h_i) correspond to elements of $Z_{(g_i, h_i)}$. We have

$$BG_T \cong LLBG \cong \coprod_i BZ_{(g_i, h_i)}. \tag{3.1}$$

We say that a cohomology class $\beta \in H^3(BG_T; \mathbb{C}^*) = H^3(LLBG; \mathbb{C}^*)$ has the *reciprocity property* if for every compact oriented 3-manifold M, whose boundary ∂M is a disjoint union of 2-tori, the cohomology class of $\prod_j b_j^* \beta$ in $H^3(Map(M, BG); \mathbb{C}^*)$ is trivial. Here $b_j : M \to S^1 \times S^1$ denotes the restriction to the j-th boundary component.

Our tool for constructing such classes is the double transgression

$$\tau^2 := \int_T \circ ev^* : H^i(BG) \to H^{i-2}(LLBG),$$

where $ev : LLBG \times T \to BG$ is the usual evaluation map and $\int_T \alpha := \alpha \,/\, [T]$ is the slant product with $[T]$–the fundamental class of the torus. We then have a *geometric reciprocity law*:

Theorem 3.2. Those cohomology classes which lie in the image of the natural map

$$\tau^2 : H^5(BG; \mathbb{C}^*) \to H^3(LLBG; \mathbb{C}^*)$$

have the reciprocity property.

Proof: Fix an oriented 3-manifold M with ∂M equal to a disjoint union of 2-tori T_j. Then for $\alpha \in H^5(BG; \mathbb{C}^*)$, we have

$$\prod_j b_j^* \tau^2 \alpha = ev^* \alpha \,/\, \sum_j [T_j].$$

Since $\sum_j [T_j]$ is homologically trivial in $H_2(M; \mathbb{Z})$, the result follows. ∎

The transgression τ^2 can be described purely in terms of group cocycles. Let us regard $\alpha \in H^5(BG; \mathbb{C}^*) \cong H^5_{group}(G; \mathbb{C}^*)$ as a 5-cocycle on G. Then from (3.1), we see that $\tau^2 \alpha$ amounts to specifying a 3 group cocycle on each subgroup $Z_{(g_i, h_i)}$. By iterating (2.3), we obtain a formula for

$$\tau^2 \alpha_{(g_i, h_i)}(l, m, n) := \tau(\tau \alpha_{g_i})_{h_i}(l, m, n),$$

namely

$$[\alpha, \alpha](g_i, h_i, l, m, n) \cdot [\alpha, \alpha](g_i, l, h_i, m, n)^{-1}$$

$$\cdot [\alpha, \alpha](g_i, l, m, h_i, n) \cdot [\alpha, \alpha](g_i, l, m, n, h_i)^{-1}$$

$$\cdot [\alpha, \alpha](l, g_i, h_i, m, n) \cdot [\alpha, \alpha](l, g_i, m, h_i, n)^{-1}$$

$$\cdot [\alpha, \alpha](l, g_i, m, n, h_i) \cdot [\alpha, \alpha](l, m, g_i, h_i, n)$$

$$\cdot [\alpha, \alpha](l, m, g_i, n, h_i)^{-1} \cdot [\alpha, \alpha](l, m, n, g_i, h_i),$$

where $l, m, n \in Z_{(g_i, h_i)}$ and $[\alpha, \alpha](g_i, h_i, l, m, n)$ denotes the "commutator"

$$\alpha(g_i, h_i, l, m, n) \cdot \alpha(h_i, g_i, l, m, n)^{-1}, \tag{3.3}$$

etc.... It can be verified directly that $\tau^2 \alpha_{(g_i, h_i)}$ defines a degree 3 cocycle on $Z_{(g_i, h_i)}$.

Notice that $\tau^2 \alpha_{(g_i, h_i)} = \tau^2 \alpha_{(h_i, g_i)}^{-1}$.

We can also formulate this 2-dimensional notion of reciprocity in a more geometric fashion using the theory of *abstract kernels* [**E-M**]. These are the objects classified by the 3-dimensional cohomology group $H^3_{group}(G; \mathbb{C}^*)$. By definition, an abstract kernel is a pair (θ, L), consisting of a non-abelian group L with center $\mathbb{C}^*$, together with a homomorphism $\theta : G \to Out(L)$. Given such (θ, L), one can ask for an extension of G by L, which has θ as its "group of operators". The obstruction is easily seen to lie in $H^3_{group}(G; \mathbb{C}^*)$ and it is shown in [**E-M**] that every degree three group cohomology class arises as an obstruction in this manner. There is an obvious notion of equivalence of abstract kernels and a natural group structure on the set of equivalence classes. The obstruction problem of realizing an abstract kernel by an extension, defines an isomorphism between $H^3_{group}(G; \mathbb{C}^*)$ and the group of equivalence classes of abstract kernels. If there exists an extension $1 \to L \to E \to G \to 1$ realizing a given abstract kernel (θ, L), then we refer to E as an *object* of the trivial abstract kernel (θ, L). Any other realization can be obtained by twisting E by a central extension of G by $\mathbb{C}^*$. We will frequently abbreviate "abstract kernel" to "kernel".

Note that for any space X, we have

$$BG_X \cong \coprod_{f \in S} BZ_f, \qquad (3.4)$$

where S is the set of conjugacy classes of homomorphisms of $\pi_1(X)$ to G and for $f \in S$, Z_f denotes the stabilizer $\{g \in G : g^{-1} \cdot f \cdot g = f\}$. We can now translate our definition of 2-dimensional reciprocity into the language of kernels. By (3.4), we can represent any class $\beta \in H^3(BG_T; \mathbb{C}^*)$ by a family of kernels $\{(\theta, L)_{(g,h)}\}$, where each kernel corresponds to a cohomology class in $H^3_{group}(Z_{(g,h)}; \mathbb{C}^*)$ and (g, h) ranges over representatives of the orbits of G acting by simultaneous conjugation of the set of commuting pairs in G. Given any oriented 3-manifold M, whose boundary is a disjoint union of 2-tori, let S be the set of conjugacy classes of homomorphisms of $\pi_1(M)$ into G. Then each $f \in S$ induces a pair $(g^{(r)}, h^{(r)})$ labelling the r-th component of ∂M. We see that β has the reciprocity property if and only if the product kernel $\prod_r (\theta, L)_{(g^{(r)}, h^{(r)})}$ is trivial when pulled back to Z_f, i.e., there exists an extension of Z_f realizing the product kernel. It is then possible to strengthen Theorem 3.2 to obtain a *canonical* trivialization of the trivial abstract product kernel on each Z_f. The canonical nature of this trivialization is implicit in the proof of Theorem 4.3.

In section 5, we will consider the transgression τ^2 applied to a degree 4 characteristic class $\alpha \in H^4(BG; \mathbb{C}^*)$. In this case, $\tau^2 \alpha \in H^2(LLBG; \mathbb{C}^*)$ is given on the level of group cohomology by specifying a 2-group cocycle $\tau^2 \alpha_{(g_i, h_i)}$ on each subgroup $Z_{(g_i, h_i)}$ (recall that $\{(g_i, h_i)\}$ is a set of representatives for the orbits of G acting on the set of commuting pairs in G). The formula for $\tau^2 \alpha_{(g_i, h_i)}(k, l) := \tau(\tau \alpha_{g_i})_{h_i}(k, l)$ (obtained by iterating the 1-dimensional transgression) is

$$[\alpha, \alpha](g_i, h_i, k, l) \cdot [\alpha, \alpha](g_i, k, h_i, l)^{-1} \cdot [\alpha, \alpha](g_i, k, l, h_i)$$

$$\cdot [\alpha, \alpha](k, g_i, h_i, l) \cdot [\alpha, \alpha](k, g_i, l, h_i)^{-1} \cdot [\alpha, \alpha](k, l, g_i, h_i), \qquad (3.5)$$

where $k, l \in Z_{(g_i, h_i)}$. As in (3.3), $[\alpha, \alpha](g_i, h_i, k, l)$ denotes the "commutator", i.e., $\alpha(g_i, h_i, k, l) \cdot \alpha(h_i, g_i, k, l)^{-1}$. Again we have $\tau^2 \alpha_{(g_i, h_i)} = \tau^2 \alpha_{(h_i, g_i)}^{-1}$.

Remark 3.6. Notice that for $\alpha \in H^4(BG; \mathbb{C}^*)$, $\tau^2 \alpha$ defines a central extension $\widetilde{G_T}$ of the category G_T by $\mathbb{C}^*$. This extension has the *reciprocity property*, in the sense that for any compact oriented 3-manifold M, whose boundary ∂M is a disjoint union of tori, the extension $\widetilde{G_{\partial M}}$ induced by the product of the extensions $\widetilde{G_T}$ on the boundary components, *splits canonically* when pulled back by the natural functor $G_M \to G_{\partial M}$. This is proved in exactly the same way as Theorem 3.2 and is a direct consequence of Proposition A.5 of the appendix.

4. Two-dimensional reciprocity laws

Let X be a two-dimensional complex-analytic space. Consider a Galois covering of X (with group G) which is unramified on some Zariski open set Z. This defines a principal G-bundle $E \to Z$. Let p be a point on an irreducible curve $C \subset X$. Choose a Riemannian metric in a neighbourhood of p (by this we mean a metric induced by some riemannian metric on some complex-analytic manifold in which this neighbourhood is embedded), and let $B_\epsilon(p)$ be the ball of radius ϵ centered at p. From the arguments given in [**M**], we know that for ϵ small enough, $\partial B_\epsilon(p)$ intersects the curve C transversally in a link K_ϵ. There is one component of K_ϵ for each branch of C through p. Choose $\delta << \epsilon$ and consider a δ-neighbourhood $N_\delta(K_\epsilon)$ of K_ϵ in $\partial B_\epsilon(p)$. The boundary $\partial N_\delta(K_\epsilon)$ is a disjoint union of 2-tori which may be assumed to lie in Z (by choosing δ, ϵ sufficiently small). Let $[\partial N_\delta(K_\epsilon)]$ be the class in $H_2(Z; \mathbb{Z})$ determined by $\partial N_\delta(K_\epsilon)$. It is not hard to see that this homology class is independent of the choice of ϵ, δ and the choice of local Riemannian metric and so we will drop the ϵ and δ from our notation. This construction was used by Parshin in his formulation of two-dimensional residue theorems [**P1**]. We will sometimes use $\phi_{(p,C)}$ to denote the homology class $[\partial N(K)]$, thereby emphasizing the dependence on the "place" (p,C).

Let $ev : Map(Z, BG) \times Z \to BG$ be the evaluation map. We use $Map^E(Z, BG)$ to denote the component containing the principal bundle $E \to Z$.

Definition 4.1. For a characteristic class $\alpha \in H^6(BG; \mathbb{Z}) \cong H^5(BG; \mathbb{C}^*)$, we define the *symbol* $< E, \alpha >_{(p,C)}$ of E at the "place" (p,C) to be the cohomology class

$$ev^*\alpha \ / \ [\partial N(K)] \in H^3(Map^E(Z, BG); \mathbb{C}^*).$$

Using (3.1), we may also view the symbol as taking values in $H^3(BAut(E); \mathbb{C}^*)$, where $Aut(E)$ denotes the group of gauge transformations of $E \to Z$. Therefore, the symbol may be represented by an abstract kernel for the group $Aut(E)$.

The symbol also admits a description in terms of group cocycles; the homology class $[\partial N(K)]$ can be represented as a disjoint union of 2-tori $\phi_i : T \to Z$, $1 \le i \le n$. Now

$$ev^*\alpha \ / \ [\partial N(K)] = \prod_i ev^*\alpha \ / \ [\phi_i],$$

and each term in the product can be computed by pulling back to the torus, i.e.,

$$ev^*\alpha \ / \ [\phi_i] = \phi_i^* \tau^2 \alpha.$$

Let $f \in Hom(\pi_1(Z), G)$ be the "holonomy representation" of the bundle E. It then follows that $ev^*\alpha \ / \ [\phi_i]$ is given by the 3-group cocycle (3.3) pulled back to Z_f.

Theorem 4.2. (Reciprocity) Let $E \to X$ be a Galois covering (with group G), unramified on some Zariski open set Z.

(1) Fix $p \in X$, then

$$\prod_C < E, \alpha >_{(p,C)}$$

is trivial in $H^3(BAut(E); \mathbb{C}^*)$, where the product is taken over all irreducible curves containing p.

(2) Fix an irreducible curve C. Then

$$\prod_p < E, \alpha >_{(p,C)}$$

is trivial in $H^3(BAut(E) : \mathbb{C}^*)$, where the product is taken over all points p lying on C.

Proof: For the proof of (1), we must first show that there are only finitely many non-trivial terms contributing to the product. This is really a topological fact. Suppose that the Zariski open set Z is the complement of finitely many irreducible curves $D_1, ..., D_k$. If the point p does not lie on any D_i, then the link of solid tori $N_\delta(K_\epsilon)$ associated to (p,C) can be chosen to lie entirely in Z. This means that the homology class $[\partial N(K)]$ is trivial and therefore so is the symbol $< E, \alpha >_{(p,C)}$. We may then suppose that p belongs to some D_i.

Now if no D_j is a component of the curve C, then the link of solid tori $N_\delta(K_\epsilon)$ can again be chosen to lie completely in Z, so that the symbol $< E, \alpha >_{(p,C)}$ is trivial. We conclude that the only non-trivial contributions to the product come from the finite number of irreducible curves $D_1, ..., D_k$.

Next we observe that the complement in $\partial B_\epsilon(p)$ of all the δ-neighbourhoods of $\partial B_\epsilon(p) \cap D_j$ is a 3-manifold M, which for sufficiently small ϵ, δ, lies entirely in Z. Therefore

$$\prod_j < E, \alpha >_{(p,D_j)} = ev^* \alpha \ / \ \sum_j [\partial N(K)_{(p,D_j)}].$$

Just as in Theorem 3.2, the result follows since $\sum_j [\partial N(K)_{(p,D_j)}]$ is homologous to zero in M.

Turning to the proof of (2), we must again show that there are only finitely many non-trivial terms in this product. As we have just seen, if no D_j is a component of C, then $< E, \alpha >_{(p,C)}$ is trivial. By decomposing C into irreducible components, we are therefore reduced to verifying the Theorem in the case C is one of $D_1, ..., D_k$ (say D_i). Now if the point $p \in D_i$ is a smooth point and does not also lie on some other D_j, then the link of solid tori $N(K)$ may be chosen to lie entirely in Z so that the symbol is trivial. We conclude that the only non-trivial contributions to the product come from the singular points of D_i and the finite number of points which lie on both D_i and some other irreducible curve D_j, $i \neq j$. Denote this finite set of points by $p_1, ..., p_r$.

Choose a Riemannian metric in some neighbourhood of C. Set $V := C - (\cup \ intB_\epsilon(p_j))$, for some small ϵ. Choose $\delta << \epsilon$ and let $\pi : N_\delta(C_{sm}) \to C_{sm}$ be a closed δ-neighbourhood of the smooth locus of C. Restricting $N_\delta(C_{sm})$ to V, we obtain a 3-manifold M, whose boundary

$$\partial M = \coprod_j \pi^{-1}(C \cap \partial B_\epsilon(p_j))$$

is a disjoint union of links of tori, each of which can be used to compute the symbol at (p_i, C), for sufficiently small ϵ, δ. We can now argue as in (1) to finish the proof.

∎

This geometric proof of Theorem 4.2 is adapted from the proof of the two-dimensional reciprocity law in class field theory given by us in [**B-M6**]. Note that the symbol $< E, \alpha >_{(p,C)}$ is a kind of "local monodromy" at (p, C).

In proving Theorem 4.2, we saw that there are is a *finite* list of pairs (p, C) for which the corresponding homology class $\phi_{(p,C)}$ is non-trivial in $H_2(Z; \mathbb{Z})$. Let S_0 be the finite set of points and S_1, the finite set of curves which occur in this list. For each $p \in S_0$, we constructed a 3-manifold M_p, whose boundary is equal to $\sum_{\{C \in S_1 : p \in C\}} \phi_{(p,C)}$. On the other hand, for each fixed $C \in S_1$, we constructed a 3-manifold M_C, with boundary equal to $\sum_{\{p \in S_0 : p \in C\}} \phi_{(p,C)}$. It follows that the smooth singular 3-chain $\sum_p M_p - \sum_C M_C$ is in fact a *cycle* γ. The homology class $[\gamma] \in H_3(Z; \mathbb{Z})$ determined by this cycle is *canonically defined*. We may therefore consider the slant product of $ev^* \alpha$ with $[\gamma]$ to obtain

Theorem 4.3. Let $E \to X$ be a Galois covering with group G, defined over a two-dimensional complex-analytic space X. Suppose that E is unramified over some Zariski open set Z. Then each cohomology class $\alpha \in H^6(BG; \mathbb{Z})$ defines a *canonical* central extension by $\mathbb{C}^*$ of the automorphism group (i.e. the group of gauge transformations) of the covering $E \to Z$.

This result is a prototype for the canonical central extension of $GL_n(O(Z))$, which we constructed in [**B-M7**] from the universal 3-rd Chern class. Here $O(Z)$ denotes the holomorphic functions on Z. The role of this extension in higher non-commutative class field theory is still somewhat mysterious.

5. Applications to knots

Let $K \subset S^3$ be a *framed* knot. Consider an ϵ-neighbourhood K^ϵ of K in S^3. The boundary of K^ϵ is a 2-torus $\partial K^\epsilon := T$. Let M be the 3-manifold with boundary which is the complement in S^3 of the interior of T, so that $\partial M = T$. A *peripheral system* $P(K)$ for the knot K is given by the knot group $\pi_1(M)$, together with a choice (up to isotopy) of a meridean and longitude on the torus T. The notion of a peripheral system is due to Fox, who used it to show that the granny knot and the square knot are inequivalent [**F**].

Fix a peripheral system for K. By a famous theorem of Papakyriakopoulos [**Pap**], the complement of the knot is aspherical. From the homology sequence associated to the pair $(\pi_1(M), \pi_1(T))$, we then have the isomorphism in group homology

$$H_3(\pi_1(M), \pi_1(T); \mathbb{Z}) \xrightarrow{\sim} H_2(\pi_1(T); \mathbb{Z})$$

induced by the boundary map ∂. The peripheral system $P(K)$ defines a pair of commuting generators a, b for $\pi_1(T)$. By convention, a will be the loop associated to the longitude, i.e., the loop which follows the core of the torus T. The fundamental class in $H_2(\pi_1(T); \mathbb{Z})$ can then be represented by the cycle $[a, b] - [b, a]$ in the bar complex (see the Appendix). The above isomorphism ∂ gives a corresponding element $[M]$ in the the relative homology group $H_3(\pi_1(M), \pi_1(T); \mathbb{Z})$. The element $[M]$ is by definition, the *fundamental class* of the peripheral system.

Let α be a 4-group cocycle representing a characteristic class in $H^4(BG; \mathbb{C}^*)$. By Proposition A.5 of the Appendix, the slant product of α with $[M]$ gives a central extension $\widetilde{G_{\partial K^\epsilon}}$ of the category $G_{\partial G^\epsilon}$ of "G-bundles on ∂K^ϵ", together with a *canonical splitting* of the extension $\widetilde{G_M}$ of G_M induced by the natural functor $G_M \to G_{\partial M}$.

Consider the finite set $\{R_l\}$ of irreducible functors from $\widetilde{G_{\partial K^\epsilon}}$ to **Vec** defined in section 2. These are the "irreducible representations" of $\widetilde{G_{\partial K^\epsilon}}$. Each R_l induces a functor $\overline{R_l} : \widetilde{G_M} \to$ **Vec** which in view of the canonical splitting, actually yields a "representation" of the category G_M itself. Let V_l denote the invariant part of this representation, i.e., if for each object $A \in G_M$, we set $V_l(A)$ equal to the subspace of $\overline{R_l}(A)$ on which $\overline{R_l}(\phi)$ acts trivially, for all $\phi \in Aut(A)$, then $V_l = \oplus_A V_l(A)$. Let $V := \oplus_l V_l$. The vector space V is a kind of "state space" and its construction is analogous to the modular functor $\Sigma \to V(\Sigma)$ described in §2. Summarizing, we have shown:

Theorem 5.1. Let G be a finite group and $\alpha \in H^4(BG; \mathbb{C}^*) \cong H^5(BG; \mathbb{Z})$. Suppose that $P(K)$ is a peripheral system for the framed knot K in S^3. Then there exists a canonical vector space $V(P(K), \alpha)$, which is intrinsically associated to the pair $(P(K), \alpha)$.

We are now interested in computing the dimension of $V(P(K), \alpha)$. Note that a G-bundle on ∂K^ϵ is determined by a pair (g, h) of commuting elements; the element g gives the holonomy around the a-loop, while h gives the holonomy around the b-loop. A G-bundle on the three manifold M is given via the holonomy representation, by an element of $Hom(\pi_1(M), G)$. The group G acts on this set by conjugation and we let $f_1, ..., f_r$ be representatives for the conjugacy classes. Let Z_{f_i} be the centralizer of f_i, as defined in section 4. We then have

$$BG_{\partial K^\epsilon} \cong \coprod_{i=1}^{r} BZ_{f_i}.$$

A central extension of G_M is given by specifying an extension of each Z_{f_i}. To describe the split extension of G_M induced from $\widetilde{G_{\partial K^\epsilon}}$, let (g_i, h_i) be the commuting pair of elements induced by restricting f_i to ∂M. Then α determines an extension $\widetilde{Z}_{(g_i, h_i)}$ with cocycle $\tau^2 \alpha_{(g_i, h_i)}$. The induced extension on Z_{f_i} is given by pulling back this cocycle via the natural inclusion $res : Z_{f_i} \to Z_{(g_i, h_i)}$. The resulting cocycle is equal to some coboundary, say $\delta\gamma_i$. The relative cohomology class in $H^2_{group}(Z_{(g_i, h_i)}, Z_{f_i}; \mathbb{C}^*)$ determined by the pair $(\tau\alpha_{(g_i, h_i)}, \gamma_i)$ is *canonical*. The set $\{(\tau\alpha_{(g_i, h_i)}, \gamma_i) : 1 \leq i \leq r\}$ describes the slant product $ev^*\alpha \,/\, [M]$.

Let $\{R_l^i\}$ be the finite set of irreducible representations of $\widetilde{Z}_{(g_i, h_i)}$ (with cocycle $\tau^2 \alpha_{(g_i, h_i)}$) and denote the character of each R_l^i by ρ_l^i. Then each representation R_l^i may be pulled back via the homomorphism res and after "rescaling", we obtain a genuine (i.e. not projective) representation of Z_{f_i}. The effect of this rescaling is to multiply the character $\rho_l^i(x)$ by a factor $\gamma_i(x)^{-1}$. By Schur orthogonality, the dimension of the invariant part of this representation is given by

$$n_{i,l} := \frac{1}{|Z_{f_i}|} \sum_{x \in Z_{f_i}} \rho_l^i(x) \cdot \gamma_i(x)^{-1}.$$

Now each irreducible functor $\overline{R}_l : G_M \to$ **Vec** contributing to the vector space $V(P(K), \alpha)$ can be described by specifying an irreducible representation of each Z_{f_i}. It follows that

$$dim\ V(P(K), \alpha) = \sum_{i,l} n_{i,l},$$

where i indexes the conjugacy classes $f_1, ..., f_r$ and for each fixed i, the irreducible projective representations of $Z_{(g_i, h_i)}$ with cocycle $\tau^2 \alpha_{(g_i, h_i)}$ are indexed by l.

In summary, we have shown:

Theorem 5.2. Let $P(K)$ be a peripheral system for a framed knot K in S^3 and α a characteristic class in $H^4(BG; \mathbb{C}^*) \cong H^5(BG; \mathbb{C}^*)$. Then with the above notation, the dimension of the vector space $V(P(K), \alpha)$ of Theorem 5.1 is given by

$$\sum_{i,l} \frac{1}{|Z_{f_i}|} \sum_{x \in Z_{f_i}} (\rho_i^l(x) \cdot \gamma_i(x)^{-1}).$$

It would be interesting to have an explicit formula for γ_i analogous to the expression (2.6) for the modular functor. This would involve finding an expression for the fundamental class of the peripheral system in terms of a set of Wirtinger generators for the knot group. Finally, it is possible to generalize the results of this section to links.

Appendix

Some constructions in group cohomology

In this section, we collect together various facts about group cohomology which were used throughout the paper.

For G a group and A a G-module, let $C_\bullet(G; A)$ denote the complex of chains $G^p \to A$ with finite support, and let $C^\bullet(G; A)$ denote the complex of cochains $G^p \to A$. For $g_1, ..., g_p \in G$ and $a \in A$, we let $[g_1, ..., g_p; a]$ be the corresponding element of $C_p(G; A)$. If $A = \mathbb{Z}$, we abbreviate $[g_1, ..., g_p; 1]$ to $[g_1, ..., g_p]$.

For two groups H, K, let A be an H-module, B a K-module and C an $H \times K$-module. Any map $f : A \otimes B \to C$ of $H \times K$-modules induces a map of complexes

$$\psi : C_\bullet(H; A) \otimes C_\bullet(K; B) \to C_\bullet(H \times K; C),$$

defined by the formula

$$\psi([g_1, ..., g_p; a] \otimes [g_{p+1}, ..., g_{p+q}; b]) = \sum_{shuffles\ \sigma} \epsilon(\sigma)\ [g_{\sigma(1)}, ..., g_{\sigma(p+q)}; f(a \otimes b)].$$

$$(A.1)$$

This map of complexes is of course the standard shuffle product in simplicial homology, followed by the map of simplicial sets $BH \times BK \to B(H \times K)$. On homology, ψ induces the usual product map $H_\bullet(H; A) \otimes H_\bullet(K; B) \to H_\bullet(H \times K; C)$.

Note that the group homology $H_\bullet(G; A)$ is also equal to the the homology of the quotient complex of $C_\bullet(G; A)$ by the degenerate subcomplex $C_\bullet^{deg}(G; A)$, which is generated by all the $[g_1, ..., g_p; a]$ having at least one g_i equal to 1. This quotient complex is called the *normalized* chain complex, and its elements are called normalized chains. In this paper, we always work with such chains.

As an application of the shuffle product ψ, we give a (normalized) representative for the canonical class (orientation class) in $H_n(\mathbb{Z}^n; \mathbb{Z})$. This representative is equal to

$$\sum_{\sigma \in S_n} [g_{\sigma(1)}, ..., g_{\sigma(n)}],$$

where $g_1, ..., g_n$ are the generators of $\mathbb{Z}^n$. This is easily obtained by induction using (A.1).

Now suppose that A, B, C are abelian groups and $f : A \otimes B \to C$ is a homomorphism. Then we have a map of complexes

$$\phi : C^\bullet(H \times K; B) \otimes C_{-\bullet}(H; A) \to C^\bullet(K; C),$$

given by the formula

$$\phi(\gamma, [g_{q+1}, ..., g_{p+q}; a])(g_1, ..., g_q) := \sum_{shuffles \ \sigma} \epsilon(\sigma) f(a \otimes \gamma(g_{\sigma(1)}, ..., g_{\sigma(p+q)})).$$
$$(A.2)$$

On cohomology, this induces the *slant product*

$$H^{p+q}(H \times K; B) \otimes H_p(H; A) \to H^q(K; C).$$

The normalized cochain complex consists of those cochains $H^p \to A$, which vanish on degenerate chains. The map (A.2) is also defined on the level of normalized complexes.

As an application, let w be a central element of G. Then there is a group homomorphism $\mathbb{Z} \times G \to G$, which maps (n, g) to $w^n \cdot g$. We can pull-back a class γ in $H^k(G; B)$ to a class in $H^k(\mathbb{Z} \times G; B)$, and take the slant product with the class $[g]$ in $H_1(\mathbb{Z}; \mathbb{Z})$. This gives a class in $H^{k-1}(G; B)$, which is represented by the normalized cocycle

$$\beta(g_1, ..., g_{k-1}) := \sum_{0 \leq i \leq k-1} (-1)^i \gamma(g_1, ..., g_i, w, g_{i+1}, ..., g_k). \qquad (A.3)$$

If we denote the centralizer of w in G by Z_w, then the same formula (A.3) defines a map $H^p(G; B) \to H^{p-1}(Z_w; B)$. For $p = 3$ and $B = \mathbb{C}^*$, this is the 1-dimensional transgression τ considered in §2.

Now if w_1, w_2 are two central elements of G, then we have a map $\mathbb{Z}^2 \times G \to G$ and so we may pull back a class $\gamma \in H^k(G; B)$ to obtain a class in $H^k(\mathbb{Z}^2 \times G; B)$. Taking the slant product with the canonical class in $H_2(\mathbb{Z}^2; \mathbb{Z})$, we obtain a class in $H^{k-2}(G; B)$ which is represented by the cocycle:

$$(g_1, ..., g_{k-2}) \mapsto \sum_{i \leq j, \sigma \in S_2} (-1)^{i+j} \epsilon(\sigma) \gamma(g_1, ..., g_i, w_{\sigma(1)}, g_{i+1}, ..., g_j, w_{\sigma(2)}, ..., g_{k-2})$$
$$(A.4)$$

Now if w_1, w_2 *commute* and $Z_{(w_1, w_2)}$ denotes their simultaneous centralizer, the same formula (A.4) defines a map $H^p(G; B) \to H^{p-2}(Z_{(w_1, w_2)}; B)$. This coincides with the 2-dimensional transgressions given in equations (3.3) and (3.5).

We now introduce relative group homology and cohomology. Recall that for any map of spaces $f : X \to Y$ and for any abelian group A, we have relative homology groups $H_j(f; A)$ (or $H_j(Y, X; A)$ when there is no ambiguity about f) and these fit into the long exact sequence

$$... \to H_j(X; A) \to H_j(Y; A) \to H_j(f; A) \to H_{j-1}(X; A) \to ...$$

Now suppose that $\theta : G \to H$ is a group homomorphism. Then θ induces a map $B\theta : BG \to BH$, and we define $H_j(\theta, A) := H_j(B\theta; A)$. We often write $H_j(H, G; A)$ instead of $H_j(\theta; A)$. We then have a long exact sequence

$$... \to H_j(G; A) \to H_j(H; A) \to H_j(\theta; A) \to H_{j-1}(G; A) \to ...$$

Note that $H_j(\theta; A)$ is the homology of the cone of the map of complexes $\theta_* : C_\bullet(G; A) \to C_\bullet(H; A)$.

Now any map $f : X \to Y$ of path-connected spaces gives rise to a commutative diagram

$$
\begin{array}{ccc}
X & \xrightarrow{f} & Y \\
\downarrow & & \downarrow \\
B\pi_1(X, x) & \xrightarrow{Bf_*} & B\pi_1(Y, f(x))
\end{array}
$$

and hence a map on relative homology groups $H_j(f; A) \to H_j(Bf_*; A)$. In particular, if M is an oriented connected manifold, with connected boundary ∂M, there is a fundamental class $\kappa \in H_n(M, \partial M; \mathbb{Z})$. We therefore obtain a *fundamental class* in $H_n(\pi_1(M), \pi_1(\partial M); \mathbb{Z})$. This was used in §5.

The relative cohomology groups are defined in a similar fashion as the cone of the map of complexes $\theta^* : C^\bullet(H; A) \to C^\bullet(G; A)$. We note the following result which underlies the reciprocity theorems of sections 2 and 5:

Proposition A.5. Let $\theta : G \to H$ be a group homomorphism. Then

(1) $H^1(G, H; A)$ is the group of characters $\chi : H \to A$ such that $\chi \circ \theta = 0$.

(2) $H^2(G, H; A)$ is the group of isomorphism classes of pairs (E, s), where $1 \to A \to E \to H \to 1$ is a central extension of H and $s : G \to E \times_H G$ is a section of the induced extension $1 \to A \to E \times_H G \to H \to 1$.

Finally, we define the relative slant product. Let $\rho : H_1 \to G$ and $F : H_2 \to H_1$ be group homomorphisms. Let $K_1 \subseteq G$ be the centralizer of the image of ρ and $K_2 \subseteq G$ be the centralizer of the image of ρF. Then for $j = 1, 2$, we have a group homomorphism $H_j \times K_j \to G$ and therefore a commutative diagram:

$$
\begin{array}{ccccc}
C^{p+q}(G; B) \otimes C_p(H_2; A) & \to & C^{p+q}(H_2 \times K_2; B) \otimes C_p(H_2; A) & \to & C^q(K_2; C) \\
\downarrow \cong & & \downarrow & & \downarrow \\
C^{p+q}(G; B) \otimes C_p(H_1; A) & \to & C^{p+q}(H_1 \times K_1; B) \otimes C_p(H_1; A) & \to & C^q(K_1; C)
\end{array}
$$

Hence we get a map of complexes from

$$C^\bullet(G; B) \otimes [Cone \; (C_{-\bullet}(H_2; A) \to C_{-\bullet}(H_1; A))]$$

to

$$Cone\ [C^\bullet(K_2; C) \to C^\bullet(K_1; C)]\ [1],$$

which induces the *relative slant product*

$$H^{p+q}(G; B) \otimes H_p(H_1, H_2; A) \to H^{q+1}(K_2, K_1; C).$$

Throughout the paper, we have applied this map frequently to the following situation: M is an oriented manifold (of dimension 2 or 3) with boundary ∂M. $H_1 = \pi_1(M), H_2 = \pi_1(\partial M)$, G is a finite group and $\rho: \pi_1(M) \to G$ is a homomorphism. The coefficients are $A = \mathbb{Z}, B = \mathbb{C}^*$ and $C = \mathbb{C}^*$. F is the homomorphism induced by the inclusion of the boundary $\partial M \to M$.

References

[**A-D-K**] E. Arbarello, C. De Concini and V. Kac, *The infinite wedge representation and the reciprocity law for loop groups,* Proceedings of Symposia in Pure Mathematics vol. 49, part 1 (1989), American Math. Soc., 171-190.

[**B-M1**] J.-L. Brylinski and D. A. McLaughlin, *A geometric construction of the first Pontryagin class,* Quantum Topology, Series on Knots and Everything, vol. **3**, World Scientific (1993), 209-220.

[**B-M2**] J.-L. Brylinski and D. A. McLaughlin, *The geometry of degree four characteristic classes and of line bundles on loop spaces I,* Duke Math. Jour. **75** No. 3 (1994), 603-638.

[**B-M3**] J.-L. Brylinski and D. A. McLaughlin, *The geometry of degree four characteristic classes and of line bundles on loop spaces II,* preprint 1994, to appear in Duke Math. Jour.

[**B-M4**] J.-L. Brylinski and D. A. McLaughlin, *Holomorphic quantization and unitary representations of the Teichmuller group,* in *Lie Theory and Geometry in Honor of Bertram Kostant,* Progress in Math. **123** (1994), Birkhauser, Boston, 21-64.

[**B-M5**] J.-L. Brylinski and D. A. McLaughlin, *The geometry of two dimensional symbols,* preprint 1995, to appear in K-theory.

[**B-M6**] J.-L. Brylinski and D. A. McLaughlin, *Multidimensional reciprocity laws,* preprint 1995, submitted to Crelle's Journal.

[**B-M7**] J.-L. Brylinski and D. A. McLaughlin, *Characteristic classes and multidimensional reciprocity laws,* Math. Res. Letts. vol. 3, No. 1 (1996), 19-30.

[**De1**] P. Deligne, *Le symbole modéré,* Publ. Math. IHES **73** (1991), 147-181.

[**De2**] P. Deligne, *Seminar on WZW Theories,* Inst. for Adv. Study, Spring 1991.

[**Di-Wi**] R. Dijkgraaf and E. Witten, *Topological gauge theories and group cohomology,* Comm. Math. Phys. **129** (1990), 393-429.

[**E-M**] S. Eilenberg and S. Mac Lane, *Cohomology theory in abstract groups II,* Ann. of Math. **48** (1947), 326-341.

[**F**] R. Fox, *On the complementary domains of a certain pair of inequivalent knots,* Kon. Nederl. Akad. van Wetenschappin, Proceedings Series A **55** (1952), 37-40.

[**K**] K. Kato, *Milnor K-Theory and the Chow group of zero cycles,* Contemp. Math. **55**, A.M.S. (1986), 241-253.

[**K-S-V**] T. Kimura, J. Stasheff and A. Voronov, *On operad structures of moduli spaces and string theory,* Comm. Math. Phys. **177** (1995), 1-15.

[**M**] J. Milnor, *Singular points of complex hypersurfaces,* Ann. of Math. Studies **61** (1968), Princeton Univ. Press.

[**Pap**] C. D. Papakyriakopoulos, *On Dehn's lemma and the asphericity of knots,* Ann. of Math., **66** (1957), 1-26.

[**P1**] A. N. Parshin, *On the arithmetic of two-dimensional schemes,* Math. USSR Izvestija **10** (1976) no. 4, 695-729.

[**P2**] A. N. Parshin, *Local class field theory,* Proc. Steklov Inst. Math. (1985), No. 3, 157-185.

[**S1**] G. Segal, *The definition of conformal field theory,* preprint; condensed version in *Differential Geometrical Methods in Theoretical Physics* (Como 1987), NATO Adv. Sci. Inst. Ser. C Math. Phys. Sci. **250** (1988), Kluwer, Dordrecht, 165-171.

[**S2**] G. Segal, Lecture at M.S.R.I., January 1990.

[**Wa**] F. Waldhausen, *On irreducible 3-manifolds which are sufficiently large,* Ann. of Math. **87** (1968), 56-88.

438   J.-L. BRYLINSKI AND D. A. MCLAUGHLIN

[**We**] A. Weil, *Sur les fonctions algébriques à corps de constantes finis,* C. R. Acad.
Sci. Paris **210** (1940), 592-594.

Addresses of authors
J-l. B: The Pennsylvania State University
Department of Mathematics
305 McAllister Bdg
University Park, PA. 16802
e-mail: jlb@ math.psu.edu

D. M.: Princeton University
Department of Mathematics
Princeton, NJ. 08544
e-mail:dennisa@ math.princeton.edu

Index

Selected Titles in This Series

(*Continued from the front of this publication*)

(See the AMS catalog for earlier titles)